Lecture Notes in Mathematics V
ISBN 978-3-540-18905-3 © Springer-Verlag Berlin He

T0253587

Jaak Peetre

Function Spaces and Application

Errata

PAGE 1
Add the following note: "**Note** A revised and expanded version of this article is going to be published in the book [P4]."
Second paragraph: Replace the word "bibliography' by "biography"

PAGE2
First paragraph: Replace the word "history" by "prehistory"

Modern era
line 1: replace "Norlund" by "the Dane Nørlund"

Stockholm period 1910-1925
line 2: After "1908" add, within dashes "– but very superficially, for the true story see [P4] – "

PAGE 3
Addresses. Personalia
Replace "Marcel Riesz address" by "One of Marcel Riesz's several addresses"

Scientific profile. Students
Penultimate line of last paragraph: Change "Cesaro" to "Cesàro" (with an accent)

PAGE 4
Last line of the paragraph of the section under consideration: after the words "ending up as" replace the text by the following: "county governor (landshövding) of, first, the county of Kristianstad (Kristianstads län), then, the county of Stockholm (Stockholms län)."

PAGE 7
Last paragraph: Replace "An" by "A".

PAGE 8
Penultimate line: Replace "N.F.Riesz" by "our sponsor the N.F.R."

PAGE 10

Add the following reference: "P4. Peetre J. et al (Editors): Correspondence of Gsta Mittag-Leffler with Marcel Riesz and Andries MacLeod. (In preparation.)"

Lecture Notes in Mathematics

Edited by A. Dold and B. Eckmann

1302

M. Cwikel J. Peetre
Y. Sagher H. Wallin (Eds.)

Function Spaces and Applications

Proceedings of the US- Swedish Seminar
held in Lund, Sweden, June 15–21, 1986

Springer-Verlag
Berlin Heidelberg New York London Paris Tokyo

Editors

Michael Cwikel
Technion, Israel Institute of Technology, Department of Mathematics
Haifa 32000, Israel

Jaak Peetre
Lund Institute of Technology, Department of Mathematics
S-221 00 Lund, Sweden

Yoram Sagher
University of Illinois at Chicago Circle, Department of Mathematics
Chicago, IL 60680, USA

Hans Wallin
University of Umeå, Department of Mathematics
S-901 87 Umeå, Sweden

Mathematics Subject Classification (1980): 41A25, 46E30, 46E35, 47B35

ISBN 3-540-18905-X Springer-Verlag Berlin Heidelberg New York
ISBN 0-387-18905-X Springer-Verlag New York Berlin Heidelberg

Printed in Germany

Printing and binding: Druckhaus Beltz, Hemsbach/Bergstr.
2146/3140-543210

Foreword

These are the proceedings of a seminar held in Lund in June 1986, devoted to function space methods in analysis. The organizers felt that recent developments in Interpolation Theory have important implications in various areas of analysis, and that further work in Interpolation would benefit from the input of new problems and applications. Thus the makeup of the seminar was considerably broader than its two predecessors (Lund 1982 and 1983). It is particularly appropriate that these seminars have taken place in Lund, where Interpolation Theory was born some 60 years ago. To further emphasize the historical origin of the subject, the collection is preceded by a historical lecture on the life of Marcel Riesz.

The organizers wish to thank Naturvetenskapliga Forskningsrådet and the N.S.F. for supporting the conference. The two non-Swedish organizers feel that they express the sentiments of all participants in thanking Jaak Peetre and Hans Wallin as well as the members of the mathematics department at Lund for their warm hospitality.

The Editors

CONTENTS.

MARCEL RIESZ IN LUND

Jaak Peetre

(Public lecture delivered on June 18, 1986 at 7 PM.)

Ladies and gentlemen,

Marcel Riesz was born on Nov. 16, 1886 in Györ, Hungary and died on Sept. 4, 1969 in Lund, Sweden. Thus this conference has also given us a convenient opportunity to celebrate the 100th anniversary of his birth.

The organizers had originally planned this lecture to be delivered by Prof. Lars Gårding. Unfortunately, Prof. Gårding is currently abroad and so, as we have not been able to materialize his spirit, I shall try to take over his role. So imagine now a much more charismatic personality. At least, I hope you will not end up by throwing tomatoes at me.

We all wish to believe that we are forerunners of our science, but is also important not to lose the contact with its glorious past. Here is an appropriate quotation: "We may say that its (mathematical history) first use to us is to put or to keep before our eyes 'illustrious examples' of first-rate mathematicians" ([We], p. 229). Therefore this talk certainly has a place within the framework of this conference.

<u>Sources.</u> There are two orbituaries by Lars Gårding [G1], [G2]; [G2] is an expanded Swedish version of [G1], with more personal recollections and a few anecdotes.

I myself once wrote a bibliography of Marcel Riesz for an Italian encyclopedia [P1]. It turned out to be a double catastrophe: 1st it was translated into Italian and 2nd it was cut by a factor 2 (in the middle). So don't read it. But it was this preoccupation with Marcel Riesz some 15 years ago that makes that I possess some of my present knowledge.

John Horvath, one of Riesz's last associates and a personal friend, has written a very detailed account of his early work [H1].

The very interesting correspondence between Marcel Riesz and the British mathematicians (read: Hardy), is analyzed at length by Dame Mary Lucy Cartwright [C].

It should also be mentioned that Gårding is now writing a book on Swedish mathematics. Below I draw heavily on his expert knowledge.

<u>Pre-history of Lund mathematics.</u> Before I start my subject proper, let me say a few words about the status of mathematics in Lund prior to Marcel Riesz's times.

You have already encountered the name Tycho Brahe (b. 1546 at Knutstorp mansion (50 km NW of Lund), d. 1601 in Prague), but he has no bearing upon the University of Lund, founded only in 1668, after the Swedish conquest in 1660. It has been maintained that the university was created mainly for geo-political reasons, to prevent the sons of the local clergy to attending Danish schools.

There are at least three names of some note in the history of mathematics connected with Lund. Characteristically, none of these were professional mathematicians.

18th century. Erlang Samuel Bring (1736 - 1798) was a professor of history. His name is attached to the Bring-Jerrard theorem, stating that an algebraic equation of the 5th degree can be brought to the normal form $x^5 + ax + b = 0$, using a suitable Tschirnhaus transform. Some people have seen in Bring a precursor of Abel's but, according to Gårding, this is a gross overestimation.

19th century. Albert Victor Bäcklund (1845-1922) was a professor of physics. His name again is connected with the Bäcklund transform, now very fashionable in non-linear p.d.e. (sine-Gordon and all that).

Turn of the century. Finally, Carl Ludwig Wilhelm Charlier (1862 - 1934) was an astronomer. Being an pioneer in stellar statistics, it is natural that he became involved in questions of probability. The Charlier polynomials form a system of "discrete" orthogonal polynomials corresponding to the Poisson distribution.

The modern era of Lund mathematics started - if we exclude a short interlude with the Dane Niels Norlund, famous for several monographs on topics such as difference equations and special functions, who was a professor here for a year or so - with Marcel Riesz being appointed a full professor in 1926. After his retirement in 1952 he spent some ten years at various places in the United States - Bloomington, Chicago and finally College Park, Maryland. When he became ill, he returned to Lund and died here.

His brother. Marcel Riesz was one out of three brothers. An elder brother was the famous F. Riesz (1880 - 1956). The third brother devoted himself to practising of law.

The general concensus among mathematicians seems to have been that F. was by far the more powerful and influential among the two. But such judgements change with time and perhaps we are now going to witnes a phase when the importance of Marcel Riesz's work is being reevaluated. Gårding tells us ([G2], p. 72) that during his last encounter with him - Riesz was then already hospitalized - his thoughts wandered to this very subject. "Remember", Riesz told Gårding, "that I am just a brightly colored copy of my brother".

Stockholm period 1910 - 1925. Marcel Riesz came to Sweden in 1911 upon the invitation of G. Mittag-Leffler. The two had met, apparently, in 1908 on the occasion of the ICM in Rome and, as Riesz's first mathematical subject was summation methods, Mittag-Leffler took a liking for the young man. Remember that Marcel Riesz was only in his 20's at the time. Riesz had begun his studies in Budapest. After a shorter stay at Göttingen, he had spent the year 1910 - 1911 in Paris.

One of his first assignments in this country was to compile an index for the first 35 volumes of Acta Mathematica, as you know, a journal founded by Mittag-Leffler himself. As a unique specimen in the mathematical literature, it contains a complete collection of portraits of all the authors!

As Mittag-Leffler was already an aged man, Marcel Riesz soon became the leader of the young generation of Stockholm mathematicians. His students from that period include names such as Harald Cramér, known for his achievements in probability, who also later became Chancellor of the Swedish universities (universitetskansler) - he died only recently aged 92 - and Einar Hille, a Swedish-American, born in New York, who had returned to his country of origin to study mathematics; Hille, apparently, was underestimated in Sweden and ultimately went back to the U.S. and ended up as a professor at Yale.

Addresses. Personaliae. Marcel Riesz address in Stockholm - Döbelnsgatan 5 - is one of the famous ones in the history of mathematics - it is here that he proves his celebrated conjugate function theorem. In Lund he soon moves to Kävlingevägen 1, to live in on the top floor of what was then and still is one of the most fashionable apartment buildings in town. I have been told that the flat was originally intended for Nils Zeilon, who became a professor in the same year as Riesz. But Zeilon had a large family so ultimately the flat turned out to be too expensive for him. One of Riesz's daughter Birgit Riesz-Larsson has after her own retirement moved back there. The other (twin) daughter Margit Riesz-Pleijel is married to my thesis advisor Åke Pleijel.

Competition. As you may know, in Sweden one has to compete for a professorship. In 1925? when Riesz applied for a professorship at Stockholm his competitors were Torsten Carleman and Harald Cramér. Riesz lost to Carleman and had to content himself with a chair at Lund. To become a professor at Lund, at the time, was considered, or at least Riesz himself considered it as an exile. Is it still?

Scientific profile. Students. For Marcel Riesz his arrival in Lund means, scientifically speaking, a watershed. Presumably through his contacts with Nils Zeilon, Riesz turned away from classical analysis and became interested in p.d.e. and became thereby the founder of the Swedish or Lund school in p.d.e. (Gårding, Hörmander etc.). Zeilon is a name which is little known in the mathematical community but Gårding has tried to reevaluate him on several occasions (see e.g. his address to the 1970 ICM [G3]). In the 1910's Zeilon wrote several penetrating studies on the singularities of hyperbolic equations but, being published in an obscure Swedish periodical, their influence has been negligible. In particular, he reached in this way an understanding of the physically important problem of double refraction in crystals, a question where Sonja Kowalewsky, incidentally another one of Mittag-Leffler's "protégés", had grossly erred (actually by misusing the Cauchy--Kowalewsky theorem!; this issue has recently again aroused some controversy; see several articles in the Intelligencer). Zeilon was also an amateur painter and painting seems to have been one of his major activities in later professorial years. In contrast to Riesz, he never had students.

As I have mentioned already, Riesz's first mathematical field had been summation methods, something that was very fashionable in the first decades of the century - remember that L. Fejér was a Hungarian and in 1901 had proved his famous theorem on the Cesaro summability of Fourier series. (It may well be that there will soon be a revival of interest in summation; this is connected with advances on the numerical (computer) side;

see e.g. [Wi] for a recent book on the subject.) But he also worked in classical function theory - the F. and M. Riesz theorem is surely well-known to everybody - and on trigonometric series, where he collaborated with Hilb on an Encyklopädie-article.

Then he invested some effort (ca. 1920) on another problem fashionable for the day, the moment problem. In this connection he did some work in which he in a way anticipated the Hahn-Banach theorem. However, by and large his achievements here fall into the shadow of men such as Carleman and Nevanlinna, generally considered to be mathematicians of greater stature.

Riesz's contribution to mathematics from the Lund (or post-Lund) period is intimately tied up with the work of his many talented students and associates. Before proceeding it will be convenient to draw a chart, depicting some general trends.

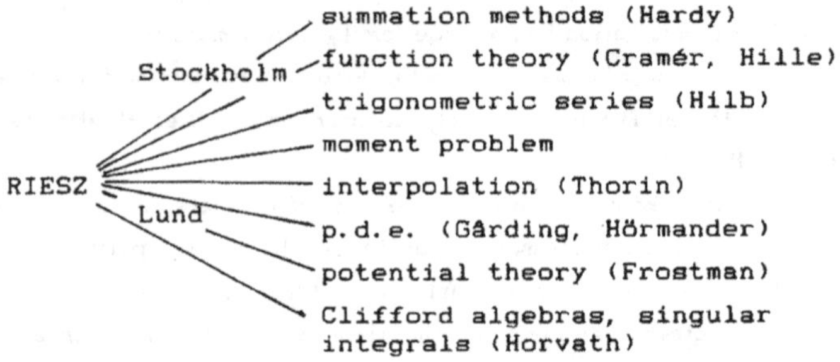

Among less known names, let me mention Carl Hylthén-Cavallius and Lennart Sandgren. The former published some original work on positive trigonometric sums, while the latter wrote on convexity and also a thesis on the Steklov eigenvalue problem; later he made a career as a civil servant, ending up as "landshövding" in "Kristianstads län".

Riemann-Liouville integral. If one looks for a common denominator in Riesz's work, there is one thing par excellence that comes to ones mind and this is the Riemann-Liouville integral of fractional integration:

$$\boxed{I^{\alpha}f(x) = \frac{1}{\Gamma(\alpha)}\int_0^x (x-y)^{\alpha-1}f(y)\,dy} \qquad (\alpha > 0),$$

which he came to investigate from various angles. He was, presumably, led to this through his previous preoccupations with summation, where he had invented what he himself termed the typical means and other people have begun to call the Riesz means. In terms of integrals, rather than series, the basic fact (on the Abelian side) is that if f is a function on the real line such that $f(x) \to A$ (as $x \to \infty$) then, for any $\alpha > 0$, also

$$\frac{\alpha}{x}\int_0^x (1-\frac{y}{x})^{\alpha-1}f(y)\,dy \to A.$$

The Riesz method of summation has several advantages compared to the older Césaro method. Riesz's early work in this area is summarized in the Hardy-Riesz Cambridge tract [HR] from 1915, which, however, due to war circumstances was mostly written by Hardy alone; the two authors happened to be in opposite camps. During the war (World War I), Riesz, still a citizen of Austria-Hungary, served at his country's embassy in Stockholm.

Returning to the Riemann-Liouville integral itself, Riesz observed that the operators I possess the semigroup property:

$$I^{\alpha}I^{\beta} = I^{\alpha+\beta}, \qquad I^{0} = \text{identity}.$$

Using analytic continuation (taking, at an intermediate stage, α to be complex, with Re α > 0) he could define I^{α} also for negative α. Especially, one has I^{-1} = derivation, whereas I^{1} = (indefinite) integration. It is curious that Gel'fand in the preface of the book [GS] maintains that Riesz in this way became a precursor of distributions or, as Gel'fand himself says, generalized functions; the name of L. Schwartz hardly occurs in [GS].

Riesz then goes on to generalizations in several variables. In particular, he defines the generalized or Riesz potentials

$$I^{\alpha}\mu(x) = c_{\alpha,n} \int |x - y|^{\alpha-n}\, d\mu(y) \qquad (0 < \alpha < n),$$

where $c_{\alpha,n}$ is a suitable "gamma factor", for an arbitrary "mass distribution" μ in R^{n} (equipped with the Euclidean metric); if $\alpha = 2$ (and $n \neq 2$) one has the Newton potential. Now formally $I^{-2} = \Delta$ = Laplacean. The subject was then followed up by Otto Frostman in his famous 1935 thesis [F], which still may serve as a most readable introduction to "modern" potential theory. Frostman was a secondary school teacher for many years; from my own boyhood I remember his characteristic profile on the yard of the Lund "Cathedral School" (Katedralskolan; incidentally one of the oldest institutions of higher learning in Scandinavia). When at last in 1952 (the very year of Riesz's retirement) he obtained a nomination as a university professor at Stockholm, his energy seemed to have been spent, and he published little after that.

In another direction, one can define operators I^{α} generalizing the Riemann-Liouville integral also in R^{n+1} equipped with the Minkowski metric. Then $I^{-2} = \Box$ = d'Alembertian. Riesz could exploit this to give a treatment of the Cauchy problem for the wave equation. Thereby he got a most elegant substitute for Hadamard's "partie finie", used by the French mathematician in his studies of general second order hyperbolic equations. Riesz's investigations in this area, dating from the 30's, were published only much later in a truly monumental Acta paper [R1].

The Riemann-Liouville integral recurs, for instance, also in a little known paper by Gårding [G4], where the author - prior to his later work on general higher order hyperbolic equations with constant coefficients - investigates certain hyperbolic equations related to Siegel's generalized upper halfplanes. This is perhaps a matter into which one ought to dig deeper, as Siegel halfplanes connect to such now popular issues as Jordan

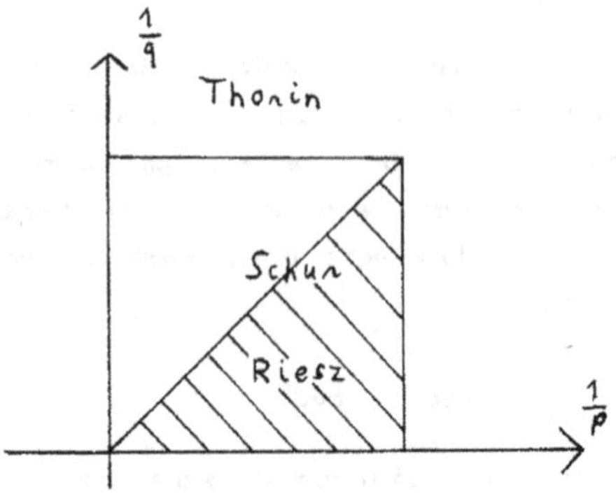

triple systems (JTS). It strikes me now also that a few years ago I pointed out a connection between Gårding's theory and the then freshly solved van der Waerden's conjecture [P2].

Interpolation (for "aficionados" only). I come now to the topic which really has served as the chief raison d'être of this meeting. The first result about interpolation in the literature is however the one by Schur (1911), which says, roughly speaking, that T: $L^1 \to L^1$, T: $L^\infty \to L^\infty$ entails T: $L^p \to L^p$ ($1 < p < \infty$), that is, what we now understand as the "diagonal" case of the Riesz-Thorin theorem. For some reason, this is a fact that never has established itself as a genuine "theorem"; it is being "rediscovered" over and over again. Riesz is of course fully aware of his predecessor, whom he duly quotes.

The road by which Riesz reaches his "convexity" theorem [R2] is curiously enough connected with the conjugate function theorem, another of Riesz's lasting contributions to Analysis (discovered in 1923 but published only in 1927 [R3]). For the real line, H denoting the Hilbert transform, it simply states that H: $L^p \to L^p$, $1 < p < \infty$. Riesz's proof of this result, reproduced many times before, goes, rougly speaking, as follows:

1) He proves it for $p = 2, 4, 8, \ldots$

2) Then he "interpolates" to get it for $2 < p < \infty$ (p not an even integer).

3) Finally, he applies duality to incorporate the case $1 < p < 2$ as well.

It is thus the middle step when generalized, which is the genesis of the whole discipline of (abstract) interpolation.

As for step 1, the case $p = 2$ is trivial. If, for instance, $p = 4$ the argument is the following. By Cauchy's theorem $\int_R f^4\,dx = 0$ where $f = u + iv$ is an analytic function in the upper halfplane with $v = Hu$ on the boundary R. Expanding and taking real parts, one gets the identity

$$\int u^4 - 6\int u^2v^2 + \int v^4 = 0.$$

From this, using again Hölder's inequality, to take care of the middle term, one then readily obtains $\|v\|_4 \le c\|u\|_4$. For p a general even integer, the proof is analogous.

Riesz tells himself how he hit on this argument. He intended to give the (trivial) case p = 2 as a problem to a student (not one of the famous ones!) in an examination and, as he suspected that the poor man did not know Plancherel's theorem, he began to think if there was not an alternative route... Whence Hardy's laconic comment ([C], p. 502): "Your student's life is not entirely without value (though I suppose he will never understand why)."

Nowadays the conjugate function theorem is subsumed in the Calderón-Zygmund theory of singular integral, thus real variable theory. But in questions of obtaining sharp estimates of the constant, the function theoretic approach, thus having its origin in Riesz's work, still is of importance; see e.g. [E1], [E2] for a recent contribution.

In [P3] I gave a "contemporary" version of Riesz's original proof of his interpolation theorem, interpreting it in terms of a suitable interpolation method, the "Riesz method". Can other known proofs of Riesz's theorem, e.g. the ones by Paley or the one by Thorin (see [Th2], appendix and [Th3]), be given a similar reinterpretation?

The extension to the whole square was open for some time and Riesz himself seems to have been of the opinion that this was not possible (probably because of an example in [R2]). Therefore, Thorin's proof [Th1] in 1939 of the full Riesz-Thorin theorem which also forms the basis for Calderón's complex method of interpolation, came as a surprise to everybody. Littlewood alludes to Thorin's achievement as "the most impudent idea in mathematics" ([L], p. 1). Olov Thorin, who is an utterly modest man, has himself described in detail the process of discovery [Th3]. Apparently, it was a cursory remark of Frostman's that triggered him off. In his 1948 thesis [Th2] Thorin independently proves also Sobolev's theorem about the L^p-continuity of the Riesz potentials (the n-dimensional version of a theorem by Hardy-Littlewood). Most of his life he worked as an actuary - Riesz too served as a consultant for Swedish insurance companies - but upon his retirement he is reported to have said: "Now at last I am free to do mathematics!" His name is well-known to probabilists and he has done important work on i.d.r.v. (cf. [BP]).

Clifford algebras. Singular integrals. In his later life Marcel Riesz worked on somewhat more esoteric subjects. I shall bypass here his preoccupations with number theory and "pure" algebra, concentrating instead on "geometry". Riesz, apparently, was always fascinated by the internal beauty of geometry, in particular, the interplay between geometry and "physics" (relativity, quantum mechanics). In particular, he tried to promote the use of Clifford numbers in Analysis, notably in connection with the Dirac equation. At the time, people did not think very highly of this. At least, his Maryland notes [R4] got a bad review [Ti]. But, as we have seen before, customs change even in such a conservative science as ours.

An 2^n-dimensional Clifford algebra C is an algebra with generators $e_1,\dots,e_n$ subject to relations $e_i^2 = \pm 1$, $e_ie_k + e_ke_i = 0$ $(i \neq k)$. Then, formally, a solution ϕ (with values in a suitable C-module) of the equation

$$\sum_{i=1}^{n} e_i \frac{\partial \phi}{\partial x_i} = 0$$

will have the property that each component is a nullsolution of the "wave" operator $\Sigma \pm\partial\phi^2/\partial x_i^2 = 0$. Whence the connection with Dirac's theory. If we add one more variable x_0, taking $e_0 = 1$, we get a generalization of the Cauchy-Riemann equations, which leads to a generalized function theory based on the notion of "monogenic" function (cf. e.g. the monograph [BDS]). Monogenic functions have recently atracted a lot of interest. For instance, they have been used in the context of singular integrals (see e.g. [McI]). The famous Riesz transforms (generalizing the Hilbert transform in the case of one variable) presumably also arose in this context (see [H2]). Later they came to be the substratum of the Fefferman--Stein-Weiss theory of generalized H^p spaces [FS], [SW].

At the '83 Lund conference Svante Janson and I suggested [JP], in an attempt to generalize Calderón's method, to use, instead of analytic functions in a strip (or a disk) in the complex plane, harmonic vector fields in an analogous domain in R^{n+1}, and similar considerations can be made more generally with Clifford numbers. The Thorin construction, in the original L^p context, basically uses only the properties of one special function, the exponential. If one could find a suitable Clifford analogue of the exponential, perhaps one could do something... I wonder what Marcel Riesz himself would have said had somebody in his lifetime pointed out to him this connection between two apparently quite separate aspects of his work.

Conclusion. Compared with other great mathematicians of the past, Marcel Riesz published relatively little and, as we have seen, often with a great delay. This has in part to do with his personality, about which I have said very little. Above all, he was a perfectionist. Let me conclude with a piece of happy news. After many years of delay, his collected works will now be published by Springer-Verlag, with Lars Hörmander as editor. The sponsors are N.F.Riesz - the same as for this conference - and the Swedish Actuarial Society, this as a tribute to his toils in actuarial mathematics.

References

BDS. Brackx, F., Delanghe, R., Sommen, F.: Clifford analysis. Research notes in mathematics 76. Boston: Pitman 1982.

BP. Bondesson, L., Peetre, J.: The classes V_a are monotone. These proceedings.

C. Cartwright, M.L.: Manuscripts of Hardy, Littlewood, Marcel Riesz and Titchmarsh. Bull. London Math. Soc. 14, 472-532 (1982).

E1. Essén, M.: A superharmonic proof of M. Riesz conjugate function theorem. Ark. Mat. 22, 241-249 (1984).

E2. Essén, M.: A generalization of the M. Riesz theorem on conjugate functions and the Zygmund LlogL-theorem to R^d, d ≥ 2. Technical report. Uppsala: 1986.

F. Frostman, O.: Potentil d'équilibre et capcité des ensembles avec quelques applications a la théorie des functions. Commun. Sém. Math. Univ. Lund 3, 1-118 (1935).

FS. Fefferman, C., Stein, E.: H^p spaces of several variables. Acta Math. 129, 137-193 (1972).

G1. Gårding, L.: Marcel Riesz in memoriam. Acta Math. 124, I-XI (1970).

G2. Gårding, L.: Marcel Riesz. In: Kungl. Fysiogr. Sällsk. Årsbok 1970, pp. 65-70. Lund : 1970.

G3. Gårding, L.: Sharp fronts and lacunas. In: Actes, Congr. International des Mathématiciens, 1/10 septembre 1970, Nice, France, t. 2, pp. 723-729. Paris: Gauthier-Villars 1971.

G4. Gårding, L.: The solution of Cauchy's problem for two totally hyperbolic linear differential equations by means of Riesz integrals. Ann. Math. 48, 785-826 (1947).

GS. Gel'fand, I.M., Shilov, G.E.: Generalized functions I. Moscow: Goz. Izd. Fiz.-Mat. Lit. 1958 [Russian].

H1. Horvath, J.: L'oeuvre mathématique de Marcel Riesz I, II. Cah. Semin. Math. Hist. 3, 83-121 (1982), 4, 1-59 (1983).

H2. Horvath, J.: Sur les fonctions conjugées a plusieurs variables. Indag. Math. 16, 17-29 (1953).

HR. Hardy, G.H., Riesz, M.: The general theory of Dirichlet series. Cambridge Tracts in Mathematics 18. Cambridge: Cambridge University Press 1915.

JP. Janson, S., Peetre, J.: Harmonic interpolation. In: Lecture Notes in Mathematics 1070, pp. 92-124. Springer: Berlin etc. 1984.

L. Littlewood, J.: Some problems in real and complex analysis. Lexington: Heath 1968.

McI. McIntosh, A.: Clifford algebras and some applications in analysis. Notes by J. Picton-Warlow.

P1. Peetre, J.: biography of Marcel Riesz. In: Biographic Encyclopedia of Scientists and Technologists. Milano: Mondadori 1974.

P2. Peetre, J.: Van der Waerden's conjecture and hyperbolicity. (Technical report). Lund: 1981.

P3. Peetre, J.: Two new interpolation methods based on the duality map. Acta Math. 143, 73-91 (1979).

R1. Riesz, M.: L'intégrale de Riemann-Liouville et le probleme de Cauchy. Acta Math. 81, 1-223 (1949).

R2. Riesz, M.: Sur les fonctions conjuguées. Math. Z. 27, 218-244 (1927).

R3. Riesz, M.: Sur les maxima des formes bilinéaires et sur les fonctionelles linéaires. Acta Math. 49, 465-497 (1927).

R4. Riesz, M.: Clifford numbers and spinors. Lecture notes. College Park: Institute for Fluid Dynamics and Applied Mathematics of the University of Maryland 1957.

SW. Stein, E.M., Weiss, G.: On the theory of harmonic functions of several variables, I. The theory of H^p-spaces. Acta Math. 103, 25-62 (1960).

Th1. Thorin, O.: An extension of a convexity theorem of M. Riesz Kungl. Fysiogr. Sällsk. i Lund Förh. 8, 166-170 (1938).

Th2. Thorin, O.: Convexity theorems generalizing those of M. Riesz and Hadamard with some applications. Comm. Sém. Math. Univ. Lund. 9, 1-58 (1948).

Th3. Thorin, O.: personal communication (Dec. 1979).

Ti. Tits, J., review of [R4]. Math. Rev. 31, 1113-1114 (1966).

We. Weil, A.: History of mathematics: why and how. In: Proc. Internat. Congr. of Mathematicians, Helsinki, Aug. 15-23, 1978, pp. 227-236. Helsinki: 1980.

Wi. Wimp, J.: Sequence transformations and their applications. New York: Academic Press 1981.

N.B. - A complete list of Marcel Riesz's scientific publications can be found in [G1]; here we have listed only those few items explicitly mentioned in the text ([HR], [R1-4]).

INTERPOLATION OF TENT SPACES AND APPLICATIONS

J. Alvarez and M. Milman
Department of Mathematics
Florida Atlantic University
Boca Raton, Florida 33431

In a series of papers [CMS], [CMS1], Coifman, Meyer and Stein have studied a new family of function spaces, the so-called "tent spaces." These spaces have proved to be useful for the study of a variety of problems in harmonic analysis. Moreover, their study has led to interesting simplifications and refinements of some basic techniques of real analysis. In particular we mention the results of [CMS1] concerning the Cauchy integral on Lipchitz curves, the natural and simple approach to the atomic decomposition and interpolation theory of H^p spaces (cf. [CMS1]) and the connection with the theory of Carleson measures (cf. [CMS1], [AM], [BJ]).

Tent spaces on product domains (cf. [AM]) have been used in [GM] to settle the problem of describing the complex interpolation spaces between H^p and BMO on product domains. We also mention the authors' work on vector valued tent spaces and their duality and interpolation properties (cf. [AM1]). An interesting consequence of this development is a generalization of the Lions-Peetre interpolation theorem of vector valued L^p spaces to the setting of vector valued H^p spaces.

These papers also provide applications to the study of maximal operators, square functions, balayages, weighted norm inequalities....

In this note we survey results concerning the interpolation theory of tent spaces and spaces of Carleson measures. The plan of the paper is as follows: the first three sections give the definition and basic properties of tent spaces and spaces of Carleson measures, §4 treats the interpolation theory of spaces of Carleson measures, §5 and §6 provide the real and complex interpolation of tent spaces. In particular §6 includes a Wolff type theorem for Quasi Banach lattices (cf. [GM1]), §7 provides an application to the interpolation theory of H^p spaces, while §8 contains a brief synopsis of tent spaces on product domains, vector valued tent spaces and further applications.

1. The T^p_q spaces

We shall work in $\mathbf{R}^{n+1}_+$ but most of the results can be stated in the more general context of homogeneous spaces. Points in $\mathbf{R}^{n+1}_+$ shall be denoted by (x,t) or (y,t), $x,y \in \mathbf{R}^n$, $t > 0$. Given $x \in \mathbf{R}^n$, let $\Gamma(x)$ denote the cone $\{(y,t)/|x-y| < t\}$. Let Ω be an open set, $\Omega \subseteq \mathbf{R}^n$, we let $T(\Omega)$ ("the tent over Ω") be the subset of $\mathbf{R}^{n+1}_+$ defined by $T(\Omega) = \{(x,t)/B(x,t) \subset \Omega\}$, where $B(x,t)$ denotes the ball with centre x and radius t.

The tent spaces T_q^p, $0 < p \leqslant \infty$, $0 < q \leqslant \infty$, are defined using two families of functionals. The A_q functionals defined by

$$A_q(f)(x) = \{\int_{\Gamma(x)} |f(y,t)|^q \frac{dydt}{t^{n+1}}\}^{1/q} \tag{1.1}$$

$$A_\infty(f)(x) = \sup_{\Gamma(x)} |f(y,t)| \tag{1.1'}$$

The C_q functionals given by

$$C_q(f)(x) = \sup\{\frac{1}{|B|}\int_{T(B)} |f(y,t)|^q \frac{dydt}{t}\}^{1/q} \tag{1.2}$$

where the sup is taken over all balls containing x.

The tent spaces T_q^p, $0 < p$, $q < \infty$ are defined by $T_q^p = \{f/A_q(f) \in L^p(\mathbb{R}^n)\}$, and we let $||f||_{T_q^p} = ||A_q(f)||_p$. The case $q = \infty$ and $p < \infty$ requires a natural modification: T_∞^p will denote the class of continuous functions in $\mathbb{R}_+^{n+1}$ such that $A_\infty(f) \in L^p(\mathbb{R}^n)$ and such that $\lim_{\varepsilon\to 0^+} ||f_\varepsilon - f||_{T_\infty^p} = 0$, where $f_\varepsilon(x,t) = f(x+\varepsilon,t)$, and $||f||_{T_\infty^p} = ||A_\infty(f)||_p$. Finally the T_q^∞ spaces $(q < \infty)$, are defined by the condition $C_q(f) \in L^\infty$ and $||f||_{T_q^\infty} = ||C_q(f)||_\infty$.

It is readily verified that the T_q^p spaces are complete and that for $1 \leqslant p$, $q \leqslant \infty$, the T_q^p are Banach spaces.

2. Spaces of Carleson measures

For measures on $\mathbb{R}_+^{n+1}$ we can also consider variants of the A_1 and C_1 functionals. For a measure w on $\mathbb{R}_+^{n+1}$ we define

$$A_1(w)(x) = \int_{\Gamma(x)} t^{-n} d|w|(y,t) \tag{2.1}$$

$$C_1(w)(x) = \sup_{B\ni x} \frac{1}{|B|}\int_{T(B)} d|w|(y,t) \tag{2.2}$$

For $0 < \Theta < 1$ the spaces V^Θ of Carleson measures of order Θ (cf. [AB]) consist of those measures w such that for some $C > 0$

$$|w|(T(\Omega)) < C|\Omega|^\Theta \quad \forall\ \Omega \text{ open in } \mathbb{R}^n \tag{2.3}$$

We let

$$||w||_{V^\Theta}\ \inf\{C/\ (2.3) \text{ holds } \forall\ \Omega \text{ open, } \Omega \subset \mathbb{R}^n\}$$

It is also convenient to define $V^0 = \{w|w$ finite measure on $\mathbb{R}_+^{n+1}\}$.

The spaces V^Θ can be also characterized in terms of the A_1 functionals (cf. [AB]):

$$V^\Theta = \{w|A_1(w) \in L(\tfrac{1}{1-\Theta},\infty)\} \tag{2.4}$$

It is shown in [CMS1] that

$$(T_\infty^1)^* = V^1 \tag{2.5}$$

The duality result (2.5) is a consequence of the basic inequality valid for Carleson measures on $\mathbb{R}_+^{n+1}$. In fact if f is a function in T_∞^1 , and w is a Carleson measure on $\mathbb{R}_+^{n+1}$, then

$$|w|\{(x,t)/|f(x,t)|>\lambda\} \leq C|\{x/A_\infty(f)(x) > \lambda\}| \; ||w||_{V^1} \tag{2.6}$$

Integrating we obtain

$$|\int f(x,t)dw(x,t)| \leq C||w||_{V^1} \; ||f||_{T_\infty^1} \tag{2.7}$$

from which one half of (2.5) follows.

To complete the proof of (2.5) one makes appeal of the atomic decomposition of functions in T_∞^1 . A function f is an "atom" if it is supported in the tent over some ball $B \subseteq \mathbb{R}^n$, with $|f(x,t)| \leq |B|^{-1}$. In [CMS1] it is shown that if $f \in T_\infty^1$ then f admits an atomic decomposition: $f = \sum_{i=1}^{\infty} \lambda_i f_i$, f_i atoms, $\lambda_i \in \mathbb{R}$, and moreover

$$||f||_{T_\infty^1} \sim \inf\{\Sigma|\lambda_i| : f = \Sigma\lambda_i f_i \text{ , } f_i \text{ atoms}\} . \tag{2.8}$$

In [AM] , [BJ] it is shown that the spaces V^Θ can be characterized as the duals of appropriate tent spaces. This requires one additional ingredient, namely, the introduction of tent spaces based on Lorentz spaces. Thus, the spaces $T_q^{r,s}$ are defined by demanding that the appropriate functional (A_q or C_q) belongs to the Lorentz space $L(r,s)$.

A typical result is

$$(T_\infty^{p,1})^* = V^{1/p} \quad , \quad 1 \leq p < \infty . \tag{2.9}$$

For a detailed study we refer to [AM] , [BJ] and [AM1] .

Finally, following Amar and Bonami [AB] we shall consider the spaces of measures W^Θ defined by

$$W^\Theta = \{w/A_1(w) \in L^{\frac{1}{1-\Theta}}\} . \tag{2.10}$$

3. Atomic decompositions and duality of tent spaces

The extension of the duality theory of §2 is provided by a generalization of (2.6) (cf. [CMS1]):

$$\int_{\mathbb{R}_+^{n+1}} |f(x,t)g(x,t)|\frac{dxdt}{t} \leq C \int_{\mathbb{R}^n} A_q(f)(x)C_{q'}(g)(x)dx \tag{3.1}$$

where $\frac{1}{q} + \frac{1}{q'} = 1$.

It is to be noted that the A_q and C_q functionals are related by the following estimates

$$||A_q(f)||_p \leq C_{p,q}||C_q(f)||_p \quad , \quad 0 < p < \infty \tag{3.2}$$

$$||C_q(f)||_p \leq C_{p,q}||A_q(f)||_p \quad , \quad q < p < \infty \tag{3.3}$$

It is then shown in [CMS1] that

$$(T_q^p)^* = T_{q'}^{p'} \quad , \quad 1 < p < \infty \quad , \quad 1 < q < \infty . \tag{3.4}$$

For $0 < p \leqslant 1$, $1 < q \leqslant \infty$, the tent spaces T^p_q can be characterized through atomic decompositions. Let $0 < p \leqslant 1$, $1 < q \leqslant \infty$, a (p,q) atom f is a function supported in the tent of a ball such that

$$||f||_{L^q(\mathbf{R}^{n+1}_+,\frac{dxdt}{t})} \leqslant |B|^{\frac{1}{q}-\frac{1}{q}} .$$

Then (cf. [CMS1]) , $f \in T^p_q$ can be written as $f = \sum_{i=1}^{\infty} \lambda_i a_i$, where the a_i are (p,q) atoms, and moreover

$$||f||_{T^p_q} \sim \inf\{(\Sigma|\lambda_i|^p)^{1/p}: f = \Sigma\lambda_i a_i\} \tag{3.5}$$

The duality theory of tent spaces for $p \leqslant 1$ is developed in [AM] and [AM1] .

4. Interpolation of spaces of Carleson measures

The interpolation theory of spaces of Carleson measures is considered in [AB] and [AM] . In [AM] the approach is the computation of the K functional for spaces of Carleson measures. On the other hand, the method of [AB] is based on a factorization theorem and has the advantage of treating the real and complex methods at the same time.

The factorization theorem of [AB] is given by

(4.1) Theorem Let w be a positive measure, $w \in V^{\Theta}$ (resp. $w \in W^{\Theta}$), $0 < \Theta \leqslant 1$, then there exists a positive Carleson measure μ and a function $f \in L(p,\infty)(\mathbf{R}^{n+1}_+,d\mu)$ (resp. $f \in L^p(\mathbf{R}^{n+1}_+,d\mu)$) where $\frac{1}{p} = 1 - \Theta$, such that $w = \mu.f$.

In fact the needed factorization is obtained as follows. Let h be defined by $h = A_1(w)$, and $\mu = A_1(\frac{1}{h})w$, $f = [A_1(\frac{1}{h})]^{-1}$, where the A_1 of $\frac{1}{h}$ is taken by considering $\frac{1}{h}$ as a measure (cf. definition (2.1)).

As a consequence we have (cf. [AB]).

(4.2) Theorem (i) $V^{\Theta} = (V^0,V^1)_{\Theta,\infty}$

(ii) $W^{\Theta} = (V^0,V^1)_{\Theta,\frac{1}{1-\Theta}} = [V^0,V^1]_{\Theta}$.

Proof. It is easy to check the inclusions

(a) $(V^0,V^1)_{\Theta,\infty} \subseteq V^{\Theta}$;(b) $(V^0,V^1)_{\Theta,\frac{1}{1-\Theta}} \subset W^{\Theta}$,(c) $[V^0,V^1]_{\Theta} \subset W^{\Theta}$.

For example we prove (a) . Let w be a measure in $(V^0,V^1)_{\Theta,\infty}$ and let Ω be an open set in $\mathbf{R}^n$. We need to show that $\exists C > 0$ independent of Ω such that

$$|w|(T(\Omega)) \leqslant C\,|\Omega|^{\Theta} . \tag{4.3}$$

Consider the functional G defined

$$G(\mu) = \int_{T(\Omega)} d\mu$$

Clearly G is a bounded linear functional, $G \in (V_0)^*$ and $G \in (V_1)^*$ with

$||G||_{(V^0)^*} \leq 1$, $||G||_{(V^1)^*} \leq |\Omega|$. It follows by interpolation that

$$G: (V^0,V^1)_{\Theta,\infty} \to C$$

$$||G||_{(V^0,V^1)_{\Theta,\infty}} \leq C_\Theta |\Omega|^\Theta$$

therefore (4.3) follows.

The converse inclusions are obtained through the factorization theorem. In fact to prove that $V^\Theta \subseteq (V^0,V^1)_{\Theta,\infty}$ we argue as follows. Let $w \in V^\Theta$ be a positive measure, factorize it as $w = \mu.f$ where $\mu \in V^1$, $f \in L(\frac{1}{1-\Theta},\infty)$ $(\mathbb{R}_+^{n+1},d\mu)$. Consider the operator $F(h) = h\mu$, clearly, F defines a bounded operator

$$F: \begin{cases} L^1(\mathbb{R}_+^{n+1},d\mu) \to V^0 \\ L^\infty(\mathbb{R}_+^{n+1},d\mu) \to V^1 \end{cases}$$

Interpolating we find

$$F: L(\tfrac{1}{1-\Theta},\infty) \to (V^0,V^1)_{\Theta,\infty}$$

in particular,

$$F(f) = f.\mu = w \in (V^0,V^1)_{\Theta,\infty} .$$

as desired.

In [AM] results of this type are obtained by computing the K functional of Peetre for the couple (V^0,V^1) .

<u>(4.4) Theorem</u> $K(t,w,V^0,V^1) \leq C\int_0^t M(A_1(w))^*(s)ds$, where M is the maximal operator of Hardy-Littlewood.

The proof of (4.4) consists in exhibiting a nearly optimal decomposition. Let $w \in V^0+V^1$, $t > 0$, and define $\Omega_t = \{x/C_1(w)(x) > M(A_1(w))^*(t)\}$, then a nearly optimal decomposition is given by $w = \mu_0 + \mu_1$, where $\mu_0 = w\chi_{T(\Omega_t)}$. For the details see [AM] theorem (4.3) .

5. Interpolation of tent spaces

In this section we consider the real and complex interpolation of T_q^p spaces, $1 \leq p \leq \infty$, $1 < q \leq \infty$. The results for the T_∞^p scale are given in [AM] (cf. also [BJ]):

<u>(5.1) Theorem</u> Let $0 < p < \infty$, then

$K(t,f,T_\infty^p,L^\infty) \approx K(t,A_\infty(f),L^p,L^\infty)$.

From (5.1) it follows readily that

$$(T_\infty^p,L^\infty)_{\Theta,r} = T_\infty^{p_\Theta,r} , \quad \frac{1}{p} = \frac{1-\Theta}{p_\Theta} . \tag{5.2}$$

The proof of (5.1) is obtained by looking at the decomposition $f = f_0 + f_1$, where $f_0 = f.\chi_{T(\Omega_t)}$, and $\Omega_t = \{x/A_\infty(f)(x) > A_\infty(f)^*(t^p)\}$.

Similar estimates had been shown first in [CMS1] for the corresponding K functionals for the couple $(T^{p_0},T_q^{p_1})$, $1 \leqslant p_0,p_1 < \infty$. In fact using these estimates, duality and Wolff's theorem, it is proved in [CMS1] that

$$(T_q^{p_0},T_q^{p_1})_{\Theta,p} = T_q^p , \frac{1}{p} = \frac{1-\Theta}{p_0} + \frac{\Theta}{p_1}, 1 < q < \infty, 1 \leqslant p_0,p_1 \leqslant \infty . \tag{5.3}$$

We propose here a different approach to (5.3) . We show that (5.3) follows from (5.2) using a factorization technique.

(5.4) Theorem (cf. [CMS1]) (i) Let $1 \leqslant q < p \leqslant \infty$, then every $f \in T_q^p$ can be factored as $f = gh$ with $g \in T_\infty^p$, $h \in T_q^\infty$, and conversely $T_\infty^p . T_q^\infty \subseteq T_q^p$.

(ii) $T_\infty^q . T_q^\infty \subseteq T_q^{q,\infty}$.

Proof. (i) The proof of $T_\infty^p . T_q^\infty \subseteq T_q^p$ follows the same argument that will be given for (ii) . On the other hand if $A_q(f) \in L^p$, then let $g(x,t) = t^{-n}\int_{|x-y|<t} (M(A_q (f)^q))^{1/q}(y)dy$. Then $g \in T_\infty^q$ and one can also check that $h = \frac{f}{g} \in T_q^\infty$.

(ii) Let $g \in T_\infty^q$, $h \in T_q^\infty$. Then,

$$C_q^q(f)(x) \leqslant C \sup_{B \ni x} \frac{1}{|B|}\int_B A_\infty^q(g)(x)dx \leqslant CM(A_\infty^q(g))$$

Therefore, $C_q(f) \in L(q,\infty)$ since $A_\infty(g) \in L^q$.

We consider now a proof of (5.3) . To simplify the notation we consider only the case $q = 2$. It is not difficult to show the inclusion

$$(T_2^{p_0},T_2^{p_1})_{\Theta,p} \subseteq T_2^p . \tag{5.5}$$

The reverse inclusion lies deeper and will be treated in separate cases. Suppose first that $1 < 2 \leqslant p_0 < p < p_1 \leqslant \infty$. Clearly we only need to consider the case $p_0 = 2$. Let $f \in T_2^p$, then using (5.4) we factorize $f = gh$, $g \in T_\infty^p$, $h \in T_2^\infty$. Next we consider the operator $P_h(\alpha) = \alpha h$. Then, $P_h: T_\infty^2 \to T_2^{2,\infty}$, $P_h: T_\infty^{p_1} \to T_2^{p_1}$. By (5.2) , $P_h: T_\infty^p \to (T_2^{2,\infty},T_2^{p_1})_{\Theta,p}$. Thus, $f \in (T_2^{2,\infty},T_2^{p_1})_{\Theta,p}$. But a simple argument proves that $(T_2^{2,\infty},T_2^{p_1})_{\Theta,p} = (T_2^2,T_2^{p_1})_{\Theta,p}$. Therefore we have proved (5.3) under the conditions $1 < 2 \leqslant p_0 < p < p_1 \leqslant \infty$. Using duality (cf. (3.4)) we obtain (5.3) under the conditions $1 \leqslant p_0 < p < p_1 \leqslant 2$. Let us also observe that using (5.5) and duality we easily see that (5.3) holds under the conditions $1 < p_0 < p < p_1 < \infty$. Thus, we may fuse these results using Wolff's theorem and obtain (5.3) without restrictions.

We consider now the complex method. We shall prove

$$[T_2^{p_0}, T_2^{p_1}]_\Theta = T_2^p , \ 1 \leq p_0 < p \leq \infty , \ \frac{1}{p} = \frac{1-\Theta}{p_0} + \frac{\Theta}{p_1} . \tag{5.6}$$

In fact everything will follow from the real interpolation result (5.3) and a Fourier type argument (cf. [M]). First we consider the space $[T_2^1, T_2^2]_\Theta$. Observe that $T_2^2 = L^2$ is of Fourier type 2, and T_2^1 is of Fourier type 1, thus

$$T_2^p = (T_2^1, T_2^2)_{\Theta,p} \subset [T_2^1, T_2^2]_\Theta , \ \frac{1}{p} = 1 - \Theta + \frac{\Theta}{2} .$$

Since the reverse inclusion is trivial we have obtained

$$T_2^p = [T_2^1 , T_2^2]_\Theta , \ \frac{1}{p} = 1 - \Theta + \frac{\Theta}{2} .$$

By duality, we get

$$T_2^p = [T_2^2, T_2^\infty]_\Theta , \ \frac{1}{p} = \frac{1-\Theta}{2} .$$

Using the formula (5.3) for $1 < p_0 < p < p_1 < \infty$ and mixed reiteration we get

$$[T_2^{p_0}, T_2^{p_1}]_\Theta = [(T_2^{r_0}, T_2^{r_1})_{\Theta_0, r}, (T_2^{s_0}, T_2^{s_1})_{\Theta_1, s}]_\Theta = (T_2^{r_0}, T_2^{r_1})_{\mu, q} = T_2^p$$

if the indices are well chosen... Thus, we are in the position to invoke Wolff's theorem for the complex method to conclude the proof of (5.6) .

For a different approach to the results of this section we refer to [CMS1] .

6. Interpolation of tent spaces: The quasi Banach space case

It turns out that the extension of (5.3) to the range of parameters $0 < p_0 < p < p_1 \leq \infty$, $1 < q < \infty$ is routine (cf. [CMS1]). Apparently that is not the case for the complex method. The reason for this difference is given by the fact that Wolff's theorem for the real method holds for quasi Banach spaces while it is an open problem to derive a complex method version of Wolff's theorem valid for quasi Banach spaces (cf. [W]).

However the tent spaces T_q^p , $0 < p < \infty$, $1 < q < \infty$ are lattices and in [GM1] we have obtained a version of Wolff's theorem for complex interpolation of quasi Banach lattices.

<u>(6.1) Theorem</u> Let $\{A_i\}_{i=1}^4$ be quasi Banach lattices such that for some $\Theta, \phi \in (0,1)$ we have $A_2 = A_1^{1-\phi} A_3^\phi$, $A_3 = A_2^{1-\Theta} A_4^\Theta$. Then, for $\Psi = \Theta/(1 - \phi + \phi\Theta)$, $\eta = \phi\Psi$, we have

(i) $A_3 = A_3^{1-\Theta/\Psi}(A_1^{1-\Psi}A_4^\Psi)^{\Theta/\Psi}$

(ii) $A_2 = A_2^{1-\Theta/\Psi}(A_1^{1-\eta}A_4^\eta)^{\Theta/\Psi}$.

<u>Proof.</u> By symmetry we only need to prove (i) . Let $f \in A_3$ then there exist $f_2 \in A_2$, $f_4 \in A_4$ such that $|f| = |f_2|^{1-\Theta}|f_4|^\Theta$, $||f||_{A_3} \sim ||f||_{A_2}^{1-\Theta}||f_4||_{A_4}^\Theta$. Us-

ing our hypothesis write $|f_2| = |f_1|^{1-\phi}|f_3|^{\phi}$, with $||f_2||_{A_3} \sim ||f_1||_{A_1}^{1-\phi}||f_3||_{A_3}^{\phi}$.

Thus $|f| = (|f_1|^{1-\Psi}|f_4|^{\Psi})^{\Theta/\Psi}|f_3|^{1-\Theta/\Psi}$ and $||f||_{A_3^{1-\Theta/\Psi}(A_1^{1-\Psi}A_4^{\Psi})^{\Theta/\Psi}} \leq C\ ||f||_{A_3}$.

The converse inclusion can be obtained in a similar fashion.

The idea to obtain a Wolff type theorem is to iterate the procedure of (6.1) . This iterative procedure is hidden in an important theorem of Stafney [S] (cf. [BL] page 82, exercise 21).

(6.2) Theorem Let $\bar{A}$ be a quasi Banach couple such that $(A_0,A_1)_{\Theta,p} = A_0$, $0 < \Theta < 1$, $p \geq 1$ with $A_0 \cap A_1$ dense in A_i , $i = 0,1$; then $A_0 = A_1$.

Now observe that for any quasi Banach lattice couple $\bar{A} = (A_0,A_1)$, $A_0^{1-\Theta}A_1^{\Theta} \subseteq (A_0,A_1)_{\Theta,\infty}$, $0 < \Theta < 1$. Combining this fact with (6.1) and (6.2) we obtain

(6.3) Theorem Let $\{A_i\}_{i=1}^{4}$ be quasi Banach lattices, $0 < \Theta$, $\phi < 1$. Suppose that the following conditions hold:

(i) $A_2 = A_1^{1-\phi}A_3^{\phi}$

(ii) $A_3 = A_2^{1-\Theta}A_4^{\Theta}$

(iii) $A_1 \cap A_4 \subset A_2 \cap A_3$ and $A_1 \cap A_4$ dense in A_2 and A_3 .

Then,

(6.4) $A_3 = A_1^{1-\Psi}A_4^{\Psi}$, $A_2 = A_1^{1-\eta}A_4^{\eta}$

where $\Psi = \dfrac{\Theta}{1-\phi+\phi\Theta}$, $\eta = \phi\Psi$.

Using (6.3) we can derive the following

(6.4) Theorem $[T_2^{p_0},T_2^{p_1}]_{\Theta} = T_2^{p}$, $0 < p_0 < p < p_1 \leq \infty$.

Proof. The case $0 < p_0 < p < p_1 < \infty$ can be obtained from the quasi Banach space version of (5.3) using the mixed reiteration theorem of [CWMS] . Combining this result with (5.6) and (6.3) gives (6.4) .

7. Interpolation of H^p spaces

The interpolation theory of T_q^p spaces can be used to derive the corresponding one for H^p spaces.

Let ϕ be a bounded function with compact support such that (i) $|\phi(x + h) - \phi(x)| \leq C(|h|/|x|)^{\varepsilon}$ for some $\varepsilon < 0$, (ii) $\int x^{\gamma}\phi(x)dx = 0$, $\forall|\gamma| < N$, where N will be specified later on. Let $\phi_t(x) = t^{-n}\phi(\frac{x}{t})$, $t > 0$. Define,

$$\pi_{\phi}(f) = \int_0^{\infty} f(\cdot,t) * \phi_t \frac{dt}{t} .$$

Let $0 < p \leqslant 1$ be given and suppose that ϕ satisfies the above conditions with $N \geqslant n[\frac{1}{p} - 1]$, then

(7.1) Theorem (cf. [CMS1]) (i) $\pi_\phi : T_2^p \to H^p(\mathbb{R}^n)$

(ii) $\pi_\phi : T_2^\infty \to BMO$.

Proof. (i) follows from the atomic decomposition of functions in T_2^p.

(ii) By duality it suffices to show that $\forall f \in T_2^\infty$, $\pi_\phi(f)$ can be paired with H^1, but

$$\int_{\mathbb{R}^n} \pi_\phi(f)(x)g(x)dx = \int f(x,t)(g * \check{\phi}_t)(x)\frac{dxdt}{t}$$

$$\leqslant C\int C_2^2(f)(x)A_2(g * \check{\phi}_t)(x)dx$$

$$\leqslant C\, ||f||_{T_2^\infty}||A_2(g * \check{\phi}_t)||_{L^1}$$

$$\leqslant C\, ||f||_{T_2^\infty}||g||_{H^1}\,.$$

It is not difficult to see that if ϕ satisfies the condition

$$-2\pi\int_0^\infty \hat{\phi}(\xi t)|\xi|e^{-2\pi|\xi|t}dt = 1\ ,\ \forall\xi$$

then

$$\pi_\phi(t\frac{\partial u}{\partial t}(x,t)) = f$$

where u is the Poisson integral of f. Thus for this choice of ϕ, π_ϕ is the inverse of the operator $f \to t\frac{\partial u}{\partial t}$ which maps $H^p \to T_2^p$ for $p \leqslant 1$.

As a consequence we have (cf. [JJ], [CMS1])

(7.2) Theorem. $[H^{p_0},H^{p_1}]_\Theta = H^p$, $0 < p_0 < p < p_1 \leqslant \infty$, $\frac{1}{p} = \frac{1-\Theta}{p_0} + \frac{\Theta}{p_1}$, where H^∞ denotes the space BMO.

Proof. We only need to show that $H^p \subsetneq [H^{p_0},H^{p_1}]_\Theta$. Let $f \in H^p$, then, if $u(x,t)$ denotes the Poisson integral of f, we have $t\frac{\partial u}{\partial t} \in T_2^p$ (cf. [CMS1]). Now by (7.1) and (6.4) $\pi_\phi : T_2^p \to [H^{p_0},BMO]_\Theta$. Thus, $\pi_\phi(t\frac{\partial u}{\partial t}) = f \in [H^{p_0},BMO]_\Theta$ as we wished to show.

8. Further results

In this section we want to briefly discuss some other recent developments in the theory of tent spaces.

a) Tent spaces on product domains

In [AM] and [GM] tent spaces on product domains are introduced. In particular [GM] treats the complex interpolation theory of these spaces and as an application the following theorem is obtained (which solves a problem of [CF]) .

(8.1) Theorem $[H^{p_0}(\mathbb{R}^2_+ \times \mathbb{R}^2_+), BMO(\mathbb{R}^2_+ \times \mathbb{R}^2_+)]_\Theta = H^p(\mathbb{R}^2_+ \times \mathbb{R}^2_+)$, $0 < p_0 < \infty$, $\frac{1}{p} = \frac{1-\Theta}{p_0}$.

b) Vector valued tent spaces

Tent spaces of vector valued functions are considered in [AM1] . In particular this paper treats the interpolation theory of these spaces and its application to the interpolation of vector valued H^p spaces. Moreover, in [AM1] the duality theory of these spaces is considered. A typical result obtained is the duality between $T^1_\infty(X)$ and the space of vector valued Carleson measures $V(X^*)$.

The interpolation theory of vector valued tent spaces can be obtained combining (6.4) with a slight generalization of Calderón's theory (cf. [C] , [R]).

References

[AM] J. Alvarez, M. Milman, Spaces of Carleson measures, duality and interpolation, Arkiv för Mat. (to aprear).

[AM1] J. Alvarez, M. Milman, Vector valued tent spaces and applications, preprint.

[AB] E. Amar, A. Bonami, Measures de Carleson d'ordre α et solutions au bord de l'equation $\bar{\partial}$, Bull. Soc. Math. France 107 (1979), 23-48.

[BJ] A. Bonami, R. Johnson, Tent spaces based on the Lorentz spaces, preprint.

[BL] J. Bergh, J. Löfström, Interpolation Spaces. Springer Verlag, 1976.

[C] A.P. Calderón, Intermediate spaces and interpolation, the complex method, Studia Math. 24 (1964), 113-190.

[CF] S.Y.A. Chang, R. Fefferman, Some recent developments in Fourier analysis and H^p theory on product domains, Bull. Amer. Math. Soc. 12 (1985), 1-43.

[CMS] R. Coifman, Y. Meyer, E.M. Stein, Un nouvel espace fonctionnel adapté a l'étude des operatéurs définis par des integrals singulièrs, Lecture notes in Math. 992 pp 1-15.

[CMS1] _______, Some new function spaces and their applications to harmonic analysis, J. Funct. Anal. 62 (1985), 304-335.

[CWMS] M. Cwikel, M. Milman, Y. Sagher, Complex interpolation of some quasi Banach spaces, J. Funct. Anal. 65 (1986), 339-347.

[GM] M. Gomez, M. Milman, Interpolation complexe des espaces H^p et théorème de Wolff, C.R. Acad. Sc. Paris 301 (1985), 631-633.

[GM1] _______, Complex interpolation of H^p spaces on product domains, preprint.

[JJ] S. Janson, P. Jones, Interpolation between H^p spaces: the complex method, J. Funct. Anal. 42 (1982), 58-80.

[M] M. Milman, Fourier type and complex interpolation, Proc. Amer. Math. Soc. 89

(1983), 246-248.

[R] N. Riviere, Interpolation theory in s-Banach spaces, Thesis, Univ. of Chicago, 1966.

[S] J.D. Stafney, Analytic interpolation of certain multiplier spaces, Pacific J. Math. 32 (1970), 241-248.

[W] T. Wolff, A note on interpolation spaces, Lecture Notes in Math. 908, Springer Verlag, 1982, pp 199-204.

TOEPLITZ LIFTINGS OF HANKEL FORMS

Mischa Cotlar
Universidad Central de Venezuela
Caracas 1040, Venezuela

Cora Sadosky
Howard University
Washington, D.C. 20059 U.S.A.

INTRODUCTION.

This paper is a survey of the basic features of a class of generalized Toeplitz forms, that appear in many different problems in analysis (see Section 1 below), and were studied in [23-35], [13], [8-11], [12], [2-7], [19], [20], [37], [55], [56], [68].

The basic property of these forms, the so-called generalized Bochner theorem, GBT, was first proved in [24], [25], where it was interpreted as a refinement of the Helson-Szegö theorem [45]. The GBT can be viewed as a Fourier representation of generalized invariant kernels or as a lifting property of Hankel kernels and forms. From the first point of view, it is closely related to general ideas of M.G. Krein [53], Iu.M. Berezanski [16] and L. Schwartz [71], concerning invariant kernels. From the second point of view, that of Hankel forms, the GBT is related to works of Coifman-Rochberg-Weiss [22], Peller-Khrushchev [65], Adamjan-Arov-Krein [1] and Ibragimov-Rozanov [48]. As a lifting property of subordinated kernels, it is a refinement of methods used by Beatrous-Burbea [15] and Burbea-Masani [21] (see also Devinatz [36]), as general tools in interpolation theory.

Considered as a lifting property of forms, the GBT is closely connected to the Nagy-Foias dilation theory [57], as well as to the Lax-Phillips scattering structures [54]. In fact, the Nagy-Foias lifting theorem can be derived from a special case of the GBT [2], and conversely, the GBT can be obtained from a special case of the Nagy-Foias lifting theorem [32], [33]. The Sarason lifting theorem [69], which is itself a special case of that of Nagy-Foias, is known to embrace basic interpolation theorems and to be equivalent to the Nehari interpolation theorem [60] (see Page [63] and N.K. Nikolskii [62], and also [32]). Similarly, the GBT finds applications in interpolation and prediction problems. Thus, the study of the generalized Toeplitz forms can be viewed as a special line of development within the Nagy-Foias dilation theory.

*Research supported in part by the National Science Foundation under Grant #DMS-8603043.

There are also connections between the GBT and the Grothendieck Fundamental Inequality [29], [68], and related questions concerning the Paley lacunary inequality and harmonizable processes [31], [33], [30], as studied by Gilbert [41], Blei [17], Fournier [40], Niemi [61] and others. Finally, the results on generalized Toeplitz kernels (GTKs) and forms are related to work of Rochberg [66], S. Janson-Rochberg-Peetre [49], I. Gohberg and H. Dym [42] [38], in ways that are still to be fully clarified.

Though the generalized Toeplitz kernels and forms were studied mainly in the one-dimensional setting of $\mathbb{Z}$ and $\mathbb{R}$, the GBT can be formulated in an ergodic setting [32], and can be applied to n-dimensional problems [35], [5]. In this survey the GBT will be considered mainly from the point of view of subordinated Hankel forms, with emphasis on the material in [28], [32], [33], [34], [13].

In Section 1 the motivation of the basic notion of generalized Toeplitz form is given, and in Section 2 it is extended to a general setting. The main theorems are stated in Section 3, where self-contained simplified proofs are sketched. Applications to interpolation and prediction problems are outlined in Sections 4 and 5.

1. WHY GENERALIZED TOEPLITZ FORMS?

The basic notions will be first recalled in the simplest situation where they arose [23], [24], [25].

Let $\mathbb{T}$ be the unit circle and $P = \{f(t) = \Sigma_{finite} \hat{f}(n)e^{int}\}$ be the set of all trigonometric polynomials defined in $\mathbb{T}$. Let $P_1 = \{f \in P; \hat{f}(n) = 0 \text{ for } n < 0\}$, $P_2 = \{f \in P; \hat{f}(n) = 0 \text{ for } n \geq 0\}$ be the sets of analytic and antianalytic polynomials, $C(\mathbb{T})$ the class of continuous functions in $\mathbb{T}$, and $H^p = H^p(\mathbb{T}) = \{f \in L^p(\mathbb{T}); \hat{f}(n) = 0 \text{ for } n < 0\}$, $1 \leq p \leq \infty$, where $\hat{f}(n) = \int_{\mathbb{T}} e^{-int} f(t)dt$. For a measure μ defined in $\mathbb{T}$, let $\mu(f) = \int_{\mathbb{T}} f d\mu$, and $\hat{\mu}(n) = \mu(e^{-int})$, $\forall n \in \mathbb{Z}$.

Every linear functional in the vector space $P, \ell: P \to \mathbb{C}$, gives rise to a sesquilinear form $B: P \times P \to \mathbb{C}$, defined by $B(f,g) = \ell(f\bar{g})$. Such forms $B \sim \ell$ are called <u>Toeplitz</u>, and are characterized by the property $B(\tau f, \tau g) = B(f,g)$, where $\tau f = e^{it}f(t)$ is the shift operator in P. Given $B \sim \ell$, B is <u>positive</u>, $B(f,f) \geq 0$, iff $\ell(f) = \int f d\mu = \mu(f)$, where $\mu \geq 0$ is a positive measure in $\mathbb{T}$. As shown in the examples below, several classical problems lead to the consideration of 2x2 matrices $(\mu_{\alpha\beta})$, $\alpha, \beta = 1,2$, whose elements are linear functionals $\mu_{\alpha\beta}: P \to \mathbb{C}$, satisfying

$$(1) \qquad \sum_{\alpha,\beta=1,2} \mu_{\alpha\beta}(f_\alpha \bar{f}_\beta) \geq 0, \quad \forall (f_1,f_2) \in P_1 \times P_2.$$

Such matrices are called <u>weakly positive</u>, $(\mu_{\alpha\beta}) \succ 0$, and we write $(\mu_{\alpha\beta}) \geq 0$ if (1) holds for all pairs $(f_1,f_2) \in P \times P$. Condition (1) implies $\Sigma_{\alpha,\beta}\, \mu_{\alpha\beta}(f_\alpha \bar{f}_\beta)\lambda_\alpha \bar{\lambda}_\beta \geq 0$ for all $\lambda_1,\lambda_2 \in \mathbb{C}$, hence (1) is equivalent to $\mu_{21} = \overline{\mu_{12}}$ and

(1a) $$|\mu_{12}(f_1\bar{f}_2)| \leq \mu_{11}(f_1\bar{f}_1)^{1/2}\, \mu_{22}(f_2\bar{f}_2)^{1/2},\ \forall (f_1,f_2) \in P_1 \times P_2.$$

This generalizes to a notion of <u>subordinated forms</u>: if three sesquilinear forms B_1,B_2,B_3 satisfy

(1b) $$|B_3(f_1,f_2)|^2 \leq B_1(f_1,f_1)B_2(f_2,f_2),\ \forall (f_1,f_2) \in P_1 \times P_2,$$

we write $B_3 \prec (B_1,B_2)$, and if (1b) holds for all $(f_1,f_2) \in P \times P$, we write $B_3 \leq (B_1,B_2)$. Finally, since $P \sim P_1 \times P_2 = P_1 \oplus P_2$, to give a sesquilinear form B: $P \times P \to \mathbb{C}$ it is enough to give the values $B(f,g)$ for $(f,g) \in P_\alpha \times P_\beta$, $\alpha,\beta = 1,2$. Therefore (1) expresses that if B is defined by $B(f,g) = \mu_{\alpha\beta}(f\bar{g})$ if $(f,g) \in P_\alpha \times P_\beta$, $\alpha,\beta = 1,2$, then $B \geq 0$. This leads to the following notion: a sesquilinear form B: $P \times P \to \mathbb{C}$ is said to be <u>generalized Toeplitz</u> if there exist four Toeplitz forms $(B_{\alpha\beta})$, $\alpha,\beta = 1,2$, such that

(1c) $$B(f,g) = B_{\alpha\beta}(f,g) \quad \text{for} \quad (f,g) \in P_\alpha \times P_\beta,\ \alpha,\beta = 1,2;$$

we write $B \sim (B_{\alpha\beta})$. It is easy to see that to give a weakly positive matrix $(\mu_{\alpha\beta}) \succ 0$ is the same as to give a triplet B_1,B_2,B_3 of Toeplitz forms $B_3 \prec (B_1,B_2)$ or a positive generalized Toeplitz form $B \geq 0$.

<u>Example (a)</u>. Consider the problem of characterizing the pairs of measures $\mu \geq 0$, $\nu \geq 0$ in $\mathbb{T}$ for which

(2) $$\int |Hf|^2 d\mu \leq M \int |f|^2 d\nu,\ \forall f \in P,$$

where H is the Hilbert transform operator and M is a fixed constant. In the case when $\mu = \nu$ and M is not prescribed, such characterization was given by Helson and Szegö [45] and reformulated by Hunt, Muckenhoupt and Wheeden [47] in terms of the A_2-condition, but neither method dealt with the case $\mu \neq \nu$. Since every $f \in P$ has an unique representation $f = f_1 + f_2$, with $f_1 \in P_1$, $f_2 \in P_2$, and since $Hf = -i(f_1 - f_2)$, (2) can be rewritten as

(2a) $$\Sigma_{\alpha,\beta=1,2}\, \mu_{\alpha\beta}(f_\alpha \bar{f}_\beta) \geq 0,\ \forall (f_1,f_2) \in P_1 \times P_2,$$

where $\mu_{11} = \mu_{22} = M\nu - \mu$, $\mu_{12} = \mu_{21} = M\nu + \mu$. Thus (2) is equivalent to $(\mu_{\alpha\beta}) \succ 0$ such that $(\mu_{\alpha\beta})$.

Example (b). If $\mu \geq 0$ is the spectral measure of a discrete stationary Gaussian process, and γ is the value of the cosine of the angle between the "past" subspace $P_2 \subset L^2(\mu)$ and the "future" subspace $P_1 \subset L^2(\mu)$, then

$$(3) \qquad |\mu(f_1\bar{f}_2)| \leq \gamma||f_1||_{2,\mu} \cdot ||f_2||_{2,\mu}, \ \forall (f_1,f_2) \in P_1 \times P_2.$$

Setting $B(f_1,f_2) = \mu(f_1\bar{f}_2)$, $B_1(f,g) = B_2(f,g) = \gamma\mu(f\bar{g})$, (3) says that $B \prec (B_1,B_2)$. Thus the prediction problems of characterizing the measure $\mu \geq 0$ for a prescribed value of γ leads to the notions (1b) and (1c).

Example (c). The Nehari theorem solves a classical moment problem by characterizing the bounded Toeplitz forms $B: P_1 \times P_2 \to \mathbb{C}$, that is, these forms defined by $B(f_1,f_2) = \mu(f_1\bar{f}_2)$, μ a finite complex measure in $\mathbb{T}$, satisfying

$$(4) \qquad |\mu(f_1\bar{f}_2)| \leq ||f_1||_2 \ ||f_2||_2, \ \forall (f_1,f_2) \in P_1 \times P_2.$$

Setting $B_1(f,g) = B_2(f,g) = \int f\bar{g}dt$, (4) becomes $B \prec (B_1,B_2)$.

Example (d). A measure $\rho \geq 0$, defined in the unit disc D, is Carleson if

$$(5) \qquad \iint_D |Pf|^2 d\rho \leq M\int_{\mathbb{T}} |f|^2 dt, \ \forall f \in P,$$

where $P: C(\mathbb{T}) \to C(D)$ is the Poisson operator. Peter Jones characterized the set of balyages of Carleson measures, i.e., $\{\rho^b = P^*\rho: \rho$ a Carleson measure$\}$, P^*: adjoint operator of P [50]. Letting $f = f_1 + f_2$, $f_\alpha \in P_\alpha$, $\alpha = 1,2$, (5) can be rewritten as

$$(5a) \qquad M\int_{\mathbb{T}}|f_1|^2dt + M\int_{\mathbb{T}}|f_2|^2dt \geq \sum_{\alpha,\beta=1,2} \iint_D (Pf_\alpha)(\overline{Pf_\beta})d\rho$$

since $\int f_1\bar{f}_2dt = 0$. As $P(f_1)\overline{P(f_2)} = P(f_1\bar{f}_2)$, (5a) implies the weaker inequality

$$(5b) \qquad M\int_{\mathbb{T}}|f_1|^2dt + M\int_{\mathbb{T}}|f_2|^2dt \geq \iint_D P(f_1\bar{f}_2)d\rho + \iint_D P(\bar{f}_1 f_2)d\rho$$

$$= \int_{\mathbb{T}} f_1\bar{f}_2 d\rho^b + \int_{\mathbb{T}} \bar{f}_1 f_2 d\rho^b.$$

Setting $\mu_{11} = \mu_{22} = Mdt$, $\mu_{12} = \mu_{21} = \rho^b$, (5b) becomes $(\mu_{\alpha\beta}) \succ 0$. In [13] it was shown that (5b) is indeed equivalent to (5), and this reduces the problem of characterizing the balyages of the Carleson measures to a special case of example (a), with $\mu = 1/2(\rho^b - Mdt)$, $\nu = 1/2(\rho^b/M + dt)$.

Example (e). Devinatz and Widom [36], [72] characterized the functions $g \in L^\infty$ for which the "Toeplitz operator" T_g defined on the closure of P in L^2 by the sesquilinear form $(T_g f_1, f_2) = \int f_1 \bar{f}_2 g dt$, $(f_1, f_2) \in P_1 \times P_2$, is invertible. As shown in [8], if g is such a function and $\mu_{11} = \mu_{22} = \lambda|g|dt$, $\mu_{12} = \overline{\mu_{21}} = gdt$, then $(\mu_{\alpha\beta}) \succ 0$ for some $0 \leq \lambda \leq 1$.

Observe that all the matrix measures $(\mu_{\alpha\beta})$ appearing in the preceding examples are weakly positive but not positive. But the basic property of such matrix measures is

(I) *For every weakly positive matrix measure* $(\mu_{\alpha\beta}) \succ 0$ *there exists a positive matrix measure* $(\nu_{\alpha\beta}) \geq 0$ *such that*

$$\mu_{\alpha\beta}(f_\alpha \bar{f}_\beta) = \nu_{\alpha\beta}(f_\alpha \bar{f}_\beta), \ \forall (f_\alpha, f_\beta) \in P \times P_\beta, \ \alpha, \beta = 1, 2. \tag{6}$$

This is the Generalized Bochner Theorem (GBT), which can be equivalently stated as:

(II) *For every positive generalized Toeplitz form* $B: P \times P \to \mathbb{C}$, *there exists a positive matrix measure* $(\nu_{\alpha\beta}) \geq 0$ *such that*

$$B(f_\alpha, f_\beta) = \nu_{\alpha\beta}(f_\alpha \bar{f}_\beta), \ \forall (f_\alpha, f_\beta) \in P_\alpha \times P_\beta, \ \alpha, \beta = 1, 2. \tag{6a}$$

(III) *If* $B_3 \prec (B_1, B_2)$, *where* B_1, B_2, B_3 *are Toeplitz forms, there exists a positive matrix measure* $(\nu_{\alpha\beta}) \geq 0$ *such that*

$$\begin{aligned} &B_\alpha(f,g) = \mu_{\alpha\alpha}(f\bar{g}), \ \forall (f,g) \in P_\alpha \times P_\alpha, \ \alpha = 1,2, \text{ and} \\ &B_3(f,g) = \mu_{12}(f\bar{g}), \ \forall (f,g) \in P_1 \times P_2. \end{aligned} \tag{6b}$$

Similar statements are valid for measures in $\mathbb{R}$. In [27], [28], [33], [67], Theorem (I) was extended to the case where the terms $\mu_{\alpha\alpha}(f_\alpha \bar{f}_\alpha)^{1/2}$ in the right hand side of (1a) are replaced by general norms $\sigma_\alpha(f_\alpha)$ (see Section 5).

It was shown in the papers quoted above that the GBT provides, among other applications, solutions and refinements for the examples (a) - (e). Since every sesquilinear form $B: P \times P \to \mathbb{C}$ is of the form $B = B_K$ where $K: \mathbb{Z} \times \mathbb{Z} \to \mathbb{C}$ is a kernel (see Section 4(B)), generalized Toeplitz forms in P correspond to generalized Toeplitz kernels, GTKs in

$\mathbb{Z}$, and the kernel terminology was used in most of the above-mentioned papers. A more general form of the GBT, as well as other examples to which it can be applied, are discussed in the next section. The solutions the GBT provides for the problems in the examples above are sketched in Section 4.

2. TOEPLITZ FORMS WITH RESPECT TO GENERAL TRANSFORMATIONS.

The definitions of the previous section will be considered now in more detail and in the light of the more general situation where P, P_1, P_2 are replaced by general vector spaces, and the shift τ by general transformations. Fix two vector spaces V_1, V_2, two vector subspaces $W_1 \subset V_1$, $W_2 \subset V_2$, and two linear isomorphisms $\tau_1: V_1 \to V_1$, $\tau_2: V_2 \to V_2$, satisfying

$$(7) \qquad \tau_1(W_1) \subset W_1,\ \tau_2^{-1}(W_2) \subset W_2.$$

Of special interest will be the case where the two following conditions are satisfied:

$$(7a) \qquad \cap_{n=1}^{\infty} \tau_1^n W_1 = \cap_{n=1}^{\infty} \tau_2^{-n} W_2 = \{0\},$$

$$(7b) \qquad \vee_{m=-\infty}^{\infty} \tau_1^m W_1 = V_1,\ \vee_{m=-\infty}^{\infty} \tau_2^m W_2 = V_2.$$

Condition (7a) means that for every $f \in W_\alpha$ the trajectory $\{\tau_\alpha^m : m \in \mathbb{Z}$ contains some element of the complement of $W_\alpha, \alpha = 1,2$, and (7b) that each element of V_α is in some such trajectory. When $V_1 = V_2$, $\tau_1 = \tau_2$, (7),(7a) and (7b) are the conditions of the Lax-Phillips scattering theory [53], and they are also satisfied in the basic case considered in Section 1, where $V_1 = V_2 = P$, $W_\alpha = P_\alpha, \alpha = 1,2$, $\tau_1 = \tau_2 = \tau$, the shift operator. But while

$$(8) \qquad P \sim P_1 \times P_2 = P_1 \oplus P_2,$$

in the scattering situation, the opposite condition, $W_1 \times W_2 \neq V_\alpha$, is of interest. By (8) sesquilinear forms B in $P \times P$ can be identified with sesquilinear forms $\overset{\#}{B}$ in $(P_1 \times P_2) \times (P_1 \times P_2) \subset (P \times P) \times (P \times P)$. Similarly, in the case of general V_α, W_α, we will be interested in forms B defined in $(V_1 \times V_2) \times (V_1 \times V_2)$, and in forms $\overset{\#}{B}$ defined in $(W_1 \times W_2) \times (W_1 \times W_2)$, and this will allow to extend the notion of generalized Toeplitz forms to the present setting.

For $\tau = 1,2$, a sesquilinear form $B_\alpha: V_\alpha \times V_\alpha \to \mathbb{C}$ (respec., $B: V_1 \times V_2 \to \mathbb{C}$) is said to be $\underline{\tau_\alpha}$-Toeplitz (respec., $(\underline{\tau_1}, \underline{\tau_2})$-Toeplitz) if $B_\alpha(\tau_\alpha f,\tau_\alpha g) = B_\alpha(f,g)$, $\forall(f,g) \in V_\alpha \times V_\alpha$ (respec., $B(\tau_1 f_1,\tau_2 f_2) = B(f_1,f_2)$, $\forall(f_1,f_2) \in V_1 \times V_2$). The restrictions of the τ_α-Toeplitz form B_α to $W_\alpha \times W_\alpha$ is again Toeplitz, but since W_2 is not τ_2-invariant, the restriction of the (τ_1,τ_2)-Toeplitz form B to $W_1 \times W_2$ is not Toeplitz, but Hankel in the following sense: a sesquilinear form $B^{\#}: W_1 \times W_2 \to \mathbb{C}$ is (τ_1,τ_2)-Hankel if $B^{\#}(\tau_1 f_1,f_2) = B^{\#}(f_1 \tau_2^{-1} f_2)$, $\forall(f_1,f_2) \in W_1 \times W_2$.

As before we write $B \leq (B_1,B_2)$ (resp., $B^{\#} \prec (B_1,B_2)$) if $|B(f_1,f_2)|^2 \leq B_1(f_1,f_1)B_2(f_2,f_2)$, $\forall(f_1,f_2) \in V_1 \times V_2$ (respec., $(f_1,f_2) \in W_1 \times W_2$). By the same argument used to show the equivalence between (1) and (1a), $B \leq (B_1,B_2)$ (respec., $B \prec (B_1,B_2)$) is equivalent to

(9) $$B_1(f_1,f_1) + B_2(f_2,f_2) + B(f_1,f_2) + \overline{B(f_1,f_2)} \geq 0$$

$$\forall(f_1,f_2) \in V_1 \times V_2 \text{ (respec., } (f_1,f_2) \in W_1 \times W_2).$$

If $B_\alpha: V_\alpha \times V_\alpha \to \mathbb{C}$ and $B: V_1 \times V_2 \to \mathbb{C}$, then, setting $B_{\alpha\alpha} = B_\alpha$, $B_{12} = B$, $B_{21}(f_2,f_1) = \overline{B_{12}(f_1,f_2)}$, the triplet B_1,B_2,B can be identified with the 2x2-matrix $(B_{\alpha\beta})$, $\alpha,\beta = 1,2$, and $(B_{\alpha\beta})$ can be identified with the sesquilinear form $\mathcal{B}: (V_1 \times V_2) \times (V_1 \times V_2) \to \mathbb{C}$, $V_1 \times V_2 = V_1 \oplus V_2$, given by

(10) $$\mathcal{B}((f_1,f_2), (g_1,g_2)) = \Sigma_{\alpha,\beta=1,2} B_{\alpha\beta}(f_\alpha,g_\beta).$$

By (10), condition $B \leq (B_1,B_2)$ says that the associated form $\mathcal{B}$ is positive.

If $B^{\#}: W_1 \times W_2 \to \mathbb{C}$, then the triplet $B_1,B_2,B^{\#}$ can be similarly associated with the sesquilinear form $\mathcal{B}^{\#}: (W_1 \times W_2) \times (W_1 \times W_2) \to \mathbb{C}$, defined by

(10a) $$\mathcal{B}^{\#}((f_1,f_2), (g_1,g_2)) = B_{11}(f_1,g_1) + B_{22}(f_2,g_2) + B^{\#}_{12}(f_1,g_2) + B^{\#}_{21}(f_2,g_1)$$

where $B^{\#}_{12} = B^{\#}$, $B^{\#}_{21} = B^{\#*}_{12}$, and $B^{\#} \prec (B_1,B_2)$ is equivalent to $\mathcal{B}^{\#} \geq 0$.

Define the linear isomorphism $\tau: V_1 \times V_2 \to V_1 \times V_2$ by $\tau(f_1,f_2) = (\tau_1 f_1,\tau_2 f_2)$ and say that the form $\mathcal{B}$ is <u>τ-Toeplitz</u> if the four forms $B_{\alpha\beta}$ are (τ_α,τ_β)-Toeplitz. Since the subspace $W_1 \times W_2$ of $V_1 \times V_2$ is not τ-invariant, the restriction $\mathcal{B}^{\#}$ of a τ-Toeplitz form $\mathcal{B}$ to $(W_1 \times W_2) \times (W_1 \times W_2)$ is not τ-Toeplitz but generalized τ-Toeplitz, in the sense that follows.

A form $\mathcal{B}^{\#}: (W_1 \times W_2) \times (W_1 \times W_2) \to \mathbb{C}$ is <u>generalized τ-Toeplitz</u> if $\mathcal{B}^{\#} \sim (B_{11},B_{22},B_{12},B_{21})$ where $B_{\alpha\alpha}$ is τ_α-Toeplitz, $\alpha = 1,2$, $B^{\#}_{12}$ is

(τ_1,τ_2)-Hankel and $\overset{\#}{B}_{21} = \overset{\#}{B}{}^*_{12}$.

As already observed when $V_\alpha = P$, $W_\alpha = P_\alpha$, the τ-Toeplitz forms $\overset{\#}{\mathcal{B}}$ can be identified with forms in $P \times P$, and it is easy to see that these coincide with the forms defined in (1c).

Now fix two positive sesquilinear forms $B_\alpha\colon V_\alpha \times V_\alpha \to \mathbb{C}$, $\alpha = 1,2$, and let

$$
\begin{aligned}
\Lambda_0 &= \{B\colon V_1 \times V_2 \to \mathbb{C};\ B \text{ is } (\tau_1,\tau_2)\text{-Toeplitz}\}\\
\overset{\#}{\Lambda}_0 &= \{\overset{\#}{B}\colon W_1 \times W_2 \to \mathbb{C};\ \overset{\#}{B} \text{ is } (\tau_1,\tau_2)\text{-Hankel}\}\\
\Lambda &= \Lambda(B_1,B_2) = \{B \in \Lambda_0;\ B \leq (B_1,B_2)\}\\
\overset{\#}{\Lambda} &= \{\overset{\#}{B} \in \overset{\#}{\Lambda}_0;\ \overset{\#}{B} \quad (B_1,B_2)\}.
\end{aligned}
\tag{11}
$$

Thus, $\overset{\#}{B} \in \overset{\#}{\Lambda}$ iff the form $\overset{\#}{\mathcal{B}}$ associated with the triplet $B_1,B_2,\overset{\#}{B}$ is a positive generalized τ-Toeplitz form, and $B \in \Lambda$ iff the form $\mathcal{B}$ associated with the triplet B_1,B_2,B is a positive τ-Toeplitz form.

For Λ any family of sesquilinear forms in $V_1 \times V_2$, and Λ any family of such forms in $W_1 \times W_2$, with $\overset{\#}{\Lambda} \supset \Lambda|_{W_1 \times W_2}$, the following interpolation problem can be posed: given $\overset{\#}{B} \in \overset{\#}{\Lambda}$, does there exist $B \in \Lambda$ such that $B = \overset{\#}{B}$ on $W_1 \times W_2$? If the answer is yes, B is called a lifting of $\overset{\#}{B}$. The pair $\Lambda,\overset{\#}{\Lambda}$ is said to have the interpolation property if every $\overset{\#}{B} \in \overset{\#}{\Lambda}$ has a lifting in Λ. In the next section we consider the case where Λ and $\overset{\#}{\Lambda}$ are given by (11), and prove that the pair $\Lambda,\overset{\#}{\Lambda}$ has the interpolation property.

3. MAIN THEOREMS.

With the notations of Section 2, the following result asserts that the pair $\overset{\#}{\Lambda},\Lambda$ defined in (11) have the interpolation property.

Theorem 1 (LIFTING PROPERTY). *If τ_1,τ_2 are linear isomorphisms satisfying (7), $B_\alpha\colon V_\alpha \times V_\alpha \to \mathbb{C}$ are τ-Toeplitz forms, for $\alpha = 1,2$, and $\overset{\#}{B}\colon W_1 \times W_2 \to \mathbb{C}$ is a (τ_1,τ_2)-Hankel form such that $\overset{\#}{B} \prec (B_1,B_2)$, then there exists a (τ_1,τ_2)-Toeplitz form, $B\colon V_1 \times V_2 \to \mathbb{C}$, such that $B \leq (B_1,B_2)$ and $B = \overset{\#}{B}$ in $W_1 \times W_2$.*

Remark 1. The lifting property tells that every positive generalized τ-Toeplitz form $\overset{\#}{\mathcal{B}}$ is the restriction of a positive τ-Toeplitz form $\mathcal{B}$.

Proof. We sketch the proof for the special case where (7b) is satisfied. Since $\overset{\#}{B} \prec (B_1,B_2)$, the form $\overset{\#}{\mathcal{B}}$ associated with $B_1,B_2,\overset{\#}{B}$, is non-negative, and therefore defines a semi-scalar product in $W_1 \times W_2$, giving rise to a Hilbert space of which $W_1 \times W_2$ is "a dense subspace", and τ is an isometric operator, with domain $W_1 \times (\tau_2^{-1}W_2)$ and range

$(\tau_1 W_1) \times W_2$. The operator τ can be extended to a unitary operator U in a larger Hilbert space, so that $U = \tau$ in the domain of τ and $U^{-1} = \tau^{-1}$ in its range. By condition (7b), every element of $V_1 \times V_2$ is of the form $(\tau_1^n f_1, \tau_2^m f_2)$, with $(f_1, f_2) \in W_1 \times W_2$, $n, m \in \mathbb{Z}$. Define $B(\tau_1^n f_1, \tau_2^m f_2) = \langle U^n(f_1,0), U^m(0,f_2)\rangle$. This definition does not depend on the representation of $g_1 \in W_1$ as $g_1 = \tau_1^n f_1$, because if $g_1 = \tau_1^{n-p} f_1'$, $f_1' \in W_1$, $p \geq 0$, then $\langle U^{n-p}(f_1',0), U^m(0,f_2)\rangle = \langle U^{n-p}U^p(f_1,0), U^m(0,f_2)\rangle = \langle U^n(f_1,0), U^m(0,f_2)\rangle$, since $(f_1,0)$ is in the domain of τ and $U = \tau$ in its domain. Using this observation and the property that, for given $g_1' = \tau_1^n f_1'$, $g_1'' = \tau_1^m f_1''$, $f_1', f_1'' \in W_1$, there is a large $r > 0$ such that $\tau_1^r g_1'$, $\tau_1^r g_1''$ and $\tau_1^r(g_1' + g_1'')$ all belong to W_1, it follows that the B thus defined is a sesquilinear form. Setting $n = m = 0$, $B = \overset{\#}{B}$ in $W_1 \times W_2$, and $B \leq (B_1, B_2)$ since U is unitary and $\|U^n(f_1,0)\| = \|\tau_1^n(f_1,0)\| = \|f_1\|$. □

Different proofs of Theorem 1 were given in [25], [26] and [13] in the special case when $V_1 = V_2 = P$, $W_\alpha = P_\alpha$, τ = shift, and it was extended to general vector spaces $V_1 = V_2$, $\tau_1 = \tau_2$ in [32], where the proof depended on the Nagy-Foias lifting theorem, and to other V_1, V_2 in [33].

If the Lax-Phillips conditions (7a), (7b) are satisfied, then for each $f_\alpha \in V_\alpha$ there is an $e_\alpha \in W_\alpha$ such that $f_\alpha = \tau_\alpha^n e_\alpha$, $\alpha = 1,2$, and $\tau_1^m e_1 \in W_1$ iff $m \geq 0$, $\tau_2^m e_2 \in W_2$ iff $m \leq 0$. Therefore we can fix two sets, $E_1 \subset W_1$, $E_2 \subset W_2$, such that each $f \in V_\alpha$ has a unique representation

(12) $$f = \tau_\alpha^{n(f)} e_\alpha, e_\alpha \in E_\alpha, \text{ and } f \in W_1 \text{ (resp., } f \in W_2) \text{ iff } n(f) \geq 0 \text{ (resp., } n(f) \leq 0).$$

From this observation and the preceding proof follows.

Theorem 2. (DILATION PROPERTY). *If τ_1, τ_2 satisfy (7), (7a), (7b), and $\overset{\#}{B} \prec (B_1, B_2)$, where B_α are τ_α-Toeplitz, $\alpha = 1,2$, and B is (τ_1,τ_2)-Hankel forms, then there is a unitary operator U defined in a space containing a subspace identifiable with $W_1 \times W_2$ and such that, for all $(e_1, e_2) \in E_1 \times E_2$, E_α as in (12),*

(12a) $$B_\alpha(\tau_\alpha^n e_\alpha, \tau_\alpha^m e_\alpha) = \langle U^{n-m} e_\alpha, e_\alpha\rangle, \quad \alpha = 1,2,$$

(12b) $$\overset{\#}{B}(\tau_1^n e_1, \tau_2^m e_2) = \langle U^{n-m} e_1, e_2\rangle, \text{ if } n \geq 0,\ m \leq 0.$$

U is called a unitary dilation of $B_1, B_2, \overset{\#}{B}$ or of its associated generalized τ-Toeplitz form B.

In the case when $V_1 = V_2 = P$, $W_\alpha = P_\alpha$, τ = shift, the dilation property was obtained as a consequence of the lifting property in [8],

[10], and in the general case, for $V_1 = V_2$, $\tau_1 = \tau_2$, in [32].

The following result is a general form of the GBT (II and III in Section 1), for which different proofs were given in [25] and [13] in the case $V_1 = V_2 = P$, etc., and in [32] for the general case $V_1 = V_2$, vector space.

Theorem 3 (FOURIER REPRESENTATION). *If τ_1, τ_2 satisfy (7), (7a), (7b) and $\overset{\#}{B} \prec (B_1, B_2)$, where B_α are τ_α-Toeplitz, $\alpha = 1,2$, and $\overset{\#}{B}$ is (τ_1, τ_2)-Hankel forms, then every $(e_1, e_2) \in E_1 \times E_2$, E_α as in (12), there is a positive matrix measure $(\nu_{\alpha\beta}) \geq 0$, $\nu_{\alpha\beta} = \nu_{\alpha\beta}(e_1, e_2)$, such that*

(12c) $$B_\alpha(\tau_\alpha^n e_\alpha, \tau_\alpha^m e_\alpha) = \hat{\nu}_{\alpha\alpha}(n - m), \ \forall (n,m) \in \mathbb{Z} \times \mathbb{Z}, \quad \alpha = 1,2,$$

(12d) $$B(\tau_1^n e_1, \tau_2^m e_2) = \hat{\nu}_{12}(n - m), \ \forall (n,m) \in \mathbb{Z}_+ \times \mathbb{Z}_-.$$

The proof follows from the dilation properties (12a), (12b) by using the spectral representation of U.

It is easy to verify that if $B \prec (B_1, B_2)$, where B_α are τ_α-Toeplitz and B is (τ_1, τ_2)-Hankel forms, then τ_α extends to a unitary operator in the pre-Hilbert space V_α, with seminorm $\sigma_\alpha(f)^2 = B_\alpha(f,f)$, and that letting $T_\alpha = P_\alpha \tau_\alpha|_{W_\alpha}$, P_α the projection of $\bar{V}_\alpha$ onto $\bar{W}_\alpha$, $\alpha = 1,2$, and $\overset{\#}{A}: W_1 \to W_2$, the operator given by the form $\overset{\#}{B}$, then $T_2\overset{\#}{A} = \overset{\#}{A}T_1$. Thus from Theorem 1 derives the following special case of the Nagy-Foias lifting theorem [58].

Theorem 1a (Nagy-Foias). *Let H_1, H_2 be two Hilbert spaces, for $\alpha = 1,2$, $W_\alpha \subset H_\alpha$ two subspaces, P_α the orthogonal projections of H_α onto W_α, and $\tau_\alpha: H_\alpha \to H_\alpha$ two unitary operators satisfying (7). If $T_\alpha = P_\alpha \tau_\alpha|_{W_\alpha}$, and if $\overset{\#}{A}: W_1 \to W_2$ is an intertwining contraction such that $T_2\overset{\#}{A} = \overset{\#}{A}T_1$, then there exists another contraction $A: H_1 \to H_2$, such that $\tau_2 A = A\tau_1$ and $\overset{\#}{A} = P_2 A|_{W_1}$.*

Conversely, Theorem 1 can be deduced from Theorem 1a, as done in [32]. In particular, we have

Corollary 1. *If Λ and $\overset{\#}{\Lambda}$ are the classes of Toeplitz and Hankel forms defined in (11), then for every $\overset{\#}{B} \in \overset{\#}{\Lambda}$ the liftings $B \in \Lambda = \Lambda(B_1, B_2)$ of $\overset{\#}{B}$ are in 1-1 correspondence with the Nagy-Foias liftings A of the intertwining contractions $\overset{\#}{A}$ of Theorem 1a.*

Remark 2. The general Nagy-Foias lifting theorem holds even if τ_1, τ_2 do not satisfy (7) but a weaker condition. It was shown in [2] that it can be deduced from a special case of Theorem 1, and the proof is based in a construction that we sketch next for the case $H_1 = H_2 = H$. Let $V_1 = V_2 = P(H) = \{f: \mathbb{Z} \to H;\ f \text{ of finite support}\}$,

$W_1 = P_+(H) = \{f \in P(H);\ \mathrm{supp}\ f \subset \mathbb{Z}_+\}$, $W_2 = P_-(H) = \{f \in P(H);\ \mathrm{supp}\ f \subset \mathbb{Z}_-\}$. To each triplet $T_1, T_2, \mathring{A}$ is in Theorem 1a, can be associated in a natural way a generalized Toeplitz form in $P_+(H) \times P_-(H)$ which is proved positive. Theorem 1 applied to this case leads to the Nagy-Foias theorem. In [55] such construction was studied in more detail and the Arsene-Ceausescu-Foias parametrization [14] of all the Nagy-Foias liftings was derived from it.

Remark 3. Observe that Theorem 2 is an extension for generalized Toeplitz forms of the classical dilation theorem of Naimark [59]. Extensions of Naimark theorem to Krein spaces (i.e., spaces with indefinite metrics) were studied in [4] and [55].

The preceding results extend to 1-parametric groups. Let V_1, V_2, $W_\alpha \subset V_\alpha$, be as before, but consider, instead of τ_1, τ_2, two continuous groups of linear transformations, $\tau_\alpha^t: V_\alpha \to V_\alpha$, $t \in \mathbb{R}$, $\alpha = 1,2$, such that

$$\tau_1^t(W_1) \subset W_1,\quad \tau_2^{-t}(W_2) \subset W_2,\quad \forall t \geq 0. \tag{13}$$

A form $B: V_1 \times V_2 \to \mathbb{C}$ is said to be (τ_1^t, τ_2^t)-Toeplitz if $B(\tau_1^t f_1, \tau_2^t f_2) = B(f_1, f_2)$, $\forall t \in \mathbb{R}$, $\forall (f_1, f_2) \in V_1 \times V_2$, and similarly are defined the τ_α^t-Toeplitz and the (τ_1^t, τ_2^t)-Hankel forms.

Theorem 4 [34]. *For $\alpha = 1,2$, let B_α be two τ_α^t-Toeplitz forms, and B a (τ_1^t, τ_2^t)-Hankel form, satisfying $B \prec (B_1, B_2)$, and assume that $B_\alpha(\tau_\alpha^t f_1, \tau_\alpha^t f_2)$, $B(\tau_1^t f_1, \tau_2^t f_2)$ are continuous in $t \in \mathbb{R}$. Then there exists a continuous (τ_1^t, τ_2^t)-Toeplitz form B such that $B \leq (B_1, B_2)$ and $B = \mathring{B}$ in $W_1 \times W_2$.*

Sketch of the proof. In each trajectory $\Gamma_\alpha = \Gamma_\alpha(f_\alpha) = \{\tau_\alpha^t f_\alpha : t \in \mathbb{R}\}$, $f_\alpha \in V_\alpha$, choose a fixed element $\xi(\Gamma_\alpha)$, and for $k = 1,2,\ldots$, let V_α^k be the vector subspace spanned by the elements of the form $\tau_\alpha^{r2^{-k}} \xi(\Gamma_\alpha)$, $r \in \mathbb{Z}$. Set $W_\alpha^k = W_\alpha \cap V_\alpha^k$, and let B_α^k, $\mathring{B}^k$ be the restrictions of B_α, $\mathring{B}$ to $V_\alpha^k \times V_\alpha^k$ and $W^k \times W_2^k$, respectively. By Theorem 1, for each k there is a $(\tau_1^{2^{-k}}, \tau_2^{2^{-k}})$-Toeplitz form B^k such that $B^k \leq (B_1, B_2)$, $B^k = \mathring{B}^k = \mathring{B}$ in $W_1^k \times W_2^k$. If $u_1, u_2, \ldots$ is a basis of the pre-Hilbert space (V_1, B_1), since $B^k(f, u_n)$ is a bounded linear form in f, of norm less than or equal to $B_2(u_n, u_n)$, passing to subsequences it is possible to obtain a form $B(f, u_n) = \lim_k B^k(f, u_n)$, of norm bounded by $B_2(u_n, u_n)$. By repeating this process, the required form B is obtained.

Remark 4. From Theorem 4 follows the analog of Theorem 1a for 1-parametric unitary groups τ_α^t, $t \in \mathbb{R}$, $\alpha = 1,2$, and conversely. Thus, also in Theorem 4 the liftings of $\overset{\#}{B}$ are in 1-1 correspondence with Nagy-Foias liftings. The general Nagy-Foias lifting theorem can also be extended to 1-parametric groups by using a special case of Theorem 1 [2].

4. APPLICATIONS.

(A) In the case when $V_1 = V_2 = P$, $W_\alpha = P_\alpha$, $\alpha = 1,2$, $\tau_1 = \tau_2 = \tau$ the shift operator in P, $P \sim P_1 \times P_2$, then Theorem 3 reduces to the GBT (I,II,III) of Section 1. Moreover, condition (6) gives $\hat{\mu}_{12}(n) = \hat{\nu}_{12}(n)$ for $n < 0$, so that $\nu_{12} - \mu_{12} = h\,dt$ for $h \in H^1$, and (I) can be stated as follows:

Corollary 2. *If $(\mu_{\alpha\beta}) \succ 0$, there exists $h \in H^1$ such that, letting $\nu_{\alpha\alpha} = \mu_{\alpha\alpha}$, $\alpha = 1,2$, $\nu_{12} = \mu_{12} + h\,dt$, then $(\nu_{\alpha\beta}) \geq 0$. Hence*

(14) $$|\mu_{12}(\Delta) + h(\Delta)|^2 \leq \mu_{11}(\Delta) \cdot \mu_{22}(\Delta), \ \forall \text{ Borel set } \Delta \subset \mathbb{T},$$

where $h(\Delta) = \int_\Delta h\,dt$. In particular, if $\mu_{\alpha\beta} = \omega_{\alpha\beta}(t)\,dt$, then

(14a) $$|\omega_{12}(t) + h(t)|^2 \leq \omega_{11}(t) \cdot \omega_{22}(t), \text{ a.e.}$$

Applied to example (a) of Section 1, Corollary 2 gives that a pair of positive measures μ,ν satisfies condition (2) iff there exists $h \in H^1$ such that

(15) $$\mu(\Delta) \leq M\nu(\Delta) \text{ and } |(M\nu + \mu - h\,dt)(\Delta)| \leq (M\nu - \mu)(\Delta), \ \forall\Delta \subset \mathbb{T}.$$

From (15) it follows that $\mu = \omega\,dt$. For the particular case when $\mu = \nu = \omega\,dt$, (15) gives the Helson-Szegö theorem, with control on the constant M, since it is equivalent to $\omega = \exp(u + Hv)$, with $u = c \log \omega/|h|$, $v = \arg h$, and $\|u\|_\infty \leq C_M$, $\|v\|_\infty \leq \frac{\pi}{2} - \varepsilon_M$ (see [26],[13]). For the case of μ and $\nu = dt$ the Lebesgue measure, by (15) it is $\mu = \omega\,dt$, $\omega \leq M$ and it is sufficient to take $h = 0$. But if instead of considering, as in the last case, $(\mu_{\alpha\beta}) \succ 0$ for $\mu_{11} = \mu_{22} = M\,dt - \mu$, $\mu_{12} = \overline{\mu_{21}} = M\,dt + \mu$, we look at $(\mu'_{\alpha\beta}) \succ 0$, for $\mu'_{11} = \mu'_{22} = M\,dt$, $\mu'_{12} = \overline{\mu'_{21}} = M\,dt + \mu$, Corollary 2 entails that $d\mu = \omega\,dt$ and $|M + \omega + h| \leq M$ a.e., from (14a). From this follows easily that $\omega = u + \tilde{v} + \text{cst.}$, for u,v bounded, $|u|^2 + |v|^2 \leq M^2$, $\tilde{v}$ the conjugate function of v, and therefore, that $\omega \in BMO$ (see [13], Proposition 10), according to the Fefferman-Stein duality theorem [39].

Letting $W_1 = e^{int} P_1$, $n > 0$, $W_2 = e^{imt} P_2$, $m < 0$, $V_1 = V_2 = W_1 \oplus W_2$, Theorem 3 characterizes the pairs of measures μ,ν satisfying (2) only for $f \in e^{int} P_1 \oplus e^{imt} P_2$. This solves the prediction problem of example (b), as well as its variants, with control over the fixed constant $\gamma = \gamma_{nm}$, the cosine of the angle between the "past" and "future" subspaces $e^{imt}P_2$ and $e^{int}P_1$. This control allows the passage to the limit when $|n| \to \infty$, $|m| \to \infty$, to obtain the prediction theorems of Helson-Sarason [44] and Ibragimov-Rozanov [48] (see [13]), and connects these results with theorems of Koosis [51].

Applied to example (c), Theorem 1 gives the Nehari theorem. In fact, by (4) there exists a Toeplitz form B' such that $B' \leq (B_1, B_2)$ and $B' = B$ on $P_1 \times P_2$. Setting $\hat{\phi}(n) = \ell(n)$ for $\ell \sim B'$, the conditions on B' say that $|\phi(t)| \leq 1$ and that $\hat{\phi}(n) = \hat{\mu}(n)$ for $n > 0$ (see [8], [33]).

Applied to example (a), Theorem 3 gives the characterization of the balyapes of Carleson measures as $d\rho^b = \phi dt$, $\phi = u + \tilde{v} + cst. \in BMO$ with an argument similar to that used for example (a) (see [13]).

In the case of example (e), Theorem 3 gives the necessary and sufficient condition for invertibility of the Toeplitz operator T_g, as $g^{-1} \in L^\infty$ and $g/|g| = \exp(i(\tilde{u} + v + cst.))$, for bounded u,v, $\|v\|_\infty < \pi/2$, u the conjugate of u, by considerations similar to those of the particular case of example (a) (for details, see [8]).

(B) Now fix two sets X_1, X_2, two subsets $E_1 \subset X_1$, $E_2 \subset X_2$, two bijections $\tau_\alpha: X_\alpha \to X_\alpha$ (or two groups of bijections $\tau^t_\alpha: X_\alpha \to X_\alpha$, $t \in \mathbb{R}$), $\alpha = 1,2$, such that $\tau_1(E_1) \subset E_1$ and $\tau_2^{-1}(E_2) \subset E_2$ (idem for τ^t_α), and set $V_\alpha = V(X_\alpha) = \{f: X_\alpha \to \mathbb{C};$ f with finite support$\}$, $W_\alpha = \{f \in V_\alpha:$ supp $f \subset E_\alpha\}$, $(\tau_\alpha f)(x) = f(\tau_\alpha x)$, $\alpha = 1,2$.

To each kernel $K: X_\alpha \times X_\beta \to \mathbb{C}$ corresponds a sesquilinear form $B_K: V_\alpha \times V_\beta \to \mathbb{C}$, defined by

$$(16) \qquad B_K(f_1,f_2) = \sum_{\alpha,\beta=1,2} \sum_{(x_1,x_2)\in X_\alpha \times X_\beta} K(x_1,x_2) f_1(x_1)\overline{f_2(x_2)},$$

and $K \leq (K_1,K_2)$ means $B_K \leq (B_{K_1}, B_{K_2})$. K is said to be Toeplitz or Hankel if B_K is so.

Theorem 1 entails the following theorem, proved in [25] for the case of $X_1 = X_2 = \mathbb{Z}$, $E_1 = \mathbb{Z}_+$, $E_2 = \mathbb{Z}_-$, and in [32] in the general case.

<u>Theorem 5</u>. (LIFTINGS OF SUBORDINATED GTKs). *If $\overset{\#}{K} \prec (K_1,K_2)$, where the K_α are τ_α-Toeplitz kernels and $\overset{\#}{K}$ is a Hankel kernel, then there exists a Toeplitz kernel $K: X_1 \times X_2 \to \mathbb{C}$, such that $K \leq (K_1,K_2)$ and $K = K$ in $E_1 \times E_2$.*

<u>Remark 5</u>. The triplet $K_1, K_2, \overset{\#}{K}$ has a Fourier representation similar to that of Theorem 3. This applies to both the cases of a pair of

isomorphisms τ_1, τ_2 or of groups τ_1^t, τ_2^t, $t \in \mathbb{R}$. The theorem holds also when $K_1, K_2 \overset{\#}{K}$ are operator-valued kernels [32],[10].

Theorem 4 furnishes the following generalization of the Bochner-Schwartz theorem to GTKs.

Theorem 6 [34]. *Let $V_1 = V_2 = S(\mathbb{R})$ the Schwartz space, $W_1 = \{f \in S(\mathbb{R});\ \operatorname{supp} f \subset \mathbb{R}_+\}$, $W_2 = \{f \in S(\mathbb{R});\ \operatorname{supp} f \subset \mathbb{R}_-\}$, $(\tau_1^t f)(x) = (\tau_2^t f)(x) = f(x-t)$, $t \in \mathbb{R}, x \in \mathbb{R}$, $B_\alpha: S \times S \to \mathbb{C}$, $\alpha = 1,2$, two continuous τ_α^t-Toeplitz forms, $\overset{\#}{B}: W_1 \times W_2 \to \mathbb{C}$ a (τ_1^t, τ_2^t)-Hankel form satisfying $\overset{\#}{B} \prec (B_1, B_2)$. Then there exists a continuous Toeplitz form $B: S \times S \to \mathbb{C}$, and a tempered matrix measure $(\mu_{\alpha\beta}) \geq 0$ in $\mathbb{R}$, such that $B \leq (B_1, B_2)$, $B = \overset{\#}{B}$ in $W_1 \times W_2$, and $B_\alpha(\hat{f}, \hat{g}) = \mu_{\alpha\alpha}(f\bar{g})$, $\forall (f,g) \in S \times S$ $\alpha = 1,2$, $\overset{\#}{B}(f_1, f_2) = \mu_{12}(f_1 \bar{f}_2)$, $\forall (f_1, f_2) \in W_1 \times W_2$.*

Remark 6. The measures μ_{11} and μ_{22}, in the preceding statement, are uniquely determined, but not so μ_{12}, and the possible μ_{12}'s are in 1-1 correspondence with the Nagy-Foias liftings of certain commutants. For $B_1 = B_2$, $\overset{\#}{B} = B_1$ in $W_1 \times W_2$, Theorem 6 reduces to the Bochner-Schwartz theorem.

Remark 7. Theorem 6 allows the refinement of results stated in [13] without complete proofs, concerning weighted inequalities and prediction problems in $\mathbb{R}$. A thorough study of prediction problems in $\mathbb{R}$, using the method of the GTKs, was done in [37].

Remark 8. The analog of Theorem 6 for kernels has recently been proved independently [20], by a different method developed in [19] and based on a generalization of Stone's spectral theorem for local semigroups. This method provides a new approach to the study of GTKs, in particular it allows to extend the GBT to kernels defined in finite intervals $[-a,a] \subset \mathbb{R}$, thus generalizing the results of M.G. Krein for Toeplitz kernels defined in finite intervals [52]. Other generalizations of this "reduced moment" type, in $\mathbb{R}$ and $\mathbb{R}^d$, $d > 1$, are given in [35].

Remark 9. It was already noted in Remark 6 that the matrix measure $(\mu_{\alpha\beta})$ appearing in Theorem 6 is not unique, and the same holds for those in Theorem 3 and 5 and in Corollary 2. More precisely, it is the μ_{12} which is not uniquely determined, and the problem arise to describe all such measures. In the case of Corollary 2, such description was given in [3]. In the situation of example (c) of Section 1, this description reduces to the classical parametrization of Adamjan-Arov-Krein [1]. Some unicity and parametrization results for the "reduced" problem are given in [19]. The parametrization was studied independently by Yamamoto in [72] by a different metod.

In [12] it was shown that, in the case of operator-valued kernels, the matrix $(\mu_{\alpha\beta})$ leads to a generalized Lax-Phillips structure, such that μ_{12} is the scattering function.

For applications of the results in Section 4 to harmonizable processes (in the sense of Bochner [18]), scattering functions and linear systems, see [30], [12], [6], [7]. Another approach to the study of GTKs was started recently in [5].

5. HANKEL FORMS SUBORDINATED TO GENERAL NORMS.

In the preceding sections, sesquilinear forms B, subordinated to a pair of fixed positive forms B_1,B_2, were systematically considered. Since B_1 and B_2, being positive, give rise to hilbertian seminorms $\sigma_\alpha(f) = B_\alpha(f,f)^{1/2}$, $\alpha = 1,2$, it is natural to consider the situation where a form B is subordinated to any pair of (non-necessarily hilbertian) seminorms.

Fix as before two vector spaces V_1,V_2, two vector subspaces $W_\alpha \subset V_\alpha$, two linear isomorphisms $\tau_\alpha: V_\alpha \to V_\alpha$, and let σ_α be a seminorm in each V_α, $\alpha = 1,2$. Let the families of Toeplitz and Hankel forms Λ_0, $\overset{\#}{\Lambda}_0$ be defined as in (11), and set now

(17)
$$\Lambda = \Lambda(\sigma_1,\sigma_2) = \{B \in \Lambda_0;\ B \leq (\sigma_1,\sigma_2)\},$$
$$\overset{\#}{\Lambda} = \overset{\#}{\Lambda}(\sigma_1,\sigma_2) = \{\overset{\#}{B} \in \overset{\#}{\Lambda}_0;\ \overset{\#}{B} \prec (\sigma_1,\sigma_2)\},$$

where $B \leq (\sigma_1,\sigma_2)$ (resp.,$\prec$) if $|B(f_1,f_2)| \leq \sigma_1(f_1$ $\forall (f_1,f_2) \in V_1 \times V_2$ (resp., $W_1 \times W_2$).

The interpolation problem for the families $\Lambda = \Lambda(\sigma_1,\sigma_2)$ and $\overset{\#}{\Lambda} = \overset{\#}{\Lambda}(\sigma_1,\sigma_2)$ is: given $\overset{\#}{B} \in \overset{\#}{\Lambda}$, does there exist $B \in \Lambda$ such that $B = \overset{\#}{B}$ in $W_1 \times W_2$? This problem was given an affirmative answer in [33] for the case $V_1 = V_2 = P$, $W_\alpha = P_\alpha$, $\tau_1 = \tau_2$ = the shift operator in P, and σ_α two seminorms defined in $C(\mathbb{T})$, that satisfy (i) $|f| \leq |g| \Rightarrow \sigma_\alpha(f) \leq \sigma_\alpha(g)$, (ii) $\sigma_\alpha(f) \leq c\,\|f\|_\infty$ (examples of such seminorms are those of the L^p, L^p-weighted and Orlicz spaces, as well as those of some Lorentz spaces). As in Theorem 1, every $\overset{\#}{B} \in \overset{\#}{\Lambda} = \overset{\#}{\Lambda}(\sigma_1,\sigma_2)$, for such σ_1,σ_2, has a lifting in $\Lambda = \Lambda(\sigma_1,\sigma_2)$. Moreover, the result is strengthened by replacing the condition $\overset{\#}{B} \prec (\sigma_1,\sigma_2)$ by a weaker condition $\overset{\#}{B} \prec\prec (\sigma_1,\sigma_2)$ (see [28], [67] for details). In particular, this leads to refinements of the Nehari theorem, where $\|f_1\|_2\ \|f_2\|_2$ in condition (4) is replaced by a more general $\sigma_1(f_1)\ \sigma_2(f_2)$ [33]. As an application, the following generalization of the Paley lacunary inequality is given.

A sequence $\{n_k\} \subset \mathbb{N}$ is said to be σ-q-Paley, where σ is a seminorm and $2 \leq q < \infty$, if, for every $f(t) = \Sigma_{n\geq 0} c_n e^{int} \in C(\mathbb{T})$, the inequality

$$(18) \qquad (\sum_k |c_{n_k}|^q n_k^{2-q})^{1/q} \leq C_q \sigma(f)$$

holds. It was proved in [33] that, if σ satisfies (i) and (ii), a sufficient condition for $\{n_k\}$ to be a σ-q-Paley sequence is that, whenever $\phi(t) = \Sigma_{k>0} v_k e^{in_k t} \in H^2(\mathbb{T})$, is defined for $n > 0$ as v_k for $n = n_k$ and as 0 for $n \neq n_k$, and $\overset{\#}{B}$ is the Hankel forms given by s(n), the subordination $\overset{\#}{B} \prec (r\sigma, r\sigma)$ holds for

$$(19) \qquad r^2 = r_q^2 = C_q (\sum_k |v_k|^{q'} n_k^{(q-2)/(q-1)})^{1/q'}.$$

From this follows that every lacunary sequence $\{n_k\}$, $n_{k+1}/n_k \geq \lambda > 1$ for all k, <u>is</u> σ_q-<u>q-Paley for</u>

$$(20) \qquad \sigma_q(f) = (\int_{\mathbb{T}} |f|^{q/2} |t|^{q-2} dt)^{2/q}, \quad q \geq 2.$$

For q = 2, (18) reduces to the classical Paley lacunary inequality [63].

It was shown in [28] that the lifting theorem for forms, defined in $P \times P$ and subordinated to general seminorms, holds if $\sigma_1(f_1)\ \sigma_2(f_2)$ is replaced in (17) by $\sigma(|f_1 + f_2|)^{1/2}$.

Using this result, in [27] and [28] the Helson-Szegö theorem was generalized for the case of L^p, $p \neq 2$, and for pairs of different measures μ, ν (see example (a) in Section 1 for p = 2). As an example of such generalization, we have the following equivalence of the Muckenhoupt A_p condition [68]: a positive weight $\omega \in A_p$ iff for every $0 \leq \phi \in L^{p/(p-2)}$, $\|\phi\| \leq 1$, there is $h = h_\phi \in H^1$ and $a = a_\phi \in L^{p/(p-2)}$, such that

$$(21) \qquad 2\phi\omega^{2/p} \leq \mathrm{Re}\, h \leq |h| \leq a\omega^{2/p}.$$

If p = 2, it is enough to pick $\phi \equiv 1$, and (21) reduces to (15), and it is again the well-known equivalence between A_2 and the Helson-Szegö condition [47]. Similar extensions to L^p were provided for the other examples of Section 1 in [28] and [68], where the weighted norm inequalities for the Poisson operator are also considered. The main theorem of [27] was recently extended by Rubio de Francia [66] to more general 2-convex norms.

In [29] and [68] the lifting theorem for forms subordinated to seminorms was used to obtain a Bochner-Eberlein-Horn [46] theorem for GTKs, and a Grothendieck type inequality for such kernels, as well as an interpretation of these results in the framework of Grothendieck's theory of bilinear forms [43]. Since Grothendieck's Fundamental Inequality is known to be closely related to the classical Paley inequality (see, for instance, [40]), the connection between the σ-q-Paley sequences and the Grothendieck inequality for GTKs is to be further explored.

REFERENCES

1. V.D. Adamjan, D.Z. Arov and M.G. Krein, Infinite Hankel matrices and generalized Carathéodory-Fejér and I. Schur problems, Funct. Anal. Appl. 2 (1968), 1-17. (Russian)

2. R. Arocena, Toeplitz kernels and dilations of intertwining operators, Integral Eqs. and Operator Thy. 6 (1983), 759-778.

3. ________, On generalized Toeplitz kernels and their relation with a paper of Adamjan, Arov and Krein, North Holland Math. Studies 86 (1984), 1-22.

4. ________, Scattering functions, linear systems, Fourier transforms of measures and unitary dilations to Krein spaces: a unified approach, Publ. Math. d'Orsay 85-02 (1985), 1-57.

5. ________, Une classe de formes invariantes par translation et le théorème de Helson et Szegö, C.R. Acad. Sci. Paris 302, I (1986), 107-110.

6. ________, Naimark's theorem, linear systems and scattering operators, to appear in J. Functional Anal.

7. ________, Generalized Toeplitz kernels and dilations of intertwinning operators, to appear in Acta Sci. Math. (Szeged).

8. R. Arocena and M. Cotlar, Generalized Toeplitz kernels, Hankel forms and Sarason's commutation theorem, Acta Cient. Venez. 33 (1982), 89-98.

9. ________________________, Generalized Toeplitz kernels and Adamjan-Arov-Krein moment problems, Op. Theory: Adv. & Appl. 4 (1982), 37-55.

10. ________________________, Dilation of generalized Toeplitz kernels and L^2-weighted problems,, Lecture Notes in Math., Springer-Verlag 908 (1982), 169-188.

11. ________________________, Generalized Herglotz-Bochner theorem and L^2-weighted problems with finite measure, in Conf. Harmonic Analysis in honor of A. Zygmund, (Eds.: A.P. Calderon, W. Beckner, R. Fefferman, P. Jones), Wadsworth Intl. Math. Ser. (1982), 258-269.

12. R. Arocena, M. Cotlar and J. León, Toeplitz kernels, scattering structures and covariant systems, North-Holland Math. Lib. 34 (1985), 1-19.

13. R. Arocena, M. Cotlar and C. Sadosky, Weighted inequalities in L^2 and lifting properties, Adv. Math. Suppl. Studies 7A (1981), 95-128.

14. Gr. Arsene, Z. Ceausescu, and C. Foias, On intertwining dilations VIII, J. Operator Thy. 9 (1983), 107-142.

15. F. Beatrous and J. Burbea, Positive-definiteness and its applications to interpolation problems for holomorphic functions, Trans. Amer. Math. Soc. 284 (1984), 247-270.

16. Iu. M. Berezanskii, Expansions in eigenfunctions of selfadjoint operators, Transl. Math. Monographs, Amer. Math. Soc., Providence, 1968.

17. R. Blei, Uniformity property for Λ(2) sets and Grothendieck's inequality, Symp. Math. 22 (1977), 321-337.

18. S. Bochner, Stationarity, boundedness, almost periodicity of random-valued functions, in Proc. Third Berkeley Symp. Math. Stat. Probl, Vol. 2, Univ. Calif. Press, Berkeley (1956), 7-27.

19. R. Bruzual, Teoría de semigrupos locales de isometrías y algunas aplicaciones, Ph.D. Thesis, U.C.V., Caracas, Venezuela, 1986.

20. ________, Local semigroups of contractions and some applications to Fourier representation theorems, submitted for publication.

21. J. Burbea and P. Masani, Banach and Hilbert spaces of vector-valued functions, their general theory and applications to holomorphy, Pitman Research Notes in Math. 90, Pitman, London, 1984.

22. R.R. Coifman, R. Rochberg and G. Weiss, Factorization theorems for Hardy spaces in several variables, Annals of Math. 103 (1976), 711-735.

23. M. Cotlar and C. Sadosky, A moment theory approach to the Riesz theorem on the conjugate function for general measures, Studia Math. 53 (1975), 75-101.

24. ________________, Transformée de Hilbert, théorème de Bochner et le problème des moments, I, II, C.R. Acad. Sci., Paris, A 285 (1977), pp. 433-435, and 661-665.

25. ________________, On the Helson-Szegö theorem and a related class of modified Toeplitz kernels, Proc. Symp. Pure Math. AMS 34 (1979), 383-407.

26. ________________, Characterization of two measures satisfying the Riesz inequality for the Hilbert transforms in L^2, Acta Cient. Venez. 30 (1979), 346-348.

27. ________________, On some L^p versions of the Helson-Szegö theorem, in Conf. Harmonic Analysis in honor of A. Zygmund, (Eds.: W. Beckner, A.P. Calderon, R. Fefferman, P.W. Jones), Wadsworth Int. Math. Series (1982), pp. 306-317.

28. ________________, Majorized Toeplitz forms and weighted inequalities with general norms, Lecture Notes in Math., Springer-Verlag 908 (1982), 132-168.

29. ________________, Vectorial inequalities of Marcinkiewicz-Zygmund and Grothendieck type for generalized Toeplitz kernels, Lecture Notes in Math., Springer-Verlag 992 (1983), 278-308.

30. ________________, Generalized Toeplitz kernels, stationarity and harmonizability, J. Anal. Math. 44 (1985), 117-133.

31. ________________, Inégalités à poid pour les coefficients lacunaires de certaines fonctions analytiques, C.R. Acad. Sci. Paris I 299 (1984), 591-594.

32. ______________, A lifting theorem for subordinated invariant kernels, J. Functional Anal. 67 (1986), 345-359.

33. ______________, Lifting properties, Nehari theorem and Paley lucunary inequality, to appear in Rev. Matem. Iberoamericana.

34. ______________, Extensions de Toeplitz de formes de Hankel généralisées, submitted for publication.

35. ______________, Extensions de Toeplitz de noyaux de Hankel pour groupes d-paramétriques, submitted for publication.

36. A. Devinatz, On extensions of positive definite functions, Acta Math. 102 (1959), 109-134.

37. M. Domínguez, Procesos continuous completamente regulares, Master Thesis, UCV, Caracas, Venezuela (1984), 1-134.

38. H. Dym, On a Szegö formula for a class of generalized Toeplitz kernels, Integral Eqs. and Operator Thy. 8 (1985), 427-431.

39 C. Fefferman and E.M. Stein, H^p spaces of several variables, Acta Math. 129 (1972), 137-193.

40. J. Fournier, On a theorem of Paley and the Littlewood conjecture, Arkiv f. Math. 17 (1979), 199-216.

41. J.E. Gilbert, Harmonic analysis and the Grothendieck's fundamental theorem, Symp. Math. 22 (1977), 393-420.

42. I. Gohberg and H. Dym, A maximum entropy principle for contractive interpolants, J. Functional Anal. 65 (1986), 83-125.

43. A. Grothendieck, Résumé dela théorie métrique des produits tensoriels topologiques, Bol. Soc. Mat. Sao Paulo 8 (1956), 1-79.

44. H. Helson and D. Sarason, Past and future, Math. Scand. 21 (1967), 5-16.

45. H. Helson and G. Szegö, A problem in prediction theory, Ann. Mat. Pura Appl. 51 (1960), 107-138.

46. R.A. Horn, Quadratic forms in harmonic analysis and Bochner-Eberlein theorem, Proc. Amer. Math. Soc. 52 (1975), 263-270.

47. R. Hunt, B. Muckenhoupt and R.L. Wheeden, Weighted norm inequalities for the conjugate function and the Hilbert transform, Trans. Amer. Math. Soc. 176 (1973), 227-252.

48. I. Ibragimov and Iu. Rozanov, Random Gaussian Processes, Springer-Verlag, Berlin and New York (1978); Mir, Moscow (1974), (in Russian).

49. S. Janson, J. Peetre and R. Rochberg, Hankel forms and the Fock spaces, U. U. D. Math., Report 96-6 (1986).

50. P.W. Jones, Carleson measures and the Fefferman-Stein decomposition of BMO(R), Ann. Math. 111 (1980), 197-208.

51. P. Koosis, Moyennes quadratiques pondérées de fonctions périodiques et de leur conjuguées harmoniques, C.R. Acad. Sci. Paris A 291 (1980), 255-257.

52. M.G. Krein, Sur le problème du prolongement des fonctions hermitiennes positives et continues, Dokl. Akad. Nauk SSSR 26 (1940), 17-22 (in Russian).

53. M.G. Krein, On hermitian operators with directed functionals, Akad. Nauk Ukrain. RSR Zbirnik Proc' Inst. Mat. (1947), 104-129 (in Russian).

54. P. Lax and R. Phillips, Scattering theory, Academic Press, New York, 1967.

55. S. Marcantognini, Dilataciones unitarias a espacios de Krein, Master Thesis, UCV, Caracas, Venezuela (1985), 1-150.

56. M.D. Morán, Parametrización de Foias-Ceausescu y nucleos de Toeplitz generalizados, Ph.D. Thesis, UCV, Caracas, Venezuela (1986).

57. B. Sz. Nagy and C. Foias, Analyse harmonique des opérateurs de l'espace de Hilbert, Masson- Akad. Kiado, Paris and Budapest, 1967; English transl. North-Holland-Akad. Kiado, Amsterdam and Budapest, 1970.

58. ______________________, Dilation des commutants, C.R. Acad. Sci. Paris A 266 (1968), 493-495.

59. M.A. Naimark, Extremal spectral functions of a symmetric operator, I.V. Akad. Nauk SSSR Ser. Mat. 11 (1947), 327-344 (in Russian).

60. Z. Nehari, On bilinear forms, Ann. Math. 68 (1957), 153-162.

61. H. Niemi, On the linear prediction problem of certain non-stationary stochastic processes, Math. Scad. 39 (1976), 146-160.

62. N.K. Nikolskii, Lectures on the shift operator, Mir, Moscow, 1980, English transl. Springer-Verlag, New York, 1985.

63. L.B. Page, Applications of the Sz. Nagy and Foias lifting theorem, Indiana Univ. Math. J. 20 (1970), 135-145.

64. R.E.A.C. Paley, On the lacunary coefficients of power series, Ann. Math. 34 (1933), 615-616.

65. V.V. Peller and S.V. Khrushchev, Hankel operators, best approximations and stationary Gaussian processes, Uspehi Mat. Nauk 37 (1982), 53-124 (in Russian).

66. R. Rochberg, Toeplitz and Hankel operators on a Paley-Wiener Space, to appear in Int. Eq. & Operator Thy.

67. J.L. Rubio De Francia, Linear operators in Banach lattices and weighted L^2 inequalities, Volume in honor of H. Triebel, Nachr., in the press.

68. C. Sadosky, Some applications of majorized Toeplitz kernels, Topics in Modern Harm. Anal., II, Ist. Alta Mat. Roma (1983), 581-626.

69. D. Sarason, Generalized interpolation in H^∞, Trans. Amer. Math. Soc. 127 (1967), 179-203.

70. L. Schwartz, Mathemática y Física Cuántica, Cursos y Sem. de Matem., Vol. 1, Fac. de Ciencias, UBA, Buenos Aires (1959), 1-266.

71. __________, Sous-espaces hilbertiens d'espaces vectoriels topologiques et noyaux associés, J. Anal. Math. 13 (1964), 115-256.

72. H. Widom, Inversion of Toeplitz matrices III, Notices Amer. Math. Soc. 7 (1960), 63.

73. T. Yamamoto, On the generalization of the theorem of Helson and Szegö, Hokkaido Math. J. 14 (1985), 1-11.

74. V.E. Katznelson, Integral respresentation of hermitian positive kernels of mixed type and generalized Nehari problem, Theoria Funktii, Funk. Anal. i Priloz. 43 (1985), 54-70 (in Russian).

Some Recent Developments in Fourier Analysis and H^p Theory on Product Domains - II

Robert Fefferman, University of Chicago

This note is intended as an update of the survey article of the same name published by Alice Chang and the author in the Bulletin of the A.M.S. in January of 1985 [1]. Since then there have been several important developments in the area of product domains, (most notably a certain covering lemma for rectangles, due to J. L. Journé), which have changed our point of view considerably.

As pointed out in [1], a certain counterexample due to Carleson in 1974 had shown that the dual space of $H^1(R^2_+ \times R^2_+)$ could not be characterized by the most straightforward generalization of the one variable duality $[(H^1(R^2_+ \times R^2_+) = \{f \in L^1(R^2) | H_1f, H_2f,$ and $H_1H_2f \in L^1(R^2)\}$. For other characterizations of this space, as well as further discussion of it, see [1])]. To be more specific, define the mean oscillation of a function $\phi(x_1,x_2)$ over the rectangle $R \subseteq R^2$ by

$$\mathrm{osc}_R(\phi) = \inf\left(\frac{1}{|R|}\int_R |\phi(x_1,x_2) - \phi_1(x_1) - \phi_2(x_2)|^2 dx_1dx_2\right)^{1/2}$$

where the inf is taken over all pairs of functions ϕ_1 and ϕ_2 which depend only on the x_1 and x_2 variables respectively. Then Carleson's counterexample shows that for a function ϕ, the condition of bounded mean oscillation,

$$\mathrm{osc}_R(\phi) < C \quad \text{for all rectangles } R$$

is much too weak to insure that ϕ belong to the dual of $H^1(R^2_+ \times R^2_+)$. In order to give a condition which is equivalent to $\phi \in H^1(R^2_+ \times R^2_+)^*$, we require some notation. Fix a function $\psi(x)$ on R^1 such that $\psi \in C^\infty(R^1)$, ψ is odd, supported in $[-1,1]$ and does not vanish identically. For an open set Ω R^2, define the Carleson region $S(\Omega) \subseteq R^2_+ \times R^2_+$ by the condition

$$(y_1,t_1,y_2,t_2) \quad S(\Omega) \quad \text{iff} \quad [y_1 - t_1, y_1 + t_1] \times [y_2 - t_2, y_2 + t_2] \subset \Omega.$$

Finally, for a function ϕ defined on R^2, let Φ be defined on $R^2_+ \times R^2_+$ by

$$\Phi(y_1,t_1,y_2,t_2) = \iint_{R^2} \phi(y_1 - z_1, y_2 - z_2)\psi\left(\frac{z_1}{t_1}\right)\psi\left(\frac{z_2}{t_2}\right)t_1^{-1}t_2^{-1}dz_1dz_2 .$$

Then $\phi \in H^1(R^2_+ \times R^2_+)^*$ iff

$$(*) \qquad \iint_{S(\Omega)} |\phi(y_1,t_1,y_2,t_2)|^2 \frac{dy_1 dt_1 dy_2 dt_2}{t_1 t_2} \leq C\cdot|\Omega|,$$

for every open set $\Omega \subseteq R^2$. (Here m denotes Lebesgue measure in the plane). As the reader recognizes immediately this is the standard characterization of $BMO(R^1)$ extended to R^2, but where the role of intervals is now played by arbitrary open sets, rather than by rectangles. The geometry of arbitrary open sets in R^2 could be quite complicated, and so the space $BMO(R^2_+ \times R^2_+)$ defined by (*) above, is best understood in terms of a certain covering lemma for rectangles. This is described in detail below.

At this point let us also mention the atomic decomposition of the spaces $H^p(R^2_+ \times R^2_+)$ for $0 < p \leq 1$. (See [1] for information concerning these spaces). An $H^p(R^2_+ \times R^2_+)$ atom is best described in terms of the collection $\mathcal{M}(\Omega)$ of maximal dyadic subrectangles of Ω. The most straightforward generalization of H^p atom to the product case would be a function $a(x_1,x_2)$ supported in a rectangle $R = I \times J$ so that for each fixed x_1, as a function of x_2 a has a certain number of moments vanishing, and the same is true if x_2 is fixed when a is considered as a function of x_1. In analogy with the classical case, we also require that $\|a\|_2 \leq |R|^{1/2-1/p}$. Unfortunately with this definition of atom, it is an immediate consequence of Carleson's counterexample that it is not possible to write an arbitrary $H^p(R^2_+ \times R^2_+)$ function f in the form

$$f = \sum_{k=1}^{\infty} \lambda_k a_k \quad \text{where the } \lambda_k \text{ are scalars}$$

satisfying $\sum|\lambda_k|^p < \infty$ and a_k are atoms. To emphasize that such a_k do not span all of H^p we shall refer to such functions as "rectangle atoms" and reserve the name atom for a function $a(x_1,x_2)$ supported in an open set $\Omega \subseteq R^2$ which can be further decomposed as $a = \sum_{R\in\mathcal{M}(\Omega)} \alpha_R$ where the α_R are supported in the double of the maximal subrectangle $R = I \times J$, where

$$\int_I \alpha_R(x_1,x_2)x_1^j dx = 0, \; j \leq N(p)$$

$$\int_J \alpha_R(x_1,x_2)x_2^j dx_2 = 0 \; j \leq N(p) \quad \text{and where}$$

$$\Big(\sum_{R\in\mathcal{M}(\Omega)} \|\alpha_R\|_2^2\Big)^{1/2} \leq |\Omega|^{1/2-1/p}$$

It is proven in [2] that with such atoms a, we do get the decomposition; i.e. given $f \in H^p$ we can write f as $f = \sum\lambda_k a_k$ with $(\sum|\lambda_k|^p)^{1/p} \leq C\|f\|_{H^p}$ and a_k H^p atoms. Conversely, if $\sum|\lambda_k|^p < \infty$ and a_k are H^p atoms, then

$$\sum\lambda_k a_k \in H^p(R^2_+ \times R^2_+) \quad \text{and} \quad \|\sum \lambda_k a_k\|_{H^p} \leq C(\sum|\lambda_k|^p)^{1/p}.$$

In order to apply the above characterizations of BMO and H^p in the product setting, one can make use of a basic geometric lemma about rectangles due to J. L. Journé.

The statement of the lemma can be understood in the following terms: If $\Omega \subseteq R^2$ is open then it may well be the case that $\sum_{R\in \mathcal{M}(\Omega)} |R| = \infty$, where $\mathcal{M}(\Omega)$ denotes the collection of all maximal dyadic subrectangles of Ω. Now, let M_s denote the strong maximal operator, and $\tilde{\Omega} = \{x \mid M_s(\chi_\Omega)(x) > 1/2\}$. Let $\delta > 0$. Suppose $\gamma_1(R)$ is defined for a dyadic subrectangle $R = I \times J \subseteq \Omega$ to be $|\hat{I}|/|I|$ where $\hat{I}$ is a maximal dyadic interval containing I so that $\hat{I} \times J \subseteq \tilde{\Omega}$. Then Journe's Lemma [3] is

$$\sum_{R\in \mathcal{M}_2(\Omega)} |R|\gamma_1^{-\delta}(R) < C|\Omega|, \quad \text{where } \mathcal{M}_2(\Omega)$$

denotes the class of dyadic subrectangles of Ω which are maximal in the x_2 direction.

Using this result, Journé introduces a class of operators in [3] which are proven to be bounded from L^∞ to BMO. The result in [3] is proven when the functions are defined on product spaces with arbitrarily many factors, i.e. on $R^{n_1} \times R^{n_2} \cdots \times R^{n_k}$. Here we shall state the results in $R^1 \times R^1$. To do so, we require some notations. Suppose that $k(x,y)$ is a kernel defined for $x,y \in R^1$ and defining the operator $Tf(x) = \int_{y\in R^1} k(x,y)f(y)dy$. Then the Calderon-Zygmund norm of T is defined as follows: Let C_0 denote the smallest positive number so that

$$\int_{|x-y|>\gamma|x-x'|} |k(x,y) - k(x',y)|dy < C_0\gamma^{-\delta} \quad \text{for all } \gamma > 2$$

and some fixed $\delta > 0$. Then $\|T\|_{CZ}$ is defined as

$$\|T\|_{L^2,L^2} + C_0 .$$

Now suppose $k(x_1,x_2,y_1,y_2)$ is a kernel on $R^2 \times R^2$ defining an operator $Tf(x_1,x_2) = \iint_{R^2} k(x_1,x_2,y_1,y_2)f(y_1,y_2)dy_1dy_2$. Let $k^{(1)}(x_1,y_1)$ denote the operator whose kernel is given as

$$[k^{(1)}(x_1,y_1)](x_2,y_2) = k(x_1,x_2,y_1,y_2)$$

Then we shall say that the operator T acting on functions on R^2 is in Journé's class, $\mathcal{J}$, iff T is bounded on $L^2(R^2)$ and

$$\int_{|x_1-y_1|>\gamma|x_1-x_1'|} \|k^{(1)}(x_1,y_1) - k^{(1)}(x_1',y_1)\|_{CZ}dy_1 < C\gamma^{-\delta} \text{ for all } \gamma > 2,$$

for all x_1, x_1',

$$\int_{|x_2-y_2|>\gamma|x_2-x_2'|} \|k^{(2)}(x_2,y_2) - k^{(2)}(x_2',y_2)\|_{CZ}dy_2 \leq C\gamma^{-\delta} \text{ for all } \gamma \geq 2,\ x_2,x_2'.$$

In [3] it is shown that every operator in the class $\mathcal{J}$ is bounded from $L^\infty(R^2)$ to $BMO(R_+^2 \times R_+^2)$. By the interpolation results in [4] and [5] it follows that such operators are bounded on $L^p(R^2)$ for $2 \leq p < \infty$. (These operators may be unbounded on the L^p spaces for the other values of p however). Using this Calderón-Zygmund type result, Journé is able to obtain a product version of the results of Calderón, Coifman and Meyer on commutator integrals.

The $H^p(R_+^2 \times R_+^2)$ theory for $0 < p \leq 1$ of the operators in $\mathcal{J}$ can be found in [6]. The class we consider there contains $\mathcal{J}$, but the results have a somewhat different flavor than those of [3]. In [6] we are interested in trying to circumvent the counterexample of Carleson, and the failure of rectangle atoms to span all of $H^p(R_+^2 \times R_+^2)$. It turns out that for an L^2 bounded linear operator it is enough to check the action of the operator on a rectangle atom in order to know that the operator acts boundedly on $H^p(R_+^2 \times R_+^2)$. To be more specific, suppose we want to show that T is bounded from $H^p(R_+^2 \times R_+^2)$ to $L^p(R^2)$ for some p, $0 < p \leq 1$. Since a rectangle atom $a(x_1,x_2)$ has a uniformly bounded H^p norm, it is clearly necessary to have

$$(**) \qquad \int_{R^2} |T(a)|^p dx \leq C$$

for some constant C, independent of a. Because rectangle atoms do not span H^p, this condition is not sufficient to imply that T is bounded from H^p to L^p. Nevertheless, a very slightly stronger condition than (**) is sufficient. To be precise, suppose that for a rectangle R and $\gamma > 0$ we denote by $\tilde{R}_\gamma$ the concentric dilate of R by a factor of γ in both directions. Let δ be an arbitrary positive number. Then we have the following theorem [6]:

Theorem: Let T be a linear $L^2(R^2)$ bounded operator. Suppose that whenever a is an $H^p(R_+^2 \times R_+^2)$ rectangle atom supported on a rectangle $R \subsetneq R^2$ and $\gamma \geq 2$ that

$$\int_{{}^c\tilde{R}_\gamma} |T(a)|^p dx \leq C\gamma^{-\delta}.$$

The T maps $H^p(R_+^2 \times R_+^2)$ boundedly onto $L^p(R^2)$.

The philosophy that it is possible to circumvent the failure of the rectangle atomic decomposition on H^p can be carried further in order to analyze product singular integrals in other ways. This is done in [7] where we obtain weighted norm inequalities for a class which is, for all practical purposes, the class $\mathcal{J}$. In so doing we also come to grips with the problem of understanding how the strong maximal operator M_s defined by

$$M_s f(x) = \sup_{x\in R} \frac{1}{|R|} \int_R |f(y)|dy,$$

(the sup is taken over all rectangles R with sides parallel to the axes which contain the point x) controls product singular integrals. Both of these are accomplished through the sharp operator, which generalizes the Fefferman-Stein sharp operator in the classical theory. Let us recall that given a function locally in L^2 and a rectangle $R \subseteq R^2$, we defined above the notion of mean oscillation of ϕ over R:

$$\mathrm{osc}_R(\phi) = \inf\left(\frac{1}{|R|} \int_R |\phi(x_1,x_2) - \phi_1(x_1) - \phi_2(x_2)|^2 dx_1 dx_2\right)^{1/2}$$

where the inf is taken over all pairs of functions ϕ_1 and ϕ_2 depending only on x_1 and x_2 respectively. We would like to define, following Fefferman and Stein in [8]:

$$\phi^{\#}(x) = \sup_{x\in R} [\mathrm{osc}_R(\phi)]$$

and then show that for a singular integral T,

$$(Tf)^{\#}(x) \leq C\, M_s(f^2)^{1/2}(x),$$

and that $\phi^{\#}(x) \in L^p(R^2)$ implies $\phi(x) \in L^p(R^2)$. Unfortunately, as is easily seen from Carleson's counterexample, the last implication of how $\phi^{\#}$ controls ϕ is false. Though this might seem to rule out the sharp operator as a useful tool in product spaces, this is not the case. Suppose that we are given an L^2 bounded linear operator T and that there exists an operator U satisfying (for some fixed $\delta > 0$)

$$\mathrm{osc}_R(Tf) \leq \gamma^{-\delta} U(f)(x) \quad \text{whenever} \quad x \in R,\ \gamma \geq 2,$$

and f is supported outside of $\tilde{R}_\gamma$. Then we call U a sharp operator for T and write $U = T^{\#}$. (If we did not require any condition on the support of f and dropped the power of γ, this would be the classical definition of $T^{\#}$.) With this definition, it is not difficult to see that for a product singular integral, T,

$$T^{\#}f = M_s(f^2)^{1/2}.$$

In [7], it is also shown that

$$\int_{R^2} S^2(Tf)\phi \, dx \leq C \int_{R^2} (I + T^{\#})(f)^2 M_s^{(3)}(\phi)dx$$

(I denotes the identity operator, S denotes the area integral), and $M_s^{(3)}$ denotes the composition $M_s \circ M_s \circ M_s$) for any L^2 bounded linear operator T. Taking T to be a singular integral gives

$$(***) \qquad \int_{R^2} S^2(Tf)\phi \, dx \leq C \int_{R^2} M_s(f^2) M_s^{(3)}(\phi)dx.$$

At this point we recall that a locally integrable function $w(x_1,x_2)$ on R^2 is said to belong to product A^p, $A^p(R^1 \times R^1)$, if and only if

$$\left(\frac{1}{|R|}\int_R w\right)\left(\frac{1}{|R|}\int_R w^{-\frac{1}{p-1}}\right)^{p-1} \leq C \quad \text{for all rectangles } R.$$

We are interested in proving weighted inequalities of the form

$$\int_{R^2} |Tf|^p w dx \leq C \int_{R^2} |f|^p w dx, \; p > 2.$$

Because the operators T may be unbounded on $L^p(R^2)$ when $p < 2$, the appropriate requirement for w turns out to be $w \in A^{p/2}(R^1 \times R^1)$ in order for the above inequality to hold. It turns out (***) implies the desired result as follows: First, it is well known that if $w \in A^p(R^1 \times R^1)$ then $\int |f|^p w \leq C \int Sf^p w$, so it suffices to let $w \in A^{p/2}(R^1 \times R^1)$, $\int_{R^2} |\phi|^{(p/2)'} w = 1$ and to estimate $\int_{R^2} S^2(Tf)\phi w dx$. By (***)

$$\begin{aligned}\int_{R^2} S^2(Tf)\phi w dx &\leq C \int_{R^2} M_s(f^2) M_s^{(3)}(\phi w) w^{-1} w dx \\ &\leq C\left(\int_{R^2} M_s^{p/2}(f^2) w dx\right)^{2/p}\left(\int_{R^2} [M_s^{(3)}(\phi w)]^{(p/2)'} w^{-(p/2)'} w dx\right)^{1/(p/2)'}.\end{aligned}$$

Since

$$\begin{aligned}& w^{1-(p/2)'} \in A^{(p/2)'}, \left(\int_{R^2} [M_s^{(3)}(\phi w)]^{(p/2)'} w^{1-(p/2)'} dx\right)^{1/(p/2)'} \leq \\ &\leq C\left(\int |\phi|^{(p/2)'} w\right)^{1/(p/2)'} \leq C,\end{aligned}$$

because M_s is bounded on $L^{(p/2)'}(vdx)$ when $v \in A^{(p/2)'}(R^1 \times R^1)$. Hence (***) becomes

$$\int_{R^2} S^2(Tf)\phi w dx \leq C \int |f|^p w dx$$

and the weighted inequalities for singular integrals are proved.

In proving these weighted norm inequalities for singular integrals T, we are assuming only that $T^{\#}f = M_s(f^2)^{1/2}$. These assumptions imply only the boundedness of T on $L^p(w)$ when $p > 2$ and $w \in A^{p/2}$. If one wants the boundedness of T on $L^p(w)$ when $1 < p < \infty$ and $w \in A^p(R^n \times R^m)$, then one has the following result, proven in [9]: Suppose T is bounded on $L^2(R^n \times R^m)$, and is an integral operator with kernel k. Suppose that for some $\delta > 0$,

$$\|k^{(1)}(x_1,y_1) - k^{(1)}(x_1',y_1)\|_{CZ} \leq C \frac{|x_1-x_1'|^{\delta}}{|x_1-y_1|^{n+\delta}} \text{ if } |x_1-x_1'| < 1/2\ |x_1-y_1|,$$

$$\|k^{(1)}(x_1,y_1) - k^{(1)}(x_1,y_1')\|_{CZ} \leq C \frac{|y_1-y_1'|^{\delta}}{|x_1-y_1|^{n+\delta}} \text{ if } |y_1-y_1'| < 1/2\ |x_1-y_1|$$

and analogous estimates hold for $k^{(2)}$. Then T is bounded on $L^p(R^n \times R^m, wdx)$ when $w \in A^p(R^n \times R^m)$, $1 < p < \infty$.

One can ask whether all of the results above hold when we consider the product theory when there are more than two factor spaces. The results of Journé in [3] are carried out in the context of functions on $R^{n_1} \times R^{n_2} \cdots \times R^{n_k}$, but those of [6] are two parameter results. The H^p theory for operators dual to those in Journé's class can be done in arbitrarily many parameters, and this is due to J. Pipher [10]. The $A^{p/2}$ weight norm inequalities also are valid in this context (H. Lin [11]). Yet not all the theory for $R^n \times R^m$ extends. Recall that in $R^3 = R^1 \times R^1 \times R^1$ an $H^1(R_+^2 \times R_+^2 \times R_+^2)$ atom is a function $a(x_1,x_2,x_3)$ supported in a rectangle $R = I_1 \times I_2 \times I_3$, having mean value 0 in each variable separately, fixing the other two, and such that $\|a\|_{L^2} \leq |R|^{-1/2}$. Consider an $L^2(R^3)$ bounded linear operator T with the property that for some $\delta > 0$, and every $\gamma > 2$

$$\int_{{}^c\tilde{R}_{\gamma}} |T(a)|dx \leq C\gamma^{-\delta}$$

whenever a is an $H^1(R_+^2 \times R_+^2 \times R_+^2)$ atom supported on R. Then according to a remarkable theorem of Journé [12]:

(1) There exists such an operator T which fails to be bounded from $H^1(R_+^2 \times R_+^2 \times R_+^2)$ to $L^1(R^3)$.

but (2) If T is a convolution operator, and $\delta > 1/8$, then T is bounded from $H^1(R_+^2 \times R_+^2 \times R_+^2)$ to $L^1(R^3)$.

(The appearance of the constant $1/8$ seems curious. It is interesting that in most cases it can be eliminated, by requiring a condition on T which says roughly that $T^{\#} = M_s$. The "decay exponent" δ occurring in the definition of $T^{\#}$ can be taken as small as is desired. For a precise statement of this, see H. Lin [11].)

Thus, if we ask when the behavior of an operator on H^1 is determined by its action on rectangle atoms, the answer indicates a fundamental difference between the theory in one or two parameters on the one hand, and three or more parameters on the other. At least to this author, this comes as a great surprise.

References

[1] S.Y.A. Chang and R. Fefferman, Some Recent Developments in Fourier Analysis and H^p-Theory on Product Domains, Bulletin of the A.M.S., Vol. 12, No. 1, 1985.

[2] S.Y.A. Chang and R. Fefferman, A Continuous Version of the Duality of H^1 and BMO on the Bidisk, Ann. of Math., (2), 112, 1980.

[3] J.L. Journé, Calderón-Zygmund Operators on Product Spaces, Revista Matematica Iberoamericana, Vol. 3, 1985.

[4] S.Y.A. Chang and R. Fefferman, The Calderón-Zygmund Decomposition on Product Domains, Amer. Journ. of Math., 104, 1982.

[5] K.C. Lin, Thesis, Univ. of California, Los Angeles, 1984.

[6] R. Fefferman, Calderón-Zygmund Theory for Product Domains: H^p Spaces, Proc. Natl. Acad. Sci. USA, Vol. 83, Feb. 1986.

[7] R. Fefferman, Harmonic Analysis on Product Spaces, to appear.

[8] C. Fefferman and E.M. Stein, H^p Spaces of Several Variables, Acta Math. 129, 1972.

[9] R. Fefferman, A^p Weights and Singular Integrals, to appear.

[10] J. Pipher, Journé's Covering Lemma and It's Extension to Higher Dimensions, to appear in Duke Journal of Math.

[11] H. Lin, Weighted Norm Inequalities for Calderón-Zygmund Operators on Product Domains, Thesis, Univ. of Chicago, 1986.

[12] J.L. Journé, Rectangle Atoms in Three Parameters, Preprint.

A UNIFIED APPROACH TO ATOMIC DECOMPOSITIONS

VIA INTEGRABLE GROUP REPRESENTATIONS

Hans G.Feichtinger and Karlheinz Gröchenig

Institut für Mathematik, Universität Wien,
Strudlhofgasse 4, A-1090 Wien, AUSTRIA.

§ 1 Introduction

The purpose of this paper is the presentation of a general approach to atomic decompositions for Banach spaces of functions (or distributions) which are (in a suitable sense) invariant under the action of a locally compact group G. More precisely, we will show that - given an irreducible, integrable representation π of a locally compact group G on a Hilbert space $\mathcal{H}$ - it is possible to construct a family of Banach spaces, related to $\mathcal{H}$. Following Peetre, who has used this terminology in a similar context (cf. Remark 5.3 below) we shall call these spaces coorbit spaces. They are parametrized in a natural way through certain solid BF-spaces (=Banach function space) Y on G . This family is reasonably well defined (independent of particular ingredients used in their description as coorbit-spaces), and it is closed under (complex) interpolation as well as duality (with natural restrictions). Finally, each of these spaces can be given an atomic characterization, i.e. given a suitable family of atoms (derived from a sufficiently nice "wavelet" under the group action) the membership in a given space is equivalent to appropriate growth or summability conditions on the coefficients in the atomic representation. From the technical point of view, various results from the theory of (square)integrable group representations as well as from the theory of Banach function spaces on locally compact groups are required. The results formulated on an abstract level are illustrated by a couple of typical applications, including Gabor- representations for functions in $L^2(\mathbb{R}^n)$ or $S_o(\mathbb{R}^n)$ or atomic characterizations of (weighted) Besov spaces as well as for weighted Bergman spaces on the upper half plane. This situation is very similar to many situations found in the literature ([CR],[RO],[RT],[FJ], [F10],[F11] ,[PE1], [TR1], [TR2], cf. § 7 for details).

Let us mention the close connection to the theory of frames as developed by Y. Meyer and collaborators for the Heisenberg and the ax+b-group ([ME1],[ME2],[DGM]). They try to store all information on a function f through coefficients $\langle g_i, f\rangle$ with a family of functions g_i which are obtained from a single (analyzing) wavelet under a group action. The problem of reconstructing the original function from these coefficients has a satisfactory solution only in special cases (tight frames). In our approach we decompose a function into basic building blocks ("synthesizing wavelets", "atoms") which are again derived from one special wavelet under a group action. In contrast to other effective approaches to atomic decompositions we have much more freedom in choosing our *mother wavelet* .

The paper ends with an outlook on further extensions and generalizations.

§2. Some notation and definitions

2.1. By L_x and R_x we mean the left and right translation operators $L_x f(y) = f(x^{-1}y)$ and $R_x f(y) = f(yx)$. $^\vee$ and $^\triangledown$ are the involutions $f^\vee(x) = f(x^{-1})$ and $f^\triangledown(x) = \bar{f}(x^{-1})$.

2.2. A *weight function* is a positive, continuous submultiplicative function w on G, $w(xy) \leq w(x)w(y)$, $\forall x,y \in G$ (cf. [REI]), a *moderate* function is a positive, continuous function m on G such that $m(yxz) \leq w(y)m(x)w(z)$, $\forall x,y,z \in G$ for some weight function w *associated* with m.

2.3. A *solid* BF-*space* $(Y, \| \ \|_Y)$ on G is a Banach space of functions which is continuously imbedded into $L^1_{loc}(G)$ and such that $|f(x)| \leq |g(x)|$ a.e. for $g \in Y$ implies $f \in Y$ with $\| f \|_Y \leq \| g \|_Y$ .In this paper we shall restrict our attention mainly to weighted L^q-spaces

$$L^q_m(G) = \{ f \in L^1_{loc}(G) |\ fm \in L^q(G) \text{ with norm } \| f \|_{q,m} = \| fm \|_q \}$$

where m is a weight or moderate function on G. Then the L^q_m are invariant under left and right translations. Moreover, $L^1_w(G)$ is a Banach convolution algebra (called *Beurling algebra*, cf. [REI]) and $L^1_w * L^q_m \subseteq L^q_m$, whereas $L^q_m * L^1_{w_2} \subseteq L^q_m$ is valid for $w_2 := w^\triangledown \cdot \Delta^{-1/q}$ (cf.[F5]), where Δ is the Haar modulus on the group G .

2.4. A useful tool for discrete descriptions are certain (arbitrary fine) partitions of unity which exist on every locally compact group (cf.[F9]).

Definition: Given any compact neighbourhood of the identity a family $\Psi = (\psi_i)_{i\in I}$ in $C^o(G)$ is called a *bounded uniform partition of unity of size* U (U-BUPU in the sequel) if the following properties hold:

(B1) $0 \le \psi_i(x) \le 1$, $\forall\, i \in I$.

(B2) There exists $(x_i)_{i\in I}$ in G such that supp $\psi_i \subseteq x_i U \quad \forall i\in I$.

(B3) $\sup_{z\in G} \#\{\, i\in I \mid z \in x_i Q \,\} \le C_Q < \infty$ for any $Q \subseteq G$ compact .

(B4) $\sum_{i\in I} \psi_i(x) \equiv 1$.

As a consequence of (B4) we have $\bigcup_{i\in I} x_i U = G$ and we shall speak of U-*density* of the family $X=(x_i)_{i\in I}$. Apparently condition (B3) means that the family $(x_i)_{i\in I}$ is not too densely spread in the group. We call it *relatively separated in* G. We shall consider the set of U-BUPUs as a net directed by inclusion of the associated neighbourhoods and write $\Psi \longmapsto \infty$ if these neighbourhoods run through a neighbourhood base of the identity.

§ 3 Square integrable representations

In this section we collect the basic facts concerning square integrable representations of lc groups. For a detailed description of the relevant facts cf. [GMP],[GP],[CA],[DM],[WA]. In the papers of A.Grossmann and his collaborators special emphasis is put on the relationship between certain integral transformations (used below) and square integrable representations.

3.1. Let $(\pi,\mathcal{H})$ be an irreducible continuous unitary representation of a lc group G on a Hilbert space $\mathcal{H}$. Given $f,g \in \mathcal{H}$ it will be convenient to call the *coefficients of the representation* $V_g(f)\colon x \to \langle \pi(x)g,f\rangle$ the *voice transform* f (*with respect to* g) (cf.[GM1,2]). π is called *square integrable* if there exists $g \in \mathcal{H}$ such that $V_g(g) \in L^2(G)$, and it is called *integrable* if $V_g(g) \in L^1(G)$ for some $g \in \mathcal{H}$. The following properties hold true for a square integrable representation π.

3.2. π is equivalent to a subrepresentation of the (left) regular representation of G on $L^2(G)$, i.e. $\mathcal{H}$ may be identified with a closed translation invariant subspace of $L^2(G)$, the intertwining operator being just a suitably renormed voice transform, cf. below.

3.3. There exists a unique positive, selfadjoint, densely defined operator A on $\mathcal{H}$ such that $V_g(g) \in L^2(G)$ iff $g \in$ dom A and the *orthogonality relations* hold:

$$\int_G \overline{V_{g_1}(f_1)(x)} V_{g_2}(f_2)(x)dx = \langle Ag_2, Ag_1\rangle \langle f_1, f_2\rangle \quad \forall f_i \in \mathcal{H},\ g_i \in \mathrm{dom}\,A,\ i=1,2. \quad (3.1)$$

As an important immediate consequence one has

$$V_{g_1}(f_1) * V_{g_2}(f_2) = \langle Ag_1, Af_2\rangle V_{g_2}(f_1) \quad \forall\ g_1, f_2 \in \mathrm{dom}\ A,\ g_2, f_1 \in \mathcal{H}. \quad (3.2)$$

For unimodular groups A is just a scalar multiple of the identity operator, in particular dom A = $\mathcal{H}$. Choosing now $g_1 = g_2 = f_2 =: g \in$ dom A and the normalization $\| Ag \| = 1$ one obtains the following *reproducing formula*, which will be of great relevance for our approach

$$V_g(f) = V_g(f) * V_g(g) \quad \forall f \in \mathcal{H}\ . \quad (3.3)$$

The operator V_g: $f \longmapsto V_g(f)$ defines an isometry from $\mathcal{H}$ into $L^2(G)$, intertwining the group actions on these Hilbert spaces, i.e. satisfying

$$V_g[\pi(x)f] = L_x V_g(f). \quad (3.4)$$

Moreover, the orthogonal projection P from $L^2(G)$ onto the range of V_g is given by $F \longmapsto F * V_g(g)$.

In the sequel we shall extensively exploit the reproducing formula and the orthogonality relations. We shall require the representations to be integrable, because in that case the convolution operator defining P extends to a larger range of function spaces on G.

3.4. For a bounded measure $\mu \in M(G)$ let $\pi(\mu)$ denote the operator on $\mathcal{H}$ which is given weakly by

$$\langle f, \pi(\mu)(h)\rangle := \int_G \langle f, \pi(x)h\rangle\, d\mu(x) \quad \text{for } f,h \in \mathcal{H} \quad (3.5)$$

Under certain circumstances $\pi(\mu)$ can also be defined for other measures/functions, eventually with other domains or target spaces. If $\pi = L$ is the regular representation, then $\pi(\mu)h = \mu * h$ for $h \in L^2(G)$.

§4. Analyzing vectors and minimal invariant Banach spaces

In this section a particular family of coorbit spaces within the Hilbert space $\mathcal{H}$ will be described. There are various reasons for discussing these spaces at first: a) they are technically easier to treat than general coorbit spaces b) they form probably the correct "pool" of "analyzing" vectors c)their dual spaces appear as the natural reservoir for the definition of general coorbit spaces.

4.1. Throughout this section we assume that we consider a fixed integrable representation $(\pi,\mathcal{H})$ of the (separable) locally compact group G. First we define a set of *admissible* (or *analyzing*) vectors.

Definition. Given a weight function we define

$$\mathcal{A}_w := \{ g \mid g \in \mathcal{H} \ , \ V_g(g) \in L^1_w(G) \}.$$

If $\mathcal{A}_w(G) \neq \{0\}$ fix some $g \in \mathcal{A}_w(G)$ and set further

$$\mathcal{H}^1_w(G) := \{ f \mid f \in \mathcal{H} \ , \ V_g(f) \in L^1_w(G)\}$$

endowed with the natural (semi)norm

$$\| f \,|\mathcal{H}^1_w\| := \| V_g(f) \,|\, L^1_w(G)\| \ .$$

4.2 Lemma. (i) $(\mathcal{H}^1_w, \| \ \|_{\mathcal{H}^1_w})$ is a π-invariant Banach space, dense in $\mathcal{H}$, on which $\pi|_{\mathcal{H}^1_w}$ acts strongly continuous.

(ii) Set $w^{\#} = w + w^{\triangledown}\Delta^{-1}$ and take $g \in \mathcal{A}_w$ arbitrary. Then

$$\mathcal{A}_w = \mathcal{A}_{w^\#} = \mathcal{H}^1_{w^\#} = \{ f \mid f = \pi(F)g, \ F \in L^1_{w^\#} \}$$

(iii) The definition of $\mathcal{H}^1_w$ is independent of the choice of $g \in \mathcal{A}_w$.

Proof. (i) is easily verified from the definitions.

ad (ii) $\mathcal{A}_w = \mathcal{A}_{w^\#}$ follows from the symmetry $V_g(g) = V_g(g)^{\triangledown}$.

If $f,g \in \mathcal{A}_{w^\#}$ (and by 3.3 in dom A) we can choose $h \in \text{dom } A \cap \mathcal{A}_w$ such that Ah is not orthogonal to Af and Ag. Then by (3.2)

$\langle Ah,Ag\rangle \ \langle Af,Ah\rangle \ V_g(f) = V_f(f) * V_h(h) * V_g(g) \in L^1_{w^\#}$, i.e. $f \in \mathcal{H}^1_{w^\#}$.

On the other hand, for every $f \in \mathcal{H}^1_{w^\#}$ the element $f_1 = \pi\big(\|Ag\|^{-2}V_g(f)\big)g$ has the same voice transform $V_g(f_1) = V_g(f) * \|Ag\|^{-2}V_g(g) = V_g(f)$ by (3.2) and therefore $f = f_1$.

Finally, that $g \in \mathcal{A}_w$, $F \in L^1_{w^\#}$ implies $\pi(F)g \in \mathcal{A}_w$ follows from

$$V_{\pi(F)g}(\pi(F)g) = F * V_g(g) * F^{\triangledown} \in L^1_{w^\#}.$$

ad(iii) If $g,g' \in \mathcal{A}_w$, then $g' = \pi(F)g$ for some $F \in L^1_{w^\#}$ and $V_{g'}(f) = V_g(f) * F^{\triangledown} \in L^1_w$. Interchanging the role of g and g' we obtain $V_g(f) \in L^1_w$ iff $V_{g'}(f) \in L^1_w$. □

4.3. We now aim at an atomic decomposition of these spaces. For that purpose we shall discretize the convolution operator T on L^1_w, $TF = F * G$, $F,G \in L^1_w$ by means of a U-BUPU $\Psi = (\psi_i)_{i\in I}$. The next lemmas serve to get information on the "approximation" operator

$$T_\Psi : F \longmapsto \sum_{i\in I} \langle \psi_i, F\rangle L_{x_i} G \tag{4.1}$$

which is composed of a coefficient mapping $F \longmapsto \langle \psi_i, F\rangle_{i\in I}$ and a convolution operator: $(\lambda_i)_{i\in I} \longmapsto \sum_{i\in I} \lambda_i L_{x_i} G == (\sum_{i\in I} \lambda_i \delta_{x_i}) * G$.

Given a family $X = (x_i)_{i\in I}$ and a weight w we use $\mathscr{w}$ for its discrete version given by $\mathscr{w}(i) := w(x_i)$, $i\in I$.

<u>4.4.Lemma.</u> i) For $f \in L^1_w(G)$ the "sequence" of coefficients $\Lambda = (\lambda_i)_{i\in I}$ given by $\lambda_i := \langle \psi_i, F \rangle$ belongs to $\ell^1_{\mathscr{w}}$. More precisely, given a fixed compact neighbourhood U_o of the identity there exists a constant C_o such that the norms of the linear operators $F \longmapsto \Lambda$ are uniformly bounded by C_o for all U_o-BUPUs.

ii) Conversely, given $G \in L^1_w(G)$ and $\Lambda = (\lambda_i)_{i\in I} \in \ell^1_{\mathscr{w}}$, (where $X=(x_i)_{i\in I}$ is any family in the group) one has $F := \sum_{i\in I} \lambda_i L_{x_i} G \in L^1_w$ (the sum being absolutely convergent in L^1_w) and there is a universal constant C_1 such that $\| F \|_{1,w} \le C_1 \| \Lambda \|_{1,\mathscr{w}}$.

iii) As a consequence the set of operators $\{ T_\Psi \}$, where Ψ runs through the family of U_o-BUPUs acts uniformly bounded on $L^1_w(G)$.

<u>Proof.</u> i) From 2.2 we conclude that

$$w(x_i) \le C_o\, w(y) \quad \text{for } y \in x_i U_o \quad \text{for } C_o := \sup_{z\in U_o} w(z) \quad) \tag{4.2}$$

Therefore $|\langle \psi_i, F\rangle| w(x_i) \le C_o \langle \psi_i, w|F|\rangle$ and

$$\| \Lambda \|_{1,\mathscr{w}} = \sum_{i\in I} |\langle \psi_i, F\rangle| w(x_i) \le C_o \sum_{i\in I} \langle \psi_i, w|F|\rangle = C_o \| F \|_{1,w} .$$

ii) follows immediately from

$$\| \sum_{i\in I} \lambda_i L_{x_i} G \|_{1,w} \le \sum |\lambda_i| \| L_{x_i} G \|_{1,w} \le \sum |\lambda_i| w(x_i) \|G\|_{1,w} = \|\Lambda\|_{1,\mathscr{w}} \|G\|_{1,w} . \square$$

<u>4.5. Lemma.</u> The net $\{ T_\Psi \}$ of U-BUPUs (directed according to inclusions of the corresponding neighbourhoods U to $\{ e \}$) is norm convergent to T (as operators on $L^1_w(G)$): $\lim_{\Psi\to\infty} ||| T_\Psi - T |||_{1,w} = 0$.

<u>Proof.</u> Given $F \in L^1_w(G)$ we can give the following estimate:

$$\| TF - T_\Psi F \|_{1,w} = \| (\sum_{i\in I} (F\psi_i - \langle \psi_i, F\rangle \delta_{x_i}) * G \|_{1,w} \le$$

$$\le \sum \| \int_{x_i U} F(y) \psi_i(y) (L_y G - L_{x_i} G)\, dy \|_{1,w} \text{ (as vector-valued integral)}$$

$$\leq \sum \int_{x_i U} |F(y)|\,\psi_i(y) \| L_y G - L_{x_i} G \| dy \leq \sum \sup_{u \in U} \| L_{x_i u} G - L_{x_i} G \| \; \langle \psi_i, |F| \rangle$$

$$\leq \sup_{u \in U} \| \; L_u G - G \; \|_{1,w} \sum_{i \in I} w(x_i) \langle \psi_i, |F| \rangle \leq \omega_U(G) \; C_o \| \; F \; \|_{1,w} \quad \text{by 4.4.ii)},$$

where $\omega_U(G) := \sup_{u \in U} \| \; L_u G - G \; \|_{1,w}$ is the modulus of continuity of G with respect to the norm $\| \;\; \|_{1,w}$. Consequently

$$|\!|\!| \; T - T_\Psi \; |\!|\!|_{1,w} \leq C_o \omega_U(G) \quad \longrightarrow \quad 0 \quad \text{for} \quad U \quad \longrightarrow \quad \{e\} \; . \; \square$$

<u>4.6.Theorem.</u> For any $g \in \mathcal{A}_w$ there exists a neighbourhood U_o of the identity and a constant $C > 0$ (both only dependent on g) such that for every U_o-dense family $(x_i)_{i \in I}$ in G any $f \in \mathcal{H}^1_w(G)$ can be written as

$$f = \sum_{i \in I} \lambda_i \pi(x_i) g \quad \text{with} \quad \sum_{i \in I} |\lambda_i| w(x_i) \leq C \| \; f \; | \mathcal{H}^1_w \| \tag{4.3}$$

The series is absolutely convergent in $\mathcal{H}^1_w(G)$. Moreover, we can choose the coefficients to depend linearly on f .

<u>Proof.</u> We may assume that $g \in \mathcal{A}_w$ is normalized by $\| \; Ag \; \| = 1$. Then convolution from the right with $G := V_g(g)$ is a bounded projection from L^1_w onto the closed(!) subspace $L^1_w * G$ (because $G*G=G$ by (3.3) and $\| \; F * G \; \|_{1,w} \leq \leq \| \; F \; \|_{1,w} \| \; G \; \|_{1,w}$) and $F \in L^1_w * G$ if and only if $F = V_g(f)$ for a unique element $f \in \mathcal{H}^1_w$ by (3.3) . Now T acts as identity on $L^1_w * G$, and therefore T_Ψ (as operator on $L^1_w * G$!) is invertible for sufficiently small neighbourhoods U of e , as a consequence of 4.4 (using Neuman's series $|\!|\!| \; T - T_\Psi |_{L^1_w * G} |\!|\!| < a < 1$ implies $|\!|\!| \; T_\Psi^{-1} \; |\!|\!| \leq (1-a)^{-1}$). We thus obtain for $F \in L^1_w * G$ or $f \in \mathcal{H}^1_w$ $V_g(f) = F = T_\Psi(T_\Psi^{-1} F) = \sum_{i \in I} \langle \psi_i, T_\Psi^{-1} F \rangle \; L_{x_i} G$, and finally

$$f = \sum_{i \in I} \langle \psi_i, T_\Psi^{-1} V_g(f) \rangle \; \pi(x_i) g \; .$$

The remaining assertions on the coefficients $\lambda_i := \langle \psi_i, T_\Psi^{-1} V_g(f) \rangle$ now follow easily from the fact that $T_\Psi^{-1} F$ belongs to $L^1_w * G$, Lemma 4.3.i) and $|\!|\!| \; T_\Psi^{-1} \; |\!|\!| \leq (1-a)^{-1} < \infty$. $\square$

Almost the same arguments apply to the decomposition of the spaces $\mathcal{H}^1_m$, for a moderate function m with associated weight function w .

As a consequence we have the following useful results (cf.[F10],[F9]).

<u>4.7.Corollary.</u> $\mathcal{H}^1_w(G)$ is the minimal π-invariant Banach space $(B, \| \;\; \|)$ in $\mathcal{H}$ with $\mathcal{A}_w \cap B \neq \{ \; 0 \; \}$ satisfying $|\!|\!| \; \pi(x) \; |\!|\!|_B = \mathcal{O}(w(x))$. In particular, $\mathcal{H}^1$ is the minimal isometrically π-invariant Banach space in $\mathcal{H}$ (satisfying $\mathcal{A}_w \cap B \neq \{ \; 0 \; \}$ for $w \equiv 1$).

4.8. Corollary. The finite linear combinations of "atoms" $\pi(x_i)g$ form a dense subspace of $\mathcal{H}_w^1$ (with respect to its norm topology).

§ 5 Basic properties of general coorbits

5.1. The dual spaces $(\mathcal{H}_w^1)'$ of the minimal Banach spaces discussed in the previous section are sufficiently big Banach spaces, invariant under the group action. Furthermore, for their elements a voice transform is well defined and satisfies $V_g(f)(x) = \mathcal{O}(1/w(x))$. Thus is makes sense to use it for the definition of a coorbit space with respect to a space Y as considered in § 3. We shall assume from now on that $(Y, \| \ \|_Y)$ is a two-sided translation-invariant solid BF-space on G such that for a suitable weight function w the following facts are true:

W1) $\mathcal{A}_w$ is non-trivial

W2) $Y * L_w^1(G)^\vee \subseteq Y$ and $\|F * G\|_Y \le \|F\|_Y \ \|G^\vee\|_{1,w}$, $F \in Y, G \in L_w^1$

Fixing $g \in \mathcal{A}_w(G)$ we define:

$$\mathcal{C}o Y := \{ f \mid f \in (\mathcal{H}_w^1)' \ , \ V_g(f) \in Y \} .$$

As a natural (semi)norm we may take $\| f \|_{\mathcal{C}oY} := \| V_g(f) \|_Y$.

5.2. The following theorem collects the basic properties of the coorbit spaces just defined. Note that the theory is of course non-empty if $Y = L^p(G)$, whenever π is an integrable representation.

Theorem. (Properties of coorbit spaces)

(i) $\mathcal{C}oY$ is a π-invariant Banach space with its natural norm.

(ii) $V_g : \mathcal{C}oY \longmapsto Y$ intertwines π and L, i.e. we have

$$V_g(\pi(x)f) = L_x V_g(f) \quad \forall \ f \in \mathcal{C}oY . \tag{5.1}$$

If L is (strongly) continuous on Y, then π is (strongly) continuous on $\mathcal{C}oY$.

iii) Assuming that we have normalized g (so that $\| Ag \| = 1$) the reproducing formula holds true, i.e. we have

$$V_g(f) = V_g(f) * V_g(g) \quad \forall \ f \in \mathcal{C}oY . \tag{5.2}$$

In fact, more is true: The convolution operator $F \longmapsto F*G$ defines a bounded projection from Y onto the closed subspace $V_g(\mathcal{C}oY)$ which is equal to $Y * G$. It follows that $\mathcal{C}o Y$ is non-trivial in our situation.

iv) $\mathcal{C}o Y = \{ \pi(F)g\ ,\ F \in Y \}$

v) $\mathcal{C}o Y$ is independent of the choice of the analyzing vector $g \in \mathcal{A}_w$ (and of course of the weight function w as long as the assumptions formulated above hold true).

<u>Proof.</u> a) $\|.|\mathcal{C}o Y\|$ is a norm: Assume that $\| f|\mathcal{C}o Y \| = 0$, i.e. $\| V_g(f) \|_Y = 0$, then $V_g(f) \equiv 0$ because Y is a solid BF-space and $V_g(f)$ is continuous. Since $\{ \pi(x)g|\ x \in G \}$ spans a dense subspace of $\mathcal{H}_w^1$ (Cor. 4.8), we obtain $f = 0$ in $(\mathcal{H}_w^1)'$ and in $\mathcal{C}o Y$ or in other words, the map V_g: $(\mathcal{H}_w^1)' \longrightarrow L^\infty_{1/w}$ injective.

b) The reproducing property: Formula (3.3) which is valid for all $f \in \mathcal{H}$ and $g \in \operatorname{dom} A$, $\| Ag \| = 1$ at first may be written as

$$\langle \int_G V_g(g)(y^{-1}x)\ \pi(x)g\ dy - \pi(x)g,\ f \rangle = 0 .$$

For $g \in \mathcal{A}_w$ the left term is in $\mathcal{H}_w^1(G)$, hence the weak density of $\mathcal{H}$ in $(\mathcal{H}_w^1)'$ implies the validity of (3.3) for all $f \in (\mathcal{H}_w^1)'$. (W2) and the idempotency of $G := V_g(g)$ assure that $T : F \longmapsto F*G$ is a bounded projection from Y onto $Y*G$.

c) If $F * G = F$ for some $F \in Y$, then $f := \pi(F)g$ is a well-defined element of $(\mathcal{H}_w^1)'$, because π is an integrable representation with nontrivial $\mathcal{A}_w$. Then $V_g(f) = F * G = F$ as desired . Assertion (iv) follows from this and the injectivity of V_g .

d) Completeness of $\mathcal{C}o Y$: Let $f_n \in \mathcal{C}o Y$ be a Cauchy sequence, then $F_n := V_g(f_n)$ is one in Y and therefore $F_n \longrightarrow F$ in Y. Now all F_n and thus F satisf $F_n * G = F_n$ and $F * G = F$ from (b) , therefore $f_n \longrightarrow f$ where $V_g(f) = F$ (by c).

e) Uniqueness: Using (W2) the independence of $\mathcal{C}o Y$ of g is settled in the same way as in Lemma 4.1.

The other properties of the theorem are immediately verified from the definitions.□

<u>5.3. Remark.</u> Following Peetre ([PE2],p.200) one could introduce the formal adjoint (or pull-back) V_g^* : $L^\infty_{1/w} \longmapsto (\mathcal{H}_w^1)'$, which turns out to satisfy $V_g^*(F) = \pi(F)g$. The natural notion of an orbit would be in this context to set

$$\mathcal{O}Y := \{ f \in (\mathcal{H}_w^1)'\ ,\ f = V_g^*(F) \text{ for some } F \in Y \} ,$$

and to put on $\mathcal{O}Y$ the following norm:

$$\| f|\mathcal{O}Y \| := \inf \{ \| F|Y \|,\ f = V_g^*(F) \} .$$

Assertion iv) in the above theorem can thus be seen as a statement about the coincidence of coorbit and (continuous) orbit spaces.

5.4. The reproducing formula characterizing $V_g(\mathcal{C}o Y)$ can also be used to proof the expected result concerning duality.

Theorem. Assume that $(Y, \| \ \|_Y)$ has an absolutely continuous norm ($\Leftrightarrow$ the Banach dual Y' coincides with the Köthe-dual $Y^{\alpha} := \{ h \in L^1_{loc} : \check{h}f \in L^1(G) \ \forall \ f \in Y\}$). which is certainly true if Y is a reflexive Banach space. Then:

$$(\mathcal{C}o Y)' = \mathcal{C}o Y^{\alpha} = \mathcal{C}o Y'.$$

In particular, $\mathcal{C}o Y$ is reflexive as a Banach space if Y is reflexive.

5.5. It is possible to show that the basic spaces $\mathcal{A}_w$, $\mathcal{H}$ and $(\mathcal{H}^1_w)'$ coincide with coorbit spaces, more precisely, $\mathcal{H} = \mathcal{C}o L^2$ and $(\mathcal{H}^1_m)' = \mathcal{C}o L^{\infty}_{1/m}$ and $\mathcal{A}_w = \mathcal{H}^1_{w^\#} = \mathcal{C}o L^1_{w^\#}$ as subsets of $\mathcal{H}$ (4.2.ii)

Altogether the above mentioned construction of coorbit spaces yields a family of G-invariant Banach spaces which is related to the given Hilbert space $\mathcal{H}$ in a similar way as the family of solid spaces $(Y, \| \ \|_Y)$ is related to $L^2(G)$. For practical reasons we shall restrict our attention from now on mainly to the spaces $\mathcal{C}o L^p_m$, where m is any moderate function, and $1 \le q \le \infty$. Because of their relationship to $\mathcal{H}$ we shall denote them by $\mathcal{H}^p_m$.

5.6. Given two equivalent representations π_1 and π_2 any given intertwining operator $T: \mathcal{H}_1 \longmapsto \mathcal{H}_2$ extends in a natural way to an (intertwining) isomorphism between corresponding coorbit spaces. Thus the spaces $(\mathcal{H}_1)^p_m$ and $(\mathcal{H}_2)^p_m$ are automatically isomorphic as Banach spaces, and in fact can be shown to be isomorphic to the spaces ℓ^p.

5.7. We want to mention here a few facts concerning the dependence of $\mathcal{C}o Y$ from the defining BF-space Y.

Theorem. i) Fixing the moderate function m the spaces $\mathcal{H}^p_m$ are monotonically increasing with p.

ii) Assume that the quotient m_1/m_2 belongs to $L^r(G)$, for some $r \le 1/p_2 - 1/p_1 > 0$. Then $\mathcal{H}^{p_1}_{m_1} \hookrightarrow \mathcal{H}^{p_2}_{m_2}$ (even compactly).

iii) $\mathcal{H}^{p_1}_{m_1} = \mathcal{H}^{p_2}_{m_2}$ if and only if $p_1 = p_2$ and $m_1 \cong m_2$, i.e. if there are two constants $0 < C_1 \le C_2$ such that $C_1 m_1(x) \le m_2(x) \le C_2 m_1(x), \forall x \in G$.

§ 6 Atomic characterizations of general coorbits

For technical reasons and in order to bring out as good as possible the basic underlying ideas we shall give only the atomic decompositions of the spaces $Y=\mathcal{H}^p_m$, because for these spaces the corresponding conditions on the coefficients in the atomic decomposition are easy to guess. Furthermore it is possible to obtain the atomic characterization for this class of spaces by working only with the limiting cases in detail and to get the general results from there (for the basic steps) by interpolation. As the examples will show, however, the special cases treated here give already a lot of relevant information concerning decompositions of modulation spaces, Besov spaces and Bergman spaces.

6.1. In preparation for the main result we have to state some observations. Fixing m we shall only work with weight functions w for which W1) and W2) hold true. Whereas for $\mathcal{H}^1_w$ any $g \in \mathcal{A}_w$ could serve as an atom we now have to restrict our attention to a smaller but still dense subspace. We say that $g \in \mathcal{A}_w$ enjoys property (W) if $G := V_g(g)$ can be written as

$$G = \sum_n R_{z_n} G_n \text{ with} \sum_n \| G_n \|_\infty w(z_n^{-1}) =: C_w(G) < \infty , \tag{6.1}$$
where all $G_n \in \mathcal{K}(G)$ have a fixed compact support Q.

Lemma: (i) The subspace $\mathcal{B}_w$ of all $g \in \mathcal{A}_w$ with property (W) is dense in $\mathcal{H}^1_w$.
(ii) For $g \in \mathcal{B}_w$ and any relatively separated subset $X = (x_i)_{i\in I}$ the sequence $(G(x_i^{-1}y))_{i\in I}$ is in $\ell^1_{w^\vee}$ for all $y \in G$.

Remark: This is actually the assertion that $V_g(g)$ belongs to a so-called Wiener-type space, namely a "right" version of $W(C^o, L^1_w)(G)$ (cf. [F1],[F2]). Whereas our original approach relied heavily on the properties of these spaces (in particular the convolution relations among them) we have tried to avoid this theory in the present paper and have chosen a more elementary exposition. However, using these spaces it would be possible to give a more precise description of *local* and *global* properties of voice transforms. In any case one cannot dispense with the theory of Wiener-type spaces if one wants to describe coorbit-spaces in full generality. Even in the general situation the examples give the impression that all interesting "atoms" $g \in \mathcal{A}_w$ belong automatically to $\mathcal{B}_w$. If G is an IN-group then $\mathcal{A}_w = \mathcal{B}_w$.

Proof of the Lemma: i) For $\varphi \in \mathcal{K}(G)$ and $g \in \mathcal{A}_w$ one has $V_{\pi(\varphi)g}(\pi(\varphi)g) = \varphi * V_g(g) * \varphi^{\triangledown}$ and thus $\pi(\varphi)g \in \mathcal{A}_w$. It is left to the reader to verify that $h := V_g(g) * \varphi^{\triangledown} \in L^1_w$ can be decomposed as $h = \sum R_{z_n} h_n$, with $h_n \in L^1_w$, $\mathrm{supp}\, h_n \subseteq Q$ (compact) and $\sum_n \| h_n^{\triangledown} \|_1 w(z_n^{-1}) < \infty$. But then

$$V_{\pi(\varphi)g}(\pi(\varphi)g) = \sum \varphi * R_{z_n} h_n = \sum_n R_{z_n}(\varphi * h_n) \ , \ \varphi * h_n \in \mathcal{K}(G) \ ,$$

$\mathrm{supp}\, \varphi * h_n \subseteq (\mathrm{supp}\, \varphi)(\mathrm{supp}\, h_n) \subseteq K$ (a compact set), and

$$\sum_n \| \varphi * h_n \|_\infty w(z_n^{-1}) \le \| \varphi \|_\infty \sum \| h_n^{\triangledown} \|_1 w(z_n^{-1}) < \infty .$$

This means $\pi(\varphi)g \in \mathcal{B}_w$. If φ runs through an approximate identity (φ_n) in $\mathcal{K}(G)$ then (using any $h \in \mathcal{A}_w$ for the norm in $\mathcal{H}^1_w$) :

$$\| \pi(\varphi_n)g - g | \mathcal{H}^1_w \| = \| V_h(\pi(\varphi_n)g - g) \|_{1,w} = \| \varphi_n * V_h(g) - V_h(g) \|_{1,w} \longrightarrow 0$$

for $n \longrightarrow \infty$, whence $\mathcal{B}_w$ is dense in $\mathcal{H}^1_w$ (since $\mathcal{A}_w$ is dense) .

ii) Given a well-spread set $X = (x_i)_{i\in I}$ with $\sup_{z\in G} \#\{ i | \ x_i \in zQ \} \le h_Q < \infty$ and $G \in \mathcal{B}_w$ with a decomposition (6.1) we obtain :

$$\sum_{i\in I} |G(x_i^{-1}y)| w(x_i^{-1}) \le \sum_n \sum_i |G_n(x_i^{-1}yz_n)| w(x_i^{-1}) = (**)$$

Fixing n the formal sum over $i\in I$ is only over the finite index set $I_y := \{ i | \ x_i \in yz_nQ \}$ and for $i \in I_y$ one has $w(x_i^{-1}) \le w(y^{-1})w(z_n^{-1})\sup_{q\in Q} w(q)$. We continue the estimate by

$$(**) \le \sum_n \sum_{i\in I_y} |G_n(x_i^{-1}yz_n)| w(y^{-1})w(z_n^{-1}) \, C_Q \le C_w(G) \cdot \# I_y \, w(y^{-1}) \, C_Q < \infty . \square$$

We can now formulate our main result concerning atomic characterizations of coorbit spaces :

6.2. Theorem. i) Given any $g \in \mathcal{B}_w$ and any relatively separated family $(x_i)_{i\in I}$ in G one has: The mapping $(\lambda_i)_{i\in I} \longmapsto \sum_{i\in I} \lambda_i \pi(x_i)g$ defines a bounded linear operator from the sequence space ℓ^p_m (over X) into $\mathcal{H}^p_m$ and the sum is unconditionally norm convergent for $1 \le p < \infty$, w^*-convergent in the case $p = \infty$.

ii) Conversely, if $g \in \mathcal{B}_w$ is given a neighbourhood U_o of the identity can be found such that for any family $(x_i)_{i\in I}$ in G which is U_o-dense there exists a constant C_o such that for each $f \in \mathcal{H}^p_m$ $(1 \le p < \infty)$ there exists a family $(\lambda_i)_{i\in I}$ satisfying

$$\| (\lambda_i)_{i\in I} | \ \ell^p_m \| \le C_o \| f | \mathcal{H}^p_m \| \ ,$$

such that f can be represented as

$$f = \sum_{i\in I} \lambda_i \; \pi(x_i)g \; .$$

Again the coefficients will depend linearly on f, showing that $\mathcal{H}^p_m$ is a retract of the weighted sequence space ℓ^p_m .

6.3. It is clear from the theorem that (fixing g) for families $(x_i)_{i\in I}$ which are both relatively separated and U_o-dense the Banach spaces $\mathcal{H}^p_m$ can be considered as a retract of the solid sequence spaces ℓ^p_m . This allows to draw many conclusions, e.g. embedding theorems (Thm. 5.7 (i), (ii)) and the following interpolation result:

6.4 Theorem. The family $\mathcal{H}^p_m$, with $1\le p\le\infty$ and m running through the family of moderate functions is closed under complex interpolation. More precisely, if $p_1<\infty$ and $\vartheta \in [0,1]$ is given then we have

$$\left[\mathcal{H}^{p_1}_{m_1} , \mathcal{H}^{p_2}_{m_2} \right]_{[\vartheta]} = \mathcal{H}^p_m , \text{ with } \frac{1}{p} = \frac{1-\vartheta}{p_1} + \frac{\vartheta}{p_2} \text{ and } m = m_1^{1-\vartheta}\cdot m_2^{\vartheta} .$$

6.5. Proof of Theorem 6.3 As in Lemma 4.4 we shall establish that (1) for $g\in\mathcal{B}_w$, $G := V_g(g), (\lambda_i)\in\ell^\infty_m$ and any relatively separated set $X=(x_i)_{i\in I}$ the function $F := \sum_{i\in I} \lambda_i L_{x_i} G$ is in L^∞_m and (2) that the operators T_Ψ , as defined in (4.1) act uniformly bounded on L^∞_m for all U_o-BUPUs Ψ . Having proved this ii) follows immediately by complex interpolation between L^1_m and L^∞_m ([BL]). In order to prove i) we use (2) to obtain

$$|||\, (T-T_\Psi)|_{L^p_m*G} \,||| \le |||\, (T-T_\Psi)|_{L^1_m} \,|||^{1/p} \; |||\, (T-T_\Psi)|_{L^\infty_m} \,|||^{1-1/p} \le C|||\, (T-T_\Psi)|_{L^1_m} \,|||^{1/p}$$

which tends to zero for $\Psi \longrightarrow \infty$. Hence $T_\Psi|_{L^p_m*G}$ is invertible and we may proceed as in Theorem 4.6 to finish the proof.

In order to verify (1) note that $m(e) \le m(x)w^\triangledown(x)$, i.e. $m^{-1}(x) \le w^\triangledown(x)$ for $x\in G$, and Lemma 6.1.ii) yields $(G(x_i^{-1}y))_{i\in I} \in \ell^1_{1/m}$ for all $y\in G$. The duality between $\ell^1_{1/m}$ and ℓ^∞_m makes $F(y) := \sum_{i\in I} \lambda_i L_{x_i} G(y)$ a well defined function, whose partial sums converge pointwise to F . This is equivalent to w^*-convergence of $\sum_{i\in E} \lambda_i \pi(x_i)g$ to f (E running through the finite subsets of I) whenever $F\in L^\infty_m$. Since the proof of $F\in L^\infty_m$ is similar to that of (2) we shall content us with a demonstration of the uniform boundedness of the family (T_Ψ) .We insert the decomposition (6.1) of G in the definition of $T_\Psi F$ and get

$$\| F \|_{\infty,m} = \sup_{y\in G} |\textstyle\sum_n \sum_i \langle \psi_i, F\rangle L_{x_i} G_n(yz_n)\ m(y)| \le$$

$$\le \textstyle\sum_n \sup_{y\in G} \sum_i \langle \psi_i, |F|\rangle\ |G_n(x_i^{-1}y)|\ m(yz_n^{-1}) \le$$

$$\le \textstyle\sum_n \| G_n \|_\infty w(z_n^{-1}) \cdot \sup_{y\in G} \left[\sum_i \langle \psi_i, |F|\rangle\ 1_Q(x_i^{-1}y)\right] m(y)$$

$$=: C_w(G) \sup_{y\in G} K(F,y)\ m(y)$$

For every $y \in G$ the sum defining $K(F,y)$ can be replaced by a sum over $I_y = \{ i \mid x_i \in yQ \}$. For $k \in \mathcal{K}(G)$ with $k(x) \equiv 1$ on QU_o we have $\sum_{i\in I_y} \psi_i \le L_y k$ and $\sup_{y\in G} K(F,y)m(y) \le \sup_{y\in G} \langle L_y k, |F|\rangle\ m(y)$

$$\le \sup_{u\in QU_o} w(u) \cdot \sup_{y\in G} \langle L_y k, |F| m\rangle \quad \text{(compare (4.2)}$$

$$\le C_Q \sup_{y\in G} \| L_y k \|_1 \| m|F| \|_\infty .$$

Altogether this yields the desired result

$$\| T_\Psi F \|_{m,\infty} \le C_w(G)\ C_Q\ \|F\|_{m,\infty} .$$

This shows that the operator norm of T_Ψ depends only on G, but not on the choice of Ψ . □

§ 7 Examples

There seems to be a rich variety of examples for the abstract theory presented above. We shall discuss here the above results for three typical Lie groups: 1. the (reduced) Heisenberg group as an example for a nilpotent group 2. the "ax+b" group, the simplest solvable Lie group , and 3. $SL(2,\mathbb{R})$ (or $SU(1,1)$) as a semisimple Lie group with discrete series. In these cases the spaces arising as coorbit spaces have been extensively studied (as Banach spaces of functions or distributions of their own right without making use of the Lie group acting on these spaces and its representation theory). Thus in various settings decomposition theorems, duality and interpolation results were obtained using rather different methods.

7.1. The Heisenberg group $\mathbb{H}_n$ and the modulation spaces

As a set (and also topologically) $\mathbb{H}_n$ can be identified with $\mathbb{R}^n \times \mathbb{R}^n \times \mathbb{T}$, endowed with the multiplication (writing $h = (x,y,t)$)

$$h_1 h_2 = (x_1, y_1, t_1)(x_2, y_2, t_2) := (x_1+x_2, y_1+y_2, t_1 t_2 e^{iy_1 x_2})$$

and the product measure dxdydt as Haar measure . Moreover, $\mathbb{H}_n$ is evidently an IN-group, hence unimodular. We shall mainly be interested in the (Schrödinger) representation of $\mathbb{H}_n$ on $\mathcal{H} = L^2(\mathbb{R}^n)$. Setting $M_y f(x) := \exp(iy\cdot x)\ f(x)$ it acts on $f \in L^2(\mathbb{R}^n)$ by

$$\pi(x,y,t)f(z) := t(L_x M_y f)(z) = te^{iy\cdot(z-x)} f(z-x) \quad \text{for } x,y,z \in \mathbb{R}^n, t \in \mathbb{T}.$$

The representation coefficients $V_g(f)$, $f,g \in L^2(\mathbb{R}^n)$ are of the form

$$V_g(f)(x,y,t) = \langle tL_x M_y g, f\rangle = \bar{t} M_y g^{\vee} * f(x) .$$

The orthogonality relations in this group are known as *Moyal's formula* and the coefficient

$$V_g(g)(x,y,e^{ix\cdot y/2}) = \int_{\mathbb{R}^n} e^{-izy}\ \bar{g}(z-x/2)\ g(z+x/2)\ dz$$

is known as the *(radar) ambiguity function* which is an important tool in applications (cf.[SCH1-3],[PA],[AT]). In this situation our conditions (in particular W1-W2)) are all satisfied for almost arbitrary weights w on $\mathbb{H}_n$. In fact, one has with $h = (x,y,t)$:

$$\int_{\mathbb{H}_n} |V_g(f)(h)|\ dh = \int_{\mathbb{R}^n}\int_{\mathbb{R}^n} |M_y g^{\vee} * f(x)|\ dxdy = \int_{\mathbb{R}^n} \| (L_y \mathcal{F}g)\mathcal{F}f\|_{\mathcal{F}L^1}\ dy < \infty$$

whenever both functions have compactly supported Fourier transforms, for example. Thus π is integrable, but the same argument shows that $\mathcal{A}_w$ is nontrivial in many other situations, e.g.if w is dominated by a product of two weights w_1,w_2 on $\mathbb{R}^n$: $w(x,y,t) \le w_1(x)w_2(y), x,y \in \mathbb{R}^n$, where w_1 is assumed to satisfy the Beurling-Domar non-quasianalyticity condition (see [REI],p.132) and w_2 is arbitrary. In particular, the spaces $\mathcal{H}^p_m$ are well defined as long as m is of the form $m(x,y,t) = m_1(y)$, where m_1 is any moderate function on $\mathbb{R}^n$. For the choice $m(x,y,t) = (1+|y|)^s$ the resulting spaces turn out to coincide with the so-called *modulation spaces* $M^s_{p,p}(\mathbb{R}^n)$ [cf.[F6],[F7],[TR3],[F10] for various descriptions and properties of the spaces $M^s_{p,q}(\mathbb{R}^n)$]. The spaces $M^s_{2,2}(\mathbb{R}^n)$ can be identified with the classical Bessel-potential spaces $\mathcal{L}^2_s(\mathbb{R}^n)$ (cf.[TR2]). Moreover, $M^s_{1,1}(\mathbb{R}^n)$ turns out to be a Banach ideal in $L^1(\mathbb{R}^n)$, which is the smallest character invariant Banach ideal in $L^1(\mathbb{R}^n)$ satisfying an estimate of the form $|||\ M_y\ ||| = \mathcal{O}[(1+|y|)^s]$ (cf. [F9],[F3]) for results in this direction). The most important special case is the Segal algebra (in the sense of Reiter) $S_o(\mathbb{R}^n)$, as introduced in [F3]. It corresponds to $\mathcal{H}^1(\mathbb{R}^n)$ in our setting. As indicated in [F4] and [F12] this space together with its dual are extremely useful tools for harmonic analysis. As the spaces $\mathcal{H}^p$(including the limiting cases $S_o(G)=\mathcal{H}^1$ and $S_o(G)'=(\mathcal{H}^1)'$) can easily be defined for arbitrary lca. groups their relevance is by no means restricted to the Euclidean case. Applying our general result on atomic decompositions we obtain representations of the following form (which should be called *Gabor type representations* following the terminology used in signal analysis, cf.[JS],[BA], because the choice $g_o(x) := \exp(-x^2)$ is among the options) : Given $s \in \mathbb{R}$ and some $g_o \neq 0$, $g_o \in M_{1,1}^{|s|}(\mathbb{R}^n)$ one has:

There exist $\alpha_o > 0$ and $\beta_o > 0$ such that for $\alpha \leq \alpha_o$ and $\beta \leq \beta_o$ there exists $C = C(\alpha,\beta) > 0$ with the following property:

$f \in M_{p,p}^{s}(\mathbb{R}^n)$ if and only if $f = \sum_{n,k} a_{n,k} L_{\beta k} M_{\alpha n} g_o$,
for some double sequence of coefficients satisfying

$$\left[\sum_{n,k} |a_{n,k}|^p (1+|n|)^{sp} \right]^{1/p} \leq C \| f | M_{p,p}^{s}(\mathbb{R}^n) \| .$$

Even for s=2 and n=1 this result goes beyond known results, in particular as far as the freedom in the choice of g_o is concerned. For another approach to atomic decompositions of modulations spaces cf.[F11] and [F10] for the special case p=1. These results extend to the setting of arbitrary lca. groups instead of $G=\mathbb{R}^n$ ($\mathbb{H}_n$ has to be replaced by $\mathbb{H}_G := G \times \hat{G} \times \mathbb{T}$ with appropriate multiplication, cf.[IG]). Another approach to the basic properties of modulation spaces (duality, interpolation, trace theorems,...) is given in [F6,7,14].

Considering the atomic characterization it is also evident that the spaces $M_{p,p}^{0}(\mathbb{R}^n)$ are invariant under the Fourier transform.

There are other, non-equivalent representations of $\mathbb{H}_n$ denoted π_k, $k \in \mathbb{Z}\setminus\{0\}$ (with $\pi_1=\pi$ above) which satisfy $\pi_k(x,y,t) = t^k L_x M_y$. However, the coorbit spaces with respect to these representations are the same as those considered above.

In contrast, the Stone-von Neumann uniqueness theorem tells us that any infinite-dimensional irreducible representation ρ of $\mathbb{H}_n$ is equivalent to some π_k. By 5.6 the isomorphism between the Hilbert spaces $L^2(\mathbb{R}^n)$ and $\mathcal{H}_\rho$ extends to isomorphism (intertwining operator) between the coorbit spaces attached to these representations. An interesting special case is the Fock representation of $\mathbb{H}_n$ on a Hilbert space of entire functions on $\mathbb{C}^n$ (cf. [JPR],[TA]) . Consequently the Banach spaces E_p, $1 \leq p < \infty$, used for a characterization of Hankel operators in the Schatten class $\mathcal{S}_p$, are isomorphic to $M_{p,p}^{0}(\mathbb{R}^n)$.

7.2. The ax+b group, Besov and potential spaces

Let $G = \mathbb{R} \times_\beta \mathbb{R}^*$ be the non-connected "ax+b"-group, with multiplication $(b,a)\,(x,y) = (ax+b,ay)$ with left Haar measure $dxdy/y^2$ and Haar modulus $\Delta(x,y) = |y|^{-1}$. We shall be concerned with the "unique" infinite-dimensional irreducible unitary continuous representation π on $L^2(\mathbb{R})$:

$$[\pi(x,y)f](t) := y^{-1/2} f\left(\frac{t-x}{y}\right) = L_x D_y f(t)$$

where $D_y f(t) = y^{-1/2}\ f(t/y)$ is the L^2-isometric dilation operator. The representation coefficients $V_g(f)$ are realized as

$$\langle \pi(x,y)g,f\rangle = \langle L_x D_y g,f\rangle = D_y g^\triangledown * f(x).$$

The proof of the integrability follows the same path as in 3.1., this time starting with any $g \in L^2(\mathbb{R})$, such that supp $\mathcal{F}g$ is compact and bounded away from zero. Then $V_g(g) \in L^1_w(G)$ for practically every well-behaved weight function on G, in particular for the powers $w_s(x,y) = |y|^{-s}$, $s \in \mathbb{R}$, of the Haar module. It follows that

$$f \in \mathcal{C}o L^p_w \quad \text{if and only if} \quad \int_{\mathbb{R}^*} \| D_y g^\triangledown * f \|_p \ |y|^{-sp} \frac{dy}{y^2} < \infty .$$

Comparing this expression with the work of Peetre [PE1] or Triebel[TR4] one recognizes this condition to be equivalent to the membership of f in the homogeneous Besov space $\dot{B}^{s-1/2+1/p}_{p,p}$(cf.[TR4] §2.2,Cor.6). The (formally) different parameters are a consequence of the normalization of the dilation operators D_y and the use of the Haar measure on G . An application of our main result gives in this situation: Given any non-zero symmetric $g \in \mathcal{B}_{w_s}$ there exist positive values $\alpha_o > 0$ and $\beta_o > 0$ such that for α,β with $0 < \alpha \le \alpha_o$ and $|\beta-1| \le \beta_o$ there exists $C = C(\alpha,\beta) > 0$ with the following property:

$$f \in \dot{B}^{s-1/2+1/p}_{p,p} \quad \text{if and only if} \quad f = \sum_{(n,k)\in\mathbb{Z}^2} a_{n,k}\, L_{k\alpha/\beta^n}\, D_{\beta^n}\, g \ \text{and}$$

$$\left[\sum_{n,k} |a_{n,k}|^p\ \beta^{nsp}\right]^{1/p} \le C \| f | \dot{B}^{s-1/2+1/p}_{p,p} \|$$

Results of this type for Besov-Triebel spaces have been given by Frazier and Jawerth ([FJ]). Their approach only allows to take elements g which have plateau-like Fourier transforms. The minimal isometrically translation and dilation invariant space in this setting is $\mathcal{C}oL^1 = \dot{B}^{1/2}_{1,1}$. It can be shown that this space is closely related to the space of "good vectors" considered by Eymard and Terp ([ET]), who work with a different, but equivalent representation of the ax+b group (cf.[F10]).

The Besov-potential spaces on $\mathbb{R}^n$ can be treated in exactly the same way, if one replaces the "ax+b"-group by the semidirect product $\mathbb{R}^n x_\beta (\mathbb{R}^+ x\ O(n))$ with Haar measure $y^{-n-1}dxdydU$, acting on $L^2(\mathbb{R}^n)$ by its natural representation π

$$[\pi(x,y,U)f](t) := y^{-n/2}f(y^{-1}U^*(t-x)) = L_x D_y T_U f(t)$$

$$\text{with } x,t \in \mathbb{R}^n,\ y\in\mathbb{R}^+,\ U\in O(n),\ f\in L^2(\mathbb{R}^n),\ T_U f(t)=f(U^* t).$$

If one uses a function g which is invariant under O(n), i.e. a radial function g the computations are simplified because then the integration over O(n) is trivial and one gets the same formulas as in the one-dimensional case.

7.3. SU(1,1) and the Bergman spaces

Let $G = SL(2,\mathbb{R})$ be the group of all real 2x2 matrices with determinant one. Every matrix $g = \begin{pmatrix} a & b\\ c & d\end{pmatrix} \in SL(2,\mathbb{R})$ is a product

$$\begin{pmatrix} 1 & b\\ 0 & 1\end{pmatrix}\begin{pmatrix} a & 0\\ 0 & a^{-1}\end{pmatrix}\begin{pmatrix} \cos\phi & \sin\phi\\ -\sin\phi & \cos\phi\end{pmatrix},\ b\in\mathbb{R}, a\in\mathbb{R}^+, 0\le\phi\le 2\pi,$$

of a unipotent, a diagonal and an orthogonal matrix (Iwasawa decomposition) and the Haar measure of G is $dg = a^{-2}dndad\phi$. $SL(2,\mathbb{R})$ has a family of square-integrable representations - the so-called discrete series - which may be realized on Hilbert spaces of analytic functions on the upper half plane $U = \{ z=x+iy \in \mathbb{C},\ y > 0 \}$. The representation spaces of the discrete series are special cases of the Bergman spaces $A^{p,\alpha}$, $1\le p\le\infty$, $\alpha>1$.

$$A^{p,\alpha} = \left\{ f \text{ holomorphic on } U:\ \| f \|^p_{p,\alpha} = \iint_U |f(x+iy)|^p\, y^\alpha \frac{dxdy}{y^2} < \infty \right\}$$

where we have used for these spaces a parameterization slightly different from that in [RO] (cf. also [LUE]). For any integer $m \ge 2$ the representation π_m is defined on $A^{2,m}$ by $\pi_m\begin{pmatrix} a & b\\ c & d\end{pmatrix} f(z) = f\left(\frac{dz-b}{-cz+a}\right)(-cz+a)^{-m}$Let us consider the particularly simple functions $f_m(z) = (z+i)^{-m}$ .One can check the following properties (cf. e.g. [LA], p.181-187, [WA],vol II,p.413): (1) $f_m \in A^{2,m}$ for all $m \ge 2$. (2) $G_m := = \langle\pi_m(.)f_m, f_m\rangle \in L^1[SL(2,\mathbb{R})]$ for all $m \ge 3$, with other words , $f_m \in \mathcal{H}^1(\pi_m)$, because $SL(2,\mathbb{R})$ is unimodular, whereas $\mathcal{H}^1(\pi_2) = \{ 0 \}$. Furthermore the functions G_m enjoy property (W) of Lemma 6.1 and thus are appropriate analyzing vectors. (3) For $m\ge 3$, $1\le p<\infty$, f_m is in $A^{p,mp/2}$, $f_2 \in A^{p,p}$ only for $1<p<\infty$, and π_m acts isometrically on $A^{p,pm/2}$. (Compare [LA],p.183,Lemma 2)

(4) The subgroups K and AN of the Iwasawa decomposition act in a very simple way on f_m:

$$\pi_m\begin{bmatrix} \cos\vartheta & \sin\vartheta \\ -\sin\vartheta & \cos\vartheta \end{bmatrix}(z+i)^{-m} = e^{-im\vartheta}(z+i)^{-m}$$

and $$\pi_m\begin{bmatrix} a & ab \\ 0 & a^{-1} \end{bmatrix}(z+i)^{-m} = \frac{a^m}{(z-(a^2b-a^2i))^m} = \frac{y_0^{m/2}}{(z-\bar{z}_0)^m} ,$$

where $z_0 = x_0+iy_0 = a^2b+ia^2$. With this knowledge about our atoms we can identify the Bergman spaces as coorbits of $L^p(SL(2,\mathbb{R}))$ under the representations π_m :

Theorem: $A^{p,pm/2} = \mathcal{C}o(L^p,\pi_m)$

This is the theorem of Coifman and Rochberg ([CR],[RO]) in a rather unfamiliar disguise. For simplicity we have only described the $A^{p,pm/2}$. If we are inclined to take into account also the projective representations π_r , $r > 2$ and to extend our theory to such representations the $A^{p,\alpha}$, $\alpha > p$ occur as $\mathcal{C}o(L^p,\pi_r)$ whereas for $A^{p,\alpha}$, $1 < \alpha \le p$, we have to deal with coorbits of weighted L^p-spaces.

The Moebius invariant function spaces in the sense of [APF1] cannot be handled within our scheme in a satisfying manner, because they arise in connection with the representation on the Dirichlet space which is equivalent to π_2 and thus is not integrable, hence our theory is not applicable . There are , however, representation coefficients of π_2 in $L^{1+\varepsilon}(SL(2,\mathbb{R}))$ for every $\varepsilon > 0$, thus it should be possible to define coorbit spaces of BF-spaces Y which contain some L^p, $p > 1$. The minimal Moebius invariant Banach space which is originally defined as an orbit $\mathcal{O}\ L^1$ can definitely not be identified with a coorbit space because $\mathcal{H}^1(\pi_2) = \{ 0 \}$. Thus we have an example where orbits and coorbits fall apart.

&8. Generalizations, extensions and aspects

8.1. Atomic decomposition of general coorbit spaces

(a) A refinement of the methods presented here, (b) bringing in the techniques of Wiener-type spaces and convolution equations between them, (c) a detailed investigation of the correspondence between solid BF-spaces and their associate discrete sequence spaces, allows us to obtain atomic decompositions of general coorbit spaces. Then the case of mixed norm spaces in the sense of [BP] on $\mathbb{H}_n$, the ax+b-group and $SL(2,\mathbb{R})$ yields Gabor-type representations for modulation spaces and generalized amalgams ([F11]), atomic decompositions for Besov- and Triebel-Lizorkin spaces which are in a sense stronger than those due to Frazier-Jawerth [FJ], because we may start our construction with any reasonable basic function, and of mixed norm Bergman spaces as treated by [RT].

8.2. Instead of T_Ψ we could have worked with another approximation operator $S_\Psi F := \sum_{i\in I} F(y_i)\ \|\psi_i\|_1\ L_{x_i} G$ which, however, is a much more delicate matter to handle. As a consequence of the usability of S_Ψ instead of T_Ψ every F with the reproducing equation $F = F * G$ is completely determined and can be reconstructed by our method if its values are known on any sufficiently fine net $(x_i)_{i\in I}$ in G. A typical situation where our method works is the related to 7.1. and concerns the reconstruction of f from a sampled *short time Fourier transform* with a given window $g \in \mathcal{S}(\mathbb{R}^n)$ (cf.[AR],[BA]).

8.3. Our examples suggest to take into consideration also projective representations (to explain the Bergman spaces for the whole range of p,α as coorbit spaces) or, equivalently, representations which are integrable only modulo the center of the group (as in 7.1. for the universal covering group of $\mathbb{H}_n$). The orthogonality relations and hence a reproducing formula are still available in a slightly modified form and therefore our theory will go through with only minor changes.

8.4. It seems to be of interest to extend the discussion to coorbit spaces of L^p-spaces with $0<p<1$ as well as other quasi-Banach spaces.

8.5. One possible approach to Hankel operators (cf. [PE2], [AFP2])in our general setting seems to be to consider them as operators of the form T_b: $f \longmapsto f - V_g^*[(b\cdot V_g(f)) * V_g(g)]$. Translated into the group setting T_b can be considered as a product-convolution operator and methods from [BS] and [F8] can be applied.

REFERENCES

AR J.B.ALLEN, L.R.RABINER: A unified approach to short-time Fourier analysis and synthesis. Proc. IEEE, Vol. 65/11 ,1558-1564 (1977) .

AFP1 J.ARAZY, S.D. FISHER, J. PEETRE: Möbius invariant function spaces. J.Reine Angew.Math. 263 ,110-145 (1986) .

AFP2 J.ARAZY, S.D.FISHER, J.PEETRE: Hankel operators on weighted Bergman spaces. Preprint Sept.1986.

AT L.AUSLANDER, R. TOLIMIERI: Radar ambiguity functions and group theory. SIAM J. Math. Anal. 16, 577-601 (1985) .

BA M.J.BASTIAANS: Signal description by means of local frequency spectrum. SPIE Vol.373, Transformation in optical signal processing,49-62 (1981) .

BP A.BENEDEK, R. PANZONE: The spaces L^p, with mixed norm. Duke Math. J. 28, 303-324 (1961) .

BL J.BERGH, J.LÖFSTRÖM: Interpolation Spaces (An Introduction), Grundl.math.Wiss. 223, Berlin - Heidelberg - New York, Springer, 1976.

BS R.C.BUSBY, H.A.SMITH: Product-convolution operators and mixed norm spaces. Trans.Amer.Math.Soc. 263 , 309-341 (1981).

CA A.L.CAREY: Square-integrable representations of non-unimodular groups. Bull.Austral.Math.Soc. 15 ,1-12 (1976).

CM T.CLAASEN, W.MECKLENBRÄUCKER: The Wigner distribution - a tool for time--frequency signal analysis. I-III. Philips J.Res.35 , 217-250, 276-300, and 372-389 (1980).

CR R.COIFMAN, R.ROCHBERG: Representation theorems for holomorphic and harmonic functions. Astérisque 77 , 11-65 (1980).

DGM I.DAUBECHIES, A.GROSSMANN, Y.MEYER: Painless nonorthogonal expansions. J.Math.Phys. 27 , 1271-1283 (1986).

DM M.DUFLO,C.C.MOORE: On the regular representation of a non unimodular locally compact group. J.Functional Anal.21 , 209-243 (1976).

ET P.EYMARD, M.TERP: La transformation de Fourier et son inverse sur le groupe des ax+b d'un corps local. In "Analyse Harmonique sur les Groupes de Lie II", pp.207-248, Lect.Notes in Math. 739, Springer. 1979.

F1 H.G.FEICHTINGER: Banach convolution algebras of Wiener's type. "Functions, Series, Operators", Proc.Conf., Budapest 1980, Coll. Soc. Janos Bolyai, North Holland ,Amsterdam (1983), 509-524.

F2 -----: Banach spaces of distributions of Wiener's type and interpolation. "Functional Analysis and Approximation" , Proc. Conf., Oberwolfach 1980, Ed. P.Butzer, B Sz.Nagy and E.Görlich, Birkhäuser-Verlag,ISNM 60, (1981), 153-165.

F3 -----: On a new Segal algebra. Monatsh.Math. 92 , 269-289 (1981).

F4 -----: Un espace de Banach de distributions tempérées sur les groupes localement compacts abéliens. C.R.Acad.Sci. Paris, Sér A 290 , no.17, 791-794(1980).

F5 -----:Gewichtsfunktionen auf lokalkompakten Gruppen, Sitzber. Österr. Akad.Wiss,Abt.II,Bd.188 (1979),451-471.

F6 -----:Modulation spaces on locally compact abelian groups, Techn.Report, Vienna, 1983.

F7 -----:A new class of function spaces. Proc.Conf. "Constructive Function Theory", Kiew 1983.

F8 -----:Compactness in translation invariant Banach spaces of distributions and compact multipliers, J.Math.Anal. Appl.102, 289-327 (1984).

F9 -----:A characterization of minimal homogeneous Banach spaces. Proc.Amer.Math.Soc. 81 , 55-61(1981).

F10 -----: Minimal Banach spaces and atomic decompositions. Publ. Math. Debrecen 33 ,167-168 (1986)(An expanded version will appear in the same journal in 1987).

F11 -----: Atomic characterizations of modulation spaces. Proc. Conf. "Constructive Function Theory", Edmonton, July 1986, to appear.

F12 -----: The appropriate frame for harmonic analysis over locally compact abelian groups. Talk presented at the ICM 86, Berkeley.

FJ M.FRAZIER, B.JAWERTH: Decomposition of Besov spaces. Indiana Univ.Math.J. 34 , 777-799 (1985).

GM1 A.GROSSMANN, J.MORLET: Decomposition of Hardy functions into square integrable wavelets of constant shape. SIAM J.Math.Anal.**15** , 723-736 (1984).

GM2 A.GROSSMANN, J.MORLET: Decomposition of functions into wavelets of constant shape and related transforms. "Mathematics and Physics, Lect. on Recent Results", World Sci.Publ. Singapore.

GMP A.GROSSMANN,J.MORLET, T.PAUL: Transforms associated to square integrable group representations I. J.Math.Phys. 26(10), 2473-2479 (1985).

GP A.GROSSMANN, T.PAUL: Wave functions on subgroups of the group of affine canonical transformations. In Proceedings "Resonances - Models and Phenomena", Bielefeld 1984, Lecture Notes in Physics 211, 128-138 (1984)..

IG J.IGUSA: Theta Functions. Grundl.math.Wiss. Bd.194, Springer Berlin - Heidelberg - New York ,1972.

JPR S.JANSON, J.PEETRE, R.ROCHBERG: Hankel forms and the Fock space. Uppsala Univ.Dept.Math. Rep. 1986/6.

JS A.J.JANSSEN: Gabor representation of generalized functions. J.Math. Anal. Appl. **83**, 377-394 (1981).

LA S.LANG: $SL_2(\mathbb{R})$, Springer,New York-Berlin-Heidelberg (1985).

LUE D.H.LUECKING: Representations and duality in weighted spaces of analytic functions. Ind.Univ.Math.J. **34**, 319-336(1985).

ME1 Y.MEYER: De la recherche petrolière á la géometrie des espaces de Banach en passant par les paraproduits. Sem.Equ.Der.Part. 1985/86, Ecole Polytechn. Paris.

ME2 Y.MEYER: Principe d'incertitude, bases hilbertiennes et algèbres d' opérateurs. Séminaire Bourbaki, 38ème année,1985/86 n° 662.

PA A.PAPOULIS: Signal analysis. McGraw-Hill Book Comp.. 1977.

PE1 J.PEETRE: New Thoughts on Besov Spaces. Duke Univ.Math.Ser. 1, Durham, 1976.

PE2 J.PEETRE: Paracommutators and minimal spaces, Proc.Conf. "Operators and Function Theory",S.C.Powers ed., NATO ASI Series.,Reidel (1985).

REI H.REITER: Classical Harmonic Analysis and Locally Compact Groups. Oxford Univ.Press, (1968).

RT F.RICCI, M.TAIBLESON: Boundary values of harmonic functions in mixed norm spaces and their atomic structure. Ann. Scuola Norm.Sup. Pisa, Ser. IV, X 1-54 (1983).

RO R.ROCHBERG: Decomposition theorems for Bergman spaces and their applications. In "Operators and Function Theory". S.C.Powers ed. NATO ASI Series. Reidel (1985).

SCH1 W.SCHEMPP: Radar reception and nilpotent harmonic analysis I-IV. C.R. Math./ Math.Reports, Acad.Sci Canada **4**, 43-48, 139-144, 219-224, 287-292 (1982).

SCH2 W.SCHEMPP: Radar ambiguity function, the Heisenberg group, and holomorphic theta series, Proc.Amer.Math.Soc. **92**, 103-110(1984).

SCH3 W.SCHEMPP: Harmonic analysis on the Heisenberg group with applications, Pitman, Boston, Mass.(1986).

TA M.TAYLOR, Noncommutative Harmonic Analysis. Math. Surveys and Monographs Nr.22. Amer.Math.Soc., Providence (1986).

TR1 H.TRIEBEL: Spaces of Besov-Hardy-Sobolev Type. Teubner Texte zur Mathematik. Teubner,Leipzig (1978).

TR2 H.TRIEBEL: Theory of Function Spaces. Akad. Verlagsges., Leipzig (1983).

TR3 H.TRIEBEL: Modulation spaces on the Euclidean n-space, Zeitschr.für Analysis und ihre.Anwendungen **2**, 443-457 (1983).

TR4 H.TRIEBEL: Characterizations of Besov-Hardy-Sobolev-spaces. A unified approach. Preprint(1986).

WA G.WARNER: Harmonic Analysis on Semisimple Lie Groups I, II. Springer Berlin - Heidelberg - New York.1972.

THE WORK OF COIFMAN AND SEMMES ON COMPLEX INTERPOLATION, SEVERAL COMPLEX VARIABLES, AND PDE'S

Richard Rochberg[1]
Department of Mathematics
Washington University
St. Louis, MO 63130
USA

I reported on some very interesting work currently being done by R. R. Coifman and S. Semmes. This write up is based on that talk. The presentation is extremely informal; the only thing I hope to do with any accuracy is give their points of view and some of their results.

1. Classical Complex Interpolation:

We first recall the theory of complex interpolation of Banach spaces in a form which suggests the generalizations which we will present. This is A. P. Calderon's theory of complex interpolation as extended by Coifman, Cwikel, Rochberg, Sagher, and Weiss. Details are in the survey [RW] and in the references there.

Let D be the unit disk in the complex plane and ∂D its boundary, the unit circle. The starting data for interpolation is a family of Banach spaces parametrized by points ζ in ∂D. That is, for each ζ in ∂D we have the Banach space $A_\zeta = (\mathbb{C}^n, N_\zeta)$, the vector space $\mathbb{C}^n$ equipped with the norm function N_ζ. (Here and throughout we will assume the Banach spaces considered are <u>finite</u> dimensional. Some of the results we discuss require a more complicated formulation for infinite dimensional spaces,

---

[1]This work supported in part by a grant from the National Science Foundation.

others havn't been studied systematically.) Some restrictions are needed on N_ζ. To focus attention on the main issues we assume more than is needed, we suppose throughout this paper that N_ζ is continuous in ζ and that the ratio of N_ζ to the Euclidean norm is bounded above and below, independently of ζ.

The interpolation construction starts with this boundary data and produces a family of Banach spaces $A_z=(\mathbb{C}^n,N_z)$ parametrized by points z in D. These spaces are called the <u>intermediate spaces</u> or the <u>interpolation spaces</u> associated with the boundary data. Families of spaces obtained this way are called <u>interpolation families</u>. When we want to indicate the dependence of the interpolation spaces on the boundary data we write $A_z=(A_\zeta)_z$.

Here are some of the basic facts about these spaces:

<u>A. Boundary Values</u>: If ζ is in ∂D and the points z are in D then $\lim_{z\to\zeta} A_z=A_\zeta$. (With our assumptions the convergence is unrestricted; in general there is non-tangential convergence a.e. $d\theta$.)

<u>B. Interpolation Theorem</u>: Suppose X is a fixed Banach space and that for z in a neighborhood of D, T_z is an analytic family of linear maps of X <u>into</u> $\mathbb{C}^n$. If for each ζ in ∂D, T_ζ has operator norm at most one as a map from X to A_ζ, then for all z in D, T_z has operator norm at most one as a map of X to A_z. In short, operator norm estimates on the boundary imply operator norm estimates inside the disk. (There is a more technical result in which the norms on the boundary are estimated by a function rather than a constant--this is another refinement which we put aside in this description.)

<u>C. Iteration Theorem</u>: Suppose we are given A_ζ and obtain the A_z by the interpolation construction. Now suppose we take a small disk D' inside D and associate to the points in $\partial D'$, the boundary of D', the Banach spaces A_w, $w\in\partial D'$. The interpolation construction associated to the disk D adapts in a completely

natural way to the disk D' and using it we can construct Banach spaces $(A_w)_z$ for z in D'. The iteration theorem states that this repetition of the construction reproduces the same spaces, i.e. $((A_\zeta)_w)_z=(A_\zeta)_z$.

D. Duality Theorem: For any Banach space X, let X^* denote the dual space; the Banach space of continuous linear functionals on X. The duality theorem states that the interpolation spaces of the duals are the duals of the interpolation spaces: $(A_\zeta{}^*)_z = (A_\zeta)_z{}^*$. A very closely related result is that an interpolation theorem also holds for analytic families of linear maps T_z out of the A_z and into a fixed space X. (More generally the interpolation theorem remains valid for maps between two interpolation families.)

E. Existence and Uniqueness: The interpolation construction produces the unique family of spaces A_z which satisfies A, B, C, and D. To show a family of spaces is an interpolation family it is enough to show that the family and the dual family both satisfy A and B.

The interpolation construction, which we will not describe here, is quite abstract. Often the intermediate spaces are actually identified by exhibiting the family of spaces which satisfy the conditions in E.

Before going further we recall the basic examples of the theory.

One dimensional spaces: If n=1 then the function N_ζ is determined by the positive function $a(\zeta)=N_\zeta(1)$. The interpolation spaces are obtained by setting $N_z(1)=a(z)$ where a is extended from ∂D to all of D so as to satisfy

(1) $$\Delta \log a(z) = 0.$$

(It is instructive to reread properties A through E for this example.)

ℓ^p spaces: Let ℓ^p denote the n-dimensional Lebesgue space; $\mathbb{C}^n$ normed by $N_p((z_1,\dots,z_n)) = (\sum|z_k|^p)^{1/p}$. If the boundary spaces A_ζ are $\ell^{p(\zeta)}$ for some function $p(\zeta)$ with $1\leq p(\zeta)<\infty$ then the spaces A_z will be $\ell^{p(z)}$ where $1/p(z)$ is the harmonic function in D which has boundary values $1/p(\zeta)$.

Inner Product Spaces: Suppose the A_ζ are inner product spaces; thus there is a function $\Omega(\zeta)$ defined on ∂D and taking values which are n by n positive definite matrices so that

$$(2) \qquad N_\zeta(v)^2 = (\Omega(\zeta)v,v) \qquad v\in\mathbb{C}^n$$

((,) denotes the Euclidean inner product on $\mathbb{C}^n$). The interpolation spaces are given by

$$N_z(v)^2 = (\Omega(z)v,v) \qquad v\in\mathbb{C}^n$$

with z in D and with Ω extended to all of D so that

$$(3) \qquad \frac{\partial}{\partial \bar z}\left\{\Omega^{-1}\frac{\partial}{\partial z}\Omega\right\} = 0.$$

If all of the $\Omega(\zeta)$ commute (for instance if n=1) then it is possible to define a functional calculus for all of the $\Omega(z)$ so that (3) is the matricial analog of (1). Otherwise (3) is rather mysterious; we will have more to say about it later.

Two boundary spaces: If the boundary spaces only take two values A_0, and A_1 then the spaces A_z will be elements of the Calderon interpolation family $[A_0,A_1]_\theta$. $A_z = [A_0,A_1]_{h(z)}$ with h(z) harmonic. Thus this theory actually is a generalization of Calderon's. This is described more fully in [RW].

Properties A through E as well as equations (1) and (3) suggest that the construction of intermediate spaces is closely related to solving a Dirichlet problem (i.e. a boundary value problem) for a differential equation for the function N_z. A and B together (actually, the version of B with general boundary estimates) are roughly equivalent to saying that whenever f(z) is a holomorphic $\mathbb{C}^n$ valued function then the function $F(z)=\log(N_z(f(z))$ is dominated in D by the harmonic function with the same boundary values. The iteration theorem shows that this type of domination can be localized. Thus on all small disks F is dominated by the harmonic function with the same boundary values as F--thus F is subharmonic. N_z is the largest function which satisfies A and such that the functions F constructed this way will be subharmonic. This is less clear but it is a consequence of D, the duality theorem.

The situation is clearest when n=1. If $N_z(1)=a(z)$ for some positive function a (we are not yet assuming (1) holds) then $F(z) = \log(a(z)) + \log|f(z)|$. If this is to be subharmonic for every holomorphic f then log(a) must be subharmonic. If the duality theorem holds then the dual norms must satisfy a similar conclusion. A direct computation shows that the dual norms are associated with the function 1/a and hence we get subharmonicity for $\log(1/a) = -\log(a)$. Thus log(a) is also superharmonic. Hence log(a) is harmonic. Since log(a) is harmonic, it is the maximal subharmonic function with the correct boundary values. In this case we see explicitly that N_z is the maximal subsolution of certain differential inequalities, and is a solution to the corresponding differential equation. This point of view is developed further in [R].

2. The New Theory:

A. Abstract Results:

Let B be the unit ball in $\mathbb{C}^k$ and ∂B its boundary. As before the starting data for the interpolation construction is a family of Banach spaces A parametrized by points of the boundary, $\zeta\in\partial B$. Coifman and Semmes have developed a construction of intermediate spaces A_z, z in B, so that (the natural analogs of) properties A, B, C, hold. Instead of D there is a

D'. Maximality Theorem: The function N_z is maximal subject to the requirement that it satisfy A and an appropriate mixture of B and C. More precisely, suppose M_z is an auxiliary family of norms on $\mathbb{C}^n$ such that $M_\zeta = N_\zeta$ for ζ in ∂B and so that the spaces $(\mathbb{C}^n, M_z)$ satisfy the interpolation theorem on sub-balls of B. Then for all v in $\mathbb{C}^n$, all z in B, $M_z(v) \leq N_z(v)$.

With this replacement for D we have the hoped for substitute for E; the family A_z is the unique family which satisfies A, B, C, and D'.

Although it is not known if the duality theorem holds for these families it is still true that the interpolation theorem is valid for maps between two interpolation families. Suppose A_z and A'_z are two such interpolation families and T_z is an analytic family of maps between them. If for ζ in ∂B, the operator norm of T_ζ as a map from A_ζ to A'_ζ is at most one, then the same operator norm estimate holds inside B.

B. The Examples:

k=1: When the base space (the ball B) has one (complex) dimension the new construction gives the same intermediate spaces as those obtained by the method in the first section. Thus we have an extension of that theory and hence also of the classical theory of Calderon.

k=k, n=1: As before, the case in which all the A_ζ are one dimensional is particularly nice. Again we have to describe how to extend the function $a(\zeta)=N_\zeta(1)$ from the boundary ∂B to the entire ball B. Again the solution is that (1) must be satisfied.

Although the analogy is striking, the conclusion is a bit surprising at first. The historical focus on the function theoretic aspect of the one variable theory might have suggested that we would replace (1) with the condition that log(a) be pluriharmonic. In fact the theory we are describing now has a bit more potential theory and a bit less function theory than one might have expected.

ℓ^p spaces: The Lebesgue spaces work exactly as before; $p(\zeta)$ is extended so that $1/p(z)$ is harmonic. Similarly mixed norm spaces (such as $\ell^p(\ell^q)$) and weighted ℓ^p spaces work exactly as one would guess from the one variable theory. (Although the reasons are quite different.)

Inner product spaces: If the A_ζ for ζ in ∂B are given by (2) then the A_z will also be inner product spaces (however this fact is quite difficult to prove). The required function $\Omega(z)$

satisfies a partial differentrial equation which generalizes (3). For functions defined on $\mathbb{C}^k$ define the differential operator $\mathfrak{D}$ by

$$\mathfrak{D}\Omega = \sum_{j=1}^{k} \frac{\partial}{\partial \bar{z}_j} \left\{ \Omega^{-1}(z) \frac{\partial}{\partial z_j} \Omega(z) \right\} .$$

The replacement for (3) is

(4) $$\mathfrak{D}\Omega = 0$$

which reduces to (3) if k=1 and which is the statement that $\log(\Omega)$ is harmonic if all the Ω commute.

C. The Construction:

Equation (1) for the unknown function a(z) is rather easy to deal with. First change variables to the new function $b = \log a$ and then study the linear equation $\Delta b=0$. This linearization trick isn't available for equation (3) and hence equation (3) remains rather mysterious. Although existence and uniqueness results are known for (3) many elementary questions remain open and an "explicit" form for the solution is not known. The analysis of (3) from the interpolation theoretic point of view uses relatively subtle function theoretic ideas based on work of Szego, Nevanlinna, Beurling and others. The difficulty in generalizing this function theoretic point of view to several variables is that some of the tools used, for instance the inner-outer factorization of bounded analytic functions, do not extend.

The linear equation $\Delta b=0$ can be solved explitiicly using the Poisson integral formula. However that is of little direct help with (1). $\Delta b=0$ can also be solved using the Perron method which consists, roughly, of constructing the maximal solution to the differential inequality $\Delta b \geq 0$ and then showing that this gives a solution to the equation. A very nice feature of this method is that it can be used directly on equation (1) (without the necessity of a linearizing change of variable) and in fact extends (although this is less obvious) to take care of (3), the analogous equation in noncommuting variables. This use of the Perron method in interpolation is the point of view which Coifman and Semmes use in $\mathbb{C}^k$. It is new even when k=1.

We now outline the Perron process which they use to

construct the intermediate spaces. Given the boundary data A_ζ, $\zeta\in\partial B$, we call a family of norming functions M_z, z in B <u>admissible</u> if:

(1) $\forall v\in\mathbb{C}^n$, $M_z(v)$ is upper semicontinuous

(2) $\forall v\in\mathbb{C}^n$, $\overline{\lim\limits_{z\to\zeta}}\, M_z(v) \leq M_\zeta(v)$, and

(3) for any holomorphic $\mathbb{C}^n$ valued function f(z) defined on B, $\log M_z(f(z))$ is subharmonic.

Any family (of norms or spaces) which satisfies (2) we call <u>logarithmically subharmonic</u>.

We set

$$\mathcal{A} = \{\text{admissible families of norming functions}\}$$

and define

$$N_z(v) = \sup\{M_z(v) : M_z\in\mathcal{A}\}.$$

This is the Perron solution to the problem of constructing interpolation spaces. It is relatively straightforward to show that these spaces have properties A, B, C, D', and E. The details of the proof are of the type expected when using an abstract Perron construction,...establishing admissibility by proving semicontinuity, constructing barriers, etc.

Condition (3) is the crucial one for studying maps into interpolation families. Suppose M_z is logarithmically subharmonic and that T_z are linear maps from a fixed Banach space X into $(\mathbb{C}^n, M_z)$. Suppose also that T_z depends analytically on z. If the norms of T_ζ, $\zeta\in\partial B$ are at most one then this will also be true for the T_z, $z\in B$. To see this pick v in $\mathbb{C}^n$ and z_0 in B. Then

$$M_{z_0}(T_{z_0}(v)) \leq \sup_{\zeta\in\partial D} M_\zeta(T_\zeta(v)) \leq \|v\|_X.$$

The first of these inequalities because $M_z(T_z(v))$ is logarithmically subharmonic and hence harmonic; the second by the operator norm estimate on the T_ζ. The <u>logarithmic</u> subharmonicity is used to extend this argument to the case of general norm estimates. Suppose that for $\zeta\in\partial D$ the norm of T_ζ is $e^{h(\zeta)}$. Let $h(z)$ be the harmonic extension of h to D. Because $M_z(T_z(v))$ is logarithmically subharmonic, $e^{-h(z)}M_z(T_z(v))$ is subharmonic. Hence we can follow the previous pattern and conclude that the norm of T_z is at most $e^{h(z)}$.

Condition (3) is at the heart of estimates of norms for maps into interpolation families. The study of mappings out of interpolation families uses the maximality that interpolation norms have because they were obtained by a Perron construction.

Suppose that for z in some neighborhood in $\mathbb{C}^k$, M_z and N_z are two (smooth) families of norms on $\mathbb{C}^n$ and $\mathbb{C}^m$ (we do not require that n=m) and T_z is an analytic family of linear maps of $\mathbb{C}^n$ to $\mathbb{C}^m$. Let h(z) be the norm of the operator T_z acting between the associated spaces;

$$h(z) = \sup\,\{N_z(T_z(v))/M_z(v):\ v\neq 0\}.$$

The family of norms M_z (and also the associated family of spaces) is called <u>logarithmically superharmonic</u> if log h(z) is subharmonic whenever N_z is an interpolation family. This is a local version of an interpolation theorem for maps <u>out of</u> the family normed by N_z. The definitions are set up so that analytic families of maps out of a logarithmically superharmonic family and into a logarithmically subharmonic family will satisfy an interpolation theorem; i.e. the operator norms of the maps will satisfy a maximum principle..

The Perron construction gives families of spaces that are logarithmically superharmonic.

<u>Lemma</u>: The family of spaces given by the Perron construction is the unique logarithmically subharmonic and logarithmically superharmonic family of spaces with the required boundary values.

One consequence of the lemma is that the families produced by the Perron construction will satisfy interplation theorems for maps into the family or out of the family. Another consequence is that a family of norms is an interpolation family, i.e. is obtained by the interpolation construction from some boundary values, if and only if it is an interpolation family in an arbitrarily small neighborhood of each point. The reason is that being logarithmically subharmonic and being logarithmically superharmonic are both local properties. This makes it plausible that being an interpolation family might be equivalent to having the norm function satisfy a differential equation. For instance, (4) is such an equation (although the equation is written for Ω rather than N_z). For k=1 such an equation is given in Section 5 of [RW]. Finally, this lemma gives an effective criterion for showing that a given family is an interpolation family.

The lemma also leads to an alternative formulation of the duality theorem. If the family of spaces A_z is logarithmically subharmonic, then the dual spaces A_z^* do not form a logarithmically superharmonic family (at least that is not known to be true except when k=1 in which case it is a consequence of the duality theorem.) It is fairly easy to check that A_z is logarithmically subharmonic if and only if the function log h(z) is subharmonic when m=1 (i.e. A_z^* is <u>weakly logarithmically superharmonic</u>.) In this context the possible duality theorem is equivalent to the claim that every weakly logarithmically superharmonic family is logarithmically superharmonic.

<u>D. The examples</u>:

When k=1 one can show that a family of spaces is an interpolation family by exhibiting $\mathbb{C}^n$ valued analytic functions f which meet certain conditions (for instance that $N_z(f(z))$ must be constant). (The introduction of such functions for the Lebesgue spaces is at the heart of Thorin's proof of the Riesz-Thorin theorem.)

For k greater than one we don't have analogs of these analytic functions. This is because the theory deals with the (generalized) harmonicity of log N_z and makes frequent use of a property of harmonic functions that doesn't extend to several variables: if log G(z) is a real valued harmonic function then $G(z)=|F(z)|$ for some holomorphic F. This fails if k>1. Instead we use an infinitesimal substitutes for this fact. The details are too lengthy to include and we will just make a few comments.

ℓ^p spaces: Suppose we are given a family of spaces $A_z=\ell^{p(z)}$ with $1/p(z)$ harmonic and we want to show that this is an interpolation family. Log subharmonicity is established by a lengthy direct computation or by a series of reductions which reduces the question to showing that $c^{1/p(z)}$ is subharmonic for positive constants c.

Suppose that we are given z_0 in B and $w=(w_i)$ in $\mathbb{C}^n$. Let $\bar{p}$ be the value at z_0 of $p(z)$. Log subharmonicity shows that the interpolation norms N_z satisfy $\|w\|_{\ell^{\bar{p}}} \leq N_{z_0}(w)$. The estimate in the other direction requires more work. Without loss of generality we can assume $\|w\|_{\ell^{\bar{p}}} = 1$. With a bit of work we can restrict attention to those M_z in $\mathcal{A}$ which only depend on the modulus of the w_i's. Once that is done we can assume all the w_j are positive. Now consider the function

$$G(z) = M_z((w_j^{\bar{p}/p(z)})).$$

As in the proof of the Riesz-Thorin theorem, we have set things up so that $G(\zeta)\equiv 1$ on ∂B. The required conclusion is that $G(z_0)\leq 1$; thus we are done as soon as we show that G is subharmonic. To get that conclusion we apply to following lemma to (a smoothed version of) $M_z((e^{\gamma_i}))$.

Lemma: Suppose F is a function on $\mathbb{C}^k\times\mathbb{C}^m$ which for all z in $\mathbb{C}^k$ and γ in $\mathbb{C}^m$ satisfies $F(z,\gamma)=F(z, \operatorname{Re}\gamma)$. Suppose further that for all holomorphic maps f of $\mathbb{C}^k$to $\mathbb{C}^m$, $F(z,f(z))$ is subharmonic. Then $F(z,h(z))$ is subharmonic for all harmonic $\mathbb{R}^m$ valued functions $h(z)$.

Informally, this lemma is the infinitesimal substitute for having a holomorphic function with modulus of the form $c^{1/p(z)}$.

Inner product spaces: Suppose $A_z=(\mathbb{C}^n,M_z)$ is a family of inner product spaces parametrized by points z in some open set in $\mathbb{C}^k$; $M_z(v)^2=(\Omega(v),v)$ for positive definite matrices $\Omega(z)$. Suppose also that Ω is smooth enough so that we can consider $\mathcal{D}\Omega$.

To say that A_z is logarithmically subharmonic is to require that a local maximum principle hold for certain functions (i.e. the functions can't have a srtict local maximum at an interior point.) Sometimes such a maximum principle is related to a second order differential inequality ("having negative curvature".) The differential operator which occurs in this case is $\mathfrak{D}$ which can be thought of as a generalization of $\Delta \circ \log$ (and hence $\mathfrak{D}\Omega$ is the negative of a generalized curvature.) The sign condition on $\mathfrak{D}\Omega$ is relative to the inner product on A_z; that is, when we say $\mathfrak{D}\Omega$ is negative definite we mean that for each z, $\mathfrak{D}\Omega$ is negative definite relative to the inner product which induces M_z.

Theorem: 1. A_z is logarithmically subharmonic if and only if $\mathfrak{D}\Omega$ is positive semidefinite.

2. A_z is logarithmically superharmonic if and only if $\mathfrak{D}\Omega$ is negative semidefinite.

Corollary: If $\mathfrak{D}\Omega \equiv 0$ then A_z is an interpolation family of inner product spaces.

Corollary: If A_z is an interpolation family of inner product spaces then $\mathfrak{D}\Omega \equiv 0$.

Proof discussion: Part 1 is proved by direct computation. Some of the proof of 2 is by direct computation. At a certain stage in the proof of 2 it is necessary to come to grips with the function theory. When k=1 then equation $\mathfrak{D}\Omega \equiv 0$ is equivalent to a factorization $\Omega = F^* F$ where F is a non-vanishing matrix valued holomorphic function (and * denotes the adjoint). (This should be viewed as part of the scheme: $\log(\circledast)$ is harmonic implies $\circledast$ is the modulus of something holomorphic.) This equivalence fails for k>1 and as with the proof for Lebesgue spaces we need an infinitesimal substitute for the existence of certain holomorphic functions. We indicate the flavor of the needed lemma by stating a special case.

Lemma: Suppose $\Omega(z_0) = I$. There is a matrix valued function F such that $\Omega = F^* F$ and at the point z_0,

$$\frac{\partial F}{\partial \bar{z}_j} = 0 \quad j=1,\ldots,k; \text{ and } \Delta F = 0.$$

If we could show directly that whenever the boundary spaces A_ζ were all inner product spaces then the resulting interpolation spaces A_z would also be inner product spaces which varied smoothly with z; we could then use the previous corollary and deduce an existence theorem for the Dirichlet problem for the equation $\mathfrak{D}\Omega\equiv 0$. However no such direct proof is known. (Finding such a proof would be a major advance.) Instead Coifman and Semmes give a direct proof of the existence to solutions for this Dirichlet problem. An appeal to the previous theorem then completes the analysis of the example. The proof of the existence theorem is relatively difficult and uses a mixture of classical ideas and convexity principles related to the interpolation theory.

(This is different from the situation for k=1. In that case it is possible to give a direct proof that the interpolation spaces are inner product spaces. Thus the interpolation scheme in one variable gives a proof of the existence of a solution to (3). Furthermore, as we noted, solutions to (3) are exactly the Ω which admit factorizations $\Omega=F^*F$. Thus the interpolation construction gives a proof of the factorization theorem of Masani and Wiener which asserts that given Ω on ∂D it is possible to find holomorphic F so that the boundary values of F^*F equal Ω.)

3. Further comments:

A. Variations:

Of course many variations are possible. Most of the results of the previous section hold if the ball B is replaced by a general domain for which the Dirichlet problem for the Laplace equation is solvable. This includes unbounded domains (if appropriate care is taken about growth of functions at infinity.)

This leads to an interesting technique for constructing families of Banach spaces in $\mathbb{R}^k$. Start with a region R in $\mathbb{R}^k$ and a family of Banach spaces A_x parametrized by the points x of ∂R. Extend these boundary values trivially to $\partial R \times i\mathbb{R}^k \subset \mathbb{C}^k$. Now perform the interpolation construction on the domain $R \times i\mathbb{R}^k$ contained in $\mathbb{C}^k$ and then restrict the resulting interpolation family A_z back to the original region R. If k=1 and R is an interval this process generates the classical Calderon spaces. For k>1 we obtain a new, and not well understood, generalization of Calderon's construction. If the given boundary spaces are inner product spaces then the resulting Ω will satisfy a real version of equation (4):

$$\sum_{j=1}^{k} \frac{\partial}{\partial x_j} \left\{ \Omega^{-1}(x) \frac{\partial}{\partial x_j} \Omega(x) \right\} = 0. \tag{5}$$

There are other variations that may be interesting but havn't been studied systematically. For example, what happens if we try to develop the theory on B using the invariant Laplacian instead of the Euclidean Laplacian in the definition of log subharmonic?

B. Curvature: A family of Banach spaces A_z defined for z in a subset R of $\mathbb{C}^k$ forms a (topologically trivial) holomorphic vector bundle with base R. If the A_z are all inner product spaces then we have a Hermitian holomorphic bundle. The curvature of this bundle is a second order differential expression which vanishes exactly when the bundle is trivial (i.e. Ω is constant in appropriate coordinates.) Negative curvature is often associated with a maximum principle for sections of the bundle. This suggests that the interpolation condition might be related to curvature. For k=1 this is true in the strongest possible sense; $\mathfrak{D}\Omega$ is a constant multiple of the curvature and the interpolation families are exactly those for which the curvature vanishes. An analogous result involving the curvature of Finsler bundles instead of Hermitian bundles holds if the A_z are general Banach spaces [R].

In higher dimensions $\mathfrak{D}\Omega$ is not the curvature, it is a contracted form of the full curvature tensor. It is similar in spirit to the Ricci tensor but different in detail; here the contraction is with respect to the base variables not the fiber variables. There is a similar, but more complicated, quantity related to curvature of Finsler bundles which can be used in formulating the theory for general Banach spaces.

For inner product space the relations between interpolation and the differential geometry of vector bundles go deeper than these comments about curvature might suggest. The interpolation construction produces the solution to certain function theoretic extremal problems and hence it is not surprising (in retrospect) that the operator $\mathcal{D}$ and equations (3), (4) and (5) show up other places in differential geometry and in mathematical physics. These relations are not well understood and we will just point to the other topics. The reader is encouraged to help in studying these fascinating connections.

When k=1 the interpolation families are exactly the vector bundles which have the assigned boundary values and minimize the absolute curvature (by making it zero). When k=2 it is not possible to solve the boundary value probolem with a vector bundle having curvature zero (because it is not possible to find a holomorphic function on B with prescribed real part on ∂B). However an attempt to minimize at appropriate mean (with respect to z) curvature again leads to the requirement that Ω satisfy (4). In this context (4) is known as the anti-self-dual Yang-Mills equation and has been greatly studied recently. (The Yang-Mills equation is often written in terms of the connection of the vector bundle rather than in terms of Ω. Also, the equation is often written in terms of the underlying coordinates of $\mathbb{R}^4$. This form of the equation is in [P]. See [A] for a general discussion.) Recently K. Uhlenbeck and S. T. Yau [UY] have studied vector bundles which satisfy

$$\mathcal{D}\Omega = cI. \tag{6}$$

(Here I is the identity matrix and c is a constant. The case c=0 is not especially trivial in their work.) They call a vector bundle a Hermitian-Yang-Mills bundle if it admits an inner product so that (the extension to general base manifolds of) (6) is satisfied. They prove that certain vector bundles over compact Kahler manifolds are Hermitian-Yang-Mills bundles. In short, they prove a global existence theorem for (the manifold version of) (6). (Uhlenbeck has told me that the methods of [UY] can be used to obtain the existence theorem for the Dirichlet problem for (4).)

A mapping between Riemannian manifolds is called <u>harmonic</u> if it is a critical point of certain energy functional. There is an extensive theory of such maps; [EL] is a introduction. We mention such things here to note that if the domain of the harmonic map is $\mathbb{R}^k$ with is usual metric and the range manifold is the space of real positive definite with it's natural metric as a symmetric space then the harmonic maps are exactly those which satisfy (5). The implications of this connection between interpolation theory and harmonic maps is not well understood. An interesting related question is to try to understand the complex equation (4) as a variational equation for some generalized energy functional.

<u>C. Questions:</u> There are lots of open questions in this area. Here are a few:

1. What are other examples of interpolation families. For instance, if B_θ is a complex interpolation family in the sense of Calderon and h(z) is a harmonic function defined in B in $\mathbb{C}^k$ and taking values between 0 and 1, will $A_z = B_{h(z)}$ be an interpolation family? (When k=1 the answer is yes (with some care to technical restrictions).)

2. What is going on with the duality theorem? Is it true? If it's not true, then what characterizes the many types of examples in which it is true? (For example, it it always true if the A_z are Banach lattices?) If the general theorem is not true then what are the theoretical properties of families of spaces dual to interpolation families?

3. Give a direct interpolation theoretic proof that the interpolation spaces generated by inner product spaces on the boundary will be inner product spaces.

4. The idea of defining interpolation spaces by a Perron construction makes sense in lots of contexts. What are some other examples in which it gives interesting results?

<u>D. Alternative approaches</u>: Z. Slodkowski has recentely given a very interesting approach to interpolation in one and several

complex variables which is related to his earlier work on analytic multifunctions [Z]. His work is certainly related in spirit to the work described here but detailed comparisons have yet to be made.

REFERENCES

[A] M. F. Atiyah, Geometry of Yang-Mills fields, Fermi Lectures at Scu. Norm. Pisa, 1978, Scu. Norm. Pisa 1979.

[EL] J. Eells and L. Lemaire, A Report on Harmonic Mappings, Bull. Lond. Math. Soc. 10 (1978) 1-68.

[P] K. Pohllmeyer, On the Lagrangian Theory of Anti-Self-Dual Fields in Four-Dimensional Euclidean Space.

[R] R. Rochberg, Interpolation of Banach Spaces and Negatively Curved Vector Bundles, Pac. J. Math 110 (1984), 355-376.

[RW] R. Rochberg and Guido Weiss, Analytic Families of Banach Spaces and Some of Their Uses, Recent Progress in Fourier Analysis, I. Peral and J.-L. Rubio de Francia eds., North-Holland, Amsterdam, 1985, 173-202.

[UY] K Uhlenbeck and S.T. Yau, On the Existence of Hermitian-Yang-Mills Connections in Stable Vector Bundles, preprint 1986.

[Z] Z. Slodkowski, Presentation at International Conference on Potential Theory and Related Topics, U. of Toledo, July, 1986.

INTERPOLATION OF QUASINORMED SPACES BY THE COMPLEX METHOD

Richard Rochberg, Anita Tabacco Vignati, Marco Vignati and Guido Weiss[1]
Department of Mathematics
Washington University
Saint Louis, MO 63130
USA

The complex method of interpolation developed by A. P. Calderón [2] was extended by R. R. Coifman, M. Cwikel, R. Rochberg, Y. Sagher, and Guido Weiss (see [3], [4], [5], and the two expository articles [10] and [11]). In very broad terms, the problem considered in this extension is the following one: Suppose Δ is a simply connected domain in the complex plane and that $\partial\Delta$ is its boundary. Assume that to each $\zeta \in \partial\Delta$ there is assigned a Banach space B_ζ, with norm N_ζ. Can one construct Banach spaces B_ζ, with norms N_ζ, associated with each interior point $z \in \Delta$ in a way that is consistent with the theory of interpolation of linear operators? The precise assumptions on the boundary spaces needed for this construction and the meaning of this question are given in the articles cited above. The main features of the theory, however, are the following: assuming that the boundary spaces are embedded in a topological vector space U the intermediate space B(z) consists of all vectors $v \in U$ for which $N_z(v) < \infty$, where

---

[1]Supported in part by NSF Grants DMS-8402191 and DMS-8200884

$$(1) \qquad N_z(v) = \inf \left\{ (\int_{\partial\Delta} [N_\zeta(F(\zeta))]^p \, P_z(\zeta) \, d\zeta)^{1/p} \right\}.$$

the infimun being taken over an appropriate class of U valued holomorphic functions satisfying $F(z) = v$. Here P_z is the Poisson kernel associated with the domain Δ and $0 < p \leq \infty$. A notable result in the theory is the fact that the norm defined by (1) is independent of the parameter p. These intermediate spaces are consistent with the theory of interpolation in the sense that if T is an operator on the spaces B_ζ with operator norms $\|T\|_\zeta \leq M$ on $\partial\Delta$ then T has operator norm $\|T\|_z \leq M$ on B_z for $z \in \Delta$ (more generally, it can be shown that $\log \|T\|_z$ is subharmonic on Δ; moreover, T can map one family $\{B_\zeta\}$ of boundary spaces into another). Other results that can be obtained are : a *Duality Theorem* (this states that if $\{B_\zeta\}$ is a boundary family, $\zeta \in \partial\Delta$, for which our construction can be applied, so is the family of dual spaces $\{B_\zeta^*\}$ and the intermediate spaces obtained from the latter are the duals of the spaces B_z); an *Iteration Theorem* (if $\Omega \subset \Delta$ is a simply connected subdomain then the intermediate space B_w, $w \in \Omega$, can be obtained by our construction from the intermediate spaces associated with $\partial\Delta$). Many other results are stated in the references cited above.

Equality (1) makes sense if the functionals N_ζ are quasinorms instead of norms (a quasinorm N differs from a norm only in the sense that Minkowski's inequality reads $N(u+v) \leq K(N(u)+N(v))$, where $K > 1$). The existence of many examples that are important in analysis (L^p spaces, weighted L^p spaces, H^p spaces, the Schatten ideals S^p for $p < 1$) is one of the motivations for considering the study of this interpolation method in the case of Quasi-Banach spaces.

Interpolation of two p-normed spaces was studied by N. Riviere in his thesis [8]. Here we work with quasinorms rather than p-norms. The two viewpoints, however, yield the same interpolation spaces (see [9]). A more recent extension of the Calderón method can be found in the work of M. Cwikel, M. Milman and Y. Sagher [6].

The extension of the complex method developed in [3], [4] and [5] to quasi-Banach spaces is not routine for, at least, two reasons. First, one cannot use duality as was done in the Banach space case. The duality theorem is not true in general in the quasi-Banach case. Second, we do not have, at least in the infinite dimensional case, a maximum principle or a related vector-valued integration that yields a mean value inequality.

One can exhibit, in fact, a rather dramatic example that illustrates some of the difficulties encountered in the quasinorm case: there exists a quasi-Banach space B of infinite dimension such that, if $B_\zeta = B$ for <u>all</u> $\zeta \in \partial\Delta$ then <u>all</u> the intermediate spaces B_z are zero dimensional, $z \in \Delta$.

It is not hard to see that the space $B = L^p(T)/H^p_o$, $0 < p < 1$, furnishes us with such an example (Δ is the unit disk and $\partial\Delta = T$ is the torus). Here H^p_o is the subspace of all the f in the classical Hardy space H^p such that $f(0) = 0$. This space was introduced by A. B. Alexandrov [1] who showed that holomorphic B-valued functions do not satisfy the maximum principle. Indeed, let $f : \Delta \longrightarrow B$ be the function whose value at $w \in \Delta$ is the analytic function $f(w)(z) = z/(z-w)$, $z \in \Delta$. Then $f(0)$ is the constant function 1 while $f(e^{i\theta}) \in H^p_o$. Thus, f determines the zero element in B for all $w \in \partial\Delta = T$ while $f(0)$ determines a non zero coset in B. From the definition (1) of the intermediate norm at $z = 0$, it is then reasonable to expect that the cosets in B determined by the constant functions have quasinorm 0. As stated above, we have not given here the precise definition of the intermediate spaces; however, it is shown in [12] and [13] that this is, indeed, the case. In fact, this argument can be extended to show that the cosets determined by the monomials $e^{-in\theta}$, $n = 1, 2, \ldots$, have zero quasinorm at $z = 0$. From this it follows that the intermediate space B_o is trivial (more generally, $B_z = \{0\}$ for all $z \in \Delta$).

Despite these negative results one can develop a useful theory of interpolation in the quasinorm case. As is the case in the Banach space case, the simplest situation occurs when the boundary spaces are finite dimensional. We give a brief description in this setting. Suppose that to each point $\zeta \in \partial\Delta$ (which we assume from now on to be the torus T and Δ to be the unit disk) we associate a space B_ζ of the form $(\mathbb{C}^n, N_\zeta)$, where N_ζ is a quasinorm on the n dimensional complex Euclidean space $\mathbb{C}^n$. We assume that the map $v \longrightarrow N_\zeta(v)$ is measurable and the quasinorm N_ζ is related to the Euclidean norm $\| \ \| = \| \ \|_{\mathbb{C}^n}$ by the inequalities

$$(2) \qquad m_1(\zeta)\|v\| \leq N_\zeta(v) \leq m_2(\zeta)\|v\|,$$

where $\log m_j(\zeta)$, $j = 1, 2$, is integrable on $\partial\Delta$. We also assume that the constants $K(\zeta)$ of the "quasi Minkowski" inequality for B_ζ form an integrable function. Under these quite general hypotheses it can be shown that:

(3) The functional N_z is a quasinorm on $\mathbb{C}^n$ for each $z \in \Delta$ and $N_z(u + v) \leq K(z)(N_z(u) + N_z(v))$ for all $u, v \in \mathbb{C}^n$, where

$$(4) \qquad K(z) = \exp \int_o^{2\pi} \log K(\zeta)\, P_z(\theta)\, d\theta$$

(recall that $\zeta = e^{i\theta}$). This estimate for $K(z)$ is best possible in general.

(5) The functional N_z is independent of p, $0 < p \leq \infty$.

If we let

(6) $$W_j(z) = \exp \int_0^{2\pi} \log m_j(\zeta)\, H_z(\theta)\, d\theta$$

for $j = 1.\ 2$, where $H_z(\zeta) = \frac{1}{2\pi} \frac{1 + ze^{-i\theta}}{1 - ze^{-i\theta}}$ is the Herglotz kernel whose real part is the Poisson kernel $P_z(\theta)$, then

(7) $$|W_1(z)|\ \|v\| \leq N_z(v) \leq |W_2(z)|\ \|v\|.$$

These last inequalities show that the statement of the iteration theorem makes sense in the finite dimensional quasinorm case. Whether it is true is an open question.

In the finite dimensional Banach space case one can show that the extremal problem posed by definition (1) has a solution. More precisely, there exists an extremal function $F = F_{z,v}$ in the class of $\mathbb{C}^n$ valued holomorphic functions over which the infimum in (1) is taken (this class is a weighted Hardy space depending on m_1) satisfying

(8) $$N_z(v) = \left(\int_0^{2\pi} [N_\zeta(F(\zeta))]^p\, P_z(\theta)\, d\theta\right)^{1/p}.$$

We have not been able to show the existence of such extremal functions in the finite dimensional quasinorm case.

Let us now pass to the general case. The first problem is to find the correct definition of an "interpolation family" of spaces B_ζ associated with the boundary $\partial\Delta$. We must also find the appropriate class of vector valued holomorphic functions involved in the infimum in (1). This is done in [12] and [13]. We shall not explain these technical difficulties here. Our main purpose is to describe the results that have been obtained. As in the finite dimensional case, the functional N_z does not depend on p. Moreover, (3) and (4) also hold in the general case. These intermediate spaces also allow us to interpolate linear operators as described above for the Banach space case. Several other such general results are obtainable. Perhaps most important, however, is the fact that this theory can be applied to function spaces that are important in analysis.

Suppose that $B_\zeta = L^{p(\zeta)}$, where $p(\zeta)$ is a positive function such that $1/p(\zeta)$ is integrable on $T = \partial\Delta$. The construction we described then shows that the intermediate spaces B_z are $L^{p(z)}$ for each $z \in \Delta$, where

$$\frac{1}{p(z)} = \int_0^{2\pi} \frac{1}{p(\zeta)} P_z(\zeta)\, d\theta. \tag{9}$$

Moreover, the functional N_z is the $L^{p(z)}$ "norm" (which is a norm if $p(z) \geq 1$ and is a quasinorm when $p(z) < 1$).

The Hardy spaces also interpolate in a natural way. The results obtained in [13], however, are technically more complicated then the one we just announced for the Lebesgue spaces. If $B_\zeta = H^{p(\zeta)}(\mathbb{R}^n)$ for $\zeta \in \partial\Delta$ and $0 < p(\zeta) \leq 1$, then $B_z = H^{p(z)}$ for $z \in \Delta$ provided $\frac{1}{p(\zeta)} \log p(\zeta)$ is integrable. If we want a result of this kind that allows $p(\zeta)$ to range above the index 1, we require that $p(\zeta)$ be bounded away from 0. We doubt that these results are best possible.

Weighted ℓ^p spaces interpolate in a way that is similar to the one described above for the L^p spaces. In this case we can consider varying weights μ_ζ on the boundary. The weights associated with the intermediate spaces are then also given in terms of Poisson integrals involving both μ_ζ and $1/p(\zeta)$ (the details are given in [13]). The Schatten ideals interpolate formally as do the Lebesgue spaces.

There are other positive results in this theory. Let us briefly mention two of them. A. P. Calderón introduced a method of interpolation of Banach lattices. This method was extended to the Banach space theory associated with simply connected domains in the complex plane by E. Hernandez [7]. One can also carry out this extension in the quasinorm case. In [3] and [4] a connection is given between the real method of interpolation of pairs of spaces and the complex method we have been describing. This connection can be extended to the quasinorm case as well. These extensions lead to rather elegant formulae. We exhibit two such results for those familiar with the notation:

(10) $$[X_{p(\zeta),q(\zeta)}]^z = X_{p(z),q(z)};$$

(11) $$[X(\zeta)(B(\zeta))]_z = [X(\zeta)]^z(B(z)).$$

Here $\{B(\zeta)\}$ and $\{X(\zeta)\}$ are appropriate interpolation families of quasi Banach spaces and lattices, respectively, $1/p(z)$ and $1/q(z)$ are the Poisson integrals of $1/p(\zeta)$ and $1/q(\zeta)$.

REFERENCES

[1] Alexandrov, A. B., Essays on the non-locally convex Hardy classes, Lecture Notes in Math. 856, Springer Verlag, (1981), pp. 1-89.

[2] Calderón, A. P., Intermediate spaces and interpolation, the complex method, Studia Math. 24 (1964), pp. 114-190.

[3] Coifman, R., Cwikel, M., Rochberg, R., Sagher, Y., Weiss, Guido, Complex interpolation for families of Banach spaces, Proc. Symp. Pure Math., Vol. 35, Part 2, A.M.S. (1979) pp. 269-282.

[4]——————————————, The complex method of interpolation of operators acting on families of Banach spaces, Lecture Notes in Mathematics 779, Springer Verlag (1980) pp. 123-153.

[5]——————————————, A theory of complex interpolation for families of Banach spaces, Advances in Math., 33 (1982) pp. 203-229.

[6] Cwikel, M.,Milman, M., Sagher, Y., Complex interpolation of some quasi-Banach spaces, preprint.

[7] Hernandez, E., Intermediate spaces and the complex method of interpolation for families of Banach spaces, to appear in the Annali Scuola Normale Pisa.

[8] Riviere, N., Interpolation theory in s-Banach spaces, Ph.D. Dissertation, Univ of Chicago (1966).

[9] Rochberg, R., A generalization of Szego's Theorem and the Power Theorem for complex interpolation, Math. Nach., to appear.

[10]Rochberg, R., Weiss, Guido, Complex interpolation of subspaces of Banach spaces, Suppl. Rend. Circ. Mat. Palermo (1981) pp. 179-186.

[11]——————————————, Analytic families of Banach spaces and some of their uses, Recent progress in Fourier series, Peral and Rubio de Francia, editors, North-Holland (1985) pp. 173-202.

[12] Tabacco Vignati, Anita, Complex interpolation for families of quasi Banach spaces, to appear in the Indiana J. of Math.

[13]——————————————, Topics in interpolation theory of quasi Banach spaces, Ph.D. Dissertation, Wasshington University St. Louis (1986)

NEW AND OLD FUNCTION SPACES

Hans Wallin
Department of Mathematics
University of Umeå
S-901 87 Umeå, Sweden

0. Introduction

In this survey we concentrate on Lipschitz spaces Λ_α and Lip_α (Section 1) and their natural analogues in L^p, Besov spaces $B_\alpha^{p,q}$ (Section 2), but we also consider function spaces defined by means of maximal functions (Section 3). We shall study these function spaces on R^n or on a subset of R^n and throughout we describe the spaces by the possibility to approximate their functions locally by polynomials of a given degree. This approach is connected to the names John-Nirenberg, Campanato, Stampacchia, Meyers, Morrey, Whitney, Brudnyi, Stein, Taibleson, G. Weiss, Krantz, Janson, Dynkin [11], DeVore-Sharpley, Greenwald, Bojarski, and many others. From Lund we could mention Peetre, Spanne, and Grevholm and from Umeå Jonsson, Sjögren (now in Göteborg), Ödlund and the author. In this survey some recent developments are discussed with emphasis on the interests of the research group on function spaces in Umeå. I shall not attempt to give detailed history and references will often be given to monographs rather than to original papers. Two main references for this paper are [19] and [9]; a good introduction to different techniques in Lipschitz spaces is [21]. A lot of further references relevant to this survey are found in [19], [9] and [21].

1. Lipschitz (Hölder) spaces

1.1. Lipschitz spaces on R^n. Let us recall the straightforward definition of $\mathrm{Lip}_\alpha(R^n)$ and $\Lambda_\alpha(R^n)$, $\alpha>0$ (α is the smoothness index). We let $[\alpha]$ denote the integer part of α and (α) the largest integer which is strictly smaller than α; Δ_h denotes the first difference with step h, $\Delta_h g(x) = g(x+h)-g(x)$; $j=(j_1,\dots,j_n)$ denotes a multiindex, D^j the corresponding partial derivative, and $|j|=j_1+\dots+j_n$.

DEFINITION 1 (Non-homogeneous Lipschitz spaces).

(a) $f \in \mathrm{Lip}_\alpha(R^n) \Leftrightarrow$

$f \in C^{(\alpha)}(R^n)$ and, for some constant M,

$$|D^j f(x)| \leq M, \text{ for } |j| \leq (\alpha),\ x \in R^n, \tag{1}$$

and

$$|\Delta_h D^j f(x)| \leq M|h|^{\alpha-(\alpha)}, \text{ for } |j|=(\alpha),\ x,h \in R^n. \tag{2}$$

We introduce the norm $\|f; \mathrm{Lip}_\alpha\| = \inf M$.

(b) $f \in \Lambda_\alpha(R^n)$: As in (a) if α is not an integer but change Δ_h to $\Delta_h^2 = \Delta_h\Delta_h$ if α is an integer.

DEFINITION 2 (Homogeneous Lipschitz spaces).

(a) $f \in \dot{\mathrm{Lip}}_\alpha(R^n)$: Omit (1) in Definition 1, (a).

(b) $f \in \dot{\Lambda}_\alpha(R^n)$: Omit (1) in Definition 1, (b).

<u>Properties</u>: 1) $\mathrm{Lip}_\alpha(R^n) \to \Lambda_\alpha(R^n)$ (continuous imbedding).

2) $\mathrm{Lip}_\alpha(R^n) \neq \Lambda_\alpha(R^n)$ if and only if α is an integer.

Zygmund (1945) discovered that in many applications Λ_α is the correct space rather than Lip_α ; as an example, this is true when we want to measure the degree of approximation of periodic functions by trigonometric polynomials of a given degree (Jackson-Bernstein-Zygmund (α integer)). Also, in the theory of function spaces, Λ_α occurs more naturally than Lip_α ; for instance, if you interpolate between C^k-spaces you get Λ_α-spaces.

1.2. <u>Mean oscillation and local polynomial approximation on R^n</u>. There are many others, equivalent definitions of Lipschitz spaces. We shall discuss only one (Theorem 2), based on local approximation by polynomials. This is often easier to work with in applications to partial differential equations and harmonic analysis (for instance Hardy spaces). Historically, BMO corresponding to the limit case $\alpha=0$ appeared a little earlier than the Campanato spaces - here denoted by $\Lambda_{\alpha;\rho}$ - corresponding to the case $\alpha>0$. By Q we denote different n-dimensional cubes in R^n with sides parallel to the coordinate axes and by $|Q|$ their Lebesgue measure.

DEFINITION 3 (Functions of bounded mean oscillation; John-Nirenberg 1961; see [24]). $f \in BMO(R^n) \Longleftrightarrow$

$f \in L^1_{loc}(R^n)$ and for every cube Q there exists a constant c_Q such that

$$\frac{1}{|Q|}\int_Q |f-c_Q|dx \le M,$$

with a constant $M=M(f)$ independent of Q.

One striking result involving BMO is the first part of the following theorem where $H^p(R^n)$ stands for Hardy spaces in R^n. We use the notation P_k for the set of all polynomials in n real variables of total degree of most k,

THEOREM 1.

(a) $(H^1(R^n))' = BMO(R^n)|P_0$, i.e. the dual of H^1 is BMO modulo constants (C. Fefferman, 1971).

(b) $(H^p(R^n))' = \dot{\Lambda}_\alpha(R^n)|P_{[\alpha]}$, where $\alpha=n(\frac{1}{p}-1)$, $0<p<1$, (T. Walsh, 1973).

DEFINITTION 4 (Mean oscillation spaces, local approximation spaces, Campanato spaces; see [6], [22]). $f \in \Lambda_{\alpha;\rho}(R^n)$, where $\alpha>0$, $1\le\rho\le\infty$ or $\alpha=0$, $1\le\rho<\infty \Longleftrightarrow$

$f \in L^\rho_{loc}(R^n)$ and, for some constant $M=M(f)$, for every cube Q with side δ there exists a polynomial $P_Q \in P_{[\alpha]}$ such that

$$\left\{\frac{1}{|Q|}\int_Q |f-P_Q|^\rho dx\right\}^{1/\rho} \le M\delta^\alpha, \text{ if } 0<\delta\le 1, \tag{3}$$

and

$$\left\{\frac{1}{|Q|}\int_Q |f|^\rho dx\right\}^{1/\rho} \le M, \text{ if } \delta=1. \tag{4}$$

We put $||f;\Lambda_{\alpha;\rho}|| = \inf M$.

Observe that (4) implies (4) for $\delta>1$ which implies (3) for $\delta>1$ with $P_Q=0$. We also note that in the case $\alpha>0$, $\rho=\infty$, (3) and (4) mean

$$\|f-P_Q\|_{L^\infty(Q)} = M\delta^\alpha \ , \text{ if } \ 0<\delta\le 1,$$

and

$$\|f\|_{L^\infty(Q)} \le M \ , \text{ if } \ \delta=1,$$

respectively. In analogy with Definition 4, $\dot{\Lambda}_{\alpha;\rho}(R^n)$ is defined by requiring (3) for $\delta>0$ and omitting (4), and the spaces $Lip_{\alpha;\rho}(R^n)$ and $\dot{Lip}_{\alpha;\rho}(R^n)$ are defined by changing $P_{[\alpha]}$ to $P_{(\alpha)}$. As an example, we observe that $\dot{\Lambda}_{0;1}(R^n) = BMO(R^n)$ and in analogy we denote $\Lambda_{0;1}(R^n)$ by $bmo(R^n)$. If in Definition 4 we take $\alpha<0$ we get Morrey spaces (see [22]). We also note that a notion of differentiability related to Definition 4 has been introduced by Calderon-Zygmund ([5], §1).

THEOREM 2.

(a) <u>If</u> $\alpha>0$, $1\le\rho\le\infty$, <u>then</u> $\Lambda_{\alpha;\rho}(R^n) = \Lambda_\alpha(R^n)$ <u>and</u> $\dot{\Lambda}_{\alpha;\rho}(R^n) = \dot{\Lambda}_\alpha(R^n)$ (essentially Campanato and Meyers, 1964).

(b) If $\alpha=0$, $1\le\rho<\infty$, <u>then</u> $\Lambda_{0;\rho}(R^n) = bmo(R^n)$ <u>and</u> $\dot{\Lambda}_{0;\rho}(R^n) = BMO(R^n)$ (essentially John-Nirenberg, 1961).

An analogous theorem holds with Λ_α changed to Lip_α. We refer to [6], [24], and [19] for proofs (see also [12]). Here we just indicate some vital ingredients in the proof.

<u>Sketch of parts of the proof</u>. 1) Taylor's formula gives: If $f \in Lip_\alpha(R^n)$, then $f \in Lip_{\alpha,\infty}(R^n)$, i.e. there exists, for any cube Q with side δ, $P_Q \in P_{(\alpha)}$ such that $|f(x)-P_Q(x)| \le M\delta^\alpha$ for $x \in Q$. If $f \in \Lambda_\alpha(R^n)$ you have to work a little for the analogous result. 2) The main part of the proof depends on the inequalities

$$(M_1) \quad \|\nabla P\|_{L^\infty(Q)} \le \frac{c}{\delta}\|P\|_{L^\infty(Q)}, \quad \text{(Markov's inequality)}$$

and, for $1\le\rho\le\infty$,

$$(M_2) \quad \|P\|_{L^\infty(Q)} \le c\left\{\frac{1}{|Q|}\int_Q |P|^\rho dx\right\}^{1/\rho}, \quad \text{(Gagliardo)}$$

for all cubes Q with side δ, all $P \in P_k$, and all positive integers k; $c=c(k)$ is a constant depending only on k.

1.3. Local approximation (mean oscillation) spaces on special sets F. The definitions in Section 1.2 may in a natural way be generalized to the case when the functions are defined on a closed subset F of R^n. Instead of using the Lebesgue measure on R^n we just need a suitable measure μ on F. Let $F \subset R^n$ be closed and μ a positive Borel measure with support F satisfying the following doubling condition where $Q(x,\delta)$ denotes the cube with center x and side δ, and the doubling constant c is independent of x and δ,

$$\mu(Q(x,2\delta)) \leq c\mu(Q(x,\delta)), \text{ for } x \in F, \ 0<\delta\leq 1.$$

It is a recent observation by Volberg-Konyagin [27] that μ exists.

DEFINITION 5. $f \in \Lambda_{\alpha;\rho}(F)$, where $\alpha>0$, $1\leq\rho\leq\infty$ or $\alpha=0$, $1\leq\rho<\infty \iff$ $f \in L^{\rho}_{loc}(\mu)$ and, for some constant $M=M(f)$, for every cube Q with center in F and side δ there exists a polynomial $P_Q \in \mathcal{P}_{[\alpha]}$ such that

$$\left\{\frac{1}{\mu(Q)} \int_Q |f-P_Q|^{\rho} d\mu\right\}^{1/\rho} \leq M\delta^{\alpha}, \quad \text{if} \quad 0<\delta\leq 1, \tag{3'}$$

and

$$\left\{\frac{1}{\mu(Q)} \int_Q |f|^{\rho} d\mu\right\}^{1/\rho} \leq M, \quad \text{if} \quad \delta=1. \tag{4'}$$

As usual, we put $||f; \Lambda_{\alpha;\rho}(F)|| = \inf M$.

Analogously, $\dot{\Lambda}_{\alpha;\rho}(F)$ is defined by requiring (3') for $\delta>0$ and omitting (4'), and by changing $\mathcal{P}_{[\alpha]}$ to $\mathcal{P}_{(\alpha)}$ we get the $Lip_{\alpha;\rho}$-spaces instead of the $\Lambda_{\alpha;\rho}$-spaces. The spaces defined depend on μ. This is avoided if we assume that F and μ have the special properties described in the following definition which may be compared to the notion of s-sets in the theory of fractal sets.

DEFINITION 6. F is a <u>d-set</u> ($0<d\leq n$, d not necessarily an integer) and μ a <u>d-measure on F</u> if $\text{supp}\, \mu=F$ and, for some constants $c_1, c_2>0$,

$$c_1\delta^d \leq \mu(Q(x,\delta)) \leq c_2\delta^d, \text{ for } x \in F, \ 0<\delta\leq 1. \tag{5}$$

If we require (5) for all $\delta>0$ we get the concepts of global d-set and global d-measure on F.

Any d-measure on F is equivalent to the restriction to F of the d-dimensional Hausdorff measure in R^n (see [19] or [16] where d-measures were introduced). The spaces in Definition 5 are, consequently, independent of μ if F is a d-set and μ a d-measure on F (different measures give equivalent norms).

Examples. 1) R^n is a global n-set and the Lebesgue measure in R^n a global n-measure in R^n.

2) If $F=\overline{D}$, where $D \subset R^n$ is a domain with boundary ∂D locally in Lip_1, then F is an n-set and ∂D an (n-1)-set.

3) If $F \subset R^1$ is the ordinary Cantor set, then F is a d-set with $d=\log 2/\log 3$.

In order to be able to extend Theorem 2 from R^n to F we assume that F is such that the analogues of (M_1) and (M_2) hold on F in the following sense.

DEFINITION 7. F preserves Markov's inequality if F is such that

$$(M_1') \qquad \max_{F\cap Q}|\nabla P| \leq \frac{c}{\delta} \max_{F\cap Q}|P|,$$

and, for $1\leq p\leq\infty$,

$$(M_2') \qquad \max_{F\cap Q}|P| \leq c \left\{\frac{1}{\mu(Q)} \int_Q |P|^p d\mu\right\}^{1/p},$$

for all $Q=Q(x,\delta)$, $x \in F$, $0<\delta\leq 1$, all $P \in P_k$, and all positive integers k; c is a constant depending only on k, n, F, and the doubling constant of μ. F preserves the global Markov inequality if (M_1') and (M_2') hold for all $\delta>0$.

For Definition 7 and the following remarks we refer to [19], or [18] and [15].

Remarks. 1) $(M_1') \Rightarrow (M_2')$.

2) $(M_1') \Longleftrightarrow \max_Q |P| \leq c \max_{F\cap Q}|P|$ for all Q and P as in Definition 7; c is independent of Q and P.

3) If $f \subset R^n$ is a d-set with $d > n-1$, then F preserves Markov's inequality.

We can now generalize Theorem 2 to spaces of functions on F by defining Lipschitz spaces on F, $\Lambda_\alpha(F)$, as the trace to F of Lipschitz spaces in R^n. Analogous generalizations hold for $\dot{\Lambda}_\alpha$, Lip_α and $\dot{\mathrm{Lip}}_\alpha$.

DEFINITION 8. $\Lambda_\alpha(F) = \Lambda_\alpha(R^n)|F = \{f\colon f=g|F, g \in \Lambda_\alpha(R^n)\}$.

THEOREM 3. *Assume that* $F \subset R^n$ *is a closed d-set preserving Markov's inequality. Then*

(a) $\Lambda_{\alpha;\rho}(F) = \Lambda_\alpha(F)$, *for* $\alpha>0$, $1\leq\rho\leq\infty$

and

(b) $\dot{\Lambda}_{0;\rho}(F)$ *is independent of* ρ, *for* $1\leq\rho<\infty$

(and denoted $\mathrm{bmo}(F)$).

The theorem consists of two parts: Firstly, that the space $\Lambda_{\alpha;\rho}(F)$ is independent of ρ (see Jonsson-Sjögren-Wallin [15] or [19]), and secondly the identification of this space for $\alpha>0$ as the trace to F of $\Lambda_\alpha(R^n)$ (see [19] and [18], Th.1.3).

Application (See [15] or [19]). One can define Hardy spaces on F, $H^p(F)$, by means of atoms and prove a duality theorem generalizing the one in R^n given in Theorem 1.

THEOREM 4. *Assume that* F *is a global d-set preserving the global Markov inequality. Then*

(a) $(H^1(F))' = \mathrm{BMO}(F)|P_0$ (by definition $\mathrm{BMO}(F) = \dot{\Lambda}_{0;1}(F)$)
and
(b) $(H^p(F))' = \dot{\Lambda}_\alpha(F)|P_{[\alpha]}$, *where* $\alpha = d(\frac{1}{p} - 1)$, $0<p<1$.

1.4. *The general case.* It is possible to describe Lipschitz spaces on any closed set F by means of local polynomial approximation. It is now convenient to work with jets (families of functions).

DEFINITION 9. $\Lambda_\alpha(F) = \Lambda_\alpha(R^n)|F = \{\{f^{(j)}\}_{|j|\leq(\alpha)}\colon f^{(j)} = (D^j g)|F,$
$g \in \Lambda_\alpha(R^n)\}$.

If F preserves Markov's inequality it may be proved that $f^{(j)}$, $0<|j|\leq(\alpha)$, are uniquely determined by $f^{(0)}$ and $\Lambda_\alpha(F)$ as just defined may be identified with $\Lambda_\alpha(F)$ as defined in Section 1.3. Now we have (see Jonsson-Wallin [19] or [17], [18]):

THEOREM 5. Let $F \subset R^n$ be any closed set and $\alpha>0$. Then $\{f^{(j)}\}_{|j|\leq(\alpha)} \in \Lambda_\alpha(F)$ if and only if $f^{(j)}$, $|j|\leq(\alpha)$, are defined on F and, for some constant M, for every cube Q with side $\delta>0$ intersecting F there exists a polynomial $P_Q \in \mathcal{P}_{[\alpha]}$ such that

(a) $|f^{(j)}(x) - D^jP_Q(x)| \leq M\delta^{\alpha-|j|}$, for $|j|\leq(\alpha)$, $x \in Q\cap F$,

(b) $|P_Q(x)| \leq M$, for $x \in Q$, $\delta=1$,

and

(c) if $Q'\subset Q$ is a cube with side $\delta'\geq\delta/2$ intersecting F, then

$$|P_Q(x) - P_{Q'}(x)| \leq M\delta^\alpha, \quad \text{for } x \in Q'.$$

We remark that Condition (c) may not in general be omitted and that we get the same space even if we use polynomials of degree larger than $[\alpha]$. By using instead $P_Q \in \mathcal{P}_{(\alpha)}$ we get an analogous description of

$$\mathrm{Lip}_\alpha(F) = \{\{f^{(j)}\}_{|j|\leq(\alpha)}: f^{(j)} = (D^jg)|F,\ g \in \mathrm{Lip}_\alpha(R^n)\}.$$

In this case (c) follows from (a) and (b). Whitney described $\mathrm{Lip}_\alpha(F)$ already in 1934 using formal Taylor expansions. When α is an integer, $\Lambda_\alpha(F) \neq \mathrm{Lip}_\alpha(F)$ if and only if $\inf\{|x-y|: x,y \in F, x\neq y\}=0$ (Wingren).

Open problem 1. (a) Interpolate between $\Lambda_\alpha(F)$ and $\Lambda_\beta(F)$ (and the same problem for $\mathrm{Lip}_\alpha(F)$). A complication is that the elements in the spaces are jets. (b) The same interpolation problem for $\{f: f=g|F, g \in \Lambda_\alpha(R^n)\}$.

Open problem 2. Show that there exists a linear, bounded extension operator

$$\{f: f=g|F,\ g \in \Lambda_\alpha(R^n)\} \to \Lambda_\alpha(R^n).$$

The same problem for Lip_α is an old problem among those who work with ideals of differentiable functions. We make some comments and refer to [4] for further references and discussion. Problem 2 has been solved in certain special cases: 1) if $n=1$ (Jonsson, Shvartsman, Dzyadyk, Shevchuk); 2) if $\alpha<2$ (Brudnyi-Shvartsman [4]). 3) if F preserves Markov's inequality ([28], [14], [29]). A solution of a variant of Problem 2 would reduce Problem 1, (b) to a problem in R^n where the solution is known. It is also unknown if there exists a linear continuous extension for the jet case $\Lambda_\alpha(F) \to \Lambda_\alpha(R^n)$ when α is an integer. For the jets Lip_α it is known to exist (Whitney).

1.5. Local approximation over oval bodies. The cubes Q (or balls) in the local polynomial approximation may be replaced by oval bodies; for instance ellipsoids with axes δ and δ^2. This is of importance in the theory of several complex variables and pseudo-differential operators [20], [23]. The balls are adapted to the singularity of the Poisson kernel for a ball in R^n but oval bodies to the singularity of the Szegö kernel. This approach to Lipschitz spaces may be used to identify the duals of Hardy spaces on the ball in $\mathbb{C}^n$ (Kranz [20]) and to study the boundary behavior of holomorphic functions on general domains in $\mathbb{C}^n$. It is an open problem to investigate the general effect of the shape of the oval bodies.

2. Integrated Lipschitz spaces (Besov or Nikol'skii-Besov spaces), $B_\alpha^{p,q}$

2.1. Besov spaces on R^n. The Besov spaces in L^p with smoothness index α and second-order smoothness q may be defined as follows where $\|\ \|_p$ is the $L^p(R^n)$-norm.

DEFINITION 10. $f \in B_\alpha^{p,q}(R^n)$, $\alpha>0$, $1\leq p,q\leq\infty \iff$

$$\|f;B_\alpha^{p,q}\| = \sum_{|j|\leq(\alpha)} \|D^j f\|_p + \sum_{|j|=(\alpha)} \left\{ \int_{R^n} \frac{\|\Delta_h D^j f\|_p^q}{|h|^{n+(\alpha-(\alpha))q}}\, dh \right\}^{1/q} < \infty,$$

if α is not an integer; if α is an integer Δ_h shall be replaced by Δ_h^2.

We note that $B_\alpha^{\infty,\infty}(R^n) = \Lambda_\alpha(R^n)$. There is also an analogue of $\dot{\Lambda}_\alpha$ which we denote by $\dot{B}_\alpha^{p,q}(R^n)$, and there are analogues of the Lip_α-spaces.

2.2. Mean oscillation and local polynomial approximation on R^n. Since 1980 various characterizations of Besov spaces in R^n have been given based on mean oscillation and local polynomial approximation giving Campanato type theorems. We shall state two of these which are, of course, closely related.

2.2.1. Mean oscillation (Dorronsoro 1985 [10]; see also Ricci-Taibleson [25]). For $f \in L^1_{loc}(R^n)$ and any cube Q in R^n there exists a unique $P_Q f \in \mathcal{P}_{[\alpha]}$ such that

$$\int_Q (f-P_Q f)x^j dx = 0, \quad \text{for} \quad 0 \le |j| \le [\alpha].$$

Put

$$\Omega_f(x,\delta) = \sup_Q \frac{1}{|Q|} \int_Q |f-P_Q f| dy,$$

where sup is over all Q containing x and having side δ.

THEOREM 6. $f \in \dot{B}_\alpha^{p,q}(R^n) \Leftrightarrow$

$$\left\{ \int_0^\infty (\delta^{-\alpha} ||\Omega_f(\cdot,\delta)||_p)^q \frac{d\delta}{\delta} \right\}^{1/q} < \infty .$$

The last expression gives an equivalent seminorm in $\dot{B}_\alpha^{p,q}(R^n)$. [10] also contains a result where the integrand in the definition of $\Omega_f(x,\delta)$ is raised to a power ρ.

2.2.2. Local polynomial approximation. Another generalization from the case $p=q=\infty$ based on local polynomial approximation is the following theorem by Jonsson-Wallin from 1980 ([18], [19]). Similar results have been stated by Brudnyi 1976 [3] and DeVore-Popov [8]. Let π be a net in R^n of cubes with mesh δ, i.e. a division of R^n into cubes Q with side δ obtained by hyperplanes orthogonal to the axes. We introduce the space $\mathcal{P}_{[\alpha]}(\pi)$ of piecewise polynomials of degree at most $[\alpha]$ associated with the net π,

$$P_{[\alpha]}(\pi) = \{n(\pi):\ n(\pi)|Q \in P_{[\alpha]},\ \text{for every}\ \ Q \in \pi\},$$

and state the theorem in discretized form $(\delta=2^{-\nu})$.

THEOREM 7. $f \in B_\alpha^{p,q}(R^n) \iff$
$f \in L^p(R^n)$ <u>and there exists</u> $(c_\nu)_0^\infty \in \ell^q$, $c_\nu \geq 0$, <u>such that for every net</u> π <u>with mesh</u> $\delta=2^{-\nu}$ $(\nu=0,1,2,\ldots)$ <u>there exists</u> $n(\pi) \in P_{[\alpha]}(\pi)$ <u>satisfying</u>

$$||f-n(\pi)||_{L^p(R^n)} \leq 2^{-\nu\alpha}c_\nu .$$

<u>The original norm in</u> $B_\alpha^{p,q}(R^n)$ <u>is equivalent to the norm given by</u>

$$||f||_{L^p(R^n)} + \inf\left(\sum_\nu c_\nu^q\right)^{1/q}$$

<u>where inf is over all possible</u> $(c_\nu)_0^\infty$.

2.3. <u>Local approximation (mean oscillation) spaces on special sets</u> F. Let $F \subset R^n$ be a closed d-set which preserves Markov's inequality, and μ a d-measure on F.

<u>Sketch of THEOREM</u> (Jonsson-Wallin [18] or [19] for a general d; Brudnyi [3] for $d=n$). The space

$$\{f:\ f=g|F,\ \ g \in B_\alpha^{p,q}(R^n)\}$$

can be described by generalizing the construction indicated in Theorem 7. The smoothness α in R^n gives smoothness $\beta=\alpha-(n-d)/p>0$ on F and the trace space is denoted by $B_\beta^{p,q}(F)$. The measure μ replaces the Lebesgue measure in R^n. An analogous result holds for $B_\alpha^{p,q}(R^n)$ replaced by the space $L_\alpha^p(R^n)$ of Bessel potentials of order α of L^p-functions.

2.4. <u>The case of a general d-set</u>. Let $F \subset R^n$ be any closed d-set and μ a d-measure on F. We shall describe some results which partly go back to 1978 [16]. For a full treatment see [19]. A new proof is presented in [30].

Sketch of THEOREM. Introduce $\beta=\alpha-(n-d)/p>0$. The trace space

$$B_\alpha^{p,q}(R^n)|F = \left\{\{f^{(j)}\}_{|j|\leq(\beta)}: f^{(j)} = (D^j g)|F,\ g \in B_\alpha^{p,q}(R^n)\right\}$$

can be described by piecewise polynomial approximation as in the case $p=q=\infty$ in Section 1.4, and analogously for $L_\alpha^p(R^n)$. As in Section 2.3 the smoothness α in R^n gives smoothness β on F and the trace space is denoted by $B_\beta^{p,q}(F)$. If F preserves Markov's inequality, $f^{(j)}$, $0<|j|\leq(\beta)$, are uniquely determined by $f^{(0)}$.

The result just described generalizes theorems by Nikol'skii (1953,$q=\infty$) - Besov (1959) - Taibleson (1964), Stein (1961), and others from $F=R^d$ to the more general sets F used here.

Application. Let $F=M$ be a submanifold of R^n of dimension d (d integer) of class C^γ, $\gamma>\beta=\alpha-(n-d)/p>0$. Jonsson [13] used the above results to describe in a more constructive way the space

$$\{f : f=g|M,\ g \in B_\alpha^{p,q}(R^n)\}.$$

He obtains somewhat more precise results then Besov-Il'in-Nikol'skii [1] who based their study on an integral representation of Sobolev type. Other treatments of Besov spaces on manifolds are given by Triebel [26] and Ciesielski-Figiel [7].

The theory above works also if F is not closed or if we work on $D \subset R^n$. It gives, for instance, results of the following type ([19], Ch. VIII): Let $D \subset R^n$ be open, α a positive integer and $W_\alpha^p(D)$ a Sobolev space in D. Then, for a general class of sets D,

$$W_\alpha^p(D)|\partial D = B_\beta^{p,p}(\partial D), \quad \text{if} \quad \beta=\alpha-1/p,\ 1<p<\infty.$$

3. Maximal functions measuring smoothness (DeVore-Sharpley spaces)

Let $D \subset R^n$ be open and $\alpha\geq0$ and introduce the sharp approximate maximal function

$$f_\alpha^\#(x) = \sup_Q \inf_{P\in P_{[\alpha]}} \frac{1}{|Q|^{\alpha/n}} \left(\frac{1}{|Q|}\int_Q |f-P|dy\right),$$

where sup is over all Q such that $x \in Q \subset D$. If $P_{[\alpha]}$ is replaced by $P_{(\alpha)}$ we get instead the <u>flat aproximate maximal function</u> $f_\alpha^b(x)$. As an example, we get $BMO(R^n)$ if $D=R^n$ by the condition $f_0^{\#} \in L^\infty(R^n)$. In 1984, DeVore-Sharpley [9] introduced the following spaces (see also [2]):

DEFINITION 11. $f \in C_\alpha^p(D)$, $\alpha>0$, $1\leq p\leq\infty$ $\iff$

$f \in L^p(D)$ and $f_\alpha^{\#} \in L^p(D)$. Introduce

$$||f||_{C_\alpha^p(D)} = ||f||_{L^p(D)} + ||f_\alpha^{\#}||_{L^p(D)}.$$

They also considered the analogous space for the maximal function f_α^b, the case $0<p<1$, and the case where the integrand is raised to ρ in the definition of the maximal functions. They proved:

THEOREM 8 [9]. (a) $C_\alpha^\infty(R^n) = \Lambda_\alpha(R^n)$.

(b) $B_\alpha^{p,p}(R^n) \to C_\alpha^p(R^n) \to B_\alpha^{p,\infty}(R^n)$.

(c) $L_\alpha^p(R^n) \to C_\alpha^p(R^n)$.

(d) $C_\alpha^p(R^n)$, $1\leq p<\infty$, <u>is neither a Besov nor a potential space</u>.

(e) $(C_\alpha^p(R^n), C_\alpha^q(R^n))_{\Theta,r} = C_\alpha^r(R^n)$, $\frac{1}{r} = \frac{1-\Theta}{p} + \frac{\Theta}{q}$, $0<\Theta<1$.

(f) $C_\alpha^p(R^n) \to C_\beta^q(R^n)$, $1\leq p\leq q\leq\infty$, $0<\beta\leq\alpha+n\left(\frac{1}{q} - \frac{1}{p}\right)$.

(g) <u>There exists a bounded linear extension operator</u> $C_\alpha^p(D) \to D_\alpha^p(R^n)$, <u>if</u> $D \subset R^n$ <u>is a domain whose boundary is locally in</u> Lip_1.

We remark that (a) follows from Theorem 2 and that (g) permits the generalization of (a)-(f) from R^n to D with $\partial D \in Lip_{1,loc}$. Many of the results by DeVore-Sharpley may be generalized from D to F preserving Markov's inequality, at least if F is also a d-set.

REFERENCES

[1] O.V. Besov, V.P. Il'in, and S.M. Nikol'skii, Integral Representations of Functions and Imbedding Theorems (2 volumes), John Wiley and Sons, New York, 1978.

[2] B. Bojarski, Sharp maximal operator of fractional order and Sobolev imbedding inequalities, Bull. Polish Acad. Sc., Math. 33, No. 1-2 (1985), 7-16.

[3] Ju.A. Brudnyi, Piecewise polynomial approximation, imbedding theorems and rational approximation, Lecture Notes in Math. 556, Springer-Verlag, Berlin, 1976, 73-98.

[4] Ju.A. Brudnyi and P.A. Shvartsman, A linear extension operator for a space of smooth functions defined on a closed subset of R^n, Dokl. Akad, Nauk 280. No. 2 (1985); English transl. in Soviet Math. Dokl. 31 (1985), 48-51. (See also Lecture Notes in Math. 1043, Springer-Verlag, Berlin, 1984, 583-585.)

[5] A.P. Calderon and A. Zygmund, Local properties of solutions of elliptic partial differential equations, Studia Math. 20 (1961) 171-225.

[6] S. Campanato, Proprietà di una famiglia di spazi funzionali, Ann. Sc. Norm. Sup. Pisa 18 (1964), 137-160.

[7] Z. Ciesielski and T. Figiel, Spline bases in classical function spaces on compact C^∞ manifolds, Part I and II, Studia Math. 76 (1983), 1-58 and 95-136.

[8] R.A. DeVore and V. Popov, Interpolation of Besov spaces (under preparation).

[9] R.A. DeVore and R.C. Sharpley, Maximal Functions Measuring Smoothness, Amer. Math. Soc., Providence, R.I., 1984.

[10] J.R. Dorronsoro, Mean oscillation and Besov spaces, Canad. Math. Bull. 28 (1985), 474-480.

[11] E.M. Dynkin, A constructive characterization of the classes of S.L. Sobolev and O.V. Besov, Proc. Steklov Inst. of Math., 1, 1983, Amer. Math. Soc., Providence, RI, 39-74.

[12] S. Janson, M. Taibleson, and G. Weiss, Elementary Characterization of the Morrey-Camapanto spaces, Lecture Notes in Math. 992, Springer-Verlag, Berlin, 1983, 101-114.

[13] A.Jonsson, Besov spaces on manifolds in R^n, Dept. of Math., Univ. of Umeå, No. 3 (1985).

[14] A. Jonsson, Markov's inequality and local polynomial approximation, These proceedings.

[15] A. Jonsson, P. Sjögren, and H. Wallin, Hardy and Lipschitz spaces on subsets of R^n, Studia Math. 80 (1984), 141-166.

[16] A. Jonsson and H. Wallin, A Whitney extension theorem in L^p and Besov spaces, Ann. Inst. Fourier 28 (1978), 139-192.

[17] A. Jonsson and H. Wallin, The trace to closed sets of functions in R^n with second difference of order $O(h)$, J. Approx. Theory 26 (1979), 159-184.

[18] A. Jonsson and H. Wallin, Local polynomial approximation and Lipschitz type conditions on general closed sets, Dept. of Math., Univ. of Umeå, No. 1 (1980); partly published as Local pol. approx. and Lipschitz functions on closed sets, Proc. Conf. Constructive Function Theory, Varna 1981, Sofia 1984, 368-375.

[19] A. Jonsson and H. Wallin, Function spaces on subsets of R^n, Mathematical Reports 2, Part 1, Harwood Academic Publ., New York, 1984.

[20] S.G. Krantz, Geometric Lipschitz spaces and applications to complex function theory and nilpotent groups, J. Funct. Anal. 34 (1979), 456-471.

[21] S.G. Krantz, Lipschitz spaces, smoothness of functions, and approximation theory, Expo. Math. 3 (1983), 193-260.

[22] A. Kufner, O. John, and S. Fučik, Function spaces, Noordhoff Inter. Publ., Leyden, 1977.

[23] A. Nagel and E.M. Stein, Lectures on Pseudo-Differential Operators, Princeton University Press, Princeton, 1979.

[24] H.M. Reimann and T. Rychener, Funktionen beschränkter mittlerer Oszillation, Lecture Notes in Math. 487, Springer-Verlag, Berlin, 1975.

[25] F. Ricci and M. Taibleson, Boundary values of harmonic functions in mixed norm spaces and their atomic structure, Ann. Sc. Norm. Sup. Pisa, (4), 10 (1983), 1-54.

[26] H. Triebel, Theory of Function Spaces, Birkhäuser, Basel, 1983.

[27] A.L. Volberg and S.V. Konyagin, There is a homogeneous measure on any compact subset in R^n, Dokl. Akad. Nauk 278, No. 4 (1984); English transl. in Soviet Math. Dokl. 30 (1984), 453-456.

[28] H. Wallin, Markov's inequality on subsets of R^n, Canad. Math. Soc., Conf. Proc. 3, Amer. Math. Soc., Providence, R.I. (1983), 377-388.

[29] P. Wingren, Lipschitz spaces and interpolating polynomials on subsets of Euclidean space, These proceedings.

[30] L. Ödlund, A Whitney type extension operator on Besov spaces, Dept. of Math., Univ. of Umeå, No. 9 (1986).

A NOTE ON CHOQUET INTEGRALS WITH RESPECT TO HAUSDORFF CAPACITY

David R. Adams
Department of Mathematics
University of Kentucky
Lexington, Ky 40506
U.S.A

0. INTRODUCTION.

Working on Euclidean n-dimensional space $\mathbb{R}^n$, we prove

Theorem A: There is a constant c depending only on α and n, $0<\alpha<n$, such that for all $\phi \in C_0(\mathbb{R}^n)$

$$\int M_0(\phi)\, dH_\infty^{n-\alpha} \leq C \int |\phi|\, dH_\infty^{n-\alpha} . \tag{*}$$

Here M_0 is the Hardy-Littlewood maximal function, $H_\infty^{n-\alpha}$ is the Hausdorff capacity of dimension $n-\alpha$ (see section 1), and the integral is to be understood in the sense of Choquet (see section 2). Inequality (*) is a "strong type" version of estimates found in [10] and [17] - which basically correspond to the case $\alpha = 0$ here. Also, (*) can be considered as an L^1-version of estimates of a similar type for L^p-integrals involving Bessel and Riesz L^p-capacities $B_{\alpha,p}$ and $R_{\alpha,p}$, $p>1$; see [2], [3], [11]. Theorem A can be extended to include all ϕ belonging to a natural space $L^1(H_\infty^{n-\alpha})$ - those quasi-continuous ϕ for which the right side of (*) is finite - which is the L^1 analogue of the L^p spaces $L^p(R_{\alpha,p})$ etc. in [2] and [3]. From Theorem A it follows that we get

Theorem B: There is a constant c depending only on α and n, $0<\alpha<n$, $\alpha =$ positive integer, such that for all $\phi \in C_0^\infty(\mathbb{R}^n)$

$$\int M_0(\phi)\, dH_\infty^{n-\alpha} \leq c \|\nabla^\alpha \phi\|_1 . \tag{**}$$

Here $\nabla^\alpha \phi$ is the α-th order gradient of ϕ. In the case $\alpha = 1$, (**) extends to the space $BV^1 =$ functions of bounded variation, giving an estimate that seems not to have been observed until now. The techniques involved in proving (*) rely strongly on the use of Riesz potentials of functions in the Hardy space H^1 and liberal uses of the H^1-BMO duality of C. Fefferman; see section 4.

Notation: $B(x,r) = \{y \in \mathbb{R}^n : |x-y|<r\}$ an open ball centered at x of radius r. $L^p, p\geq 1$, denotes the usual Lebesgue spaces with norms $\|\cdot\|_p$. H^1 is the Hardy space of functions in L^1 whose Riesz transforms $R_j f$ also belong to L^1. The norm is

$$\|f\|_{H^1} = \|f\|_1 + \sum_{j=1}^{n} \|R_j f\|_1 .$$

BMO = space of functions of bounded mean oscillation on $\mathbb{R}^n$; the norm is

$$\|U\|_* = \sup_Q |Q|^{-1} \int_Q |U - U_Q|\, dx$$

where the supremum is taken over all "coordinate cubes" in $\mathbb{R}^n$; U_Q = integral average of U over Q, and $|Q|$ denotes the Lebesgue n-measure of Q. The well known H^1-BMO duality is that BMO is the dual of H^1. Also we need the smooth dense class $\mathbf{S}_{00}$ = all $\phi \in \mathbf{S}$, the Schwartz class of rapidly decreasing C^∞ functions, for which the Fourier transform of ϕ has compact support disjoint from the origin. $C_0(\mathbb{R}^n)$, $C_0^\infty(\mathbb{R}^n)$ are respectively, the continuous, the infinitely differentiable, functions with compact support on $\mathbb{R}^n$. $\mathbf{M}$ denotes the "Radon measures" on $\mathbb{R}^n$ (locally finite regular signed Borel measures), $\mathbf{M}^+(K)$ refers to those elements of $\mathbf{M}$ that are non-negative and have their support in K. $\mathbf{L}^1$ are those elements of $\mathbf{M}$ for which the total variation measure $|\mu| = \mu^+ + \mu^-$ is totally finite on $\mathbb{R}^n$. In fact when there is no confusion, we shall write $\|\mu\|_1$ for $|\mu|(\mathbb{R}^n)$. $\mathbf{L}^{1,d}$, $0<d\leq n$, is the Morrey space of elements from $\mathbf{M}$ for which

$$|||\mu||| \equiv \sup_{x;r>0} r^{-d}\,|\mu|(B(x,r)) < \infty .$$

Also, $\operatorname{supp}(\mu)$ is used to denote the support of μ. The symbol "$\sim$" is read "is comparable to" and means that there are two positive finite constants such that the ratio of the two quantities under examination are bound above and below by these constants; $A \sim B$ *iff* $c_1 A \leq B \leq c_2 A$. And the letter c will generally refer to a generic constant, independent of said (or implied) variables.

Finally, the potentials used are those of M. Riesz:

$$I_\alpha * f(x) \equiv \int |x-y|^{\alpha-n} f(y)\,dy$$

with integration over $\mathbb{R}^n$ with respect to Lebesque n-measure, and the maximal functions used are

$$M_\alpha f(x) \equiv \sup_{r>0} r^{\alpha-n} \int_{B(x,r)} |f(y)|\,dy .$$

1. HAUSDORFF CAPACITY.

Let $E \subset \mathbb{R}^n$ and consider the collection of all countable open (closed) covers of E by balls. If $\{B_j\}$, $j=1,2,\ldots$, is such a cover, i.e. $\cup_j B_j \supset E$, and if the radius of the ball B_j is r_j, then we shall use the term, the *Hausdorff capacity* of E (of dimension d, $0<d\leq n$), for the quantity

$$H_\infty^d(E) \equiv \inf \sum_j r_j{}^d \tag{1}$$

where the infimum is over all such covers of E. If we had required the radii of the covering balls in (1) all to satisfy $r_j \leq \epsilon$, for some positive number ϵ, then we would have written $H_\epsilon^d(E)$. The *classical Hausdorff measure* of E (of dimension d) is the limit of $H_\epsilon^d(E)$ as $\epsilon \to 0$. It will be denoted simply by $H^d(E)$. It is well known that H^d is a metric outer measure in the sense of Cartheodory on subsets of finite H^d-measure. However, H^d is generally not finite, whereas H_∞^d is, at least on bounded sets. Moreover, H^d and H_∞^d have the same null sets. Thus H_∞^d is more amenable to estimation and these estimates can then be used to recover "almost everywhere" statements involving Hausdorff measures.

Several of the well known properties of Hausdorff capacities will be needed in what follows. These we will list below. First of all, however, a comment should be made concerning the name we have chosen. Hausdorff capacity is refered to in the literature by various names, none of which have become standard. The present one is based on the fact that H^d_∞ is, in fact, a capacity in the sense of N. Meyers ([14], page 257). This involves checking the simple properties: H^d_∞ is a monotone, countably subadditive set function on the class of all subsets of $\mathbb{R}^n$ which vanishes on the empty set. Furthermore, it is an *outer capacity* in the sense that

$$H^d_\infty(E) = \inf H^d_\infty(G) \tag{2}$$

where the infimum is over all open sets $G \supset E$.

An extremely useful variant of Hausdorff capacity is the dyadic version of H^d_∞, denoted here by $\tilde{H}^d_\infty$. The difference is that the ball-covers have now been replaced by dyadic cube-covers and the sum in (1), by the corresponding sum of the edge lengths of the cubes raised to the d-th power.

Properties of Hausdorff capacities needed include:

P_1: $\tilde{H}^d_\infty \sim H^d_\infty$ (constants depend only on n);

P_2: if $E_j \uparrow E$, then $\lim_j \tilde{H}^d_\infty(E_j) = \tilde{H}^d_\infty(E)$;

P_3: if $K_j \downarrow K$, K_j compact, $\lim_j \tilde{H}^d_\infty(K_j) = \tilde{H}^d_\infty(K)$;

P_4: $\tilde{H}^d_\infty(E_1 \cup E_2) + \tilde{H}^d_\infty(E_1 \cap E_2) \le \tilde{H}^d_\infty(E_1) + \tilde{H}^d_\infty(E_2)$;

P_5: $H^d_\infty(K) \sim \sup\{\|\mu\|_1 : \mu \in \mathbf{M}^+(K)\ \&\ |||\mu|||_d \le 1\}$, K compact.

The proof of P_1 and P_2 can be found in [16] (see chapter 2, section 7). P_3 is trivial and also holds for H^d_∞. Notice that P_2 and P_3 imply that $\tilde{H}^d_\infty$ is a capacity in the sense of Choquet (see [7]) and that all analytic, and hence all Borel sets, are $\tilde{H}^d_\infty$ capacitible. Property P_4 is often referred to as strong subadditivity; the basic argument needed here is contained in [10]. P_5 is O. Frostman's theorem; see [5] page 7.

2. CHOQUET INTEGRALS.

If $C(\cdot)$ is an outer capacity in the sense of N. Meyers, then we define the *Choquet integral* of $\phi \in C_0(\mathbb{R}^n)^+$ (with respect to the capacity C) as the Riemann integral $\int_0^\infty C[\phi > \lambda]\, d\lambda$. We will, however, continue to use the traditional notation, $\int \phi\, dC$, for this integral. Similar interpretations can be given for $\int \phi^p\, dC$, $0 < p < \infty$, and $\int_E \phi\, dC$, $E \subset \mathbb{R}^n$. The space $L^1(C)$ is now defined as the completion of $C_0(\mathbb{R}^n)$ with respect to the functional $\int |\phi|\, dC$. It is now a standard excercise to show that in fact $L^1(C)$ consists of the C-quasi continuous functions u on $\mathbb{R}^n$ for which $\int |u|\, dC < \infty$; i.e. for each $\epsilon > 0$ there is an open set G such that $C(G) < \epsilon$ and u restricted to the complement of G is continuous there.

An important property of the Choquet integral that will be crucial for what follows is the sublinearity of $\int |\phi|\, dC$. The key result in this direction was proved by Choquet [6] and later simplified and refined by Topsøe and Anger [4]. This result states that $\int |\phi|\, dC$ is sublinear (in ϕ) if and only if C is strongly subadditive. Thus due to P_4, we have that $\int |\phi|\, d\tilde{H}^d_\infty$ is sublinear on $C_0(\mathbb{R}^n)$, and with respect to this norm $L^1(\tilde{H}^d_\infty)$ is a Banach space (under the usual identification). Actually, R. Fefferman in [10] gives a direct proof of sublinearity for a Hausdorff capacity (called "entropy") which works in the present setting. And because of P_1, we see that $L^1(H_\infty^d)$ can be considered a quasi-Banach space with respect to the quasi-norm $\int |\phi| \, d\, H_\infty^d$.

3. DUALITY.

Here we identify the dual to $L^1(H_\infty^d)$.

Proposition 1: The dual of the space $L^1(H_\infty^d)$ is the Morrey space $\mathbf{L}^{1,d}$.

proof.

The duality is given by $l(u) = \int u \, d\, \mu$, $u \in L^1(H_\infty^d)$ and $\mu \in \mathbf{L}^{1,d}$. Such an l is clearly well defined since for all μ-measurable E,

$$|\mu|(E) \leq |||\mu||| \cdot H_\infty^d(E) . \tag{3}$$

Hence

$$|l(u)| \leq |||\mu|||_d \cdot \int |u| \, d\, H_\infty^d ,$$

so l is in the dual space and $||l|| \leq |||\mu|||_d$. For the converse, note that since $C_0(\mathbb{R}^n) \subset L^1(H_\infty^d)$, l must be given by: $l(\phi) = \int \phi \, d\, \mu$, $\phi \in C_0(\mathbb{R}^n)$, for some $\mu \in \mathbf{M}$. But then for any $\psi \in C_0(\mathbb{R}^n)$,

$$\begin{aligned} \left|\int \psi \, d\, |\mu|\right| &\leq \int |\psi| \, d\, |\mu| = \sup \left\{\int \phi \, d\, \mu : \phi \in C_0(\mathbb{R}^n) \ \&\ |\phi| \leq |\psi|\right\} \\ &\leq \sup \left\{||l|| \cdot \int |\phi| \, d\, H_\infty^d : \phi \in C_0(\mathbb{R}^n) \ \&\ |\phi| \leq |\psi|\right\} \\ &\leq ||l|| \cdot \int |\psi| \, d\, H_\infty^d . \end{aligned}$$

Thus if $\psi = 1$ on $B(x,r)$ and $\psi = 0$ on $B(x, r+\epsilon)^C$, then

$$|\mu|(B(x,r)) \leq ||l|| \cdot H_\infty^d \left(B(x, r+\epsilon)\right) = ||l|| \cdot (r+\epsilon)^d .$$

Hence $|||\mu|||_d \leq ||l||$. □

Corollary: If $f(x) \geq 0$ is lower semi-continuous on $\mathbb{R}^n$, then

$$\int f \, d\, H_\infty^d \sim \sup \left\{ \int f \, d\, \mu : \mu \in \mathbf{L}_+^{1,d} \ \&\ |||\mu|||_d \leq 1 \right\} . \tag{4}$$

proof. The cononical map of the Banach space $L^1(\tilde{H}_\infty^d)$ into its second dual has norm

$$\int |u| \, d\, \tilde{H}_\infty^d = \sup \left\{\left|\int u \, d\, \mu\right| : |||\mu|||_d \leq 1\right\}$$

for $u \in L^1(\tilde{H}_\infty^d)$. For a non-negative lower semi-continuous f, we approximate from below by a sequence $\{\phi_j\} \subset C_0(\mathbb{R}^n)^+$, $\phi_j \uparrow f$ as $j \to \infty$. Then

$$\int \phi_j \, d\, \tilde{H}_\infty^d = \sup_\mu \int \phi_j \, d\, \mu \leq \sup_\mu \int f \, d\, \mu .$$

From P_2, we get $\tilde{H}_\infty^d [\phi_j \geq \lambda] \to \tilde{H}_\infty^d [f > \lambda]$, hence

$$\int f \, d\, \tilde{H}_\infty^d \leq \sup_\mu \int f \, d\, \mu .$$

The result now follows from P_1 and estimate (3). □

4. POTENTIAL THEORY.

In this section, we want to investigate a capacity type functional that is naturally associated with the space of M. Riesz potentials of the Hardy H^1 functions. Because of technicalities, we actually look only at Riesz potentials of the dense class $\mathbf{S}_{00}$. The pointwise definition of $I_\alpha * f$, $f \in H^1$, is treated later on. The functional is

$$R_\alpha(\phi) = \inf\{\|f\|_{H^1}: f \in \mathbf{S}_{00} \ \& \ I_\alpha * f \geq \phi \text{ on supp } (\phi)\},$$

and its dual functional

$$r_\alpha(\phi) = \sup\{\textstyle\int \phi \, d\mu : \mu \in \mathbf{M}^+(\text{supp}(\phi)) \ \& \ \| I_\alpha * \mu \|_* \leq 1\}.$$

Here ϕ is a non-negative bounded function with compact support. All such functions will be denoted by $\mathbf{O}$. If $\phi = 1_K$, the characteristic function of the compact set K, then we will write $R_\alpha(K)$ and $r_\alpha(K)$ instead of $R_\alpha(1_K)$ etc.. We show

Proposition 2: Suppose $\phi \in \mathbf{O}$ and that its restriction to its support is continuous there and strictly positive, then

$$R_\alpha(\phi) \sim r_\alpha(\phi), \tag{5}$$

(with constants independent of ϕ).

proof. Set $M_\phi = \{\nu \in \mathbf{M}^+(\text{supp}(\phi)) : \int \phi \, d\nu = 1\}$ and $\mathbf{F} = \{f \in \mathbf{S}_{00} : \|f\|_{H^1} \leq 1\}$, and consider the functionals

$$R_\alpha'(\phi) = \left[\sup_{\mathbf{F}} \inf_{\mathbf{M}_\phi} \int I_\alpha * f \, d\nu\right]^{-1}$$

$$r_\alpha'(\phi) = \left[\inf_{\mathbf{M}_\phi} \sup_{\mathbf{F}} \int I_\alpha * f \, d\nu\right]^{-1}.$$

Notice since ϕ is bounded below away from zero on supp(ϕ), $\mathbf{M}_\phi$ is a compact subset in the vague topology of $\mathbf{M}$ and that both $\mathbf{M}_\phi$ and $\mathbf{F}$ are convex. These facts plus the linearity and continuity of the Lagrangian $\int I_\alpha * f \, d\nu$ insure that the minimax theorem of K. Fan [8] applies and yields: $R_\alpha'(\phi) = r_\alpha'(\phi)$. We conclude the proof by showing that $R_\alpha(\phi) \sim R_\alpha'(\phi)$ and $r_\alpha(\phi) \sim r_\alpha'(\phi)$. To do this, first notice that $r_\alpha'(\phi) \geq 0$ since $0 \in \mathbf{F}$. Also, $r_\alpha'(\phi) < \infty$, for if not then there would be a sequence of measures $\{\nu_j\} \subset \mathbf{M}_\phi$ such that $\sup \int I_\alpha * f \, d\nu_j \to 0$ as $j \to \infty$ and hence, by the H^1-BMO duality, $\|I_\alpha * \nu_j\|_* \to 0$. From [1] (see Propositions 3.3 and 3.4), we have:

Lemma 1: Let $0 < \alpha < n$, then

(a) if f is such that $I_\alpha * f$ is locally integratable, then there is a constant c independent of f such that

$$\|I_\alpha * f\|_* \leq c \, \|M_\alpha(f)\|_\infty ;$$

(b) if $f \geq 0$ and $\int (1+|x|)^{-\alpha-n} I_\alpha * f(x) \, dx < \infty$, then there is a constant c independent of f such that

$$\|M_\alpha(f)\|_\infty \leq c \cdot \|I_\alpha * f\|_* .$$

Hence from part (b), we have $\|\|\nu_j\|\|_{n-\alpha} \to 0$. Now there is a subsequence $\{\nu_j'\}$ that converges vaguely to a measure $\nu \in \mathbf{M}_\phi$. But by the convergence in $\mathbf{L}^{1,n-\alpha}$, $\nu(B)=0$ for all balls, contradiction.

To see $r_\alpha' \sim r_\alpha$, first suppose $r_\alpha(\phi)>0$, then there is a nontrivial measure $\nu \in \mathbf{M}^+(\mathrm{supp}(\phi))$ such that $\|I_\alpha * \nu\|_* \leq 1$, hence for $f \in \mathbf{F}$ and $\mu = (\int \phi \, d\nu)^{-1} \cdot \nu$

$$\int I_\alpha * f \, d\mu \leq c(\int \phi \, d\nu)^{-1}$$

by the H^1-BMO duality. So $r_\alpha'(\phi) \geq c \int \phi \, d\nu$. For the other inequality, suppose $r_\alpha'(\phi)>0$, then there is a measure $\nu \in \mathbf{M}_\phi$ such that for some $\epsilon>0$

$$r_\alpha'(\phi) < \left[\sup_{\mathbf{F}} \int I_\alpha * f \, d\nu\right]^{-1} + \epsilon$$

$$\leq c \, \|I_\alpha * \nu\|_*^{-1} + \epsilon$$

again by the H^1-BMO duality. So if $\mu = \|I_\alpha * \nu\|_*^{-1} \cdot \nu$, we get

$$r_\alpha(\phi) \geq \int \phi \, d\mu = \|I_\alpha * \nu\|_*^{-1} > c(r_\alpha'(\phi) - \epsilon) .$$

The result for R_α, R_α' follows in a similar manner. □

Proposition 3:

For all compact K, $R_\alpha(K) \sim H_\infty^{n-\alpha}(K)$, (with constants independent of K).

proof. With $\phi = 1_K$ in Proposition 2, we merely need to apply Lemma 1 (parts (a) and (b)) and the Frostman result P_5. □

Since the functions $I_\alpha * f$, $f \in \mathbf{S}_{00}$, are signed, it is not obvious that the set functions R_α are countably subadditive (and in fact may not be!). However, they are comparable to a set function that is.

Proposition 4: There is a constant $c>0$ such that for all $\phi \in C_0(\mathbb{R}^n)^+$,

$$R_\alpha(\phi) \leq c \cdot \int \phi \, d H_\infty^{n-\alpha} . \tag{6}$$

proof. For $\phi \in C_0(\mathbb{R}^n)^+$ and $\epsilon>0$, set $\phi^\epsilon(x) = \phi(x) + \epsilon$ whenever $x \in \mathrm{supp}(x)$, and zero otherwise. Then Proposition 2 applies to ϕ^ϵ and gives

$$R_\alpha(\phi^\epsilon) \leq c \cdot \int \phi^\epsilon \, d H_\infty^{n-\alpha}$$

by virtue of (3) and Lemma 1 (part a). Thus

$$R_\alpha(\phi) \leq c \int \phi \, d H_\infty^{n-\alpha} + c \cdot \epsilon \cdot H_\infty^{n-\alpha}(\mathrm{supp}(\phi)) ,$$

and the result follows. □

5. MAXIMAL FUNCTION ESTIMATES.

Our goal in this section is to derive estimates for the Hardy-Littlewood maximal function $M_0(U)$ of a function $U \in L^1(H_\infty^{n-\alpha})$. To do this we first prove

Proposition 5: There is a constant (depending only on α and n) such that for $0<\alpha<n$,

$$\int M_0(I_\alpha * \phi)\, d\, H_\infty^{n-\alpha} \le c \cdot \|\phi\|_{H^1} \tag{7}$$

for all $\phi \in \mathbf{S}_{00}$.

proof. Take $0<\beta<\alpha$ and write $I_\alpha * \phi = c \cdot I_\beta * I_{\alpha-\beta} * \phi$. Since the integral averages satisfy

$$\fint_{B(x,r)} |y|^{\alpha-n}\, dy \le c|x|^{\alpha-n},$$

for some constant c, it follows that

$$M_0(I_\alpha * \phi) \le c \cdot I_\beta * |I_{\alpha-\beta} * \phi|.$$

Thus

$$\begin{aligned}
\int M_0(I_\alpha * \phi)\, d\, \mu &\le c \int |I_{\alpha-\beta} * \phi| \cdot I_\beta * \mu\, dx \\
&= c \int \phi \cdot I_{\alpha-\beta} * (a \cdot I_\beta * \mu)\, dx \\
&\le c\, \|\phi\|_{H^1} \cdot \|I_{\alpha-\beta} * (a \cdot I_\beta * \mu)\|_* \\
&\le c\, \|\phi\|_{H^1} \cdot \|M_{\alpha-\beta}(I_\beta \cdot \mu)\|_\infty
\end{aligned}$$

where $a = \mathrm{sgn}(I_{\alpha-\beta} * \phi)$, and we have used Lemma 1 (parta) together with the H^1-BMO duality. Thus if $\mu \in \mathbf{L}_+^{1,n-\alpha}$, then it easily follows that $I_\beta * \mu \in \mathbf{L}^{1,n-\alpha+\beta}$. So

$$\int M_0(I_\alpha * \phi)\, d\, \mu \le c \cdot |||\mu|||_{n-\alpha} \cdot \|\phi\|_{H^1}.$$

The conclusion now follows from (4). □

Proposition 6: For $0<\alpha<n$, there is a constant c (depending only on α and n) such that for all $\phi \in C_0(\mathbb{R}^n)^+$

$$\int M_0(\phi)\, d\, H_\infty^{n-\alpha} \le c \cdot R_\alpha(\phi). \tag{8}$$

proof. Take an $f \in \mathbf{S}_{00}$ such that $I_\alpha * f \ge \phi$ on $\mathrm{supp}(\phi)$, then $\mathbf{M}_0(\phi)(x) \le \mathbf{M}_0(I_\alpha * f)(x)$ for all x. Hence from (7), we get

$$\int M_0(\phi)\, d\, H_\infty^{n-\alpha} \le c \cdot \|f\|_{H^1},$$

which gives the result. □

Putting together (6) and (8) gives Theorem A. Notice that (8) and (6) also give

$$R_\alpha(\phi) \sim \int |\phi|\, d\, H_\infty^{n-\alpha} \tag{9}$$

for all $\phi \in C_0(\mathbb{R}^n)$.

6. ESTIMATES IN TERMS OF L^1 DERIVATIVES.

In this section, α = positive integer $< n$. The following estimate for $C_0^\infty(\mathbb{R}^n)$ functions can be found in [13], page 24. In the case $\alpha = 1$, it depends on the "boxing inequality" of W. Gustin (see also [9]). The case $\alpha > 1$ can be reduced to the $\alpha = 1$ case.

Lemma 2: There is a constant c such that for all $\phi \in C_0^\infty(\mathbb{R}^n)$,

$$\int |\phi| \, d\mu \leq c \cdot |||\mu|||_{n-\alpha} \cdot \|\nabla^\alpha \phi\|_1 . \tag{10}$$

The quantity $\nabla^\alpha \phi$ denotes the vector of all derivatives of ϕ of order α, and $|\nabla^\alpha \phi|$, the Euclidean length of that vector, $\|\nabla^\alpha \phi\|_1$, the L^1-norm of $|\nabla^\alpha \phi|$.

From (4) applied to (10), we easily get

Proposition 7: There is a constant c such that for all $\phi \in C_0^\infty(\mathbb{R}^n)$,

$$\int |\phi| \, d H_\infty^{n-\alpha} \leq c \cdot \|\nabla^\alpha \phi\|_1 . \tag{11}$$

It is now clear that (11) plus Theorem A yields Theorem B. Notice also the (11) implies

$$H_\infty^{n-\alpha}(K) \leq c \cdot \Gamma_\alpha(K) \tag{12}$$

where K is a compact set in $\mathbb{R}^n$ and

$$\Gamma_\alpha(K) = \inf\left\{ \|\nabla^\alpha \phi\|_1 : \phi \in C_0^\infty(\mathbb{R}^n)^+ \ \&\ \phi \geq 1 \text{ on } K \right\}.$$

The reverse of inequality (12) is also true. It says

$$\Gamma_\alpha(K) \leq c \cdot H_\infty^{n-\alpha}(K) \tag{13}$$

for all compact K, and follows from the two elementary facts: (a) $\Gamma_\alpha(B(x,r)) \leq c\, r^{n-\alpha}$, for all x, and (b) Γ_α is countably subadditive, i.e. $\Gamma_\alpha(\cup_j K_j) \leq \sum_j \Gamma_\alpha(K_j)$. Observe, that the reason (b) is elementary is that the "test functions" in the definition of Γ_α are taken to be non-negative. This is not possible for the potentials that are used to define R_α.

7. EXTENDING THE ESTIMATES.

It is clear from the definition of $L^1(H_\infty^{n-\alpha})$ and the lower semi-continuity of M_0 that Theorem A extends to all of $\phi \in L^1(H_\infty^{n-\alpha})$. Consequently, one gets by standard arguments that the limit

$$\lim_{r \to 0} |B(x,r)|^{-1} \int_{B(x,r)} \phi(y)\, dy \ \text{ exists } H_\infty^{n-\alpha} \text{ - a.e. } x . \tag{14}$$

The values obtained in (14) can then be taken as a $H^{n-\alpha}$-precise representative of $\phi \in L^1(H_\infty^{n-\alpha})$. In a like manner, we can easily extend (7) from $\mathbf{S}_{00}$ to all of H^1 by taking a sequence $\{f_k\} \subset \mathbf{S}_{00}$ for which $f_k \to f$ in H^1-norm. Clearly $I_\alpha * f_k \to I_\alpha * f$ in L^1-loc; indeed, recall the estimate

$$\int_B I_\alpha * f\, dx \le c\,|B|^{\alpha/n}\,\|f\|_1$$

for any $f \in L^1_+$ and B = ball. A precise representative can again be defined via (14). Also, these same ideas can be applied to extend Theorem B from $C_0^\infty(\mathbb{R}^n)$ to the Sobolev space $\dot{W}^{\alpha,1}(\mathbb{R}^n)$ - the locally integrable functions on $\mathbb{R}^n$ which have distributional derivatives of order α belonging to $L^1(\mathbb{R}^n)$.

For functions in BV^α - the locally integrable functions on $\mathbb{R}^n$ which have distributional derivatives of order α belonging to $\mathbf{L}^1(\mathbb{R}^n)$ - the situation is different tho it is still possible to extend the inequality (7) to all $\phi \in BV^\alpha$. However, it is clear that BV^α does not belong to $\mathbf{L}^1(H_\infty^{n-\alpha})$ since BV^α functions are not $H_\infty^{n-\alpha}$ quasi-continuous (the characteristic function of a ball is in BV^1 but is not H_∞^{n-1} quasi-continuous). For $\alpha = 1$, the following lemma appears in [12]. It can be extended to include higher order derivatives with out difficulty.

Lemma 3: If $U \in BV^\alpha$, then there is a sequence $\{\phi_k\} \subset C^\infty(\mathbb{R}^n)$ such that $\|\nabla^\alpha \phi_k\|_1 \to \|\nabla^\alpha U\|_1$ and $\phi_k \to U$ in L^1-loc., as $k \to \infty$.

We use this lemma to extend (7) by taking

$$\psi \in C_0^\infty(\mathbb{R}^n),\ \psi(x) = 1 \text{ for } |x| \le 1 \text{ and } \psi(x) = 0 \text{ for } |x| \ge 2\,.$$

From Theorem B, we have

$$\int M_0(\phi_k \psi_R)\, d\, H_\infty^{n-\alpha} \le c\,\|\nabla^\alpha(\phi_k \psi_R)\|_1\,, \tag{15}$$

where $\psi_R(x) = \psi(x/R)$, $R \ge 0$. The right side of (15) can be estimated by

$$c \sum_{j=0}^{\alpha} R^{j-\alpha} \int_{|x|<2R} |\nabla^j \phi_k|$$

which does not exceed

$$c \sum_{j=0}^{\alpha} \|\nabla^j \phi_k\|_{n/(n-\alpha+j)}$$

which in turn does not exceed a constant multiple of $\|\nabla^\alpha \phi_k\|_1$ by the Sobolev embedding theorem. Passing to the limit in (15) now gives the desired extension of Theorem B.

Finally, we can make one additional observation regarding the class $\mathbf{L}^{1,n-1}$. It is well known (see [12] or [15]) that $\mathbf{L}^{1,n-1} \subset (BV^1)^*$ = dual of BV^1. However, $\mathbf{L}_+^{1,n-1}$ is *not* vaguely sequentially complete in $(BV^1)^*$. If so, then let ν_h, $h > 0$, be the mollified measure $\nu \in \mathbf{L}_+^{1,n-1}$. Notice that $|||\nu_h|||_{n-1} \le c\,|||\nu|||_{n-1}$ with c independent of h. Hence there is a subsequence $\{\nu_h'\}$ that converges vaguely as $h \to 0$ to an element of $\mathbf{L}^{1,n-1}$, which must in fact be ν. Thus

$$\int f\, d\,\nu_h' \to \int f\, d\,\nu$$

for each $f \in BV^1$. But this implies that $f_h' \to f$ weakly in $L^1(H_\infty^{n-1})$. Hence there will be a sequence of convex combinations of $\{f_h'\}$ which converge strongly in $L^1(H_\infty^{n-1})$ to f. Hence f will be quasi-continuous-contradiction.

8. BIBLIOGRAPHY.

[1] D. Adams, A note on Riesz potentials, *Duke Math. J.* **42** (1975), 765-778.

[2] ________, Sets and functions of finite capacity, *Ind. U. Math. J.* **27** (1978), 611-627.

[3] ________, L^p-capacity integrals with some applications, *Proc. Symp. Pure Math. A.M.S.* **35** (1979), 359-367.

[4] B. Anger, Representations of capacities, *Math. Ann.* **229** (1977), 245-258.

[5] L. Carleson, Selected problems in exceptional sets, Van Norstrand 1967.

[6] G. Choquet, Theory of capacities, *Ann. Inst. Fourier,* **5** (1953), 131-295.

[7] ________, Forme abstraite du théorème de capacitabilitié, loc.cit. **9** (1959), 83-89.

[8] K. Fan, Minimax theorems, *Proc. Nat. Acad. Sci. U.S.A.* **39** (1953), 42-47.

[9] H. Federer, Geometric measure theory, *Springer-Verlag,* 1969.

[10] R. Fefferman, A theory of entropy in Fourier analysis, *Adv. Math.* **30** (1978), 171-201.

[11] K. Hansson, Imbedding theorems of Sobolev type in potential theory, *Math. Scand.* **45** (1979), 77-102.

[12] V. Maz'ya, Sobolev spaces, *Springer-Verlag,* 1985.

[13] V. Maz'ya, T. Shaposhinkova, Theory of multipliers in spaces of differentiable functions, *Pitman Press,* 1985.

[14] N. Meyers, A theory of capacities for potentials of functions in Lebesgue classes, *Math. Scand.* **26** (1970), 255-292.

[15] N. Meyers, W. Ziemer, Integral inequalities of Poincaré and Wirtinger type for BV functions, *Amer. J. Math,* **99** (1977), 1345-1360.

[16] C. Rogers, Hausdorff measures, *Cambridge U. Press* 1970.

[17] F. Soria, Characterizations of classes of functions generated by blocks and associated Hardy spaces, *Ind. U. Math. J.* **34** (1985), 463-492.

BESOV NORMS OF RATIONAL FUNCTIONS*

J. Arazy, University of Haifa, Haifa, Israel
S.D. Fisher, Northwestern University, Evanston, IL. 60201
J. Peetre, University of Lund, Lund, Sweden

Let Δ be the open unit disc, $\{z: |z| < 1\}$, in the complex plane. Let $B^p, 1 < p \leq \infty$, be the space of those analytic functions g on Δ determined by the finiteness of the seminorm

$$\|g\|_{B^p} = \|g'\|_{L^p(\Delta, d\Sigma)}, \qquad d\Sigma(z) = (1-|z|^2)^{-2}\, dA(z).$$

The space B^1 consists of those analytic functions h which have the form $h(z)=\Sigma\lambda_k\phi_k$ where ϕ_k is a Möbius function mapping Δ onto Δ and $\{\lambda_k\}$ is a summable sequence of complex numbers. A finite Blaschke product F of degree n is an analytic function of the form

$$F(z)= \lambda \prod_1^n (a_k-z)/(1-\bar{a}_k z) \tag{1}$$

where λ is a constant of absolute value one and a_k is a point of Δ. This note gives a proof of the following theorem.

Theorem. <u>There are constants α_p and β_p depending only on p such that</u>

$$\alpha_p n^{1/p} \leq \|F\|_{B^p} \leq \beta_p n^{1/p}, \quad 1 \leq p \leq \infty \tag{2}$$

<u>for every finite Blaschke product F of degree n.</u>

In particular, the constants α_p and β_p are independent of both the degree and the location of the zeros of F. We also give a partial extension of (2) to rational functions with no poles on the closed disc, $\{z: |z| \leq 1\}$.

<u>Proof of the theorem</u>. We prove the estimates at p=1 and at p=∞ and then make use of interpolation. The upper bound at p=∞ is elementary since the

---

* Research supported in part by the National Science Foundation

B^∞ norm is dominated by the sup norm. This yields

$$\|F\|_{B^\infty} \le \|F\|_{H^\infty} = 1. \tag{3}$$

The lower bound at p=1 is also easy. The B^1 norm is larger than the $W^{1,1}$ norm since B^1 is the minimal Möbius invariant space; see[1]. Hence,

$$\|F\|_{B^1} \ge \|F\|_{W^{1,1}} = n \tag{4}$$

since the $W^{1,1}$ norm of a function F gives the length of the image of the unit circle T under F; see [1]. There are two proofs of the upper bound for p=1. The first is based on the theory of Hankel operators and a theorem of V.V. Peller. Let f be an analytic function on Δ and let H_f be the Hankel operator on the Hardy space H^2 defined by

$$H_f(g) = (I-P)\overline{f}g \tag{5}$$

where P is the orthogonal projection of $L^2(T,d\theta)$ onto H^2. By a theorem of V.V. Peller [5] the operator H_f is trace class if and only if its symbol f is in B^1 and, if this is so, then the trace norm of H_f is equivalent to the B^1 norm of f. We apply this with f=F, a Blaschke product of degree n. This then yields an upper bound on the B^1 norm of F of a constant times $tr(H_F)$. However, F is rational of degree n and so by Kronecker's theorem [3] the operator H_F has rank n. Thus, writing s_k for the k<u>th</u> singular number of the operator we have

$$tr\,(H_F) = \sum_1^n s_k(H_F) \le n\, s_1(H_F) = n\,\|H_F\|$$

where the last norm is the operator norm of H_F on H^2. This operator norm of H_F is no more than the sup norm of F which, of course, is one. This establishes the inequality

$$\|F\|_{B^1} \le c\,n \tag{6}$$

where c is an absolute constant and finishes the first proof of (6). The second proof of (6), due to Don Marshall and used here with his kind permission, is direct and entirely elementary. The first step is to note that (6) holds when n=1. This is immediate from the definition of B^1; see [1]. Further, it is known that the B^1 norm of an analytic function f is equivalent to the L^1 norm of f'' over Δ with respect to area measure plus two linear functionals; see [1]. Let $\{F_j\}$ be the degree one factors of F. We have the following formula for F":

$$F'' = \sum_1^n F\, F_j''/F_j - \sum_1^n F(F_j'/F_j)^2 + (F')^2/F \tag{7}$$

A simple change of variables argument gives

$$\int_\Delta |F'|^2/|F|\, dA = \int_{F(\Delta)} (1/|w|)\, dA(w) = 2n. \tag{8}$$

To complete the proof of (6) use the fact that $|F/F_j|$ is bounded by 1 on Δ, that the result holds for n=1, that F has n factors, and that (7) and (8) hold. This completes Marshall's proof of (6).

Finally, we use the Möbius invariant pairing

$$\langle f,g\rangle = \int_\Delta f'\,\overline{g'}\, dA$$

which effects the duality $(B^1)^* = B^\infty$, see [1], to obtain

$$\|F\|_{B^\infty} \|F\|_{B^1} \geq |\langle F,F\rangle| = \int_\Delta |F'|^2\, dA = n.$$

The last equality results from the fact that the integral gives the area of the image of Δ under F, counting multiplicities. Together with (6), this gives

$$\|F\|_{B^\infty} \geq 1/c \tag{9}$$

where c is the constant from (6) and so is independent of both F and n. The inequalities (3),(4),(6), and (9) then establish (2) for the values p=1 and p=∞. The spaces B^p are an interpolation scale, see [2], and this establishes the right-hand inequality in (2) for $1 < p < \infty$. We again make use of the Möbius invariant pairing this time for the duality $(B^p)^* = B^q$, $1/p + 1/q = 1$. Unlike the case p=1 this is only an isomorphism and not an isometry.Thus, there is a constant c_p with

$$c_p \|F\|_{B^p} \|F\|_{B^q} \geq |\langle F,F\rangle| = n, \quad 1/p + 1/q = 1.$$

This and the right-hand inequality in (2) then give the left-hand inequality in (2). The proof of the Theorem is finished.

Remark. Let F be a rational function of degree n with no poles in the closed disc $\{z: |z| \leq 1\}$. The inequalities

$$\|F\|_{B^1} \leq cn\, \|F\|_{H^\infty} \quad \text{and} \quad \|F\|_{B^\infty} \leq \|F\|_{H^\infty} \tag{10}$$

obtained above when F is a Blaschke product of degree n carry over verbatim to this more general setting. These yield the inequality

$$\|F\|_{B^p} \leq C\, n^{1/p}\, \|F\|_{H^\infty}\, ,\ 1 \leq p \leq \infty \tag{11}$$

where C is an absolute constant. The lower bounds obtained for Blaschke products, however, do not carry over and, indeed, are false. For instance, there is not a constant c with

$$\|f\|_{B^\infty} \geq c\, \|f\|_{H^\infty} \tag{12}$$

valid for all rational functions of degree n with no poles in the closed disc. This fact is easily demonstrated using lacunary series. It is known that if

$$f(z) = \sum_{k=1}^{\infty} a_k\, z^{2^k}$$

then the B^∞ norm of f is comparable to $\sup\{|a_k| : k \geq 1\}$ while the H^∞ norm is comparable to $\sum |a_k|$. By taking $a_k=1$, $k=1,\ldots,\log_2(n)$ and $a_k=0$ for the other indices, we obtain a sequence of polynomials, $\{p_n\}$, of respective degrees n with uniformly bounded B^∞ norms and H^∞ norm about $\log_2(n)$. Thus, (12) cannot hold.

REFERENCES

1. J. Arazy, S.D. Fisher, and J. Peetre, Mobius invariant function spaces, J. reine angew. Math. **363** (1985), 110-145

2. J. Bergh and J. Lofstrom, *Interpolation Spaces An Introduction* , Grundlehren 223, Springer-Verlag, Berlin, 1976

3. F.R. Gantmacher, *The Theory of Matrices*, vol.II, Chelsea, New York, 1959

4. J. Peetre, Hankel operators, rational approximation, and allied questions in analysis, Second Edmundton Conference on Approximation

Theory, Canadian Mathematical Society Conference Proceedings, **3** (1983), 287-332

5. V.V. Peller, Hankel operators of class S_p and their applications (rational appproximation, Gaussian processes, the problem of majorizing operators), Math USSR Sbornik **41** (1982), 443-479

6. R. Rochberg, Trace ideal criteria for Hankel operators and commutators, Indiana Math. J. **31** (1982), 913-925

FUNCTIONS OF BOUNDED MEAN OSCILLATION AND HAUSDORFF-YOUNG TYPE THEOREMS[(i)].

Jöran Bergh
Department of Mathematics
Chalmers University of Technology
S-412 96 Göteborg, Sweden

Bounded mean oscillation, BMO, is a notion which plays a central role in contemporary studies of boundary value problems for partial differential equations and related questions.

The inequality of Hausdorff-Young and its generalization together with lacunary techniques for the Fourier transform are likewise frequently used tools in analysis.

We will show (in Section 2) that the space BMO is in a way a better endpoint for Fourier transform interpolation than the usual L_∞. This approach yields a Hausdorff-Young type theorem which contains Paley's sharper version of the classical theorem, in addition to results for lacunary Fourier series. The BMO endpoint estimate is based on a characterization of the functions in BMO with positive Fourier coefficients due to C. Fefferman, see [S&S].

Our discussion will be conducted only for the case of the one-dimensional torus, $\mathbb{T}$. The results, however, are valid in dimension n; in both $\mathbb{T}^n$ and $\mathbb{R}^n$, mutatis mutandis. Here the BMO endpoint result is due to Sledd and Stegenga [S&S].

Moreover, in Section 1, we will characterize the functions in BMO which have decreasing Fourier coefficients. This result can be deduced from C. Fefferman's characterization mentioned above, but we

---

(i) This paper is an outline. A full report is in preparation. Our research has been partially supported by a grant from the Swedish Natural Science Research Council.

will give a proof along classical, more elementary lines. We also give a third simple proof, using the fact that the space of analytic BMO functions is inbedded in the Bloch space. Our results here settle a limiting case, which has been an open question for a long time. See, e.g., Boas' book [Bo].

O. DEFINITIONS AND PRELIMINARY RESULTS:

$$L_p = \{f; \quad \|f\|_p := (\int_0^{2\pi} |f(x)|^p dx)^{1/p} < \infty\}$$

$$\ell_p = \{c = (c_\nu)_{\nu=-\infty}^{\infty}; \|c\|_p := (\Sigma_{\nu=-\infty}^{\infty} |c_\nu|^p)^{1/p} < \infty\}$$

$$L_{pq} = \{f; \|f\|_{pq} := (\int_0^{\infty} (t^{1/p} f^*(t))^q dt/t)^{1/q} < \infty\}$$

f^* denotes "the decreasing rearrangement of f"

$$\ell_{pq} = \{c; \|c\|_{pq} := (\Sigma_{\nu\geq 0} (\nu^{1/p} c_\nu^*)^q \nu^{-1})^{1/q} < \infty\}$$

PROPOSITION 0.1 *The "real interpolation space "(cf. e.g., [B&L])*

$$(L_{p_0}, L_{p_1})_{\theta,q} = L_{pq} \quad (\textit{equivalent norms})$$

when $1/p = (1-\theta)/p_0 + \theta/p_1$, $0 < \theta < 1$. *This also holds if* BMO *is substituted for* L_{p_1} *and* $p_1 = \infty$; [R&S]. *Moreover, it is true if* ℓ *is substituted for* L *see, e.g.,* [B&L].

THEOREM 0.2 (C. Fefferman) [S&S]

Let $f \sim \Sigma_{\nu\geq 0} c_\nu e^{i\nu x}$, *with* $c_\nu \geq 0$ *for all* $\nu \geq 0$. *Then the norm on* f *in* BMO:

$$\|f\|_{BMO} := \sup_I |I|^{-1} \int_I |f(x) - f_I| dx$$

is equivalent to a norm on the coefficients:

$$\|c\|_* := \sup_{N\geq 1} (\sum_{k\geq 1} (\sum_{\nu=kN}^{(k+1)N-1} |c_\nu|)^2)^{1/2}.$$

Here I is an interval of length $|I|$ and $f_I = |I|^{-1} \int_I f(x)dx$ is the mean value of f over I.

Theorem 0.2 has a well-known consequence:

COROLLARY 0.3 *Let* $f \sim \Sigma_{\nu \geq 0} c_\nu e^{i\nu x}$. *Then holds*

$$\| f \|_{BMO} \leq C \| c \|_{*}.$$

PROOF. Decompose c ($\| c \|_{*}$ finite) into its real and imaginary parts, and these in their turn into their positive and negative parts. This yields a decomposition of f into a sum of four terms, all belonging to BMO by Theorem 0.2. Thus f is in BMO. (We owe this argument to Peter Sjögren.) □

THEOREM 0.4 *Let* $f \sim \Sigma_{\nu \geq 0} c_\nu e^{i\nu x}$. *Then*

$$\| f \|_{\mathcal{B}} \leq C \| f \|_{BMO}.$$

Here $\mathcal{B}$ is the Bloch space:

$$\mathcal{B} = \{f(z) \text{ analytic in } |z|<1;\ \| f \|_{\mathcal{B}} := \sup_{|z|<1} (1-|z|)|f'(z)| < \infty\}.$$

In this theorem, the assumption that $c_\nu = 0$ for $\nu < 0$ is, of course, relevant. The theorem can be found, e.g., in [Bae].

1. MONOTONE COEFFICIENTS.

It is well known (cf. e.g., [Z]) that if $f \sim \Sigma_{\nu \geq 0} c_\nu e^{i\nu x}$ (i) then Paley's sharpening of the Hausdorff-Young inequality:

(1) $$\| f \|_{p'} \leq C \| c \|_{pp'} \qquad (1<p\leq 2,\ 1/p + 1/p' = 1)$$

is sharp in the sense that if c is decreasing then we also have

(2) $$\| c \|_{pp'} \leq C \| f \|_{p'}\ ,$$

i.e., then the norms are equivalent, and we have a characterization. The first inequality is classically obtained by real interpolation from the two endpoint results:

(i) For simplicity only, we assume below that $c_\nu = 0$ for $\nu < 0$.

$$(3)\qquad \begin{aligned} \| f \|_\infty &\le C \| c \|_1 \\ \| f \|_2 &\le C \| c \|_2 . \end{aligned}$$

(Choose $1/p = (1-\theta)/1 + \theta/2$ and $q = p'$ in Proposition 0.1.)

However, the limiting case $p = 1$ has been open: for $p' = \infty$, the inequality (1) is false. As an example, take $c_\nu = \nu^{-1}$. Then $c \in \ell_{1\infty}$ but $f(x) = \ln(1-e^{ix})$ is not in L_∞ (but f is in BMO). The following is a result for the case $p = 1$:

THEOREM 1.1 *Let* $f \sim \sum_{\nu \geq 0} c_\nu e^{i\nu x}$ *with* c_ν *positive and decreasing. Then* f *is in* BMO *if and only if* $\sup_{\nu \geq 0} \nu c_\nu$ *is finite*:

$$\| f \|_{BMO} \le C \| c \|_{1\infty}$$

$$\| c \|_{1\infty} \le C \| f \|_{BMO} .$$

PROOF. The "if" part is well-known, and follows directly from, e.g., Theorem 0.2. The following estimates give the "only if" part:

$$\| f \|_{BMO} \geq \| \operatorname{Re} f \|_{BMO} \geq \frac{n}{\pi} \int_0^{\pi/n} |\operatorname{Re} f(x) - \operatorname{Re} f_{\pi/n}| dx \geq$$

$$\geq \frac{n}{\pi} \int_0^{\pi/n} (\operatorname{Re} f(x) - \operatorname{Re} f_{\pi/n}) \cos nx \, dx =$$

$$= \frac{n}{\pi} \int_0^{\pi/n} \operatorname{Re} f(x) \cos nx dx = \frac{n}{\pi} \sum_{\nu \geq 1} c_\nu \int_0^{\pi/n} \cos \nu x \cos nx \, dx \geq$$

$$\geq \frac{n}{\pi} \sum_{1 \leq \nu \neq n} \frac{\nu c_\nu}{n^2 - \nu^2} \sin \frac{\nu\pi}{n} =$$

$$= \frac{n}{\pi} \sum_{\nu=1}^{n-1} \left\{ \frac{\nu c_\nu}{n^2 - \nu^2} + \frac{(\nu+n) c_{\nu+n}}{(\nu+n)^2 - n^2} - \frac{(\nu+2n) c_{\nu+2n}}{(\nu+2n)^2 - n^2} + \cdots \right\} \sin \frac{\nu\pi}{n} \geq$$

$$\geq \frac{n}{\pi} \sum_{\nu=1}^{n-1} \frac{\nu c_\nu}{n^2 - \nu^2} \sin \frac{\nu\pi}{n} \geq \frac{n}{\pi} c_n \sum_{\nu=1}^{n-1} \frac{\nu}{n^2 - \nu^2} \sin \frac{\nu\pi}{n} \geq C n c_n ,$$

where the monotonicity of (c_n) is used in the two last estimates but one. □

We will now give an in a sense slightly stronger result; stronger in view of Theorem 0.4, but with an additional analyticity assumption.

THEOREM 1.2 *Let* $f \sim \sum_{\nu>0} c_\nu e^{i\nu x}$ *with* c_ν *positive and decreasing. Then* f *is in* $\mathcal{B}$ *if and only if* $\sup_{\nu \geq 0} \nu c_\nu$ *is finite.*

PROOF. The "if" part follows from Theorem 0.3 and Theorem 1.1 (and is well-known). The following estimates give the "only if" part:

$$\| f \|_{\mathcal{B}} \geq [z=1-1/N] \geq N^{-1} \sum_{\nu=N}^{2N} \nu c_\nu (1-1/N)^{\nu-1} \geq$$

$$\geq C \sum_{\nu=N}^{2N} c_\nu \geq C\, N c_{2N},$$

with an arbitrary integer $N > 1$; $C > 0$. □

There is another connection in this context between BMO and $\mathcal{B}$ for univalent (even p-valent) functions:

THEOREM 1.3 *Let* $f(z)$ *be analytic and p-valent in* $|z|<1$. *Then holds*

$$\| f \|_{\mathcal{B}} \leq C \| f \|_{BMO}$$

$$\| f \|_{BMO} \leq C \| f \|_{\mathcal{B}}$$

This theorem may also be found in, e.g., [Bae].

2. HAUSDORFF-YOUNG TYPE THEOREMS

As we noted above, the Hausdorff-Young type theorems are obtained classically by interpolation from the following two endpoint estimates.

$$\|f\|_\infty \le C\|c\|_1$$

where $f \sim \sum_{\nu\geq 0} c_\nu e^{i\nu x}$ (i)

$$\|f\|_2 \le C\|c\|_2$$

We propose to use the estimate (Corollary 0.3)

$$\|f\|_{BMO} \le C\|c\|_*$$

instead of $\|f\|_\infty \le C\|c\|_1$. Our endpoints are thus:

$$\text{(1)} \qquad \begin{aligned} &\|f\|_{BMO} \le C\|c\|_* \\ &\|f\|_2 \le C\|c\|_2 . \end{aligned}$$

Interpolating these, we get, e.g.:

THEOREM 2.1 *Let* $f \sim \sum_{\nu\geq 0} c_\nu e^{i\nu x}$. *Then holds*

$$\text{(2)} \qquad \|f\|_{p'} \le C\|c\|_{(\ell_*,\ell_2)_{\theta,p'}}$$

where θ *and* p' *are the same as in* (1.3) *and* $\ell_* = \{c;\ \|c\|_* < \infty\}$.

The estimate (2) is stronger than the estimate (1.1). This follows from two things. Firstly, $\ell_1 \subset \ell_*$ which implies that

$$\ell_{pp'} = (\ell_1,\ \ell_2)_{\theta,p'} \subset (\ell_*,\ \ell_2)_{\theta,p'} .$$

Secondly, for lacunary sequences $c(c_\nu = 0$ for $\nu \neq 2^k$, $k = 1,2,\dots)$, we have $\|c\|_* \sim \|c\|_2$. This gives $\|c\|_{(\ell_*,\ell_2)_{\theta,p'}} \sim \|c\|_2$. for these sequences. On the other hand, such a lacunary sequence need not belong to $\ell_{pp'}$. We note thus that $(\ell_*,\ell_2)_{\theta,p'}$ is not a rearrangement invariant space.

(i) Again, for simplicity only, we assume below that $c_\nu = 0$ for $\nu = 0$.

We also note that there are well-known corresponding results for a dyadic decomposition of the Fourier transform: the Beurling-Herz type spaces, cf. [P].

CONCLUDING REMARKS

We lack an explicit description of, i.a., the interpolation space $(\ell_*,\ell_2)_{\theta,p'}$. A natural guess might be that the norm of a sequence c given by the expression

$$(3) \qquad \sup_{N\geq 1} \Big(\sum_{k\geq 0} \Big(\sum_{\nu=kN}^{(k+1)N-1} |c_\nu|^p \Big)^{2/p} \Big)^{1/2}$$

could work as the right hand side in (2). However, the example $c_\nu = \nu^{-1/p}$ disproves this guess: the expression (3) is finite, but, if $f \sim \sum_{\nu\geq 1} \nu^{-1/p} e^{i\nu x}$ then $f \notin L_{p'}$, which follows from (1.1), (1.2) and the fact that $(\nu^{-1/p})_{\nu\geq 1} \notin \ell_{pp'}$.

In a subsequent paper, we will discuss the interpolation of the couple (ℓ_*,ℓ_2). In addition, we will discuss other possible endpoint estimates, modelled on analogues of the Bloch space.

REFERENCES.

[B&L] Bergh, J. & Löfström, J., Interpolation spaces, Grundlehren der Mathematik 223, Springer, Berlin, 1976.

[Bae] Baernstein A, Analytic functions of bounded mean oscillation in Brannan DA & Clunie JG (ed), Aspects of contemporary complex analysis, Academic Press, New York, 1980.

[Bo] Boas, R.P., Integrability theorems for trigonometric transforms, Ergebnisse der Mathematik 38, Springer, Berlin, 1967.

[P] Peetre, J., New thoughts on Besov spaces, Duke Univ. Math. series 1, Duke University, Durham, 1976.

[R&S] Rivière, N. & Sagher Y., Interpolation between L^∞ and H^1, the real method, J. Functional. Anal. 14, 401-409 (1973).

[S&S] Sledd, W.T. & Stegenga, D.A., An H^1 multiplier theorem, Ark. Mat. 19, 265-270 (1981).

[Z] Zygmund, A., Trigonometric series, Cambridge Univ. Press, Cambridge, 1968.

REMARKS ON LOCAL FUNCTION SPACES

Bogdan Bojarski, University of Warsaw
PL-00-901 Warszawa, POLAND

I. Let $\Omega \subseteq R^n$, be a connected open set. By a local function space we mean a function space equipped with a certain collection of local semi--norms (or quasinorms) depending on a family F of subsets of the domain Ω of definition of the elements in the function space under consideration. These seminorms are usually defined by various averaging processes, in general combined with differential or difference operations or even operations of integral type e.g. Fourier transform or fractional integration, and related with local approximations or local representations of the functions considered by families of elements of some finite dimensional space e.g. polynomials of some fixed degree, splines of some finite element approximation method etc.. The typical example is supplied by Lebesgue spaces $L^r_{loc}(\Omega)$ or Sobolev spaces $W^{k,r}_{loc}(\Omega)$ for $r \geq 1$ (or even $r > 0$), k-integer and Ω an open subset (domain) in R^n. In what follows we will restrict for simplicity our discussion to these cases only. Analogous concepts can be elaborated also if combined with the local "pointwise" norms of the maximum type: $\max_{x \in \Omega} |f|$.

The concept of the local function space given above is in a sense by intention not precise enough. We hope that some examples in this introductory part will serve to clarify the idea. In part 2 and 3 of the paper we give more precise definitions related with the case of Sobolev spaces $W^{1,p}(\Omega)$ and Lebesgue spaces $L^p(\Omega)$. We remark that the term local function space in our definition does not coincide with the its standard meaning. It is not related with partition of unity. In fact the adopted definition of local function space has a semi-global character: it tries to capture the interplay between local properties of the functions and the global geometric structure of the domain where the functions "are living".

In the case of Lebesgue spaces $L^r_{loc}(\Omega)$, taking for F the set of all measurable, relatively bounded subsets E of Ω, the basic collection of seminorms $\|f\|_{E,q} = (\fint_E |f|^q dx)^{\frac{1}{q}}$, $E \subset \Omega$ $|E| > 0$, $1 \leq q \leq r$ is usually considered. Here the barred integral denotes the average $⨍_E f dx \equiv \frac{1}{|E|}\int_E f dx \equiv f_E$ and $|E|$ is the Lebesgue measure. Few other seminorms are also introduced in the literature for various purposes: (see [10],[14],[19],[3],[30]).

We recall some of them. The sharp L^q local norms $\|f\|^{\#}_{E,q} = (⨍_E |f-f_E|^q dx)^{1/q}$ or, more generally, the best polynomial approximation norms $\mathbb{E}^q_l(f;E) = \inf_{P\in\mathbb{P}_l} (⨍_E |f-P|^q dx)^{\frac{1}{q}}$, where the inf is taken over the set $\mathbb{P}_l$ of polynomials of degree $\leq l-1$, l - integer, $1 \leq q \leq r$, appear in the study of Sobolev spaces $W^{r,l}(\Omega)$. Obviously $\mathbb{E}^q_{1o}(f;E) \sim \|f\|_E$, where the equivalence sign $A(E) \sim B(E)$ for two set functions A and B means $C^{-1} A(E) \leq B(E) \leq C\, A(E)$ for some absolute constant C.

There are some elementary classical inequalities for these seminorms. The most important are the Hölder inequalities

$$(⨍_E |f|^q dx)^{1/q} \leq (⨍_E |f|^p dx)^{1/p} \tag{1.1}$$

valid for all measurable sets $E \subset \Omega$ and $1 \leq q \leq p \leq r$.

However a much more refined and rich theory arises if the geometry of the domain Ω, where the function spaces are defined is taken into account. In the simplest case, if the consideration of the seminorms $\mathbb{E}^q_l(f;E)$ is restricted to regular subsets of Ω. e.g. cubes with coordinate parallel axes or balls contained in Ω, the geometry of the cube and cube or ball configurations and the rich structure of coverings of Ω by families of cubes, balls, rectangles or other "model" geometrical figures, comes into play.

In what follows, for simplicity we shall fix our attention on cubes and we let Q denotes any open cube in R^n. If $\delta \geq 1$, then δQ is the cube with the same center as Q, expanded by the factor δ. We recall the local Sobolev imbedding inequality

$$(⨍_Q |f(x)-f_Q|^{p^*} dx)^{\frac{1}{p^*}} \leq (⨍_Q ⨍_Q |f(x)-f(y)|^{p^*} dxdy)^{1/p^*} \leq$$

$$C(n,p)\, r(Q)\, (⨍_Q |\nabla f|^p dx)^{1/p} \tag{1.2}$$

valid for $f \in W^{p,1}_{loc}(\Omega)$, $1 < p \leq n$ and any cube Q strictly contained in

Ω. Here $r(Q)$ - is the side length of Q, $p^* = \frac{np}{n-p}$ and the constant $C(n,p)$ is the same for all cubes $Q \subset \Omega$. For k positive integer $\nabla^k f = \{D^\alpha f, |\alpha| = k\}$ is the collections of all partial derivatives of f of order k. The in equality (1.2) is much deeper than (1.1) and it implies in particular the imbedding $W^{1,p}(Q) \subset L^{p^*}(Q)$. While (1.1) is a measure theoretic inequality, (1.2) is basically geometric: if for a general open $\tilde{Q} \subset \Omega$ we set $r(\tilde{Q}) = \operatorname{diam} \tilde{Q}$ then (1.2) holds with a constant C independent of f only if $\tilde{Q}$ satisfies some special geometric conditions. It is known that the class of John domains [28] has this property.

For $f \in W^{k,p}_{loc}(\Omega)$ the local Sobolev inequality we write in the form

$$\mathbb{E}^{p^*}_{k-1}(f,Q) \leq C(n,p,k)\, r^k(Q) \left(\fint_Q |\nabla^k f|^p dx\right)^{1/p} \tag{1.3}$$

with $p^* = \frac{np}{n-kp}$ and $p < \frac{n}{k}$.

The inequalities (1.1), (1.2) and (1.3) appear as "natural" inequalities satisfied by seminorms of the considered local function space. Requiring that the local seminorms on the function space satisfy some conditions, usually in the form of inequalities, new interesting function classes can be defined. We recall some examples.

1. A function f is in $BMO(\Omega)$ - bounded mean oscillation - if $f \in L^1_{loc}(\Omega)$ and

$$\sup_Q \fint_Q |f - f_Q| dx = \sup_Q \|f\|^{\#}_{Q,1} = \|f\|_{*,\Omega} < \infty \tag{1.4}$$

for all cubes $Q \subset \Omega$.

Under the norm $\|f\|_{*,\Omega}$, $BMO(\Omega)$ is a Banach space of functions modulo constants [21],[26]. The basic property of functions in $BMO(\Omega)$ is expressed by the famous John-Nirenberg lemma, which implies, in particular, that $BMO(\Omega) \subset L^p_{loc}(\Omega)$ for any $p > 1$[21].

2. Let p,q,K be real, $p > o$, $q \neq o$, $p > q$, $K \geq 1$. We say that $f \in L^p_{loc}(\Omega)$ is in the class $B^p_q(K)$ if the inequality

$$\left(\fint_Q |f|^p dx\right) \leq K \left(\fint_Q |f|^q\right)^{1/q} < +\infty \tag{1.5}$$

holds for all cubes $Q \subset \Omega$.

(1.5) is a form of the averaged Harnack property. It excludes "big" local oscillations of f. In the extreme case $p = +\infty$, $q = -\infty$

(1.5) reduces to the uniform Harnack inequality $\max_Q |f| \le K \inf_Q |f|$ for each $Q \subset \Omega$. A non-negative measurable function f is said to satisfy the reverse Hölder inequality with exponent p if f is in the class $B_q^p(K)$ for some K and $q < p$. (K and q depending on f). For non--negative weights the union $\cup_{K\ge 1} B^1_{1/(1-p)}(K)$, $p > 1$, is the famous Muckenhoupt class A_p [24]. F.Gehring in [14] considers the class $B_1^\delta = \cup_{K\ge 1} B_1^\delta(K)$ for some $\delta > 1$. The basic fact about reverse Hölder inequalities is the "self-improving" property [14] [24], which we formulate in the following form: If (1.5) holds for all coordinate parallel subcubes $\tilde{Q}$ of a given cube Q then there exist constants $r = r(n,p,q,k)$, $r > p$ and $C = C(n,p,q,k) \ge 1$ such that

$$(\fint_Q |f|^r dx)^{1/r} \le C (\fint_Q |f|^p dx)^{1/p} \tag{1.6}$$

In particular $f \in L^r(Q)$ for some $r > p$.

The class $B_q^p(K)$, as well as the unions $B_q^p \equiv \cup_{K\ge 1} B_q^p(K)$ do not form linear subspaces of $L^p_{loc}(\Omega)$.

While the "direct" Hölder inequality (1.1) holds for arbitrary measurable subsets of Ω the reverse inequality (1.5) is required to be true for cubes only. This is quite crucial from the geometric point of view.

Essentially equivalent class $B_q^p(K)$ arises if in (1.5) instead of cubes we admitt balls or, more generally, quasiconformal images of cubes with a uniformly bounded dilatation,[5] [27],e.g. quasiballs. For the class $A_{m,M}$ of functions satisfying the inequality $m \le |f(x)| \le M$ a.e. in Ω for some fixed positive contants m,M (1.5) holds for arbitrary measurable sets $E \in \Omega$ (with $K = \frac{M}{m}$). It is easy to see that if we would require (1.5) to be true for averages taken over arbitrary measurable subsets of Ω the classes $B_q^p(K)$ would reduce to the classes $A_{m,M}$. Analogously we easily see that the condition

$$\|f\|^{\#}_{E,1} \le C < \infty \tag{1.4'}$$

for arbitrary measurable $E \subset \Omega$ and fixed C implies that the function f is essentially bounded. On the other hand it is well known that the inclusion $L^\infty(\Omega) \subset BMO(\Omega)$ is proper [26].

3. Let $\delta > 1$ and let $F_\delta(\Omega) \equiv F_\delta$ denote the family of all cubes Q

Q such that the cube $\delta Q \subset \Omega$. If instead of (1.5) we require the inequality

$$(\fint_Q |f|^p dx)^{1/p} \leq K(\fint_{\delta Q} |f|^q dx)^{1/q} \tag{1.7}$$

to hold for all cubes in F_δ then we say that the function $f \in WB_q^p(K)$ or f satisfies the weak reverse Hölder inequality. (reverse Hölder inequality with change of support).

Functions in the class $WB_q^p(K)$ also enjoy the self-improving property i.e. the inequality

$$(\fint_Q |f|^r dx)^{1/r} \leq C(\fint_{\delta Q} |f|^p dx)^{1/p} \tag{1.6'}$$

holds for each cube $Q \in F_\delta(Q_o)$ and $f \in WB_q^p(K) \cap L^p(Q_o)$ with the constants $r = r(n,p,q,K,\delta)$ and C depending on δ also. Here Q_o is a fixed cube in Ω.

The weak reverse Holder inequality (1.7) is much more flexible and has much broader range of applications then the inequality (1.5). Applications to p.d.e. have been studied by M.Giaquinta and others [13], [5]. In [15] the weak reverse inequality has been applied to quasiregular (non 1-1 generalisations of quasiconformal) mappings. The proof of the estimates (1.6') is given in [13]. Another independent approach see [5].

4. An important modification of the inequality (1.4), defining the BMO class, has been introduced by Gurov and Rešetniak [16],[17],[27]. Here we discuss some extension of the Gurov-Rešetniak class, which we call $WGR(\varepsilon,\delta)$, (weak Gurov-Rešetniak class).

Definition. Let $\varepsilon > o$ and $o < \delta \leq 1$, $q \geq 1$. A function $f \in L^q_{loc}(\Omega)$ is in $WGR(\varepsilon,\delta,q)$ if the inequality

$$(\fint_{\delta Q} |f - f_{\delta Q}|^q dy)^{1/q} \leq \varepsilon \fint_Q |f| dy \tag{1.8}$$

holds for all cubes $Q \subset \Omega$.

While inequality (1.4) gives the uniform bound for the averaged local oscillation of the function f around its average value on each cube $Q \subset \Omega$. (1.8) estimates the averaged local oscillation of f on a smaller cube δQ (for $\delta < 1$) relative to the averaged absolute value of f on the cube Q.

Of interest is the asymptotic behaviour of the class $WGR(\varepsilon)$ for

$\varepsilon \to 0$, since for big values of ε (if $\delta = 1$ for $\varepsilon = 2$) (8) holds for any function in $L^1_{loc}(\Omega)$. Gurov and Resetniak studied the case $\delta = 1$. The weak classes $WGR(\varepsilon,\delta,1)$ were discussed in [18],[19],[4]. A general result about WGR classes is contained in the following theorem.

Theorem. There exist a constant $C = C(n,q)$, depending on the dimension n and q only, such that $WGR(\varepsilon,\delta,q) \subset L^p_{loc}(\Omega)$ for $p < \frac{C}{\varepsilon}$, provided ε is small. Moreover, for constants b_1 and b_2 depending on $(\delta_o,\delta,n,p,q,\varepsilon)$ we have (for $\delta_o < \delta$) the uniform estimates

$$\left(⨍_{\delta_o Q} |f|^p dy\right)^{1/p} \leq b_1 ⨍_{Q} |f|\, dy$$

and

$$\left(⨍_{\delta_o Q} |f-f_{\delta_o Q}|^p dy\right)^{1/p} \leq b_2 \left(⨍_{_oQ} |f-f_{\delta_o Q}|^q dy\right)^{1/q}$$

Some sketch of the proof is given in [4]. See also [19]. It relies essentially on the result of I.Wik [35].

5. Instead of best local approximation norms $\mathbb{E}^q_1(f;Q)$ it is sometimes more convenient to use the local norms related with averaging of formal Taylor expansions [9],[3],[30]

$$T^{q,k}(f;Q) = \left(⨍_Q |f-T^k_Q f|^q dy\right)^{1/q} \tag{1.9}$$

where

$$T^k_Q f(x) = ⨍_Q T^k f(y;x)\varphi_Q(y)\, dy \tag{1.9'}$$

are projectors of Sobolev spaces $W^{1,r}_{loc}(\Omega)$, $1 \geq k$ on polynomials of degree $\leq k$ and $T^k f(y;x)$ is the formal Taylor expansion of order k

$$T^k f(y;x) = \sum_{|\alpha|=0}^{k} \frac{1}{\alpha!} \frac{\partial^\alpha f(y)}{\partial y^\alpha} (x-y)^\alpha$$

$\{\varphi_Q\}$ is a given family of functions, $\varphi_Q \in C^\infty_o(Q)$, normalized by the condition $⨍_Q \varphi dx = 1$. A convenient choice of the family $\{\varphi_Q\}$ can be made by translation and dilation from a fixed function φ_{Q_o}, where Q_o is the standard unit cube centered at the origin of R^n. Some examples of the use of the seminorms (1.9) and the projectors (9') are given in [3],[7],[9],[30] (See also the references quoted there). In the references quoted above the seminorms (1.1), (1.4) or

(1.9) are used to produce the corresponding maximal functions, generalizing the classical concept of the maximal operator of Hardy--Littlewood [32]. After suitable normalizations with a weight factor of type $|Q|^{-\lambda}$, for some $\lambda \geq 0$, the corresponding maximal operators can be used to describe the local differentiability properties of the functions considered [30],[7],[3]. The maximal operators produce the "pointwise" versions of the geometric inequalities for seminorms defining the considered function classes. In this connection it is an interesting question to study what geometric properties of the function classes under consideration can be recovered from the information on the behaviour of the corresponding maximal functions.
Another useful operation connected with the study of local seminorms is related with the concept of variations of the seminorms, see [7]. The maximal operators as well as the variations can be used to produce interesting examples of function spaces of Campanato-Morrey type. Also Lorentz type spaces fall in this class [7].
These introductory remarks are intended to give some flavour of the wide spectrum of geometric and analytical questions related with the concept of the local (or rather semilocal) function space. In what follows we give a more precise and rather detailed presentation how these ideas can be used to discuss the fundamental Sobolev imbedding theorems. We show that elementary geometric considerations allow to pass from local to global embeddings. One of our results can be interpreted as "factorisation" of the Sobolev imbedding inequality into a local and well known analytical inequality and a lemma describing the "integration" or "globalisation" of the local inequalities, intimately related with the semiglobal geometric structure of the domain where the function is defined. We remark that this approach gives the global Sobolev imbedding inequality for a rather general class of domains containing the John-domains (see [28]).

II. A function $f:\Omega \to R$ belongs to the local Sobolev approximation class: loc $SA^p_l(\Omega)$ if there exist constants $C = C(n,l)$ and $\hat{C}$ and for each (open) cube $Q \subset \Omega$ there exist a polynomial $P_Q(f) \equiv P(f) \subset \mathbb{P}_{l-1}$ such that the inequality

$$\left(\int\!\!\!\!-_Q |f-P_Q(f)|^{p^*}dx\right)^{1/p^*} \leq C(n,l)\,(r(Q))^l \left(\int\!\!\!\!-_Q |\nabla^l f|^p dx\right)^{\frac{1}{p}} \quad (pl < n) \tag{2.1}$$

holds and the polynomial $P_Q(f)$ satifies the estimate

$$|P_Q(f)(x)| \leq \hat{C} \fint_Q |f|\,dx \qquad x \in Q$$

The function f is in $SA_1^p(\Omega)$ if for some constant $C = C(\Omega)$ there exist a polynomial $P(f) \subset \mathbb{P}_{1-1}$ such that the Sobolev imbedding inequality holds.

$$\left(\int_\Omega |f - P(f)|^{p^*} dx\right)^{1/p^*} \leq C(\Omega) \left(\int_\Omega |\nabla^1 f|^p dx\right)^{1/p} \tag{2.2}$$

Both in (2.1) and (2.2) it is understood that all integrals exist. Introducing the local class $\text{loc } L^r(\Omega)$, $r > o$, meaning $f \in \text{loc } L^r(\Omega)$ if for each cube $Q \subset \Omega$ the integral $\int_Q |f|^r dx$ exists and is finite, we can say that in (2.1) we assume $|\nabla^1 f| \in \text{loc } L^p(\Omega)$ and $f \in \text{loc } L^1(\Omega)$. The Sobolev imbedding inequality implies then that $f \in \text{loc } L^{p^*}(\Omega)$ and (2.1) holds. As is well known for arbitrary bounded $\Omega \subset R^n$ $C_o^\infty(\Omega) \subset SA_1^p(\Omega)$ (with $P(f) \equiv 0$)

If instead of cubes we consider balls $B \subset \Omega$ we obtain, in general, another class. In many cases it is more convenient to use the "weak" local classes (see [4] [5]). For $\delta > 1$ we say that $f \in \text{Loc}_\delta\, SA_1(\Omega)$ if for each cube $Q \subset \Omega$, such that $\delta Q \subset \Omega$ the inequality

$$\left(\fint_Q |f - P_Q(f)|^{p^*} dx\right)^{1/p^*} \leq C(n,1,\delta)\, r^1(Q) \left(\fint_{\delta Q} |\nabla^1 f|^p dx\right)^{1/p} \tag{2.3}$$

holds. For $\delta > 1$, the "cubical" and "spherical" weak local classes $\text{loc}_\delta SA_1$ are essentially equivalent. For simplicity, in what follows, we mainly consider the case $\delta = 1$.

The classical local Sobolev imbedding theorem may be stated in the form [15][23] [31].

Theorem (2.1). Let Ω be an arbitrary domain in R^n. Then $W^{1,p}(\Omega) \subset \text{loc } SA_1^p(\Omega)$.

Remark. For $\delta > 1$ we have $W^{1,p}_{loc}(\Omega) \subset \text{loc}_\delta SA_1^p(\Omega)$ or even as sets $W^{1,p}_{loc}(\Omega) \equiv \text{loc}_\delta SA_1^p(\Omega)$.

If Ω is a cube Q_o, then clearly $W^{1,p}(Q_o) = \text{loc } SA_1^p(Q_o) = SA_1^p(Q_o)$.

The global Sobolev imbedding theorem states that for domains with sufficiently smooth boundary, say $C^{o,1}$ domains, bounded Lipschitz domains or domains satisfying the Sobolev cone condition ([15],[23],[31]) the Sobolev space $W^{1,p}(\Omega)$ coincides with $SA_1^p(\Omega)$.

Let g be a fixed function from $\text{loc } L^p(\Omega)$, $g \geq 0$. A function

$f \in \mathrm{loc}\, L^{p^*}(\Omega)$ is in the local approximation space $\mathrm{loc}\, A^p_{l,\delta}(\Omega)$ $\delta > 1$ if for each cube Q, $\delta Q \subset \Omega$, there exist a polynomial $P_Q(f) \equiv P_Q$ such that the inequality

$$(⨍_Q |f-P_Q|^{p^*}\, dx)^{1/p^*} \leq C(r(Q))^l (⨍_{\delta Q} g^p dx)^{1/p} \tag{2.4}$$

holds. The constant C in (2.4) does not depend on f and the cube Q (although it may depend on δ and vary with Ω if we consider a family of domains Ω). In (2.4) $p^* = \frac{np}{n-lp}$ and we assume that $n > lp$. The polynomial $P_Q(f)$ is assumed to satisfy the estimate

$$|P_Q(f)(x)| \leq \hat{C} ⨍_Q |f|\, dx \qquad x \in Q \tag{2.5}$$

for some constant $\hat{C}$ independent of f and Q.
There exist classes of domains, decomposable in cubes (a countable family of cubes) in a special way, defined by some geometrical conditions, for which the local inequality (2.4) imply a global inequality of type (2.2). We describe here one class of these domains, implicitely introduced in [6], by J.Boman, in connection with the study of the class of very strongly elliptic systems of P.D.E. See also a modification of the Boman's definition in [20]

Definition. An open set $\Omega \subset R^n$ is said to satisfy a chain condition $\mathbb{F}(\delta,B)$, $\delta > 1$, $B \geq 1$ if there exists a covering $\mathbb{F}$ of Ω by open cubes such that

$$\sum_{Q \in F} \chi_{\delta Q}(x) \leq B\chi_\Omega(x) \qquad x \in R^n \tag{2.6)i}$$

(2.6)ii) There exists a distinguished cube $Q_0 \in \mathbb{F}$ which can be connected with every cube $Q \in \mathbb{F}$ by a chain of cubes $Q_0, Q_1 \dots Q_s \equiv Q$ from $\mathbb{F}$ such that for each $\nu = 0,1,\dots,s-1$

$$Q \subset BQ_\nu$$

and

$$C^{-1} \leq \frac{|Q_\nu|}{|Q_{\nu+1}|} \leq C \qquad \nu = 0,1,\dots,s-1 \tag{2.6)iii}$$

$$|Q_\nu \cap Q_{\nu+1}| \geq C^{-1} \max(|Q_\nu|, |Q_{\nu+1}|)$$

where the constant C depends on n only.

Notice that the length s of the connecting chain is not assumed to be (uniformly) bounded in $\mathbb{F}$. It is proved in [6] that the bounded John domains satisfy the chain condition. Thus, [28], the domains in this class may have a very irregular boundary (e.g. they may be snowflake type domains, [23].

Theorem 2.2. Let the domain Ω satisfy the chain condition $\mathbb{F}(\delta,B)$. If the function $f \in \mathrm{Loc}\, L^{p^*}(\Omega)$ satisfies the inequality (2.4) for all cubes $Q \in \mathbb{F}$ and the function $g \in L^p(\Omega)$ then there exists a constant $\hat{C}$ depending on n,p,δ, B and C only, such that the global inequality

$$(\int_\Omega |f-P(f)|^{p^*}dx)^{1/p^*} \leq \hat{C}(\int_\Omega |g|^p)^{\frac{1}{p}} \tag{2.7}$$

holds.

Corollary. If the domain Ω satisfies the chain condition and the l-th gradient $\nabla^l f$ of a function f belonging to the Sobolev space $W^{l,p}_{loc}(\Omega)$ is in $L^p(\Omega)$ then the global Sobolev inequality (2.2) holds. In short, we have

$$\mathrm{loc}\ SA^p_l(\Omega) \cap L^{l,p}(\Omega) = SA^p_l(\Omega) \equiv W^{l,p}(\Omega) \tag{2.8}$$

Here $L^{l,p}(\Omega)$ is the class of distributions in Ω with $\nabla^l f \in L^p(\Omega)$. The corollary implies, in view of Boman's theorem, the global Sobolev inequality (2.2) for a rather general class of domains, see [6]. It is a very interesting problem to find necessary and sufficient geometric conditions on the domain Ω, which insure the validity of the theorem 2.2 or it's analogue. For these domains we would clearly have $W^{l,p}(\Omega) \equiv SA^l_p(\Omega)$ and they could be considered as "natural" domains for the study of Sobolev function spaces.

Sketch of the proof of theorem 2.2. Since $p^* = \frac{np}{n-lp}$ the inequality (2.4) takes the form

$$(\int_Q |f-P_Q|^{p^*}dx)^{1/p^*} \leq \hat{C}(\int_{\delta\Omega} g^p dx)^{1/p} \tag{2.9}$$

with $\hat{C}$ easily controlled by C,δ,n.

In the case $l = 1$ the polynomials $P_Q(f)$ in (2.4) are constants and the calculations in [6] or [20] go through with minor changes. In Boman's paper [6] and in [20] the case $p^* = p$ was considered, what corresponds to the Poincaré type inequality. For $l > 1$ some modifications of the estimates are necessary. However it can be checked, with some care, using the Bernstein-Markov type inequalities for polynomials in the averaged L^p form as in [2],[8], together with the conditions (2.5 iii),that also in this case the local inequalities (2.4) can be summed up to (2.6) with an explicitly calculable control of the constant $\hat{C}$. In both cases $l = 1$ and $l > 1$ a convenient tool in carrying the estimates is the lemma 3.3 in [6] going back to Strömberg and Torchinsky [33].

Remark. The definitions above and theorem (2.2) hold (with rather obvious changes) if instead of the Lebesgue measure dx we use a measure $d\mu$ satisfying a doubling condition or a measure with density w, $d\mu = w\,dx$, in a Muckenhoupt class [24],[26].

III. Let φ be a real valued function in $L^1(\Omega)$, $\int_\Omega \varphi dx = 1$. The domain Ω is said to be a $W_1(\varphi)$ domain if there exist a constant C, depending on Ω and φ, such that for each $f \in W^{1,1}(\Omega)$ the inequality

$$|f(x) - \int_\Omega \varphi(y) f(y)\,dy| \leq C \int_\Omega \frac{|\nabla f(y)|\,dy}{|x-y|^{n-1}} \tag{3.1}$$

holds for almost each $x \in \Omega$.
Ω is said to be a W_1 - domain if it is a $W_1(\varphi)$ - domain for some φ.
For $1 < l < n$, Ω is said to be a $W_l(\varphi)$ - domain if there exist a constant $C_l(\Omega,\varphi)$ such that for each $f \in W^{l,1}(\Omega)$ there exist a polynomial $P_{f,\varphi}(x) \in \mathbb{P}_{l-1}$ such that the inequality

$$|f(x) - P_{f,\varphi}(x)| \leq C_l \int_\Omega \frac{|\nabla^l f(y)|}{|x-y|^{n-1}}\,dy \tag{3.2}$$

holds a.e. in Ω.
The notation $P_{f,\varphi}(x)$ implies some constructive process of the choice of a polynomial $P_{f,\varphi}$ depending on the normalized weight $\varphi(x)$ (see [3],[7],[9],[10],[30],[31].)
Notice that, in particular, (3.2) implies that $P_{f,\varphi}(x)$, as an operator on $f \in W^{l,1}(\Omega)$, is always a projector on the finite dimensional space $\mathbb{P}_{l-1} \subset W^{l,1}(\Omega)$ i.e. $P_{f,\varphi}(x) \equiv f(x)$ for $f \in \mathbb{P}_{l-1}$.
A slightly more general definition could also be introduced admitting in (3.1), instead of the linear functional $\int_\Omega \varphi f\,dx$, a constant C_f depending on f in an unspecified way. We skip over the somewhat

delicate problem of discussing the relations of the two definitions. The constructive definition has an obvious advantage.

The inequality (3.2) states that the function $f(x)$ can be approximated globally in Ω by a polynomial of order $\leq l-1$ with the error controlled by the Riesz potential of order l of the l-th gradient $\nabla^l f$ in some uniform way.
The fundamental Hardy-Littlewood-Sobolev theorem on fractional integration implies that in a W_l domain Ω the Sobolev imbedding is valid: for $n > lp$ $W^{l,p}(\Omega) \subset L^q(\Omega)$ with $q = \frac{np}{n-lp}$. Moreover we have the estimate.

$$(\int_\Omega |f(x) - P_f(x)|^q dx)^{1/q} \leq C_l K (\int_\Omega |\nabla^l f|^p dy)^{1/p} \tag{3.3}$$

with a constant $K = K(n,p,l)$ independent of Ω.
It is well known that a convex bounded domain Ω in R^n is a $W_1(\varphi)$-
-domain for each bounded φ, with the constant $C(\Omega,\varphi) \leq \sup|\varphi| \cdot \frac{d^n}{n}$,

where $d = \operatorname{diam}\Omega$ is the diameter of Ω. In particular if E is a measurable subset of Ω, $|E| > 0$, then ([15],[9])

$$|f(x) - \int_E f(y)dy| \leq \frac{d^n}{n|E|} \int_\Omega \frac{|\nabla f(y)|}{|x-y|^{n-1}} dy. \tag{3.4}$$

A starlike domain with respect to a ball $B \subset \Omega$ or a domain satisfying a cone-condition is a W_1-domain. Also bounded uniform domains, John domains [27] and,more generally,domains of the class S introduced in [28],[29], are W_1-domains.

The snowflake or the Koch curve studied in the theory of fractals [23] is an example of a W_1 domain whose boundary is very irregular. Examples of domains for which inequalities of type (3.3) do not hold go back to O.Nikodym [25]. See also [23].

Theorem 3.1.Each W_1-domain is a W_l-domain. Moreover the constant C_l in (3.2) can be estimated from above in terms of C in (3.1)

Proof. We use an idea of Resetniak from [29]. The calculations involved are implicitely contained in the paper [31] of Sobolev. Given $x \in \Omega$ we define

$$\psi(x,\xi) = \sum_{|\alpha| \leq l-1} \frac{(x-\xi)^\alpha}{\alpha!} D^\alpha f(\xi) \tag{3.5}$$

for an $f \in W^{l,1}(\Omega)$. Because of density it suffices to assume f smooth. Differentiating (3.5) we get

$$\frac{\partial\psi}{\partial\xi_i}(x,\xi) = \sum_{|\alpha|=l-1} \frac{(x-\xi)^\alpha}{\alpha!} D^{\alpha+\delta_i} f(\xi) \tag{3.6}$$

where δ_i is the multi index $(0,\ldots,1,0,\ldots,0)$ with i-th component equal 1. It is crucial that in (3.6) all derivatives of f of order $\leq l-1$ cancel. As a function of ξ, $\psi(x,\xi) \in W^{1,1}(\Omega)$. Applying (3.1) we get

$$|\psi(x,\xi) - \int_\Omega \varphi(\eta)\psi(x,\eta)\,d\eta| \leq \hat{C} \int_\Omega \frac{(\sum_i |\frac{\partial\psi}{\partial\eta_i}(x,\eta)|^2)^{1/2}}{|\xi-\eta|^{n-1}} \, d\eta$$

For $\xi = x$ we get

$$|f(x) - P_{f,\varphi}(x)| \leq \hat{C} \int_\Omega \frac{|\nabla^l f(\eta)\,d\eta|}{|x-\eta|^{n-1}} \tag{3.8}$$

since in (3.6) only derivatives of f of order $|\alpha+\delta_i| = l$ appear and $\psi(x,x) = f(x)$. The polynomial $P_{f,\varphi}(x)$ in (3.7) is given by the formula

$$P_{f,\varphi}(x) = \int_\Omega \sum_{|\alpha|\leq l-1} \frac{(x-\eta)^\alpha}{\alpha!} D^\alpha f(\eta)\varphi(\eta)\,d\eta \tag{3.8}$$

For $x \in \Omega$ it has the estimate

$$|P_{f,\varphi}(x)| \leq \tilde{C} \|f\|_{W^{l-1,1}(E)}, \quad E = \operatorname{supp} \varphi \tag{3.9}$$

where $\tilde{C}$ depends on n, l, $\sup|\omega|$ and $d = \operatorname{diam}(\Omega)$.
If $\varphi \in C_0^\infty(\Omega)$, or $C_0^{l-1}(\Omega)$ we can differentiate by parts in (3.8) and obtain the estimate

$$|P_{f,\varphi}(x)| \leq \tilde{\tilde{C}} \int_E |f|\,dx \tag{3.10}$$

with $\tilde{\tilde{C}}$ depending on n, l and d and $\sup|D^\beta\varphi|$ for $|\beta| \leq l-1$.
The theorem expresses the fact that if the geometry of the domain Ω allows any function $f \in W^{1,p}(\Omega)$ to be globally approximated in C^0 norm by a constant in a uniform way controlled by the Riesz-potential of order 1 of the gradient $|\nabla f|$ then each function in $W^{l,p}(\Omega)$, $l > 1$, can be globally uniformly approximated by a polynomial of order $\leq l-1$ in a way controlled by a Riesz - potential of order l of the gradient $|\nabla^l f|$.
The inequalities (3.9) and (3.10) contain additional information on the approximating polynomials. Together with Markov type inequalities for polynomials [2] [8] they can be used to obtain various Sobolev type

estimates for f in $W^{k,q}(\Omega)$, $k < l$ in terms of $\|\nabla^l f\|_{L^p(\Omega)}$ and $\int_E |f|$ for appropriate values of q (see [3],[9],[23],[1],[31])
Another interpretation of inequalities (3.1), (3.2) and the theorem may be given in terms of Whitney topology in the space of (measurable) functions $f(x)$, $x \in \Omega$ [34]. They state, that, in a proper convex Whitney neighbourhood of a given $f \in W^{l,1}(\Omega)$, defined by the Riesz-potential of the gradient $\int_\Omega \frac{|\nabla^l f(\eta)|\,d\eta}{|x-\eta|^{n-1}}$ and absolute constants C, $\hat{C}$ there always exist polynomials of degree $\leq l-1$ (or constants). These polynomials may be required to satisfy additional estimates (3.9) (3.10) (by change of $\hat{C}$). We remark also that the formulas of type (3.8) give useful local approximations of functions in Sobolev spaces by polynomials [2],[7] [9],[10],[28],[30] with rather good approximating properties.

REFERENCES

[1] R.A. ADAMS, Sobolev spaces, New York; San Francisco; London 1975, 268 p.

[2] B. BOJARSKI, Remarks on Markov's Inegualities and Some Properties of Polynomials, Bull. of the Pol.Acad.of Sci. vol.33, No 7-8, 1985.

[3] B. BOJARSKI, Sharp maximal operator of fractional order and Sobolev imbedding inequalities, Bull.Pol.Ac.: Math., 33 (1985), 7 - 16

[4] B. BOJARSKI, Remarks on stability of inverse Hölder inequalities and quasiconformal mappings, Ann.Acad.Sci.Fenn., 10 (1985)

[5] B. BOJARSKI, T. IWANIEC, Analytical foundations of the theory of quasiconformal mappings in R^n, ibid. 8(1983) 257-324

[6] J. BOMAN, L^p-estimates for very strong elliptic systems-Preprint, to appear in J.Analyse Math.

[7] Ju.A. BRUDNYJ, Prostranstva opredelaemyje s Pomoŝĉu lokalnyh priblizenii. Trudy Mosk.Mat. Obscestva, 1971, t.24

[8] Ju.A. BRUDNYJ, M.I.GANSBURG, On an extremal problem for polynomials of n variables, Izviestia AN USSR, 37 (1973), 344-355.

[9] V.I. BURENKOV, Sobolev's integral representation and Taylor's formula, Proc.Steklov Inst. 131(1974), 33-38

[10] S. CAMPANATO, Proprieta di una famiglia di spazi funzionali, Ann. Scuola Norm.Pisa. 18 (1984), 137-160

[11] R. COIFMAN, C. FEFFERMAN, Weighted norm inequalities for maximal functions and singular integrals. Studia Math., 51 (1974), 241-250

[12] M. GIAQUINTA, Multiple integrals in the calculus of variations and non-linear elliptic systems, Preprint No 443, SFB 72, Bonn 1981

[13] M. GIAQUINTA, G. MODICA, Regularity results for some classes of higher order non-linear elliptic systems. J. für Reine u.Angew. Math. 311/312, 1979, pp.145-169

[14] F.W. GEHRING, The L^p-integrability of the partial derivatives of a quasiconformal mapping. Acta Math. 130, 1973, 265-277.

[15] D. GILBARG, N.S. TRUDINGER, Elliptic Partial Differential Equations of Second Order, Springer-Verlag,Berlin Heidelberg N. Tokyo 1983

[16] L.G. GUROV, The stability of Lorentz transformations, Estimates for the derivatives. - Dokl.Akad.Nauk SSSR 220. 1975, 273-276 (Russian)

[17] L.G. GUROV, Yu.G.RESETNYAK, A certain analogue of the concept of a function with bounded mean oscillation. Sibirsk.Mat.Zh.17, No 3. 1976, 540-546 (Russian)

[18] T. IWANIEC, Some aspects of partial differential equations and quasiregular mappings. - Proc. of the Int. Congress of Math. Warsaw, August 1983, PWN-North-Holland. Warsaw, 1984

[19] T. IWANIEC, On L^p-integrability in PDE's and quasiregular mappings for large exponents - Ann.Acad.Sci.Fenn. Ser A.I.Math 7, 1982, 301-322

[20] T. IWANIEC, C.A. NOLDER, Hardy-Littlewood inequality for quasiregular mappings in certain domains in R^n, Ann.Acad.S. Fenn, Ser.A.I.Math., Vol. 10, 1985, 267-282

[21] F. JOHN, L.NIRENBERG, On functions of bounded mean oscillation, Comm Pure App. Math., 141 (1961)

[22] O.A. LADYZHENSKAYA, N.N. URAL'TSEVA, Linear Quasilinear Elliptic Equations. Moscow, Izdat. "NAUKA" 1964 (Russian)

[23] V.G. MAZJA, Prostranstva S.L.Soboleva, Izdat. Lening. Univ. Leningrad 1985

[24] B. MUCKENHOUPT, Weighted norm inequalities for the Hardy Maximal function, Trans. AMS, 65 (1972), 207-226.

[25] O. NIKODYM, Sur une classe de fonctions considérée dans le probleme de Dirichlet, Fundam.Mat. 1933, vol.21, p.129-150.

[26] H.M. REIMANN, T. RYCHENER, Funktionen beschränkter mittlerer Oszillation, Lec. Notes ,No 487, Springer, 1975

[27] Ju.G. RESETNIAK, Stability theorems in geometry and analysis, Science, Novosibirsk, 1982.

[28] Ju.G. RESETNIAK , Integralnyje predstavlenia differencirujemyh funkcii v oblstiah s negladkoi granicej, Sibir.Math.Zurnal, t.21. no 6. 1980 p.108-116.

[29] Ju.G. RESETNYAK, Zamecania ob integralnyh predstavleniah differencujemyh funkcii mnogih peremènnyh, Sibir.Mat.Zurnal, t.25, No 5, 1984

[30] R. SHARPLEY, R. DEVORE Maximal function as measure of smoothness Memoirs of the AMS 293, Providence 1984,

[31] S.L. SOBOLEV, Some applications of functional analysis, Izdat. Leningrad Gos.Univ,. Leningrad 1950

[32] E. STEIN, Singular integrals and differentiability properties of functions. Princeton Univ.Press (1970)

[33] J.O. STÖMBERG, A.TORCHINSKY, Weights, sharp maximal functions and Hardy spaces, Bull. Amer.Math.Soc.(N.S.) 3, 1980, 1053-1056

[34] H. WHITNEY, Analytic extensions of differentiable functions defined in closed sets, Trans.Am.Math.Soc. 36 (1934), 63-89

[35] I.WIK, A comparison of the integrability of f and Mf with that of f*. - Preprint series 2, Dept.Math., Univ. of Umea, 1983

The classes V_a are monotone

Lennart Bondesson and Jaak Peetre

Swedish University of Agricultural Sciences
S-90 183 Umeå, Sweden

University of Lund
S-221 00 Lund, Sweden

The classes V_a, where a is a positive number, were defined in [6] in connection with a problem on interpolation spaces originating from Foiaş and Lions [3] (1961). A function f on $(0,\infty)$ is in V_a iff it admits the representation

$$f = k_a * \mu,$$

including, if we are generous, also limiting cases of such functions. Here μ is a positive measure on $(0,\infty)$ and $*$ denotes convolution with respect to the multiplicative structure, with the kernel

$$k_a(x) \overset{\text{def}}{=} \frac{x}{(1+x^a)^{1/a}}.$$

Explicitly:

$$f(x) = \int_0^\infty \frac{1}{(1+(\frac{y}{x})^a)^{1/a}}\,\mu(dy).$$

Special case: $a = 1$ - Pick (or Loewner) functions. Limiting case: $a = \infty$ - concave functions. (The letter V is for Swedish "vikning", convolution. One of the authors being a probabilist, we write $\mu(dy)$, not $d\mu(y)$.)

In this note, answering a question posed in [6], we show that

$$V_a \subset V_b \quad \text{if} \quad a \le b.$$

In [6] this was proved only for $b/a = n$ = integer. Notice also that the limiting case $b = \infty$ is trivial: the functions in any V_a are concave. (As $dk_a/dx = (1+x^a)^{-1/a-1}$, k_a certainly is concave for each a.)

The proof depends on some ideas about generalized gamma convolutions (GGC), a notion introduced by Olof Thorin [7], [8] in 1977 and subsequently studied by many authors (see e.g. [2] and the references given there.) Accordingly we start in Sec. 1 by recalling the rudiments about GGC (nothing deep is really required).

1. GGC. For the participant of this seminar, to whom the name of Thorin is so familiar, it may be interesting now to hear about some of his lasting contributions in a quite different area of mathematics. The following is just a paraphrase of what is on p. 32-33 of [7].

A distribution F(x) on $[0,\infty)$ is said to be a GGC if its Laplace transform $\int_0^\infty e^{-sx}\, dF(x)$ can be written as

$$f(s) = \exp(-as + \int_0^\infty \log\frac{1}{1+\frac{s}{y}}\, U(dy))$$

with $a \geq 0$ and U a positive measure. For instance, if (gamma distribution with parameters y and u)

$$F(x) = \frac{y^u}{\Gamma(u)} \int_0^\infty t^{u-1} e^{-yt}\, dt$$

then

$$f(s) = \frac{1}{(1+\frac{s}{y})^u}.$$

Therefore gamma distributions are special GGC's and a general GGC may be pictured as a translate (by the amount of a) of a limit of convolutions (with respect to the additive structure!) of finitely many gamma distributions. Obviously any GGC is infinitely divisible but the point is that many concrete distributions may be shown to be GGC's, thus exhibiting their infinite divisibility.

2. The proof. To prove that $V_a \subset V_b$ if $a \leq b$ it is sufficient to prove that k_a is in V_b if $a \leq b$, i.e. that $k_a = k_b * \mu$. Explicitly:

$$(1) \qquad \frac{x}{(1+x^a)^{1/a}} = \int_0^\infty \frac{\frac{x}{y}}{(1+(\frac{x}{y})^b)^{1/b}}\, \mu(dy).$$

Set $s = x^b$, $t = y^b$. Then we get instead

$$(2) \qquad \frac{1}{(1+s^\alpha)^{1/a}} = \int \frac{1}{(1+\frac{s}{t})^{1/b}}\, \mu^*(dt)$$

with $\alpha \overset{def}{=} a/b \leq 1$ and $\mu^*(dt) \overset{def}{=} \mu(dy)/y$. We claim that the left hand side of (2) is the Laplace transform of a GGC.

This is easy. Indeed, setting $\varphi(s) = (1 + s^{\alpha})^{-1/a}$ we get

$$\frac{\varphi'(s)}{\varphi(s)} = -\frac{\alpha}{a}\frac{s^{\alpha-1}}{1+s^{\alpha}} = -\int_0^{\infty}\frac{u(t)}{t+s}\,dt$$

where

$$u(t) = \frac{1}{\pi}\frac{\alpha}{a}\frac{\sin\alpha\pi\, t^{\alpha-1}}{1+t^{2\alpha}+2t^{\alpha}\cos\alpha\pi}.$$

This can be seen either by the general theory (cf. [6]) or else directly by applying contour integration in one form or other. Integrating we get

$$(3)\qquad \frac{1}{(1+s^{\alpha})^{1/a}} = \exp\left\{\int_0^{\infty}\log\left[\frac{1}{1+\frac{s}{t}}\right]u(t)\,dt\right\},$$

which, apparently, is the Laplace transform of a GGC, since $u(t) \geq 0$.

The point is now that the right hand side may be viewed as the limit of finite products

$$\prod_{k=1}^{n}\frac{1}{(1+\frac{s}{t_{kn}})^{u_{kn}}}\quad\text{with}\quad\sum_{k=1}^{n}u_{kn} = \frac{1}{b}.$$

Each such finite product can be written as an integral

$$\int_0^{\infty}\frac{1}{(1+\frac{s}{t})^{1/b}}M_n(dt)\quad\text{with}\quad M_n(dt)\geq 0,$$

as can be readily seen; we defer the detail to the following section. This settles the case of finite products. It is now easy to carry out a passage to the limit and conclude that the left hand side of (3) too has such an integral representation, thus establishing (2) and so (1); for details see [1], p. 128, Theorem 3.1.

3. An integral formula.

What we have in mind is the formula

$$\frac{1}{(s+t_1)^{u_1}(s+t_2)^{u_2}} = \frac{1}{B(u_1,u_2)}\cdot\int_0^1\frac{(1-\lambda)^{u_1-1}\lambda^{u_2-1}}{(s+t_1(1-\lambda)+t_2\lambda)^{u_1+u_2}}\,d\lambda.$$

(Here $(B(u_1,u_2)$ is Euler's beta function.) For the proof it suffices to take $s = 0$ and to make the substitution

$$\mu = \frac{(1-\lambda)t_1}{t_1(1-\lambda) + t_2\lambda}, \quad 1 - \mu = \frac{\lambda t_2}{t_1(1-\lambda) + t_2\lambda},$$

taking also into account that

$$d\lambda = \frac{d\mu}{\mu(1-\mu)}.$$

In general, one gets by induction

$$\frac{1}{(s + t_1)^{u_1} \cdots (s + t_n)^{u_n}} =$$

$$= \frac{1}{B(u_1, \ldots, u_n)} \cdot \int \frac{\lambda_1^{u_1 - 1} \cdots \lambda_n^{u_n - 1}}{(s + t_1\lambda_1 + \ldots + t_n\lambda_n)^{u_1 + \ldots + u_n}} \, dArea,$$

where we integrate over the portion of the hyperplane $\lambda_1 + \ldots + \lambda_n = 1$ which lies in the positive "2^n-ant" $\lambda_1 \geq 0, \ldots, \lambda_n \geq 0$. ($B(u_1, \ldots, u_n)$ is the multivariate generalization of the function $B(u_1, u_2)$.) This, presumably, goes back to the days of Euler and Dirichlet (for an explicit reference in a mathematical context (the case n = 2 only), see [4], p. 168).

However, it is also possible to give a probabilistic proof. Let X_i be independent gamma variables with the frequency functions

$$\frac{t_i^{u_i - 1}}{\Gamma(u_i)} e^{-t_i x}, \quad x > 0$$

(and Laplace transform

$$\frac{1}{(1 + \frac{s}{t_i})^{u_i}}.)$$

We have

$$\sum_{i=1}^{n} X_i = \{\sum c_i \cdot \frac{Z_i}{\sum Z_i}\} \sum Z_i,$$

where $c_i = 1/t_i$ and the $Z_i = X_i \cdot t_i$ are gamma variables with scale parameter 1. But it is well-known (and easy to see) that ΣZ_i has a gamma distribution with form parameter Σu_i and scale parameter 1, and, in addition, is stochastically independent of

$$\left(\frac{Z_1}{\Sigma Z_i}, \frac{Z_2}{\Sigma Z_i}, \ldots, \frac{Z_n}{\Sigma Z_i}\right) = (Y_1, \ldots, Y_n),$$

which has the so-called Dirichlet distribution (cf. e.g. [5], pp. 231-234)

$$f_{Y_1, \ldots, Y_{n-1}}(y_1, \ldots, y_{n-1}) =$$

$$= \frac{1}{B(u_1, \ldots, u_n)} y_1^{u_1-1} y_2^{u_2-1} \ldots (1 - y_1 - y_2 - \ldots - y_{n-1})^{u_n-1}.$$

This leads to the previous result if we pass to Laplace transforms and take account of the independence.

References

1. Bondesson, L.: On generalized gamma and generalized negative binomial convolutions, Part I and II. Scandinavian Actuarial J. 1979, 125-146 and 147-166.

2. Bondesson, L.: New results on generalized gamma convolutions and the B-class. Scandinavian Actuarial J. 1984, 197-209.

3. Foias, C., Lions, J.-L.: Sur certains espaces d'interpolation. Acta Szeged 22, 262-282.

4. Hirshman, I.I., Widder, D.V.: The convolution transform. Princeton: Princeton University Press 1955.

5. Johnson, N.L., Kotz, S.: Distributions in statistics. Continuous multivariate distributions. New York: Wiley 1972.

6. Peetre, J.: On Asplund's averaging method - the interpolation (function) way. In: Constr. Theory of Functions '84, Varna, 1984, pp. 664-671. Sofia: Publishing House of the Bulg. Acad. Sci. 1984.

7. Thorin, O.: On the infinite divisibility of the Pareto distribution. Scandinavian Actuarial J. 1977, 31-40

8. Thorin, O.: On the infinite divisibility of the lognormal distribution. Scandinavian Actuarial J. 1977, 121-148.

HARDY-SOBOLEV SPACES AND BESOV SPACES

WITH A FUNCTION PARAMETER

Fernando Cobos
Departamento de Matemáticas
Universidad Autónoma de Madrid
28.049-Madrid-España

Dicesar Lass Fernandez
Instituto de Matemática
Universidade Estadual de Campinas
13.081-Campinas-S.P.-Brasil

0. INTRODUCTION

Besov spaces $B^{\rho}_{p,q}$ with a function parameter ρ arise naturally by interpolation with function parameter between classical Sobolev spaces H^s_p. They have been studied by C. Merucci in [8] and [9], where some properties of Sobolev spaces H^{ρ}_p with a function parameter ρ have been also described.

The present paper can be seen as a follow-up of those investigations.

We start by introducing the Hardy-Sobolev spaces $F^{\rho}_{p,q}$ with a function parameter ρ. These spaces are extensions of Sobolev spaces H^{ρ}_p, namely $F^{\rho}_{p,2} = H^{\rho}_p$.

In order to be complete and to show the relationships between $F^{\rho}_{p,q}$ and $B^{\rho}_{p,q}$ spaces, we describe some of the already known properties of $B^{\rho}_{p,q}$ spaces. Some new ones are also given.

Next we study the interpolation properties of those generalized Hardy-Sobolev and Besov spaces. We obtain formulas that complement and improve the early ones of C. Merucci. At this point, our results are gotten via interpolation between certain sequence spaces.

Finally, we turn our attention to the UMD-property. The class of Banach spaces with that property has been extensively studied recently (see e.g. D.L. Burkholder [4],[5] and J. Bourgain [2]), but except for the spaces ℓ_p, L_p, the Schatten classes S_p $(1<p<\infty)$, and some generalizations of these spaces (see [7]), only a few more examples of Banach spaces having that property are known. In the last section we show that the Hardy-Sobolev spaces $F^{\rho}_{p,q}$ and Besov spaces $B^{\rho}_{p,q}$ have also this property.

In this paper we content the real parameters p and q in the range $[1,\infty]$. The investigations on the spaces $F^{\rho}_{p,q}$ and $B^{\rho}_{p,q}$ with $0<p,q<\infty$ will appear in a forthcoming paper now in preparation.

1. FUNCTION PARAMETERS AND SPACES $\ell^{\rho}_{q}(E)$

1.1. In what follows, $\mathcal{B}$ denotes the class of all continuous functions $\rho:(0,\infty) \longrightarrow (0,\infty)$ with $\rho(1)=1$ and such that

(1) $$\bar{\rho}(t) = \sup_{s>0} \frac{\rho(ts)}{\rho(s)} < \infty \quad \text{for every } t > 0.$$

The function $\bar{\rho}$ is then Lebesgue measurable and sub-multiplicative ($\bar{\rho}(ts) \leq \bar{\rho}(t)\bar{\rho}(s)$).

1.2. The Boyd indices $\alpha_{\bar{\rho}}$ and $\beta_{\bar{\rho}}$ of the function $\bar{\rho}$ are defined by

(1) $$\alpha_{\bar{\rho}} = \inf_{1<t<\infty} \frac{\log \bar{\rho}(t)}{\log t} = \lim_{t\to\infty} \frac{\log \bar{\rho}(t)}{\log t},$$

(2) $$\beta_{\bar{\rho}} = \sup_{0<t<1} \frac{\log \bar{\rho}(t)}{\log t} = \lim_{t\to 0} \frac{\log \bar{\rho}(t)}{\log t}.$$

The indices $\alpha_{\bar{\rho}}$ and $\beta_{\bar{\rho}}$ are real numbers with the following properties (see [3])

(3) $$\alpha_{\bar{\rho}} < 0 \text{ if and only if } \int_1^\infty \bar{\rho}(t)\frac{dt}{t} < \infty, \text{ i.e., } \bar{\rho} \in L_1^*(1,\infty),$$

(4) $$\beta_{\bar{\rho}} > 0 \text{ if and only if } \int_0^1 \bar{\rho}(t)\frac{dt}{t} < \infty, \text{ i.e., } \bar{\rho} \in L_1^*(0,1).$$

1.3. Sometimes we shall work with the subset $\mathcal{B}''$ of $\mathcal{B}$ which consists of all functions $\rho \in \mathcal{B}$ such that ρ is C^∞ on $[1,\infty)$ and, moreover, satisfies

(1) $$t^m|\rho^{(m)}(t)| \leq c_m \rho(t), \quad 1 \leq t < \infty$$

for every $m \in \mathbb{N}$ (see [8]).

<u>1.4. DEFINITION.</u> Assume $1 \le q \le \infty$ and $\rho \in \mathcal{B}$. If E is a Banach space, the space $\ell_q^\rho(E)$ consists of all sequences $(a_j)_{j\ge 0}$, with $a_j \in E$ and such that $(\rho(2^j)\|a_j\|_E)_j \in \ell_q$.

We equip $\ell_q^\rho(E)$ with the norm

(1) $$\|(a_j)_j\|_{\ell_q^\rho(E)} = \|(\rho(2^j)\|a_j\|_E)_j\|_{\ell_q} .$$

When E is the scalar field, we write simply ℓ_q^ρ.

Clearly, $\ell_q^\rho(E)$ is a Banach space. The special case $\rho(t) = t^s$ ($s \in \mathbb{R}$) gives the usual space $\ell_q^s(E)$ (see [1] and [13]).

2. THE SPACES $F^\rho_{p,q}$ AND $B^\rho_{p,q}$

2.1. Let N be a fixed positive integer. A sequence $(\phi_j)_{j\ge 0}$ in the Schwartz class $S(\mathbb{R}^d)$ is said to be a system of test functions if it satisfies the following conditions

(1) $$\operatorname{supp} \hat{\phi}_j \subset \{t : 2^{j-N} \le |t| \le 2^{j+N}\} , \quad j=1,2,\ldots,$$
$$\operatorname{supp} \hat{\phi}_0 \subset \{t : |t| \le 2^N\} ;$$

(2) $$|\hat{\phi}_j(t)| \ge c_\varepsilon > 0 \quad \text{if} \quad (2-\varepsilon)^{-N} 2^j \le |t| \le (2-\varepsilon)^N 2^j , j=1,2,\ldots,$$
$$|\hat{\phi}_0(t)| \ge c_\varepsilon > 0 \quad \text{if} \quad |t| \le (2-\varepsilon)^N ;$$

(3) $$|D^\beta \hat{\phi}_j(t)| \le c_\beta 2^{-|\beta| j} \quad \text{for all} \quad \beta \in \mathbb{N}^d \text{ and } j=1,2,\ldots;$$

(4) $$\sum_{j=0}^\infty \hat{\phi}_j(t) \ge c > 0.$$

Here $\hat{\phi}_j$ is the Fourier transform of ϕ_j.

<u>2.2. DEFINITION.</u> Suppose $\rho \in \mathcal{B}$ and let $(\phi_k)_k$ be a system of test functions. We define, for $1 < p,q < \infty$,

(1) $$F^\rho_{p,q} = F^\rho_{p,q}(\mathbb{R}^d) = \{ f \in S'(\mathbb{R}^d) : (\phi_k * f)_k \in L_p(\ell_q^\rho)\} ;$$

and, for $1 \le p,q \le \infty$, we set

(2) $$B^\rho_{p,q} = B^\rho_{p,q}(\mathbb{R}^d) = \{ f \in S'(\mathbb{R}^d) : (\phi_k * f)_k \in \ell_q^\rho(L_p)\} .$$

We equip the spaces $F^{\rho}_{p,q}$ and $B^{\rho}_{p,q}$ with the norms

(3) $$\|f\|_{F^{\rho}_{p,q}} = \|(\phi_k * f)_k\|_{L_p(\ell^{\rho}_q)} = \| \|(\phi_k * f)_k\|_{\ell^{\rho}_q} \|_{L_p}$$

(4) $$\|f\|_{B^{\rho}_{p,q}} = \|(\phi_k * f)_k\|_{\ell^{\rho}_q(L_p)} = \|(\|\phi_k * f\|_{L_p})_k\|_{\ell^{\rho}_q}$$

respectively.

2.3. For $\rho(t)=t^s$, $s\in\mathbb{R}$, we get the usual Hardy-Sobolev and Besov spaces: $F^{\rho}_{p,q} = F^{s}_{p,q}$, $B^{\rho}_{p,q} = B^{s}_{p,q}$.

The next two statements can be proved in a similar way as in the usual parameter case (see e.g. [1,p.151] and [13,p.172]).

2.4. The spaces $F^{\rho}_{p,q}$ and $B^{\rho}_{p,q}$ do not depend on the particular system of test functions used to define them.

2.5. THEOREM. Let $\rho\in\mathcal{B}$. Then the following holds

(i) $F^{\rho}_{p,q}$ is a retract of $L_p(\ell^{\rho}_q)$ for $1< p,q<\infty$.

(ii) $B^{\rho}_{p,q}$ is a retract of $\ell^{\rho}_q(L_p)$ for $1\leq p,q\leq\infty$.

2.6. COROLLARY. The spaces $F^{\rho}_{p,q}$ and $B^{\rho}_{p,q}$ are complete.

3. THE OPERATOR J^{ρ} AND THE SPACES H^{ρ}_p

In order to show some lifting properties, we shall need the generalized Bessel (or potential) operator:

3.1. THE OPERATOR J^{ρ}. Suppose $\rho\in\mathcal{B}''$ and $g\in S'(\mathbb{R}^d)$. Then $\rho(1+|.|^2)^{1/2}g \in S'(\mathbb{R}^d)$. The Bessel operator J^{ρ} is defined by

(1) $$J^{\rho}f = \mathfrak{F}^{-1}\rho(1+|.|^2)^{1/2}\mathfrak{F}f , \quad f \in S'(\mathbb{R}^d).$$

Here $\mathfrak{F}^{-1}$ is the inverse of the Fourier transform $\mathfrak{F}$ on $S'(\mathbb{R}^d)$. It is clear that J^{ρ} is a linear bijective operator from $S'(\mathbb{R}^d)$ onto $S'(\mathbb{R}^d)$, such that $(J^{\rho})^{-1} = J^{1/\rho}$ and $J^{\rho}(S(\mathbb{R}^d)) = S(\mathbb{R}^d)$.

Proceeding as in [13,p.180], one can establish

3.2. THEOREM. Let $\rho \in \mathcal{B}$ and $\sigma \in \mathcal{B}''$. Then J^{σ} is an isomorphism

(i) from $F^{\rho}_{p,q}$ onto $F^{\rho/\sigma}_{p,q}$, $1 < p,q < \infty$

(ii) from $B^{\rho}_{p,q}$ onto $B^{\rho/\sigma}_{p,q}$, $1 \leq p,q \leq \infty$.

Now we shall show that the spaces H^{ρ}_{p} introduced by C. Merucci [8] are particular cases of the spaces $F^{\rho}_{p,q}$. We shall first recall

3.3. DEFINITION. Assume $\rho \in \mathcal{B}''$ and $1<p<\infty$. The space H^{ρ}_{p} is defined by

(1) $$H^{\rho}_{p} = H^{\rho}_{p}(\mathbb{R}^d) = \{f \in S'(\mathbb{R}^d) : \|f\|_{H^{\rho}_{p}} = \|J^{\rho} f\|_{L_p} < \infty\} .$$

3.4. THEOREM. Assume $\rho \in \mathcal{B}''$ and $1 < p < \infty$. Then

$$F^{\rho}_{p,2} = H^{\rho}_{p}.$$

PROOF. According to Theorem 3.2(i) and [8, Prop.7], we only need consider the case $\rho(t) = t^0 = 1$. In this situation $F^0_{p,2} = L_p = H^0_p$, which gives the result. //

4. EMBEDDING PROPERTIES AND RELATIONSHIPS

4.1. THEOREM. Assume $\rho \in \mathcal{B}$. Then, the following continuous embeddings holds:

(1) $F^{\rho}_{p,q_0} \subset F^{\rho}_{p,q_1}$, $1 < p < \infty$, $1 < q_0 \leq q_1 < \infty$;

(2) $B^{\rho}_{p,q_0} \subset B^{\rho}_{p,q_1}$, $1 \leq p \leq \infty$, $1 \leq q_0 \leq q_1 \leq \infty$;

(3) $B^{\rho}_{p,q} \subset F^{\rho}_{p,q}$, $1 < q \leq p < \infty$;

(4) $F^{\rho}_{p,q} \subset B^{\rho}_{p,q}$, $1 < p \leq q < \infty$;

(5) $B^{\rho}_{p,1} \subset F^{\rho}_{p,q} \subset B^{\rho}_{p,\infty}$, $1 < p , q < \infty$.

Moreover, if $\rho, \sigma \in \mathcal{B}$ are such that $\sigma/\rho \in L^{*}_{1}(1,\infty)$, we have

(6) $B^{\rho}_{p,\infty} \subset B^{\sigma}_{p,1}$, $1 \leq p \leq \infty$,

and

(7) $F^{\rho}_{p,q_0} \subset F^{\sigma}_{p,q_1}$, $1 < p,q_0,q_1 < \infty$.

PROOF. (1) and (2) follow by monotonicity; (3) and (4) by Minkowski's inequality; (5) follows from (2), (3) and (4); (6) is known and due to

C. Merucci [9, Prop.8.1.5]. Let us establish (7). Since $\sigma/\rho \in L_1^*(1,\infty)$, it is easy to see that

$$\sum_{n=0}^{\infty} \sigma(2^n)/\rho(2^n) < \infty .$$

Then (7) follows from

$$\|f\|_{F^{\sigma}_{p,q_1}} \le \{\int \{\sum_n \sigma(2^n)|\phi_n * f|\}^p dx\}^{1/p}$$

$$= \{\int \{\sum_n (\sigma(2^n)/\rho(2^n))\rho(2^n)|\phi_n * f|\}^p dx\}^{1/p}$$

$$\le \{\sum_n \sigma(2^n)/\rho(2^n)\} \|f\|_{F^{\rho}_{p,q_o}} .$$

The proof is complete. //

In order to state our second embedding theorem, which is actually a regularity result, let us recall that if $m \ge 0$ is an integer then $C^m = C^m(\mathbb{R}^d)$ denotes the completion of $S(\mathbb{R}^d)$ in the norm

$$\|f\|_{C^m} = \sum_{|\alpha| \le m} \|D^{\alpha} f\|_{L_\infty} .$$

If $0<r=m+s$, $0<s<1$, then $C^r = C^r(\mathbb{R}^d)$ is defined as the collection of all $f \in C^m$ such that $D^{\alpha} f \in \mathrm{Lip}\, s$, for all α with $|\alpha|=m$.

<u>4.2. THEOREM.</u> Let $1<p<\infty$ and $r \ge 0$. If $\rho \in \mathcal{B}$ and $\beta_{\bar{\rho}} > d/p+r$, then

(1) $\qquad B^{\rho}_{p,\infty} \subset C^r$

and, for every q, $1<q<\infty$,

(2) $\qquad F^{\rho}_{p,q} \subset C^r$.

<u>PROOF.</u> Step 1. Put $\sigma(t)=t^{d/p+r}$ and $\zeta(t)=\sigma(t)/\rho(t)$. Then $\alpha_{\bar{\zeta}} = d/p+r-\beta_{\bar{\rho}} < 0$ and consequently $\zeta \in L_1^*(1,\infty)$. Hence, according to Theorem 4.1(6), we have

$$B^{\rho}_{p,\infty} \subset B^{\sigma}_{p,1} = B^{d/p+r}_{p,1} .$$

This proves the embedding (1), because $B^{d/p+r}_{p,1} \subset C^r$ (see [13, Thm.2.8.1/(c),(d)] .

Step 2. Since $F^{\rho}_{p,q} \subset B^{\rho}_{p,\infty}$, we get (2) from (1). //

5. INTERPOLATION THEOREMS

In this section we shall show the interpolation properties of the spaces $B^{\rho}_{p,q}$ and $F^{\rho}_{p,q}$, complementing earlier results of C. Merucci [8, Thms. 12 and 13]. Our approach is different from that of [8], and it is based on the interpolation properties of the sequence spaces $\ell^{\rho}_{q}(E)$.

Let us start recalling the definition of the real interpolation space with function parameter (see [10],[8],[11]).

Let (A_0,A_1) be a compatible couple of normed spaces, let $1\leq q\leq\infty$ and let $\rho\in\mathcal{B}$. The space $(A_0,A_1)_{\rho,q}$ consists of all $x\in A_0+A_1$ which have a finite norm

$$\|x\|_{\rho,q} = \left(\int_0^\infty (\rho(t)^{-1}K(t,x))^q dt/t\right)^{1/q} \quad \text{if } 1\leq q<\infty$$

$$\|x\|_{\rho,q} = \sup_{t>0} (\rho(t)^{-1}K(t,x)) \quad \text{if } q=\infty$$

where $K(t,x)$ is the functional of J. Peetre, defined by

$$K(t,x) = \inf\{\|x_0\|_{A_0} + t\|x_1\|_{A_1} : x=x_0+x_1,\ x_0\in A_0,\ x_1\in A_1\}.$$

For $\rho(t)=t^{\theta}$ $(0<\theta<1)$ we get the classical real interpolation space $((A_0,A_1)_{\theta,q}, \|\ \|_{\theta,q})$ (see [1],[13]).

We shall first prove an interpolation formula between spaces $\ell^{s}_{q}(E)$, where s is a real parameter.

5.1. THEOREM. Let E be a Banach space. Suppose that $\rho\in\mathcal{B}$ and $1\leq q_0,q_1,q\leq\infty$. If $s_0,s_1\in\mathbb{R}$ with $s_1<\beta_{\bar\rho}\leq\alpha_{\bar\rho}<s_0$ and $\gamma\in\mathcal{B}$ is given by $\gamma(t) = t^{s_0/(s_0-s_1)}/\rho(t^{1/(s_0-s_1)})$, then

$$(1) \qquad (\ell^{s_0}_{q_0}(E),\ \ell^{s_1}_{q_1}(E))_{\gamma,q} = \ell^{\rho}_{q}(E),$$

with equivalence of norms.

PROOF. Using the tecniques of [13, Thm.1.18.2/Step 1] it is not hard to verify that

$$(2) \qquad (\ell^{s_0}_{\infty}(E),\ell^{s_1}_{\infty}(E))_{\gamma,q} \subset \ell^{\rho}_{q}(E).$$

On the other hand, the indices of the function $\bar{\gamma}$ are

$$0 < \frac{s_0}{s_0-s_1} - \frac{\alpha_{\bar{\rho}}}{s_o-s_1} = \beta_{\bar{\gamma}} \leq \alpha_{\bar{\gamma}} = \frac{s_0}{s_0-s_1} - \frac{\beta_{\bar{\rho}}}{s_0-s_1} < 1.$$

Hence

$$\int_0^\infty \frac{\bar{\gamma}(t)}{\max\ (1,t)} \frac{dt}{t} < \infty.$$

From this fact it is easily derived that

$$\sum_{j=-\infty}^{\infty} \min\ (1,2^{-js})\ \bar{\gamma}(2^{js}) < \infty\ .$$

This estimate and similar arguments as in [1,Thm.5.6.1] allow us to get

$$(3) \qquad \ell_q^\rho(E) \subset (\ell_1^{s_0}(E), \ell_1^{s_1}(E))_{\gamma,q}.$$

Combining (2) and (3) we conclude that

$$\ell_q^\rho(E) = (\ell_{q_0}^{s_0}(E), \ell_{q_1}^{s_1}(E))_{\gamma,q}\ . \qquad //$$

Now, we can establish

5.2. THEOREM. Let $1 \leq q_0, q_1, q \leq \infty$. Assume that $\gamma, \rho_0, \rho_1 \in \mathcal{B}$ and put $\rho(t) = \rho_0(t)/\rho_1(t)$ and $\zeta(t) = \rho_0(t)/\gamma(\rho(t))$. If $0 < \beta_{\bar{\gamma}} \leq \alpha_{\bar{\gamma}} < 1$ and $\beta_{\bar{\rho}} > 0$ (or $\alpha_{\bar{\rho}} < 0$), then $\zeta \in \mathcal{B}$ and

$$(1) \qquad (\ell_{q_o}^{\rho_0}(E), \ell_{q_1}^{\rho_1}(E))_{\gamma,q} = \ell_q^\zeta(E)\ ,$$

with equivalence of norms.

PROOF. The fact that $\zeta \in \mathcal{B}$ is known (see [6, Thm.5.3]). Choose s_0 and s_1 such that

$$s_1 < \min(\beta_{\bar{\rho}_0}, \beta_{\bar{\rho}_1}, \beta_{\bar{\zeta}}) \leq \max(\alpha_{\bar{\rho}_0}, \alpha_{\bar{\rho}_1}, \alpha_{\bar{\zeta}}) < s_0\ .$$

If $\gamma_i(t) = t^{s_0/(s_0-s_1)}/\rho_i(t^{1/(s_0-s_1)})$, $i=0,1$, then, by Theorem 5.1, we have

$$\ell_{q_i}^{\rho_i}(E) = (\ell_1^{s_0}(E), \ell_1^{s_1}(E))_{\gamma_i,q_i}\ , \quad i = 0,1.$$

Therefore, applying the reiteration theorem [9,Thm.4.3.1] and Theorem 5.1

once again, we obtain (1). //

As a consequence of Theorem 5.2 we have the following Interpolation Theorem for the spaces $B^{\rho}_{p,q}$ and $F^{\rho}_{p,q}$.

5.3. THEOREM. Assume that $\gamma,\rho_0,\rho_1 \in \mathcal{B}$, and put $\rho(t)=\rho_0(t)/\rho_1(t)$ and $\zeta(t)=\rho_0(t)/\gamma(\rho(t))$. Then, if $0<\beta_{\bar\gamma}\le\alpha_{\bar\gamma}<1$ and $\beta_{\bar\rho}>0$ (or $\alpha_{\bar\rho}<0$), we have

(1) $$(B^{\rho_0}_{p,q_0}, B^{\rho_1}_{p,q_1})_{\gamma,q} = B^{\zeta}_{p,q} \qquad (1\le p,q_0,q_1,q\le\infty);$$

and

(2) $$(F^{\rho_0}_{p,q_0}, F^{\rho_1}_{p,q_1})_{\gamma,q} = B^{\zeta}_{p,q} \qquad (1<p,q_0,q_1<\infty,\ 1\le q\le\infty).$$

PROOF. Applying Theorem 2.5(ii) and Theorem 5.2 with $E=L_p$, we obtain formula (1).
On the other hand, according to Theorem4.1(5), we have that

(3) $$B^{\rho_i}_{p,1} \subset F^{\rho_i}_{p,q_i} \subset B^{\rho_i}_{p,\infty}, \quad i=0,1.$$

Therefore, formula (2) follows from (3) and (1), because the left hand of (1) does not depend on q_0 nor q_1. //

5.4. REMARK. Let E be a quasi-Banach space and let $0<q_0,q_1,q\le\infty$. Replacing $\ell_1^{s_i}(E)$ in Theorem 5.1(3) by $\ell_r^{s_i}(E)$ $(0<r<\min(q_0,q_1,q))$, one obtains that Theorem 5.1 (and, thus, Theorem 5.2) is also true in this more general case (see as well [12,Cor.4.3]). Furthermore, the method of retraction still works in the quasi-Banach case (we only need to replace L_p by H_p, see [14,Thm.2.2.10]). Therefore, Theorem 5.3 holds for all non-negative p,q_0,q_1,q.

In what follows, we derive some results complementing Theorem 5.3. We first state a consequence of [7,Lemma 1].

5.5. PROPOSITION. Let (E_0,E_1) be a compatible couple of Banach spaces, let $1\le q<\infty$ and let $\rho,\gamma\in\mathcal{B}$ with $0<\beta_{\bar\gamma}\le\alpha_{\bar\gamma}<1$. Then

(1) $$(\ell^{\rho}_{q}(E_0), \ell^{\rho}_{q}(E_1))_{\gamma,q} = \ell^{\rho}_{q}((E_0,E_1)_{\gamma,q}).$$

5.6. REMARK. Proposition 5.5 is also a special case of [12,Prop.3.2]. This last reference contains as well concrete descriptions of the spaces

$(\ell^{\rho}_q(E_0),\ell^{\rho}_q(E_1))_{\gamma,p}$ for the case $q \neq p$.

<u>5.7. DEFINITION.</u> Let $1\leq p,q\leq\infty$, $\rho,\sigma,\Psi\in\mathcal{B}$, let E be a Banach space and let $\Lambda^q(\Psi)$ [resp. λ^{σ}_q] be the Lorentz-Marcinkiewicz function [resp. sequence] space (see [8, §2] and [6, §2]). We set

(1) $$B^{\rho}_{\Psi,q} = \{f\in S'(\mathbb{R}^d) : \|f\|_{B^{\rho}_{\Psi,q}} = \|(\|\phi_j * f\|_{\Lambda^q(\Psi)})_j\|_{\ell^{\rho}_q} < \infty\}.$$

(2) $$\lambda^{\sigma,\rho}_q(E) = \{(a_j)\subset E : \|(a_j)\|_{\lambda^{\sigma,\rho}_q(E)} = \|(\|\rho(2^j)a_j\|_E)_j\|_{\lambda^{\sigma}_q} < \infty\}.$$

When $\rho(t) = 1$, we simply write $\lambda^{\sigma}_q(E)$.

(3) $$B^{\rho,\sigma}_{p,q} = \{f\in S'(\mathbb{R}^d) : \|f\|_{B^{\rho,\sigma}_{p,q}} = \|(\phi_j * f)\|_{\lambda^{\sigma,\rho}_q(L_p)} < \infty\}.$$

As a consequence of Proposition 5.5, Theorem 2.5(ii), and [8, Thm.3], we obtain

<u>5.8. THEOREM.</u> Let $1\leq p_0\neq p_1\leq\infty$, $1\leq q<\infty$, let $\rho,\gamma\in\mathcal{B}$ with $0<\beta_{\bar\gamma}\leq\alpha_{\bar\gamma}<1$ and put $\Psi(t) = t^{1/p_0}/\gamma(t^{1/p_0-1/p_1})$. Then

(1) $$(B^{\rho}_{p_0,q},B^{\rho}_{p_1,q})_{\gamma,q} = B^{\rho}_{\Psi,q}.$$

We end this section with

<u>5.9. THEOREM.</u> Let $1\leq q\leq\infty$, $1\leq q_0\neq q_1\leq\infty$, let $\rho,\gamma\in\mathcal{B}$ with $0<\beta_{\bar\gamma}\leq\alpha_{\bar\gamma}<1$, and put $\Psi(t) = t^{1/q_0}/\gamma(t^{1/q_0-1/q_1})$. Then

(1) $$(\ell^{\rho}_{q_0}(E),\ell^{\rho}_{q_1}(E))_{\gamma,q} = \lambda^{\Psi,\rho}_q(E).$$

Moreover, if $1\leq p\leq\infty$, we also have

(2) $$(B^{\rho}_{p,q_0},B^{\rho}_{p,q_1})_{\gamma,q} = B^{\rho,\Psi}_{p,q}.$$

<u>PROOF.</u> Step 1. By adapting the proof of [1, Thm.5.2.1], it can be seen that

$$(\ell_{q_0}(E),\ell_{q_1}(E))_{\gamma,q} = \lambda^{\Psi}_q(E).$$

On the other hand, the operator T given by $T((a_j)) = (\rho(2^j)a_j)$ is an isomorphism from $\ell^{\rho}_{q_i}(E)$ into $\ell_{q_i}(E)$, i=0,1. Whence

$$T:\ (\ell^{\rho}_{q_0}(E),\ell^{\rho}_{q_1}(E))_{\gamma,q} \longrightarrow \lambda^{\Psi,\rho}_{q}(E)$$

is also an isomorphism. This proves the desired equality.

Step 2. Equality (2) is a corollary of (1). //

6. THE UMD-PROPERTY

In this last section, we shall show that if $1<p,q<\infty$ then $F^{\rho}_{p,q}$ and $B^{\rho}_{p,q}$ have the UMD-property.

6.1. DEFINITION. Let $1<p<\infty$. A Banach space E is said to have the unconditionality property for martingale differences (UMD-property, for short) if E-valued martingale difference sequences are unconditional in $L_p([0,1];E)$.

Next we state a stability result for UMD spaces.

6.2. PROPOSITION. Assume that E and F are Banach spaces such that E has the UMD-property and F is a retract of E. Then F is also a UMD space.

PROOF. Let $S\in L(F,E)$ and $R\in L(E,F)$ such that $R\circ S = I_F$. Using the geometrical characterization of Banach spaces with the UMD-property proved by D.L. Burkholder in [4], we can find a symmetric biconvex function $\zeta : E\times E \longrightarrow \mathbb{R}$ such that $\zeta(0,0)>0$ and

$$\zeta(x,y) \leq \|x+y\| \quad \text{if} \quad \|y\| \geq 1.$$

Define $\bar{\zeta}: F\times F \longrightarrow \mathbb{R}$ by

$$\bar{\zeta}(x,y) = \frac{1}{\|R\|\ \|S\|}\,\zeta(\|R\|S(x)\ ,\ \|R\|S(y)).$$

The function $\bar{\zeta}$ has the same properties as ζ. Therefore Burkholder's characterization implies that F is a UMD space. //

Now we are in a position to establish

6.3. THEOREM. Assume $\rho\in\mathcal{B}$ and $1<p,q<\infty$. Then the spaces $F^{\rho}_{p,q}$ and $B^{\rho}_{p,q}$ have the UMD-property.

PROOF. Step 1. The space ℓ^{ρ}_{q} is a UMD space because it is isomorphic to ℓ_q. On the other hand, it is well known that if E has the UMD-property and $1<p<\infty$, then $L_p(E)$ also has the UMD-property. Hence $L_p(\ell^{\rho}_q)$ is a

UMD space. Finally, since $F^{\rho}_{p,q}$ is a retract of $L_p(\ell^{\rho}_q)$, it follows from Proposition 6.2 that $F^{\rho}_{p,q}$ has the UMD-property.

Step 2. It was shown in [7] that the UMD-property is stable by the (γ,q)-interpolation method if $1<q<\infty$. Whence, according to Step 1 and Theorem 5.3(2), we get that $B^{\rho}_{p,q}$ has the UMD-property. //

6.4. REMARK. It can be easily checked by using [7,Thm.8], that the spaces $B^{\rho}_{p,1}$ and $B^{\rho}_{p,\infty}$ $(1<p<\infty)$ fail to have the UMD-property.

Acknowledgement

We thank the referee for suggesting several improvements to the first version of this paper.

REFERENCES

1. J. Bergh and J. Löfström, Interpolation Spaces, An Introduction, Springer, Berlin-Heidelberg-New York 1976.

2. J. Bourgain, Some remarks on Banach spaces in which martingale difference sequences are unconditional, Ark. Mat. 22(1983) 163-168.

3. D.W. Boyd, The Hilbert transform on rearrangement-invariant spaces, Canadian J. Math. 19(1967) 599-616.

4. D.L. Burkholder, A geometrical characterization of Banach spaces in which martingale difference sequences are unconditional, Ann. Probability 9(1981) 997-1011.

5. D.L. Burkholder, A geometrical condition that implies the existence of certain singular integrals of Banach-space-valued functions, Proc. Conf. Harmonic Analysis in Honour of A. Zygmund (Chicago 1981), Wadsworth, Belmont 1983.

6. F. Cobos, On the Lorentz-Marcinkiewicz operator ideal, Math. Nachr. 126(1986) 281-300.

7. F. Cobos, Some spaces in which martingale difference sequences are unconditional, Bull. Acad. Polon. Sci. 34(1986) to appear.

8. C. Merucci, Applications of interpolation with a function parameter to Lorentz, Sobolev and Besov spaces, Proc. Lund Conf. 1983, Lecture Notes in Math. 1070, 183-201.

9. C. Merucci, Interpolation réelle avec fonction paramètre. Dualité, reiteration et applications. Thèse d'Etat, Nantes 1983.

10. J. Peetre, A Theory of Interpolation of Normed Spaces, Notas de Matemática Nº39, IMPA, Rio de Janeiro 1968.

11. L.E. Persson, Interpolation with a parameter function, Math.Scand. to appear.

12. L.E. Persson, Real interpolation between cross-sectional L^p-spaces in quasi-Banach bundles, Technical Report 1, Luleå (1986).

13. H. Triebel, Interpolation Theory, Function Spaces, Differential Operators, North-Holland, Amsterdam-New York-Oxford 1978.

14. H. Triebel, Spaces of Besov-Hardy-Sobolev Type, Teubner, Leipzig 1978.

AN EXTENSION OF FOURIER TYPE TO QUASI-BANACH SPACES

Michael Cwikel and Yoram Sagher
University of Illinois at Chicago
Box 4348, Chicago IL 60680

Abstract: The inclusion $(B_0,B_1)_{\theta,p} \subset [B_0,B_1]_\theta$ which is known to hold for couples of Banach spaces (B_0,B_1) where p depends on the Fourier types of B_0 and B_1 is generalized to the case of quasi Banach spaces. It turns out that when $p < 1$ the natural extension of Fourier type p is p-normability.

It is well known that real and complex interpolation spaces are related by the inclusions

$$(B_0,B_1)_{\theta,1} \subset [B_0,B_1]_\theta \subset (B_0,B_1)_{\theta,\infty}$$

which hold for all $\theta \in (0,1)$ and all compatible couples of Banach spaces (B_0,B_1) . (See [1] Theorem 4.7.1 p. 102, or [5] Theorem 1.1 p. 29.) In fact the ideas used in [5] were later refined in [9] to yield the sharpened estimates

$$(B_0,B_1)_{\theta,p} \subset [B_0,B_1]_\theta \subset (B_0,B_1)_{\theta,p'} \qquad (1)$$

in the case where the spaces B_0 and B_1 are of Fourier type p_0 and p_1 respectively, where p_0 and p_1 are in $[1,\infty]$, $1/p = (1-\theta)/p_0 + \theta/p_1$ and $1/p + 1/p' = 1$. We recall that a Banach space B is of Fourier type p for some $p \in [1,2]$ if the Fourier transform of strongly measurable B valued functions on $\mathbb{R}$ defines a bounded map from $L^p(B)$ to $L^{p'}(B)$. The notion of Fourier type has been further investigated and applied in [6, 7] where there is also some discussion of the Fourier type with respect to other groups.

In this note we shall consider the question of what analogues of (1) can be expected to hold for the case where one or both of the spaces B_j is a quasi Banach space. A brief remark about this problem can be found in [10] (p. 262). For couples of quasi Banach spaces, in the light of some difficulties which arise in constructing complex interpolation spaces (see e.g. the discussion on pp. 340-341 of [2]), it is convenient to define and consider interpolation of quasinorms on $B_0 \cap B_1$ rather than interpolation of the spaces B_0 and B_1 themselves. We shall do this using the same definitions and notation as in [2]. Thus instead of seeking to obtain inclusions of a form similar to (1) we should seek rather to establish inequalities of the form

$$C \|b\|_{(B_0,B_1)_{\theta,q}} \leq \|b\|_{[B_0,B_1]_\theta} \leq C' \|b\|_{(B_0,B_1)_{\theta,p}} \tag{2}$$

for fixed positive C and C' and all elements b in $B_0 \cap B_1$, where $0<p\leq q\leq\infty$. Our result here is that the inequality on the right in (2) holds whenever $1/p = (1-\theta)/p_0 + \theta/p_1$ and B_j is either of Fourier type p_j when $p_j \geq 1$ or is p_j-normed when $0 < p_j < 1$.

It should be pointed out that the left inequality in (2) may fail for all choices of $C>0$ if the maximum modulus principle does not hold for B_0 or B_1 valued analytic functions. (Cf. [10, 13].)

In fact, for the purposes of this discussion we must define a notion of Fourier type in a slightly different way to that of [9], thus immediately raising the question of whether the two definitions might in fact be equivalent. This problem is considered in a forthcoming paper by Janson [4] together with some similar questions raised by Milman in [7].

Definition 3 We shall say that a Banach space B is of periodic Fourier type p for some $p \in [1,2]$ if for some absolute constant C and for all sequences $\{b_n\}_{n=-\infty}^{\infty}$ of elements of B, with only finitely many non zero elements,

$$\left(\frac{1}{2\pi} \int_0^{2\pi} \|\Sigma_{n=-\infty}^{\infty} e^{int} b_n\|_B^{p'} dt\right)^{1/p'} \leq C\left(\Sigma_{n=-\infty}^{\infty} \|b_n\|_B^p\right)^{1/p} \tag{4}$$

Exactly as for the case of Fourier type as defined in [9], we see immediately that all Banach spaces are of periodic Fourier type 1 and all Hilbert spaces are of periodic Fourier type 2, with C=1. Consequently, by interpreting (4) as an estimate for an operator from $\ell^p(B)$ into $L^{p'}(B)$, we can also see that if B is a subspace of an L^p

or $L^{p'}$ space or is any other "θ-Hilbertian space" (as defined in [11]), where $1/p = 1 - \theta/2$, then B is of periodic Fourier type p .

In fact for our purposes it is convenient to use a generalization of the notion introduced in Definition 3.

<u>Definition 5</u> We shall say that a quasi Banach space B is of <u>periodic Fourier type</u> p,q for some p and q in $(0,\infty]$ if, for some absolute constant C and all B valued sequences $\{b_n\}_{n=-\infty}^{\infty}$ with only finitely many non zero elements,

$$\left(\frac{1}{2\pi} \int_0^{2\pi} \|\Sigma_{n=-\infty}^{\infty} e^{int} b_n\|_B^q \, dt \right)^{1/q} \leq C\left(\Sigma_{n=-\infty}^{\infty} \|b_n\|_B^p\right)^{1/p} \qquad (6)$$

<u>Remarks</u>

(i) In this definition it may be necessary to take $\|\cdot\|_B$ to be an r-norm for some r>0 which is equivalent to the (possibly different) quasi norm used to define B. For an r-norm it is easy to see that $\|\Sigma_{n=-\infty}^{\infty} e^{int} b_n\|_B$ is a continuous function of t. (There are examples of quasi norms for which this expression may be discontinuous or even non measurable. In the latter case the integral in (6) would have to be interpreted as an upper integral.)

(ii) Periodic Fourier type p,q of course implies periodic Fourier type $\tilde{p},\tilde{q}$ for all $\tilde{p}\leq p$ and $\tilde{q}\leq q$.

(iii) If B is r-normed then it is clearly of periodic Fourier type r,∞, with C=1 . Conversely, if B is of periodic Fourier type r,∞ then it is r-normed: This is obvious for C=1 and if C>1 then B has an equivalent r-norm defined by

$$\|b\| = \inf \left(\Sigma_n \|b_n\|_B^r\right)^{1/r}$$

where the infimum is taken over all finite sequences of elements $b_n \in B$ satisfying $\Sigma_n b_n = b$. The case C<1 is trivial since then $\|b\|_B=0$ for all b. Thus we can also observe that there are no non trivial spaces of periodic Fourier type r,∞ for r>1. (By homogeneity considerations we can also disregard Fourier type p,q for certain other values of p and q . See [4].)

We can now state our main result.

<u>Theorem 7</u> Let (B_0,B_1) be a compatible couple of quasi Banach spaces and suppose that for $j = 0,1$, B_j is of periodic Fourier type p_j,q_j for some p_j and q_j in $(0,\infty]$. Then for all $b \in B_0 \cap B_1$ and $0 < \theta < 1$,

$$\|b\|_{[B_0,B_1]_\theta} \leq C \|b\|_{(B_0,B_1)_{\theta,p}}$$

where $1/p = (1-\theta)/p_0 + \theta/p_1$ and the constant C depends only on θ , p_0, and p_1 .

The following lemma will be used in the proof of Theorem 7.

<u>Lemma 8</u> If $b \in B_0 \cap B_1$ and if p_0 , $p_1 \in (0,\infty]$ satisfy $1/p = (1-\theta)/p_0 + \theta/p_1$, then there exists a sequence $\{b_\nu\}_{\nu=-\infty}^{\infty}$ of elements in $B_0 \cap B_1$, all but finitely many of them zero, such that $b = \Sigma_{\nu=-\infty}^{\infty} b_\nu$, and $\|\{e^{(j-\theta)\nu} b_\nu\}\|_{\ell^{p_j}(B_j)} \leq C \|b\|_{(B_0,B_1)_{\theta,p}}$ for $j = 0,1$, where the constant C depends only on p_0, p_1 and θ .

<u>Remark</u>: Analogues of this lemma also hold for other interpolation methods. See for example Lemma 1.5 on p. 137 of [8] and the work of Janson and of Ovčinnikov referred to there.

<u>Proof of the Lemma</u>: By the equivalence theorem ([1, 3, 12]) there exist sequences $\{v_{0\nu}\}_{\nu=-\infty}^{\infty}$ and $\{v_{1\nu}\}_{\nu=-\infty}^{\infty}$ in B_0 and B_1 respectively, such that $a = v_{0\nu} + v_{1\nu}$ for all ν and

$$\|\{e^{(j-\theta)\nu} v_{j\nu}\}\|_{\ell^{p_j}(B_j)} \leq C_1 \|b\|_{(B_0,B_1)_{\theta,p}} \quad \text{for } j = 0,1 . \tag{9}$$

For some integer N we define new sequences $\{b_{0\nu}\}_{\nu=-\infty}^{\infty}$ and $\{b_{1\nu}\}_{\nu=-\infty}^{\infty}$ by setting

(i) $b_{j\nu} = v_{j\nu}$ for $-N \leq \nu \leq N$ and $j = 0,1$,

(ii) $b_{0\nu} = b$ and $b_{1\nu} = 0$ for $\nu > N$, and

(iii) $b_{1\nu} = b$ and $b_{0\nu} = 0$ for $\nu < -N$.

If we choose N sufficiently large then $\left(\Sigma_{\nu>N}\left(e^{-\theta\nu}\|b\|_{B_0}\right)^{p_0}\right)^{1/p_0}$ and $\left(\Sigma_{\nu<-N}\left(e^{(1-\theta)\nu}\|b\|_{B_1}\right)^{p_1}\right)^{1/p_1}$ (or the corresponding suprema if p_0 and/or p_1 are/is infinite) can both be made arbitrarily small, and thus sufficiently small to ensure, via (9), that $\|\{e^{(j-\theta)\nu}b_{j\nu}\}\|_{\ell^{p_j}(B_j)} \leq 2C_1\|b\|_{(B_0,B_1)_{\theta,p}}$ for $j = 0,1$. Finally, as in standard proofs, let $b_\nu = b_{1,\nu-1} - b_{1,\nu} = b_{0,\nu} - b_{0,\nu-1}$. As may readily be checked, $\{b_\nu\}_{\nu=-\infty}^{\infty}$ has all the required properties. ■

<u>Remark</u> This lemma enables Theorem 3 of [2] pp. 341-342 to be proved without using Gagliardo completions.

<u>Proof of Theorem 7:</u> Given any $b \in B_0 \cap B_1$ apply Lemma 8 to obtain $b = \Sigma_{|\nu|\leq N}\, b_\nu$ where $\|\{e^{(j-\theta)\nu}b_\nu\}\|_{\ell^{p_j}(B_j)} \leq C\,\|b\|_{(B_0,B_1)_{\theta,p}}$. Set $f(z) = \Sigma_{|\nu|\leq N}\, e^{(z-\theta)\nu}b_\nu$. This is clearly an element of $\mathcal{G}(B_0,B_1)$ as defined in [2] p. 340. Thus (cf. [2] p. 341), letting $w_0(\theta,y)$ and $w_1(\theta,y)$ be the Poisson kernels for the strip $\{z \mid 0 \leq \operatorname{re} z \leq 1\}$, we have

$$\|b\|_{[B_0,B_1]_\theta} = \|f(\theta)\|_{[B_0,B_1]_\theta}$$
$$\leq \exp\Big(\int_{-\infty}^{\infty}\log\|f(iy)\|_{B_0} w_0(\theta,y)dy + \int_{-\infty}^{\infty}\log\|f(1+iy)\|_{B_1} w_1(\theta,y)dy\Big)$$
$$\leq \Big\{\int_{-\infty}^{\infty}\Big(\|f(iy)\|_{B_0}^q w_0(\theta,y) + \|f(1+iy)\|_{B_1}^q w_1(\theta,y)\Big)dy\Big\}^{1/q},$$

where $q = \min(q_0,q_1)$,

$$= \Big\{\Sigma_{n=-\infty}^{\infty}\int_{2\pi n}^{2\pi(n+1)}\|f(iy)\|_{B_0}^q w_0(\theta,y) + \|f(1+iy)\|_{B_1}^q w_1(\theta,y)\, dy\Big\}^{1/q}$$
$$= \Big\{\Sigma_{n=-\infty}^{\infty}\int_{0}^{2\pi}\|f(iy)\|_{B_0}^q w_0(\theta,y+2\pi n) + \|f(1+iy)\|_{B_1}^q w_1(\theta,y+2\pi n)\, dy\Big\}^{1/q}$$
$$\leq C\Big\{\int_0^{2\pi}\|f(iy)\|_{B_0}^q + \|f(1+iy)\|_{B_1}^q\, dy\Big\}^{1/q}$$

where the constant C satisfies

$$C = C(\theta) \leq \sup_{0\leq y\leq 2\pi}\Big\{\Sigma_{n=-\infty}^{\infty} w_0(\theta,y+2\pi n) + w_1(\theta,y+2\pi n)\Big\}^{1/q}$$

and the latter sum is easily seen to be a bounded function of y because of the rapid decay of $w_j(\theta,y)$ as $|y| \to \infty$.

Since B_0 and B_1 are of periodic Fourier types p_0,q and p_1,q respectively, we deduce that

$$\|b\|_{[B_0,B_1]_\theta} \le C_2 \left\{ \|\{e^{-\theta\nu}b_\nu\}\|_{\ell^{p_0}(B_0)} + \|\{e^{-(1-\theta)\nu}b_\nu\}\|_{\ell^{p_1}(B_1)} \right\}$$

This completes the proof. ■

REFERENCES

[1] Bergh, J., Löfström, J.: Interpolation spaces. An Introduction. Grundlehren der mathematische Wissenschaften 223, Berlin-Heidelberg-New York, Springer 1976.

[2] Cwikel, M., Milman, M., Sagher, Y.: Complex interpolation of some quasi-Banach spaces. J. Functional Analysis 65, 339-347 (1986).

[3] Holmstedt, T.: Interpolation of quasi-normed spaces. Math. Scand. 26, 177-199 (1970).

[4] Janson, S.: On Fourier types of Banach spaces. (in preparation).

[5] Lions, J. L., Peetre, J.: Sur une classe d'espaces d'interpolation. Inst. Hautes Etudes Sci. Publ. Math. 19, 5-68 (1964).

[6] Milman, M.: Fourier type and interpolation. Proc. Amer. Math. Soc. 89, 246-248 (1983).

[7] Milman, M.: Complex interpolation and geometry of Banach spaces. Ann. Mat. Pura Appl. 136, 317-328 (1984).

[8] Nilsson, P.: Interpolation of Banach lattices. Studia Math. 82, 135-154 (1985).

[9] Peetre, J.: Sur la transformation de Fourier des fonctions a valeurs vectorielles. Rend. Sem. Mat. Padova 42, 15-26 (1969).

[10] Peetre, J.: Locally analytically pseudo-convex topological vector spaces. Studia Math. 73, 253-262 (1982).

[11] Pisier, G.: Some applications of the complex interpolation method to Banach lattices. J. Analyse Math. 35, 264-281 (1979).

[12] Sagher, Y.: Interpolation of r-Banach spaces. Studia Math. 26, 45-70 (1966).

[13] Tabacco Vignati, A.: Dissertation, Department of Mathematics, Washington University, St. Louis MO, 1986.

τ-MODULI AND INTERPOLATION

L.T. Dechevski
Institute of Mathematics
Bulgarian Academy of Sciences
Sofia 1090, Bulgaria

0. Introduction

Let $\Omega \subset \mathbb{R}$ be an interval, $L_p(\Omega) = L_p(\Omega, dx)$ be the usual real-valued Lebesgue spaces, $0<p \leq \infty$.

It is well known that the integral modulus of smoothness of the function f

$$\omega_k(f\ ;\ t)_p := \sup_{0<|h|\leq t} \|\Delta_h^k f \mid L_p(\Omega_{k,h})\|, \quad k\in\mathbb{N},\ 0 < p < \infty ,$$

$$\omega_k(f\ ;\ t)_\infty := \sup_{0<|h|\leq t} \ \sup_{x\in \Omega_{k,h}} |\Delta_h^k f(x)| , \text{ where}$$

$$\Delta_h^k f(y) := \sum_{n=0}^{k} (-1)^{k+n} \binom{k}{n} f(y+nh) , \quad y\in\Omega_{k,h} ,$$

$$\Omega_{k,h} := \{x : x,\ x+kh \in \Omega\},\ h\in\mathbb{R},$$

is equivalent to an appropriate K-functional, $1 \leq p \leq \infty$. This fact is known to allow a unified approach to a wide variety of problems in interpolation and approximation theory, with far-going consequences.

Integral moduli are, however, insufficient when treating many important problems in numerical analysis. A typical example is error estimation for the numerical solution of boundary problems for differential equations by means of difference schemes or finite elements. In this case, the error function (i.e., a network norm of its for difference schemes and a L_p-norm for finite elements) cannot be estimated via integral moduli. To this end, average moduli of smoothness (or τ-moduli) can be successfully used (see [15]):

$$\tau_k(f\ ;\ t)_p = \| \omega_k(f,\cdot;\ t) \mid L_p(\Omega) \|, \quad 0<p \leq \infty ,\ t > 0,\ k \in\mathbb{N},$$

$\omega_k(f,x\ ;\ t) := \sup\{|\Delta_h^k f(y)| : y,\ y+kh \in [x-(kt)/2, x+(kt)/2] \cap \Omega\},\ x \in \Omega.$

As far as we know, τ-moduli were first introduced by Sendov and Korovkin some twenty years ago and, since then, they have been finding a broad and ever-increasing range of applications in approximation theory and numerical analysis (see,e.g., [15] for references). In view of these applications, it is of considerable interest to study interpolation of spaces induced by the τ-moduli. This is the aim of our communication.

In [12] Popov considered some counterparts of Besov spaces called A-spaces. Here is a modification of Popov's definition:

$$A_{pq}^s(\Omega) := \{f : f \in L_p(\Omega),\ \| f | A_{pq}^s(\Omega) \| < \infty\}\ ,\ 0 < p,q \leq \infty\ ,\ s > 0\ ,$$

$$\| f | A_{pq}^s(\Omega) \| := \| f | L_p(\Omega) \| + \left(\int_0^1 (t^{-s}\ \tau_k(f\ ;\ t)_p)^q \frac{dt}{t}\right)^{1/q},\ s<k,\ k \in \mathbb{N}.$$

It is intriguing that methods yielding closedness of Besov spaces under complex and real interpolation, somewhat fail when applied to A-spaces. As far as we know, it is not known at present whether in the general case A-spaces are closed under real and complex interpolation or not. In Section 2 we prove that in some cases Besov spaces and A-spaces are isomorphic (cf. also [6]). In Section 3 we prove that τ-moduli are equivalent to K-functionals of the type $K(t,f\ ;\ X_{0,t}, X_{1,t})$ for appropriate spaces $X_{0,t}, X_{1,t}$. Then, in Section 4, instead of interpolation of A-spaces, we study interpolation of some space sums, considering the simple embeddings

$$(A_0 + B_0, A_1 + B_1)_{\theta,p} \supset (A_0, A_1)_{\theta,p} + (B_0, B_1)_{\theta,p} \tag{0.1}$$

$$(A_0 + B_0, A_1 + B_1)_{[\theta]} \supset (A_0, A_1)_{[\theta]} + (B_0, B_1)_{[\theta]} \tag{0.2}$$

for appropriate θ, p, A_j, B_j, $j = 0,1$. (Here $A \supset B$ (or $A \subset B$) means "B is linearly and continuously embedded in A" (or vice versa)). This allows us to interpolate,e.g., between estimates of the type (T-sublinear operator):

$\| Tf | L_{p_i} \| \leq c_i(t_i^{k_i} \| f | L_{q_i} \| + \tau_{k_i}(f\ ;\ t_i)_{q_i}),\quad i = 0,1,$ for appropriate f.

In Section 5, applying Peetre's idea of quasilinearizability to the pairs $(X_{0,t}, X_{1,t})$ and the monotonicity of norms in certain Banach lattices, we prove some "inverse" results showing that "$\supset$" in (0.1,2) can be replaced by "$\rightleftharpoons$" ($A \rightleftharpoons B$, if the quasiseminormed spaces A, B are isomorphic or, in other words, they coincide as sets and have equivalent quasiseminorms). Our "inverse" results imply a peculiar example of non-commutative interpolation of space intersections by both the real and complex methods.

Our proofs can be extended to the case of multivariate moduli, too, but for the sake of simplicity the univariate case is preferred throughout.

1. Preliminaries

In the sequel Ω is an interval; "$\sim$" denotes equivalence between quasiseminorms; c denotes a positive constant (in an embedding or equivalence relation) which may vary from line to line; $p(\theta)$: $1/p(\theta) = (1-\theta)/p_0 + \theta/p_1$, $0 < p_0, p_1 \leq \infty$, $0 < \theta < 1$; the function f: Domf $= \Omega$, Codf$\subset \mathbb{R}$ will be termed measurable, if f is μ-measurable, where μ is the usual Lebesgue measure: $\mu(dx) = dx$; for $t > 0$, A - quasinormed space,

$$tA = \{ a : a \in A \ , \ \| a \|_{tA} := t \| a \|_A \} .$$

For the properties of τ-moduli we refer to [15, 6].

For the definitions and properties of Lorentz spaces $L_{pq}(\Omega)$, $0 < p, q \leq \infty$; Sobolev spaces $W_p^k(\Omega)$, $1 \leq p \leq \infty$, $k \in \mathbb{N}$; Besov spaces $B_{pq}^s(\Omega)$, Triebel-Lizorkin spaces $F_{pq}^s(\Omega)$, $0 < p, q \leq \infty$, $s > 0$, we refer to [1, 16, 17, 8].We shall omit Ω in the denotations whenever this causes no obscurity.

For the properties of vector-valued sequence spaces

$$l_p(I;A) := \{ a : a = \{ \alpha_\nu \}_{\nu \in I}, \alpha_\nu \in A, \ \| a | l_p(I;A) \| := (\sum_{\nu \in I} \| \alpha_\nu \|_A^p)^{1/p} < \infty \},$$

where $0 < p \leq \infty$, $I \subset \mathbb{Z}$, A is a quasinormed space, see [1, 16].

For the definition and properties of the Steklov-means $f_{k,t}$ for a measurable function f, $k \in \mathbb{N}$, $0 < t \leq \mu(\Omega)/k$, see [15, 2, 14].Notice that

$$| f(x) - f_{k,t}(x) | \leq c_k \, \omega_k(f,x;2t) ,$$

$$| f_{k,t}^{(n)}(x) | \leq c_k \, t^{-n} \, \omega_n(f,x;2kt) \ , \ n = 1, \ldots, k, \ x \in \Omega \text{ (see [15])}.$$

We introduce the spaces:

$BM(\Omega) := \{ f : \text{Domf} = \Omega \ , \text{Codf} \subset \mathbb{R}, \text{ f-measurable, bounded everywhere on } \Omega, \ \| f | BM(\Omega) \| := \sup \{ | f(x) | : x \in \Omega \} < \infty \}$;

$BM_{loc}(\Omega) := \{ f : \text{Domf} = \Omega, \text{ Codf} \subset \mathbb{R} \ , \text{ f-measurable}, \ \forall F \subset \Omega, \ F \text{ - compact} \Rightarrow f|_F \in BM(F) \}$;

$$A_{p,t}(\Omega) := \{ f : f \in BM(\Omega), \ \| f | A_{p,t}(\Omega) \| := (\int_\Omega S(t, |f|;x)^p dx)^{1/p} < \infty \} ,$$

where

$S(t,f;x) := \sup \{ f(y) : y \in [x-t, x+t] \cap \Omega \}$ (Upper Baire's function);

$\dot{W}_{p,t}^k(\Omega) := \{ f : \text{Domf} = \Omega \ , \ \text{Codf} \subset \mathbb{R}, \ f^{(k)} \in BM(\Omega), \ \| f | \dot{W}_{p,t}^k(\Omega) \| :=$

$= \| f^{(k)} | A_{p,t}(\Omega) \| < \infty \}$, $0<p \leqq \infty$, $t>0$, $k \in \mathbb{N}$.

Obviously,

$\forall \delta : 0<\delta \leqq 1$, $\forall \Delta$ - interval, $\exists \{\Delta_\nu\}_{\nu \in I}$, $I \subset \mathbb{Z}$, Δ_ν - intervals (1.1)

such that: a) $\bigcup_{\nu \in I} \Delta_\nu = \Delta$, $\Delta_{\nu_0} \cap \Delta_{\nu_1} = \emptyset$, $\nu_0 \neq \nu_1$; b) $\mu(\Delta_{\nu_0}) = \mu(\Delta_{\nu_1})$, $\nu_0, \nu_1 \in I$; c) if $\mu(\Delta) = \infty$, then $\mu(\Delta_\nu) = \delta$, $|I| = \infty$; if $\delta \leqq \mu(\Delta) < \infty$, then $I : \delta \leqq \mu(\Delta_\nu) = \mu(\Delta)/|I| < 2\delta$, $\nu \in I$; if $\mu(\Delta) < \delta$, then $|I| = 1$ $\Delta_\nu = \Delta$.

It is not difficult to prove that, if $0 < p \leqq \infty$, $0 < \delta \leqq 1$, $\{\Delta_\nu\}_{\nu \in I}$ are as in (1.1), then

$$\Big(\sum_{\nu \in I} \delta \, \| f | BM(\Delta_\nu) \|^p \Big)^{1/p} \sim \| f | A_{p,\delta}(\Delta) \| \tag{1.2}$$

with constants of equivalence independent of δ. This can be shown to imply that, if $0 < p \leqq \infty$, $0 < t_0 \leqq t_1 \leqq 1$, then

$$t_1^{1/p} \, t_0^{-1/p} A_{p,t_0} \hookrightarrow A_{p,t_1} \hookrightarrow A_{p,t_0} \quad , \tag{1.3}$$

with embedding constants independent of t_0, t_1 .

For $1 \leqq p \leqq \infty$, $k \in \mathbb{N}$, (see [3]) $\exists c > 0 : \forall t : 0 < t \leqq 1$, $\forall f \in W_p^k$,

$$\| f | A_{p,t} \| \leqq \| f | L_p \| + c \tau_1(f;t)_p \leqq \| f | L_p \| + ct \| f' | L_p \| \leqq c \, \| f | W_p^k \| \tag{1.4}$$

2. Isomorphism between Besov and A-spaces

In [6] it was proved via Marchaud-type inequalities that $A^s_{pq}(\Omega) \neq B^s_{pq}(\Omega)$, $1 \leqq p,q \leqq \infty$, $s > 1/p$, Ω - interval; and for $s \leqq 1/p$ these spaces were shown to be essentially diverse. Using a different method, here we include the quasinormed case by proving

Theorem 2.1. $A^s_{pq}(\mathbb{R}) \neq B^s_{pq}(\mathbb{R})$, $0 < p,q \leqq \infty$, $s > 1/p$.

Proof (Outline): Let $\varphi \in S(\mathbb{R})$ (Schwartz's space) :

$\operatorname{supp} \varphi = \{ \xi : 1/2 \leqq |\xi| \leqq 2\}$; $\varphi(\xi) > 0$, $1/2 < |\xi| < 2$;

$\sum_{k=-\infty}^{\infty} \varphi(2^{-k}\xi) = 1$, $\xi \neq 0$ (cf. [1]).

$\varphi_k(x) := F^{-1}(\varphi(2^{-k}\cdot))(x)$, $k \in \mathbb{N}$; $\varphi_0(x) := 1 - \sum_{k=1}^{\infty} \varphi_k(x)$, $x \neq 0$, $\varphi_0(0) := 1$.

Here F^{-1} is the inverse Fourier transform.

For $j = 0,1,\ldots$, $a > 0$ and $f \in S'(\mathbb{R})$ (the space of tempered distributions), consider the maximal function (see [11]) $\varphi_j^a f$:

$$(\varphi_j^a f)(x) := \sup \Big\{ \frac{|\varphi_j * f(x-y)|}{1 + |2^j y|^a} : y \in \mathbb{R} \Big\}, \ x \in \mathbb{R},$$

where $\varphi_{j*}f$ is the convolution of $\varphi_j \in S$ and $f \in S'$ over $\mathbb{R}$.

We shall make use of the following facts:

$$\| f \| := \| f | L_p \| + \Big(\sum_{k=0}^{\infty} 2^{skq} \tau_m(f;2^{-k})_p^q\Big)^{1/q} , \quad 0 < s < m, \ m \in \mathbb{N} \tag{2.1}$$

is an equivalent quasinorm in A^s_{pq} (the proof is the same as with integral moduli and Besov spaces).

For $0 < p \leq \infty$, $s > \max\{1, 1/p - 1\}$, $\exists c > 0$: $\forall f \in B^s_{pq}$,

$$\| f | L_p \| \leq c \| f | B^s_{pq} \| \text{ (see [17])} ; \tag{2.2}$$

$$\| f \|_1 := \| f | L_p \| + \Big(\sum_{k=0}^{\infty} 2^{skq} \omega_m(f;2^{-k})_p^q\Big)^{1/q} , \tag{2.3}$$

$$\| f \|_2 := \Big(\sum_{j=0}^{\infty} 2^{jsq} \| \varphi^a_j f | L_p \|^q\Big)^{1/q} \tag{2.4}$$

are equivalent quasinorms in B^s_{pq} for $1/p < a < s < m$ (see [17]);

$\exists c > 0$: $(\varphi^a_j f)(x-z) \leq c(1+|2^j z|^a)(\varphi^a_j f)(x)$, $f \in S'$, $j = 0,1$, $x, z \in \mathbb{R}$;

$\exists c > 0$: $(\Delta^m_h \varphi_{j*} f)(x) \leq c 2^{(j-k)m} (\varphi^a_j f)(x)$, $f \in S'$, $j = 0,1, \ldots, k$, $m \in \mathbb{N}$, $0 < |h| \leq 2^{-k}$ (see [17]); the constants of embedding and equivalence in all relations listed above do not depend on the concrete choice of φ. The last two properties of $\varphi^a_j f$ and the definition of τ-moduli imply:

$$\tau_m(\varphi_{j*}f;2^{-k})_p \leq c \min\{\max\{1, 2^{(j-k)a}\}, 2^{(j-k)m}\} \| \varphi^a_j f | L_p \| \tag{2.5}$$

The embedding $A^s_{pq} \subsetneq B^s_{pq}$ is easy, in view of (2.1, 3) and the relation $\omega_m(f;\delta)_p \leq \tau_m(f;\delta)_p$, $\delta > 0$.

In order to prove $B^s_{pq} \subsetneq A^s_{pq}$, consider $f \in B^s_{pq}$, $0 < p < 1$, $0 < q < \infty$ (for $1 \leq p \leq \infty$ and/or $q = \infty$ the proof has to be modified in an obvious way).

(2.1, 3-5) imply

$$\sum_{k=0}^{\infty} 2^{skq} \tau_m(f;2^{-k})_p^q \leq c \sum_{k=0}^{\infty} 2^{skq} \Big(\sum_{j=0}^{\infty} \tau_m(\varphi_{j*}f;\ 2^{-k})_p^p \Big)^{q/p} \leq$$
$$\leq c \sum_{k=0}^{\infty} 2^{skq} \Big(\sum_{j=0}^{k} 2^{(j-k)mp} \| \varphi^a_j f | L_p \|^p \Big)^{q/p} +$$
$$+ c \sum_{k=0}^{\infty} 2^{skq} \Big(\sum_{j=k+1}^{\infty} 2^{(j-k)ap} \| \varphi^a_j f | L_p \|^p \Big)^{q/p} \leq c \| f \|_2^q .$$

The last inequality can be proved by analogy to the corresponding part of Theorem 2.5.12 in [17]. The proof is completed by applying (2.2).

Remark 2.1. Theorem 2.1 and the results from [6] show, of course, that A-spaces are closed under real and complex interpolation for $s > 1/p$. For $s_0 \leq 1/p_0$ and/or $s_1 \leq 1/p_1$, however, little is known about $(A^{s_0}_{p_0 q_0},$

$A^{s_1}_{p_1 q_1})_{\theta,p}$, $(A^{s_0}_{p_0,q_0}, A^{s_1}_{p_1 q_1})_{[\theta]}$. See more comments on this in Remark 4.6.

3. Equivalence of τ-moduli to K-functionals

In [12] Popov considered some functional characteristics equivalent to the τ-modulus. The most remarkable of them was the so-called one-sided K-functional which proved to have important applications to one-sided approximation. In this Section we find equivalent classical Peetre K-functionals.

Theorem 3.1. Let $0 < t \leq 1$, $k \in \mathbb{N}$.

a) Let $0 < p \leq \infty$. Then,

$$\tau_k(f;t)_p \sim K(t^k, f; A_{p,t}, \dot{W}^k_{p,t}) , \quad f \in A_{p,t} + \dot{W}^k_{p,t}$$

b) Let $1 \leq p \leq \infty$. Then,

$$t^k \| f | A_{p,t} \| + \tau_k(f;t)_p \sim t^k \| f | L_p \| + \tau_k(f;t)_p \sim$$
$$\sim K(t^k, f; A_{p,t}, W^k_p) , \quad f \in A_{p,t} .$$

The constants of equivalence are independent of t.

Proof (Outline): The proof is an analogue of the corresponding proof for integral moduli. Let us outline, e.g., the proof of b).

$$t^k \| f | L_p \| + \tau_k(f;t)_p \leq t^k \| f | A_{p,t} \| + \tau_k(f;t)_p$$

is obvious. Properties of τ-moduli, $A_{p,t}$, (1.4) and $0 < t \leq 1$ imply:

$$t^k \| f_o | A_{p,t} \| + \tau_k(f_o;t)_p \leq c_k \| f_o | A_{p,t} \| ,$$
$$t^k \| f_1 | A_{p,t} \| + \tau_k(f_1;t)_p \leq c_k t^k (\| f_1 | A_{p,t} \| + \| f_1^{(k)} | L_p \|) \leq$$
$$\leq c_k t^k \| f | W^k_p \| , \text{ for } f_o + f_1 = f, \ f_o \in A_{p,t} , \ f_1 \in W^k_p .$$

It remains to prove the existence of $c > 0$: $\forall f \in A_{p,t}$,

$$K(t^k, f; A_{p,t}, W^k_p) \leq c(t^k \| f | L_p \| + \tau_k(f;t)_p) .$$
$$K(t^k, f; A_{p,t}, W^k_p) = K(t,f) \leq c(K(t, f-f_{k,t}) + K(t, f_{k,t})) \leq$$
$$\leq c(\| f-f_{k,t} | A_{p,t} \| + t^k(\| f_{k,t} | L_p \| + \| f^{(k)}_{k,t} | L_p \|)) \leq$$
$$\leq c(\tau_k(f;2t)_p + t^k(\| f | L_p \| + \| f_{k,t} - f | L_p \| + t^{-k} \omega_k(f;t)_p)) \leq$$
$$\leq c(t^k \| f | L_p \| + \tau_k(f;t)_p + t^k(1 + t^{-k}) \omega_k(f;t)_p) \leq$$
$$\leq c(t^k \| f | L_p \| + \tau_k(f;t)_p) .$$

Remark 3.1. It can be proved that

$$A_{p,t} + t^k \dot{W}^k_{p,t} \neq X := \{ f : f = \bar{f} \text{ a.e. on } \Omega, \ \bar{f} \in BM_{loc}, \ \| f \|_X :=$$
$$:= \tau_k(\bar{f};t)_p < \infty \} .$$

Remark 3.2. Further versions of theorem 3.1 can be obtained by introducing "inhomogeneous" analogues of $\dot{W}^k_{p,t}$ or "homogeneous" analogues of

W_p^k . We do not go into details.

4. Interpolation - "direct" results

In this section we consider interpolation between estimates of the type: $\exists c_i > 0 : \forall f \in BM_{loc}$,

$$\| Tf|X_i \| \leq c_i (d_i t_i^{k_i} \| f|L_{q_i} \| + \tau_{k_i}(f;t_i)_{q_i}) , \quad i = 0, 1, \qquad (4.1.)$$

where (X_0, X_1) is a compatible pair of quasinormed spaces, T is a sublinear operator, $0 < q_0, q_1 \leq \infty$, $d_i \geq 0$, $i = 0,1$.

Our main tool is:

Lemma 4.1. a) Let A_i, B_i, $i = 0,1$, be quasinormed spaces, such that the pairs $(A_0 + B_0, A_1 + B_1)$, (A_0, A_1) , (B_0, B_1) are compatible. Let $0 < \theta < 1$, $0 < q \leq \infty$. Then,

$$(A_0 + B_0, A_1 + B_1)_{\theta,q} \supsetneq (A_0, A_1)_{\theta,q} + (B_0, B_1)_{\theta,q} .$$

b) Under the assumption of a), let A_i, B_i be Banach spaces. Then,

$$(A_0 + B_0, A_1 + B_1)_{[\theta]} \supsetneq (A_0,A_1)_{[\theta]} + (B_0,B_1)_{[\theta]} .$$

Proof: a) and b) can be proved analogously. We prove a).

It is easy to show that, for A, B, C-quasinormed spaces, $A \hookrightarrow C$ and $B \hookrightarrow C$ imply $A + B \hookrightarrow C$.

Since the real method $K_{\theta,q}$ is an interpolation method,

$$A_i + B_i \supsetneq A_i, \quad i = 0,1 \quad \Rightarrow \quad (A_0 + B_0, A_1 + B_1)_{\theta,q} \supsetneq (A_0, A_1)_{\theta,q} ,$$

$$A_i + B_i \supsetneq B_i , \quad i = 0,1 \quad \Rightarrow \quad (A_0 + B_0, A_1 + B_1)_{\theta,q} \supsetneq (B_0, B_1)_{\theta,q} .$$

Therefore, a) is true.

Remark 4.1. Our initial proof of Lemma 1 was longer. The present "express" proof was indicated to the author by professor Jaak Peetre.

Remark 4.2. Lemma 4.1 holds true for (quasi-)seminormed spaces, too (e.g., in case that A_i are as in the lemma and the kernels of the (quasi-)seminorms in B_i, $i = 0,1$, coincide as sets, and are closed in B_i , $i = 0,1$. This is the case in one of our applications below, where the kernels coincide and are finite-dimensional.

In order to apply Lemma 4.1 to (4.1), we need characterizations of real and complex interpolation spaces between $A_i = A_{p_i,t_i}$, $B_i = t_i^{k_i} \dot{W}^{k_i}_{p_i,t_i}$ or $B_i = t_i^{k_i} W^{k_i}_{p_i}, t_i > 0$, $i = 0,1$. In the latter case for B_i, these spaces are well-known (Sobolev spaces in partial cases, Triebel-Lizorkin spaces in general (see, e.g., [17,8])).

In view of (1.3), in order to study $(A_{p_0,t_0}, A_{p_1,t_1})_{\theta,p(\theta)}$, $t_0 \leq t_1$, it suffices to study $(A_{p_0,t}, A_{p_1,Nt})_{\theta,p(\theta)}$, where $t = t_0$, $N = [t_1/t_0]$, $[\alpha]$ is the integer part of $\alpha > 0$.

Instead of Ω-arbitrary interval, we shall consider in detail Ω of the type $[a,b)$ or $(-\infty,b)$, $-\infty<a,b\leq\infty$, because the notations can be made simpler than in the general case.

Theorem 4.1. Let Ω be of the described type, $0<p\leq\infty$, $0<t\leq 1$, $N\in\mathbb{N}$; then, there exist $c_0\in[\frac{1}{2},1)$, $I\subset\mathbb{Z}$, $I_1:I_1\subset\mathbb{Z}$, $N/2<|I_1|<2N$, depending on t, Ω, N, but not on p, such that:

$$A_{p,t}(\Omega)\rightleftharpoons t^{1/p}\, l_p(I;l_p(I_1;BM[-c_0t,c_0t))),$$

$$A_{p,Nt}(\Omega)\rightleftharpoons (Nt)^{1/p}\, l_p(I;l_\infty(I_1;BM[-c_0t,c_0t))),$$

with constants of equivalence depending on p only.

Proof (Outline): For $\Delta:=\Omega$ and $\delta:=Nt$, let I and $\{\Delta_\nu\}_{\nu\in I}$, be as in (1.1). For every $\nu\in I$, for $\Delta:=\Delta_\nu$, $\delta:=t$, consider $I_\nu=I_1$ and $\{\Delta_{\mu\nu}\}_{\mu\in I_1}$, defined as in (1.1). Evidently, $\{\{\Delta_{\mu\nu}\}_{\mu\in I_1}\}_{\nu\in I}$ is as in (1.1), with $\Delta:=\Omega$, $\delta:=t$. Consider $l(\Omega,t,N)$ - the linear space with elements $a=\{\{\alpha_{\mu\nu}\}_{\mu\in I_1}\}_{\nu\in I}$, $\operatorname{Dom}\alpha_{\mu\nu}=[-c_0t,c_0t)$, $\operatorname{Cod}\alpha_{\mu\nu}\subset\mathbb{R}$, where c_0 is chosen such that the length of $[-c_0t,c_0t)$ is the same as the length of every $\Delta_{\mu\nu}$, $\mu\in I_1$, $\nu\in I$.

We define the linear operator J and P as follows:

$J:\operatorname{Dom} J=BM_{loc}(\Omega)$, $\operatorname{Cod} J=l(\Omega,t,N)$,

$(Jf)_{\mu\nu}(\eta):=f(\eta+\xi_{\mu\nu})$,

where $\eta\in[-c_0t,c_0t)$, $\xi_{\mu\nu}$ is such that $\Delta_{\mu\nu}-\xi_{\mu\nu}=[-c_0t,c_0t)$, $\mu\in I_1$, $\nu\in I$.

$P:\operatorname{Dom} P=l(\Omega,t,N)$, $\operatorname{Cod} P=BM_{loc}$,

$$(Pa)(x):=\sum_{\nu\in I}\sum_{\mu\in I_1}\alpha_{\mu\nu}(x-\xi_{\mu\nu})\chi_{\mu\nu}(x),$$

where $x\in\Omega$, $\chi_{\mu\nu}$ is the characteristic set function of $\Delta_{\mu\nu}$.

The proof is completed by a straightforward procedure of checking that P and J are appropriate isomorphisms. (Recall that (1.2) holds true).

Setting N = 1 and applying interpolation theorems for vector-valued sequence spaces, we obtain:

Lemma 4.2. a) Let $0<p_0,p_1\leq\infty$, $0<t\leq 1$. Then,

$$(A_{p_0,t},A_{p_1,t})_{\theta,p(\theta)}\rightleftharpoons A_{p(\theta),t}.$$

b) Let $1\leq p_0,p_1\leq\infty$, $0<t\leq 1$. Then,

$$(A_{p_0,t},A_{p_1,t})_{[\theta]}\rightleftharpoons A_{p(\theta),t}.$$

Remark 4.3. In case of arbitrary Ω, the proof of Theorem 4.1 is to be modified by introducing more general sequence spaces (see [16]). Lemma 4.2 looks the same.

Lemma 4.3. a) Let $0<p_0,p_1\leq\infty$, $0<t\leq 1$, $k\in\mathbb{N}$. Then,

$$(\dot{W}^k_{p_0,t},\dot{W}^k_{p_1,t})_{\theta,p(\theta)}\rightleftharpoons \dot{W}^k_{p(\theta),t}.$$

b) Let $1\leq p_0,p_1\leq\infty$, $0<t\leq 1$, $k\in\mathbb{N}$. Then,

$(\dot{W}^k_{p_0,t}, \dot{W}^k_{p_1,t})_{[\theta]} = \dot{W}^k_{p(\theta),t}$.

<u>Proof (Outline):</u> Let $BM^k_{loc}(\Omega) := \{f : \text{Dom } f = \Omega, \text{Cod } f \subset \mathbb{R}, f^{(k)} \in BM_{loc}(\Omega)\}$.

J: Dom $J = BM^k_{loc}$, Cod $J \subset BM_{loc}$,

$(Jf)(x) := f^{(k)}(x)$, $x \in \Omega$.

Let $x_0 \in \Omega$. P : Dom $P \supset BM_{loc}$, Cod $P \supset BM^k_{loc}$,

$$(Pg)(x) := \int_{x_0}^{x} \int_{x_0}^{t_k} \dots \int_{x_0}^{t_2} g(t_1)dt_1 \dots dt_k , \quad x \in \Omega .$$

It can be checked that $\dot{W}^k_{p,t}$ is a retract of $A_{p,t}$ with retraction J and coretraction P. (It is sufficient that the composition PJ restores f up to an element of the k-dimensional kernel of the quasiseminorm in $\dot{W}^k_{p,t}$).

<u>Remark 4.4.</u> $A_{p,t}$ is complete, since it is isomorphic to an l_p-space ($BM[-c_0t, c_0t)$ is a Banach space - see [5]). $\dot{W}^k_{p,t}$ is also complete, since it is a retract of a complete space.

Theorem 4.1, Lemmas 4.1-3 imply a variety of interpolation results about (4.1). For simplicity we shall consider the case:

$X_i = L_{p_i}$, $0 < p_i \leq \infty$, $i = 0,1$.

<u>Corollary 4.1.</u> Let $0 < t \leq 1$, $0 < p_0, p_1, q_0, q_1 \leq \infty$, $k \in \mathbb{N}$, T be a sub-linear operator: Dom $T = BM_{loc}$, Cod $T \subset L_{p_0} + L_{p_1}$, such that $\exists c_i > 0$:

$\forall f \in BM_{loc}$, $\| Tf | L_{p_i} \| \leq c_i \tau_k(f ; t)_{q_i}$, $i = 0,1$.

Then, $\exists c > 0$: $\forall f \in BM_{loc}$,

$\| Tf | L_{p(\theta),q(\theta)} \| \leq c\, \tau_k(f ; t)_{q(\theta)}$.

Applying simultaneously Lemmas 4.1-3 a) and b) yields

<u>Corollary 4.2.</u> Let $0 < t \leq 1$, $1 \leq p_0, p_1, q_0, q_1 \leq \infty$, $k \in \mathbb{N}$, T be a linear operator: Dom $T = BM_{loc}$, Cod $T \subset L_{p_0} + L_{p_1}$, such that $\exists c_i > 0$:

$\forall f \in BM_{loc}$,

$\| Tf | L_{p_i} \| \leq c_i \, (t^k \| f | A_{q_i,t} \| + \tau_k(f ; t)_{q_i})$, $i = 0,1$

Then, $\exists c > 0$: $\forall f \in BM_{loc}$,

$\| Tf | L_{p(\theta),\min\{p(\theta),q(\theta)\}} \| \leq c(t^k \| f | L_{q(\theta)} \| + \tau_k(f ; t)_{q(\theta)})$.

<u>Corollary 4.3.</u> Let $0 < t_0 \leq t_1 \leq 1$; $1 \leq p_0, p_1 \leq \infty$, $1 < q_0, q_1 < \infty$; $k_0, k_1 \in \mathbb{N}$, $(1-\theta)k_0 + \theta k_1 = k(\theta)$, $k(\theta) \in \mathbb{N}$. Let T be a linear operator: Dom $T = BM_{loc}$, Cod $T \subset L_{p_0} + L_{p_1}$, such that $\exists c_i > 0$: $\forall f \in BM_{loc}$,

$\| Tf | L_{p_i} \| \leq c_i (t_i^{k_i} \| f | A_{q_i,t_i} \| + \tau_{k_i}(f ; t_i)_{q_i})$, $i = 0,1$.

Then, $\exists c > 0$: $\forall f \in BM_{loc}$,

$\| Tf | L_{p(\theta)} \| \leq cK(t_0^{(1-\theta)k_0}, f; N^{\theta/q_1} t^{1/q(\theta)} l_{q(\theta)}(I ;$

$$ {}_{1_{q_o/(1-\theta)}(I_1;BM[-c_o t, c_o t]))}, X_{q(\theta)} \text{ where } X_{q(\theta)} := (W^{k_o}_{q_o}, A_{q_1,t_1})_{[\theta]} + t_1^{\theta k_1} W^{k(\theta)}_{q(\theta)}; \quad (4.2) $$

where $N := [t_1/t_o]$. If there exist $c_1, c_2 > 0$, absolute constants, such that $c_1 t \leq t_o, t_1 \leq c_2 t$, then the right side in (4.2) is not greater than

$$ t^{(1-\theta)k_o+\theta k_1} \| f | L_{q(\theta)} \| + \tau_{k(\theta)}(f\ ;\ t)_{q(\theta)} . $$

Remark 4.5. Applying Lemma 4.1 a) instead of b), one obtains an analogue of Corollary 4.3 where discrete Lorentz spaces appear (see [16]). We do not go into details.

Remark 4.6. Corollary 4.3 raises some problems that have no analogue associated with integral moduli and Besov spaces. Let

$t_o = t^{3/2}$, $t_1 = t^{1/2}, 0 < t \leq 1$. Is $(A_{p_o,t_o}, A_{p_1,t_1})_{[\frac{1}{2}]}$ isomorphic to $A_{p(\frac{1}{2}),t}$, with constants of equivalence independent of t? (Note that $t = t_o^{1-\theta} t_1^{\theta}$ for $\theta = 1/2$). Is $K(t,f) := K(t,f\ ;\ (A_{p_o,t_o}, A_{p_1,t_1})_{[\frac{1}{2}]}, X_{p(\frac{1}{2})})$ equivalent to $K(t,f\ ;\ A_{p(\frac{1}{2}),t}, W^k_{p(\frac{1}{2})}) \sim t^k \| f | L_{p(\frac{1}{2})} \| + \tau_k(f\ ;\ t)_{p(\frac{1}{2})}$ with constants of equivalence independent of t? Do A-spaces, $s \leq 1/_p$, coincide with their counterparts, where the τ-modulus is replaced by K(t,f)? Our conjecture is that the answers to all these questions are negative.

More generally, we hazard the conjecture that there exist K-functionals K(t,f), the spaces in which depend on t, such that the method for interpolation from this section yields essentially more precise "direct" results than standard interpolation of the corresponding counterparts of Besov spaces where the integral modulus is replaced by K(t,f).

5. Interpolation - "inverse" results

Our aim here is to show that the results from the previous section are precise in the sense that the inverse of the embedding in Lemma 4.1 also holds true in some cases relevant to the K-functionals in Section 3.

Theorem 5.1. Let $0 < t \leq 1$, $0 < \theta < 1$, $0 < p \leq \infty$. Then,

$$ \text{a)}\ (A_o + B_o, A_1 + B_1)_{\theta,p} \rightleftarrows (A_o, A_1)_{\theta,p} + (B_o, B_1)_{\theta,p} \quad (5.1) $$

$$ \text{b)}\ (A_o + B_o, A_1 + B_1)_{[\theta]} \rightleftarrows (A_o, A_1)_{[\theta]} + (B_o, B_1)_{[\theta]} \quad (5.2) $$

where: $A_i := A_{p_i,t} (0 < p_i \leq \infty$ for a); $1 \leq p_i \leq \infty$ for b)),

$B_i := t^k \dot{W}^k_{p_i,t} (0 < p_i \leq \infty$ for a); $1 \leq p_i \leq \infty$ for b)), or

$B_i := t^k W^k_{p_i}$ $(1 \leq p_i \leq \infty)$,

or $A_i := L_{p_i}$ $(1 \geq p_i \leq \infty)$,

$B_i := t^k W^k_{p_i} \quad (1 \leq p_i \leq \infty)$,

and p_o, p_1 are such that $p = p(\theta)$; $i = 0, 1$; $k \in \mathbb{N}$.

Proof (Outline): We prove a) first. All cases considered are proved analogously. Let, e.g., $A_i = A_{p_i,t}$, $B_i = t^k \dot{W}^k_{p_i,t}$, $i = 0, 1$.

Let $0 < r \leq \infty$. Properties of $f_{k,t}$ yield:

$$\tau_k(f\ ;\ t)_r \sim K(t^r, f\ ;\ A_{r,t}, \dot{W}^k_{r,t}) \sim \| V_o(t) f | A_{r,t} \| + t^k \| V_1(t) f | \dot{W}^k_{r,t} \| ,$$

where $V_o(t) : f \mapsto f - f_{k,t}$, $V_1(t) : f \mapsto f_{k,t}$, f - measurable.

The monotonicity of $\| \cdot | A_{p,t} \|$ implies

$$\| f - f_{k,t} | A_{r,t} \| + t^k \| f^{(k)}_{k,t} | A_{r,t} \| \sim \| \, |f - f_{k,t}| + t^k |f^{(k)}_{k,t}| \, | A_{r,t} \| .$$

We define $V_{k,t} : f \mapsto |f - f_{k,t}| + t^k |f^{(k)}_{k,t}|$, f - measurable. $V_{k,t}$ is a sublinear operator: for $\lambda_1, \lambda_2 \in \mathbb{R}$, f_1, f_2 - measurable, $x \in \Omega$,

$$(V_{k,t}(\lambda_1 f_1 + \lambda_2 f_2))(x) \leq \lambda_1 (V_{k,t} f_1)(x) + \lambda_2 (V_{k,t} f_2)(x) .$$

The above facts apparently imply that $\| V_{k,t} \cdot | A_{r,t} \|$ is an equivalent quasiseminorm in $A_{r,t} + t^k \dot{W}^k_{r,t}$ with constants of equivalence independent of t.

Let $f \in (X_o, X_1)_{\theta, p(\theta)}$, where $X_i := A_{p_i,t} + t^k \dot{W}^k_{p_i,t}$, $i = 0, 1$.

$$K(\xi, f\ ;\ X_o, X_1) = \inf_{f=f_o+f_1} (\| f_o | X_o \| + \xi \| f_1 | X_1 \|) \sim$$

$$\sim \inf_{f=f_o+f_1} (\| V_{k,t} f_o | A_{p_o,t} \| + \xi \| V_{k,t} f_1 | A_{p_1,t} \|) \geq$$

$$\geq \inf \{ \| F'_o | A_{p_o,t} \| + \xi \| F'_1 | A_{p_1,t} \| \ :\ V_{k,t} f \leq F'_o + F'_1 ;\ F'_o, F'_1 \geq 0 \} =$$

$$= \inf \{ \| F_o | A_{p_o,t} \| + \xi \| F_1 | A_{p_1,t} \| \ :\ V_{k,t} f = F_o + F_1 ;\ F_o, F_1 \geq 0 \} \geq$$

$$\geq \inf_{V_{k,t} f = F_o + F_1} (\| F_o | A_{p_o,t} \| + \xi \| F_1 | A_{p_1,t} \|) =$$

$$= K(\xi, V_{k,t} f\ ;\ A_{p_o,t}, A_{p_1,t}) \tag{5.3}$$

The second equality in (5.3) is not obvious. For any admissible couple (F'_o, F'_1) we shall construct an admissible couple (F_o, F_1) such that:

$$\| F_o | A_{p_o,t} \| + \xi \| F_1 | A_{p_1,t} \| \leq \| F'_o | A_{p_o,t} \| + \xi \| F'_1 | A_{p_1,t} \| .$$

Let $\varphi := F'_o + F'_1 - V_{k,t} f$. Evidently, $\varphi \in BM(\Omega)$, $0 \leq \varphi \leq F'_o + F'_1$ on Ω , $\varphi(x) = 0$ for $x \in \Omega^o := \{ y : y \in \Omega,\ F'_o(y) = F'_1(y) = 0 \}$. For i=0,1 we define:

$$F_i(x) := F'_i(x)(1 - \varphi(x)/(F'_o(x) + F'_1(x))) , \quad x \in \Omega \setminus \Omega^o ;$$

$$F_i(x) := F'_i(x) = 0 , \quad x \in \Omega^o .$$

It is not difficult to check that $F_i \in BM(\Omega)$, $0 \leq F_i \leq F'_i$ on Ω and, therefore, $\| F_i | A_{p_i,t} \| \leq \| F_i' | A_{p_i,t} \|$. Besides, $F_o + F_1 = V_{k,t} f$.

(5.3) and Lemma 4.1 imply

$$\| f | (X_o, X_1)_{\theta,p(\theta)} \| \sim \| V_{k,t} f | (A_{p_o,t}, A_{p_1,t})_{\theta,p(\theta)} \| \sim \| V_{k,t} f | A_{p(\theta),t} \| \sim$$

$$\sim \| f | A_{p(\theta),t} + t^k \dot{W}^k_{p(\theta),t} \| .$$

b) is a consequence of a) and the well-known result of Karadzov (see, e.g., [1]): if $1 \leq p_o, p_1 \leq \infty$, $p = p(\eta)$, $\theta = (1-\eta)\theta_o + \eta\theta_1$, then

$$((X_o, X_1)_{\theta_o,p_o}, (X_o, X_1)_{\theta_1,p_1})_{[\eta]} = (X_o, X_1)_{\theta,p(\eta)},$$

for appropriate X_o, X_1, θ_o, θ_1, η.

Note that in the case $B_j = t^k W^k_p$ there is an embedding $B_j \hookrightarrow A_j$, therefore (5.1,2) are trivial. However, our proof implies that the constants of equivalence in (5.1,2) do not depend on t, which is not trivial.

Remark 5.1. As a "by-product" of Theorem 5.1 one obtains examples of non-commutative interpolation of space intersections (cf. [9,10]). Let $A_i = A_{p_i,t}(\mathbb{R})$, $B_i = t^k \dot{W}^k_{p_i,t}(\mathbb{R})$, $1 \leq p_i \leq \infty$, $i = 0,1$. The space of all infinitely often differentiable functions defined over $\mathbb{R}$, with finite support, is a subspace of $A_o \cap A_1 \cap B_o \cap B_1$ and proves to be dense in A_o, A_1, B_o, B_1, $(A_o,A_1)_{\theta,p}$, $(B_o,B_1)_{\theta,p}$, $A_o + B_o$, $A_1 + B_1$, $1 \leq p < \infty$. Therefore, by a simple completion and duality argument (see [1]),

$$(A_o' \cap B_o', A_1' \cap B_1')_{\theta,p'} \neq (A_o', A_1')_{\theta,p'} \cap (B_o', B_1')_{\theta,p'} \qquad (5.4)$$

holds true, where "$\cap$" denotes "intersection", A' is the dual of the (semi-)normed space A, $p' : 1/p(\theta) + 1/p' = 1$, $p(\theta) < \infty$.

Another non-trivial example of (5.4) is $A_i = L_{p_i}(\mathbb{R})$, B_i - the homogeneous version of $W^k_{p_i}(\mathbb{R})$ (see [1,17]). In this case, if $1 < p_i < \infty$, $i = 0,1$, then A_i, B_i are reflexive (see, e.g., [17]) and, therefore,

$$(A_o' \cap B_o', A_1' \cap B_1')_{[\theta]} \neq (A_o', A_1')_{[\theta]} \cap (B_o', B_1')_{[\theta]}$$

is true, too. It seems interesting to look for possible extensions and generalizations of the proof of Theorem 5.1 against the more general background of quasilinearizable pairs and Banach lattices.

6. Remarks

Theorem 3.1 is, of course, a simple analogue (at least with respect to the proof) of the corresponding results for integral moduli (cf. [7]). However, it has important new applications to numerical analysis (apart from the results in Sections 4, 5): rather precise network-norm error estimates of the numerical solutions of boundary problems for linear differential equations in terms of norms of the equation coefficients and boundary value functions in A-spaces and some other function spaces may be obtained. Such estimates are to appear in [4] (for an earlier, different, less general method, see [13,3]).

Theorem 3.1. and the completeness of $A_{p,t}$ and $\dot{W}^k_{p,t}$ may be used in proving Marchaud-type inequalities for τ-moduli for $0<p<1$ (similarly to [7]). However, in contrast to the case $1 \leq p \leq \infty$, these inequalities somewhat fail to imply Theorem 2.1.

Corollary 4.3 can be generalized for every p,q between p_0, p_1 and q_0, q_1, respectively, if we introduce " τ-moduli of fractional order s", of the form $K(t^s, f ; A_{p,t}, F^s_{p2})$, $s > 1/p$. These " τ-moduli" appear in a natural way in some of the above-mentioned error estimates, too. They can be computed by the operations usually included in computation of functional moduli, since they can be shown to be equivalent to $\| f - f_{k,t} | A_{p,t} \| + t^s \| f_{k,t} | F^s_{p2} \|$, $s < k$, $k \in \mathbb{N}$, and the norm in $F^s_{p2}(\Omega)$ can be computed by calculating finite differences, integration and finding suprema (see [8]).

The results from Section 3 were previously announced as a part of our talk at the A.Haar Memorial Conference, Budapest 1985 (unpublished).

Acknowledgement

The author would like to thank professors Jaak Peetre and Vasil A. Popov for their useful remarks and valuable pieces of advice.

References

1. Bergh, J., Löfström, J.: Interpolation spaces. An introduction. Grundlehren 223. Berlin - Heidelberg - New York: Springer 1976.
2. Brudnyĭ, Ju.A.: Approximation of functions of n variables by quasi-polynomials. Izvest. Akad. Nauk SSSR, ser. mat., 564-583 (1970) (Russian).
3. Dechevski, L.T.: Network-norm error estimates of the numerical solution of evolutionary equations. Serdica 12, pp. 53-64 (1986).
4. Dechevski, L.T.: Precise error estimates for the numerical solution of differential equations I - IV (in preparation).
5. Dunford, N., Schwartz, J.T.: Linear operators V.1: general theory. New York, London: Interscience 1958.
6. Ivanov, K.G.: On the behaviour of two moduli of functions. Compt. Rend. Acad. Bulg. Sci. 38, 5, pp. 539-542 (1985).
7. Johnen, H., Scherer, K.: On the equivalence of the K-functional and moduli of continuity and some applications. In: Proceedings of a Conference on Constructive Theory of Function of Several Variables. Oberwolfach'76. Lecture Notes 571, pp. 119-140. Berlin - Heidelberg - New York: Springer 1977.
8. Kaljabin, G.A.: Theorems about extensions, multipliers and diffeomorphisms for generalized Sobolev-Liouville classes in Lipschitz-graph domains. Trudy Mat. Inst. Akad. Nauk SSSR, 172, 173-186 (1985) (Russian).

9. Peetre, J.: A theory of interpolation of normed spaces. Notas de Matematica 39, pp. 1 - 86 (1968).

10. Peetre, J.: Über den Durchschnitt von Interpolationsräumen. Arch. Math. (Basel) 25, 511 - 513 (1974).

11. Peetre, J.: On spaces of Triebel-Lizorkin type. Ark. Mat. 13, pp. 123 - 130 (1975).

12. Popov, V.A.: Average moduli and their function spaces. In: Proceedings of the International Conference on Constructive Function Theory, Varna'81. Sofia: Bulg. Acad. Sci. 1983.

13. Popov, V.A., Dechevski, L.T.: On the error of numerical solution of the parabolic equation in network forms. Compt. Rend. Acad. Bulg. Sci. 36, 4, pp. 429 - 432 (1983).

14. Sendov, Bl.: Modified Steklov function. Compt. Rend. Acad. Bulg. Sci. 36, 3, 315 - 317 (1983) (Russian).

15. Sendov, Bl., Popov, V.A.: Average moduli of smoothness. Bulgarian Mathematical Monographs, v.4. Sofia: Bulg. Acad. Sci. 1983. (Bulgarian).

16. Triebel, H.: Interpolation theory, function spaces, differential operators. Math. Libr. 18. Amsterdam - New York - Oxford: North Holland 1978.

17. Triebel, H.: Theory of function spaces. Monographs in Math. 78. Basel - Boston - Stuttgart: Birkhäuser 1983.

Interpolation Spaces and Non-linear Approximation

R.A. DeVore
Department of Mathematics
University of South Carolina
Columbia, South Carolina

V.A. Popov
Bulgarian Academy Of Science
Sofia, Bulgaria

One of the central problems of approximation theory is to characterize the set of functions which have a prescribed order of approximation by a given method of approximation. The classical results in this subject go back to the turn of the century and the work of D. Jackson and S. Bernstein, who among other things showed that a continuous 2π-periodic function f is approximated in the uniform norm to an order $O(n^{-\alpha})$, $0<\alpha<1$, by trigonometric polynomials of degree n if and only if f is in Lip α. Many other results of this type have been obtained for various other types of approximation including algebraic polynomials and splines with fixed knots.

In the 1960's, the subject of approximation turned in large part to the study of non-linear methods. However, until recently, there was no really concrete characterization of the approximation spaces for the two most important types of non-linear approximation, namely, approximation by rational functions, and approximation by splines with free knots. One of the main purposes of this paper is to show how recent results of P. Petrushev [Pt, Pt_1] and P. Pekarskii [Pe_1] can be used to give characterizations of the approximations spaces for these two methods of approximation. In certain cases, the description of these approximation spaces is very concrete (they are Besov spaces).

Suppose we are given a sequence G_n of sets which will be used to approximate a function f in the metric of some space X. We let $E_n(f) := E_n(f)_X$ denote the error of approximation of f by the elements of G_n. We are interested in characterizing those f which have a given order of approximation. For example, we would like to describe the set, A_∞^α, $0<\alpha$, which consists of all f which have an order of approximation $E_n(f) = O(n^{-\alpha})$, $n\to\infty$. More general approximation classes are the A_q^α, $\alpha, q>0$, which consist of all functions f for which

$$(1.1) \qquad \sum_{n=1}^{\infty} [n^\alpha E_n(f)]^q \, n^{-1}$$

is finite. The primary index here is α which shows that the order of approximation is like $n^{-\alpha}$ while the q gives a finer gradation.

During the 1960's, much information [B-P, B, Bu] was given about the classes A_q^α(splines of order k) for free knot spline approximation in the space $L_p[0,1]$, $1\leq p\leq\infty$ (p is fixed and we do not indicate the dependence of the approximation spaces on p). In this type of approximation, G_n is the set of all piecewise polynomials of degree < k which have at most n pieces on [0,1]. We mention in particular, the results of Bergh and Peetre [B-P] for approximation in the space C[0,1]. They introduce a space V_σ defined by means of a generalized variation and show that the spaces A_q^α are the real interpolation between C and V_σ. While this result gave important information about A_q^α, it was not completely satisfactory since there was no description of the resulting interpolation spaces.

Recently, P. Petrushev [Pt_1] studied free knot spline approximation in $L_p[0,1]$, $p>0$ and proved two fundamental inequalities for this type of approximation (see §5) which are the anolgues of the Jackson and Bernstein inequalities for trigonometric polynomials. From these inequalities, it follows that the spaces A_q^α(splines of order k) can be characterized as the real interpolation spaces between L_p and a Besov space $B_\sigma^\lambda(L_\sigma)$, where λ is any real number larger than α and σ is determined by the relation $1/\sigma = \lambda+1/p$. Here, the Besov spaces are defined using the modulus of smoothness of the function (see §5). Petruscev went further and anounced that for $0<\alpha<k$,

(1.2) A_σ^α(splines of order k) $= B_\sigma^\alpha(L_\sigma)$, when $1/\sigma= \alpha+1/p$ and $1\leq p<\infty$.

Yu. Brudnyi also announced this result at a conference in Kiev in 1983 but again with no indication of proof. Of course (1.2) is very interesting because it gives a classical description of the approximation spaces for these values of the parameters. We should mention that the first explicit characterization of this type for non-linear approximation was the famous result of V. Peller [Pl] for approximation on the unit disc in the metric BMO by rational functions.

The research in this paper came about because it was not clear to us how the identity (1.2) followed from the Bernstein and Jackson inequalities, since in particular we did not know the interpolation spaces between L_p and B_σ^λ. This led us to investigate more thoroughly the connections between interpolation spaces and approximation spaces like the A_q^α and in particular to determine the role of the Jackson and Bernstein inequalities in such questions.

It turns out that it is possible to develop a systematic theory (§2,3) for deciding when approximation spaces are interpolation spaces and to establish precisely the role of Jackson and Bernstein inequalities in this process. This is of course not unexpected since such results for linear approximation (see for example Butzer-Scherer [B-S]) are well known and as we have already mentioned, the techniques of interpolation have already been used for specific cases of non-linear approximation. We want also to mention that Peetre and Sparr [P-S] also have developed such a theory based on their E functional. In a sense (see §2,3), our approach is equivalent to theirs (we make this clearer in §3) but for the applications in approximation, our approach seems more natural.

In §2,3, we introduce approximation spaces for a rather general form of approximation and give some of their simple properties. In particular, we show that such approximation spaces are always interpolation spaces and we also show that such spaces always satisfy Bernstein and Jackson inequalities. The familiar reader will find this material rather "old hat" although the organization of these ideas may be useful.

Our main result given in §4 is what we call an extrapolation theorem. If we are given a family of interpolation spaces X_q^α (i.e. this family is invariant under the real method of interpolation) then our extrapolation theorem gives a condition which guarantees that the X_q^α are interpolation between some space X and one of the X_p^γ. The proof of this extrapolation theorem is quite simple and yet this theorem has important application to approximation theory.

Our main application of the extrapolation theorem is to free knot spline approximation and the interpolation of Besov spaces. For example, we use it to show that for any $0<p<\infty$,

$$(1.3) \qquad B_\sigma^\alpha(L_\sigma) = (L_p, B_\gamma^\beta(L_\gamma))_{\alpha/\beta,\sigma}$$

where α and σ and likewise β and γ are related as in (1.2). As a consequence, we obtain (1.2). The extrapolation theorem was our first proof of (1.3). Subsequently, we have heard of two other ways to obtain (1.3), at least with some restrictions on the parameters. Michael Cwickel has shown us how to use retracts and the Littlewood-Paley theory for a proof of (1.3) provided $\sigma,p>1$. Additionally, Bjorn Jaewerth informed us at this conference that he has used an atomic decomposition to prove (1.3) when the Besov spaces $B_\sigma^\alpha(L_\sigma)$ are defined via Fourier transforms (this definition agrees with ours when $\sigma\geq 1$). We should mention that the case when $\sigma<1$ in (1.3) is most important for non-linear approximation.

It is also possible to use (1.3) to describe the approximation spaces A_q^α(rational) for rational approximation. In fact, there are inequalities between rational and free knot spline approximation due to Petrushev [Pt] and Pekarskii [Pe_1] which show that (Theorem 5.2) A_q^α(rational) = A_q^α(splines) provided $1<p<\infty$.

Some additional applications of the extrapolation theorem are given in §5. Finally, in §6, we mention some related results and problems.

2. Approximation spaces. We wish to begin with a general theory whose main aim is to describe approximation spaces as interpolation spaces. We shall denote by X, the space in which approximation will take place. We wish to let X be as general as possible, in particular, we shall not assume that X is linear or normed. Instead, we follow Peetre and Sparr [P-S] and require only that X be what they call a quasi-normed Abelian group. This means that X is an Abelian group under addition with a neutral element denoted by 0 and for each $f \in X$, there is defined $||f|| := ||f||_X$ with the properties:

(2.1) i) $||f|| \geq 0$,

ii) $||-f|| = ||f||$

iii) $||f+g||^\mu \leq ||f||^\mu + ||g||^\mu$, for all $\mu>0$ sufficiently small.

It follows from iii) that $||0|| = 0$. However, there may be other f with $||f|| = 0$. Property iii) is equivalent to

(2.1) iv) $||f+g|| \leq c\,(||f|| + ||g||)$,

for some constant $c \geq 1$, (see [B-L,p.59]).

We let $G_o := \phi$ and let G_n, n=1,2,... denote a sequence of sets whose elements will be used in the approximation. We wish to assume as little as possible about the structure of the sets G_n. For example, we do not want to assume that G_n is a linear space since this is too restrictive for our intended applications. Instead, we shall assume that the sets G_n have the following properties:

(2.2) i) $0 \in G_n$, n=1,...,

ii) $G_n \subset G_{n+1}$, n=0,1,...,

ii) $G_n \pm G_m \subset G_{c(n+m)}$, n,m= 0,1,.... , with c an absolute constant.

iii) $\cup G_n$ is dense in X.

This is satisfied by all of the non-linear families mentioned above. For convenience, we shall also assume that

(2.2) iv) each $f \in X$ has a best approximation from G_n, n=1,2,.... .

This assumption eliminates some technical difficulties and could be replaced by weaker assumptions.

We let $E_n := E_n(f) := E_n(f)_X$ denote the error of approximation

$$E_n := \inf_{g \in G_n} ||f-g||, \qquad n=0,1,\ldots,$$

In particular $E_o(f)=||f||$.

We shall call a sequence of sets (G_n) which satisfies the conditions (2.2) a normal approximating family. Such sequences were first introduced by Petrushev and Popov [P-P, Chapter 3]. For such a sequence, we let A_q^α denote the set of $f \in X$ such that

$$(2.3) \qquad N(f) := \left(\sum_{n=0}^{\infty} [2^{n\alpha} E_{2^n}(f)]^q \right)^{1/q}$$

is finite. As usual, here and in the sequel, the l_q norm is replaced by the l_∞ norm when $q=\infty$. Since E_m is monotone non-increasing, we obtain an equivalent norm if we let the sum in (2.3) run over all $n \geq 0$ and introduce the weight n^{-1} as we did in (1.1). The set A_q^α is a normed Abelian group and N is a quasi-norm on A_q^α. However, there is another quasi-norm which is much more useful for our purposes which we now describe.

For each $f \in X$, we let $S_n := S_n(f)$ be a best approximation to f from G_{2^n}. S_n is generally not unique, so we fix S_n once and for all. In addition, we let $S_{-1} := 0$ and define $T_n := T_n(f) := S_n - S_{n-1}$. We shall use the space $l_q^\alpha(X)$ of sequences (a_n) of functions from X and its norm:

$$(2.4) \qquad ||(a_n)||_{l_q^\alpha(X)} := \left(\sum_{o}^{\infty} [2^{\alpha n} ||a_n||]^q \right)^{1/q},$$

THEOREM 2.1. <u>The following is an equivalent quasi-norm in</u> A_q^α:

$$(2.5) \qquad N_{q,\alpha}(f) := ||(T_n f)||_{l_q^\alpha(X)}.$$

<u>Proof</u>. From (2.1) iv), we have $||T_n|| \leq c[E_{2^n} + E_{2^{n-1}}]$. Therefore, applying the $l_q^\alpha(X)$ norm to (T_n) gives that $N_{q,\alpha}(f) \leq c\, N(f)$ with N(f) as in (2.3). To reverse this inequality, we note that $T_n \to f$, $n \to \infty$ by (2.1) iii). Hence, we have for $\mu > 0$ sufficiently small:

$$(2.6) \qquad E_{2^n}^\mu \leq \sum_{n+1}^{\infty} ||T_k||^\mu.$$

With $\beta := \alpha/2$, we write $||T_k||^\mu = 2^{-k\beta\mu}\, 2^{k\beta\mu}\, ||T_k||^\mu$ in (2.6) and then apply Holder's inequality to find

$$N(f)^q \le c \sum_{n=o}^{\infty} 2^{\alpha nq} \Big(\sum_{n+1}^{\infty} ||T_k||^\mu \Big)^{q/\mu} \le c \sum_{n=o}^{\infty} 2^{n\alpha q}\, 2^{-n\beta q} \sum_{n+1}^{\infty} 2^{k\beta q} ||T_k||^q .$$

A change in the order of summation shows that the right side of this inequality is less than $c\, N^q_{\alpha,q}(f)$ as desired.

Peetre and Sparr [P-S] have used a different approach to define approximation spaces. Namely, they consider a quasi-abelian group $Y \subset X$ and they introduce the E-functional

$$(2.7) \qquad E(f,t) := E(f,t,X,Y) := \inf_{||g||_Y \le t} ||f-g||_X .$$

Their approximation spaces are then defined as the set of all $f \varepsilon X$ for which The expression in (2.3) is finite when $E_{2^n}(f)$ is replaced by $E(f,2^n)$. Both approaches (theirs and ours) lead to the same approximation spaces.

For example, if $Y := \cup G_n$ with $||g||_Y := \inf\{n: g \varepsilon G_n\}$ then $E_n(f) = E(f,n)$ and therefore their approach gives our approximation spaces A^α_q with this choice. On the other hand, if we are given X and Y, we can take $G_n := \{g \varepsilon Y: ||g|| \le n\}$ and again obtain $E(f,n) = E_n(f)$, so that our approach gives their approximation spaces as well.

In the next section, we shall give various properties of the A^α_q spaces. Some of these can be found in [P-S].

3. Jackson and Bernstein inequalities. It has been understood for some time that one of the keys to characterizing the approximation spaces A^α_q in concrete cases is to establish certain inequalities known as the Jackson and Bernstein inequalities.

We suppose that for some (for the present fixed) number $\lambda > 0$, we have a subset X^λ of X which is itself an Abelian group with a quasi-norm $||.||_{X^\lambda}$. We ask whether the following inequalities are valid

$$(3.1) \qquad E_n(f) \le c\, n^{-\lambda} ||f||_{X^\lambda}, \quad f \varepsilon X^\lambda \quad \text{(Jackson inequality)},$$

$$(3.2) \qquad ||g||_{X^\lambda} \le n^\lambda ||g||_X, \quad g \varepsilon G_n,\ n=1,2,\dots \quad \text{(Bernstein inequality)}.$$

We now show that if these two inequalities are satisfied for some space X^λ, then the approximation spaces A^α_q, $0<\alpha<\lambda$, can be characterized as interpolation spaces between X and X^λ. Such results are well known for linear approximation (see [B-S]) and are implicit in the paper of Peetre-Spaar, although we have found no exact formulation as given in the next theorem. We let $(A,B)_{\theta,r}$ denote the interpolation spaces generated by the K method for the pair (A,B).

THEOREM 3.1. If X^λ satisfies the Jackson and Bernstein inequalities, then for $0<\alpha<\lambda$ and $q>0$, we have $A_q^\alpha = (X,X^\lambda)_{\alpha/\lambda,q}$.

Proof. We let $K(f,t) := K(f,t,X,X^\lambda)$ be the K-functional for (X,X^λ). We have for appropriate constants c_1, $c_2 > 0$, and for $n=0,1,\ldots$:

$$(3.3)\qquad c_1\, E_{2^n}(f) \le K(f,2^{-n\lambda}) \le c_2\Big(E_{2^n}(f) + 2^{-n\lambda}\Big(\sum_{k=0}^{n} 2^{k\lambda\mu} E^\mu_{2^k}\Big)^{1/\mu}\Big).$$

Indeed, to prove the left inequality, we use property (2.1)iv) and the Jackson inequality to find that for each r in X^λ,

$$E_{2^n}(f) \le c\Big(||f-r|| + E_{2^n}(r)\Big) \le c\Big(||f-r|| + 2^{-n\lambda}\, ||r||_{X_\lambda}\Big).$$

Taking an infimum over all r gives the left inequality in (3.3).

For the right inequality, we have from Bernstein's inequality that for any $\mu>0$ sufficiently small:

$$K(f,2^{-n\lambda}) \le ||f-S_n|| + 2^{-n\lambda}\, ||S_n||_{X^\lambda} \le E_{2^n} + 2^{-n\lambda}\Big(\sum_{0}^{n} ||T_k||^\mu_{X^\lambda}\Big)^{1/\mu}$$

$$\le E_{2^n} + c\, 2^{-n\lambda}\Big(\sum_{0}^{n} 2^{k\lambda\mu} E^\mu_{2^k}\Big)^{1/\mu},$$

as desired.

To complete the proof of the theorem, we note that because $K(f,t)$ is bounded $(X^\lambda\subset X)$, the q-th power of the norm in the space $(X,X^\lambda)_{\theta,q}$ equivalent to

$$(3.4)\qquad \sum_{n=0}^{\infty}\Big(2^{n\alpha}\, K(f,2^{-n\lambda})\Big)^q .$$

From the left inequality in (3.3), we get that the norm of f in A_q^α is smaller than the right side of (3.4). For the converse, we fix β satisfying $\alpha<\beta<\lambda$ and we write $2^{k\lambda\mu} = 2^{k(\lambda-\beta)\mu}\, 2^{k\beta\mu}$ and then apply Holder's inequality to find that the series on the right side of (3.3) is less than

$$2^{n(\lambda-\beta)}\Big(\sum_{0}^{n} 2^{k\beta q} E^q_{2^k}\Big)^{1/q} .$$

It follows from this and (3.3) that the sum in (3.4) is less than a constant multiple of

$$(3.5) \qquad \sum_{n=o}^{\infty} 2^{n\alpha q} E^q_{2^n} + \sum_{n=o}^{\infty} 2^{n(\alpha-\beta)q} \sum_{k=o}^{n} 2^{k\beta q} E^q_{2^k}.$$

Changing the order of summation, we see that (3.5) is less than a constant multiple of the q-th power of the norm in A^{α}_{q}. This proves the theorem.

It is rather easy to see that the spaces A^{λ}_{q} satisfies Jackson and Bernstein inequalities.

LEMMA 3.2. *If $q,\lambda>0$, then $X^{\lambda}:=A^{\lambda}_{q}$ satisfies the Jackson and Bernstein inequalities.*

Proof. If $f\varepsilon X^{\lambda}$, then from the monotonicity of E_m, we have that $m^{\lambda}E_{2m} \le ||f||_{X^{\lambda}}$ which is the Jackson inequality. On the other hand, if $g\varepsilon G_{2^n}$, then $E_m(g)=0$ for $m\geq 2^n$. Hence,

$$||g||^q_{X^{\lambda}} = \sum_{o}^{n} [2^{m\lambda} E_{2^m}(g)]^q \le c\, 2^{n\lambda q} ||g||^q_X .$$

which is the Bernstein inequality for g in G_{2^n}. The Bernstein inequality for all m=1,2,... follows easily from this.

From Lemma 3.2 and Theorem 3.1, it follows that approximation spaces are always an interpolation family. Namely, we have

COROLLARY 3.3. *If $q,\lambda,r>0$, and if $0<\alpha<\lambda$, then we have*

$$A^{\alpha}_{q} = (X, A^{\lambda}_{r})_{\alpha/\lambda, q}.$$

In particular, from the reiteration theorem for interpolation, we have the following result of Peetre-Sparr [P-S].

COROLLARY 3.4. *For any $q,\alpha_1,q_1,\alpha_2,q_2>0$, and $\alpha=\theta\alpha_1+(1-\theta)\alpha_2$, with $0<\theta<1$, we have*

$$(A^{\alpha_o}_{q_o}, A^{\alpha_1}_{q_1})_{\theta,q} = A^{\alpha}_{q}$$

The above results show that approximation spaces are interpolation spaces. The converse also holds. Namely, if Y is an Abelian subgroup of X which is a quasi-normed space then the interpolation spaces $(X,Y)_{\theta,q}$ are approximation spaces for the sets $G_n=\{g\varepsilon Y: ||g||_Y \le n\, ||g||_X\}$. In fact, in view of Theorem 3.1, we only have to verify that the Jackson and Bernstein inequalities hold for Y. But the Bernstein inequality is built into the definition of G_n. For the Jackson inequality, we write $||f||_Y =: \gamma\ ||f||_X$ for an appropriate γ. Then $E_m(f) = 0$ for $m>\gamma$ because $f \in G_m$. While if $m \le \gamma$, then $E_m(f) \le ||f||_X = \gamma^{-1}\ ||f||_Y \le m^{-1}\ ||f||_Y$.

4. An extrapolation theorem. In the previous section, we have seen that the concepts of interpolation spaces and approximation spaces are identical. We want to exploit this to give a sort of extrapolation theorem for interpolation.

We call a family of spaces X_q^α, $0<q$: $\alpha_0<\alpha\le\alpha_1$ an interpolation family if they are invariant under interpolation, i.e. if $(X_p^\alpha, X_r^\beta)_{\theta,q} = X_q^\gamma$ when $\gamma = \theta\alpha + (1-\theta)\beta$. Of course typically such a family comes from interpolation between a pair of spaces (X,Y). Namely, the spaces $X_q^\alpha := (X,Y)_{\alpha,q}$ are an interpolation family.

Given an interpolation family, we are interested in when there is a sort of limiting space X, as $\alpha\to 0$, for this family. We shall see that this is the case if the X_q^α satisfy Jackson and Bernstein inequalities with respect to X. While the proof of this fact is rather simple, it has several interesting applications.

In most of our applications, we shall not have a doubly indexed family of spaces but instead will have for each value of α, one space X^α. We shall say that (X^α) is selection of an interpolation family $(X^\alpha\ )$, if for each $0<\alpha<\alpha_0$, there is a $q=q(\alpha)$ such that $X^\alpha = X_q^\alpha$.

To formulate our extrapolation theorem, we suppose that G_n are approximation sets with the usual properties and A_q^α are its approximation spaces. We begin with the following simple lemma (see [P-S]).

LEMMA 4.1. *If X^λ is a quasi-normed sub Abelian group of X which satisfies the Jackson and Bernstein inequalities (3.1) and (3.2), then we have the continuous embeddings:*

$$A_\mu^\lambda \subset X^\lambda \subset A_\infty^\lambda, \tag{4.1}$$

provided μ is sufficiently small.

Proof. The right inequality is immediate from the Jackson inequality. For the left inequality, we use that $f \in X$ can be represented as $f = \sum_0^\infty T_k$ with our usual notation. Since $||.||_{X^\lambda}$ is a quasi-norm, for $\mu>0$ sufficiently small, we have from the Bernstein inequality

$$||f||^\mu_{X^\lambda} = ||\ \sum_0^\infty T_k\ ||^\mu_{X^\lambda} \le \sum_0^\infty ||T_k||^\mu_{X^\lambda} \le c\sum_0^\infty [2^{k\lambda}||T_k||]^\mu,$$

as desired.

THEOREM 4.2. (Extrapolation theorem). Let X^λ, $0<\lambda<\lambda_o$ be a selection of the interpolation family (X_q^λ) and suppose that X^λ, $0<\lambda<\lambda_o$ satisfy Jackson and Bernstein inequalities for some approximation sets G_n, $n=0,1,\ldots$ and some space X. Then,

$$(4.2) \qquad (X,X_r^\lambda)_{\theta,q} = X_q^{\lambda\theta} = A_q^{\lambda\theta} \text{ for all } r,q>0 \text{ and } 0<\lambda<\lambda_o;\ 0<\theta<1.$$

Proof. Let A_q^α be the approximation spaces for G_n. From Lemma 4.1, we have that $A_\mu^\lambda \subset X^\lambda \subset A_\infty^\lambda$, provided $\mu>0$ is sufficiently is sufficiently small. Since the approximation spaces are interpolation spaces, they satisfy the reiteration theorem. Likewise for the spaces X_r^λ. Hence, if $0<\theta<1$,

$$A_q^\lambda = (A_r^\alpha,A_s^\beta)_{\theta,q} = (X^\alpha,X^\beta)_{\theta,q} = X_q^\lambda$$

for all $q>0$ and $0<\alpha<\beta\le\lambda_o$ with $\lambda = \theta\alpha + (1-\theta)\beta$. The theorem now follows from Corollary 3.2.

5. Applications. We want now to apply the above development to the study of the degree of non-linear approximation. We let $B_q^\alpha(L_\sigma)$, $\alpha,\sigma,q>0$, denote the Besov spaces on the interval $\Omega:=[0,1]$. Here, we use the definition of Besov spaces with the modulus of smoothness $\omega_k(f,t)_\sigma$ of f in $L_\sigma(\Omega)$. Namely, the "quasi-norm" in $B_q^\alpha(L_\sigma)$ is given by

$$|f|_{B_q^\alpha(L_\sigma)} := \left(\int_0^\infty \left(t^{-\alpha}\omega_k(f,t)_\sigma\right)^q \frac{dt}{t} \right)^{1/q},$$

where $k:=[\alpha]+1$.

When the value of σ is fixed, the Besov spaces are an interpolation family. In fact for $\sigma\ge1$, the $B_q^\alpha(L_\sigma)$, $0<\alpha<k$ are the interpolation spaces for L_σ and the Sobolev space W_σ^k. When $\sigma<1$, this is shown in [D-P]. This latter paper also describes the interpolation of Besov spaces for other values of the parameters.

A special case of the general interpolation for Besov spaces is very important in non-linear approximation. We let L_p, $0<p\le\infty$ be the space where the approximation is to take place. So p is fixed. We then relate the parameters, α,σ, and q by the equations

$$(5.1) \qquad 1/\sigma = \alpha + 1/p$$
$$q = \sigma.$$

Hence, we are talking about a one dimensional scale of spaces $B^{\alpha} := B^{\alpha}_{\sigma}(L_{\sigma})$. In [D-P], it is shown that B^{α} is a selection from an interpolation family. Namely, if $0<\alpha<\beta$, then $(B^{\alpha}, B^{\beta})_{\theta,q} = B^{\gamma}$ provided $\gamma = \theta\alpha + (1-\theta)\beta$ and $1/q = \gamma + 1/p$.

The spaces B^{α} appear naturally in non-linear approximation. For example, P. Petrushev [Pt] has recently shown that these spaces satisfy the Jackson and Bernstein inequalities for free knot spline approximation in $L_p(\Omega)$. Namely, if $0<p<\infty$, then

$$\text{(5.3) i)} \quad s_n(f)_p \leq c\, n^{-\alpha} \, ||f||_{B^{\alpha}}$$

$$\text{ii)} \quad ||S||_{B^{\alpha}} \leq c\, n^{\alpha} \, ||S||_p, \quad \text{for all } S \in S_n.$$

As was observed by Petrushev, it follows from (5.3), that the approximation spaces A^{α}_q(splines) for this type of approximation are the same as the interpolation spaces $(L_p, B^{\beta})_{\alpha/\beta,q}$(see Theorem 3.1). From our extrapolation theorem (Theorem 4.2), we obtain the following characterization for free knot spline approximation

THEOREM 5.1. _The approximation spaces_ A^{α}_q(splines), $0<\alpha<k$ _for free knot approximation by splines of order_ k _are characterized by_

$$\text{(5.4)} \qquad A^{\alpha}_q = (L_p, B^{\beta})_{\alpha/\beta,q}, \quad \text{whenever } 0<\alpha<\beta<k \text{ and } q>0.$$

In particular,

$$\text{(5.5)} \qquad A^{\alpha}_{\sigma} = B^{\alpha} = B^{\alpha}_{\sigma}(L_{\sigma}), \quad \text{whenever } 1/\sigma = \alpha + 1/p \text{ and } 0<\alpha<k.$$

The above results do not apply when $p=\infty$, that is for approximation L_{∞} or in BMO. In fact, in these cases, the approximation classes cannot be characterized in terms of Besov spaces. Indeed, there is not a Bernstein inequality for such spline approximation. For example, a characteristic function of an interval is in L_{∞} and BMO but not in $B^{1/p}_p$ for any $p>0$. Since, as we noted earlier in Lemma 3.2, approximation spaces always satisfy Jackson and Bernstein inequalities, the $B^{1/p}_p$ cannot be approximation spaces.

It is possible however to characterize A^{α}_q in terms of certain spaces which are closely related to Besov spaces when we assume that the approximating splines are continuous. We let $s^*_n(f) := s^*_{n,k}(f)$ be the error in approximating f in the uniform norm on Ω by continuous splines of degree $< k$ which have n free knots. For this type of approximation, P. Petrushev [Pt_1] has given Jackson and Bernstein inequalities in terms of the spaces B^{α}_*, $\alpha>1$, which consists of all absolutely continuous functions f for which f' is in $B^{\alpha}_{1/\alpha}(L_{1/\alpha})$. That is,

(5.6) i) $s^{*}_{n}(f)_C \leq c\, n^{-\alpha}\, ||f||_{B^{\alpha}_{*}}$,

ii) $||S||_{B^{\alpha}_{*}} \leq c\, n^{\alpha}\, ||S||_C$.

The spaces B^{α}_{*}, $\alpha>1$ are invariant under interpolation because the corresponding Besov spaces $B^{\alpha-1}_{1/\alpha}(L_{1/\alpha})$ are. Hence, it follows from our extrapolation theorem and from (5.6) that for any $1<\beta\leq k$ and all $q>0$ and $0<\alpha<\beta$ we have

$$(5.7) \qquad A^{\alpha}_{q}(\text{continuous splines}) = (C, B^{\beta}_{*})_{\alpha/\beta, q}.$$

In addition,

$$(5.8) \qquad A^{\alpha}_{1/\alpha}(\text{continuous splines}) = B^{\alpha}_{*} \quad \text{for } \alpha>1.$$

The exact same results hold when the approximation is in the space BMO instead of C. That is, the Jackson and Bernstein inequalities (5.6) hold with BMO in place of C. Therefore (5.8) also holds with C replaced by BMO. As a consequence, we have the following interpolation theorem which is of independent interest:

$$(5.9) \qquad (C, B^{\alpha}_{*})_{\theta,q} = (BMO, B^{\alpha}_{*})_{\theta,q}, \quad \text{for all } 0<\theta<1;\ q>0 \text{ and } \alpha>1.$$

It is also possible to characterize the classes of functions A^{α}_{q}(rational) when $1<p<\infty$ (V. Peller [P_1] has solved the case BMO). For this, we use connections between the errors of rational and spline approximation. Petrushev [Pt] has shown that for $1\leq p<\infty$,

$$(5.10) \qquad r_n(f)_p \leq c\, n^{-\lambda} \sum_{m=0}^{n} (m+1)^{\lambda-1} s_{m,k}(f)_p,$$

provided $k>\lambda$. This gives an estimate for the error in rational approximation once the spline approximation error is known. In the other direction, Pekarskii [Pe_1] has given the following inequality for $1<p\leq\infty$:

$$(5.11) \qquad s_{n,k}(f)_p \leq c\, n^{-k} \left(\sum_{m=0}^{n} m^{-1}[m^k r_m(f)]^{\sigma} \right)^{1/\sigma}$$

with $\sigma := (k+1/p)^{-1}$.

From these two inequalities, it is rather easy to obtain the following rather remarkable result.

THEOREM 5.2. *For approximation in* $L_p(\Omega)$ *with* $1<p<\infty$, *we have for any* $\alpha,q>0$,

(5.12) $$A_q^\alpha(\text{rational}) = A_q^\alpha(\text{splines of order } k)$$

provided $\alpha<k$. *In particular,*

(5.13) $$A_\sigma^\alpha(\text{rational}) = A_\sigma^\alpha(\text{splines of order } k) = B_\sigma^\alpha(L_\sigma),$$

when $1/\sigma = \alpha + 1/p$.

Proof. It follows from (5.10) that if $s_{n,k}(f)_p = O(n^{-\alpha})$ then $r_n(f)_p = O(n^{-\alpha})$ and the inequality (5.11) gives the reverse implication. Hence, the classes $A_\infty^\alpha(\text{rational}) = A_\infty^\alpha(\text{splines of order } k)$. Since each of these families of spaces $A_q^\alpha(\text{rational})$ and $A_q^\alpha(\text{splines of order } k)$ is an interpolation family, (5.12) follows. The inequality (5.13) then follows from this and (5.5).

6. *Remarks*.

6.1. It still remains an open problem to characterize the spaces $A_\infty^\alpha(\text{splines})$, if possible, in terms of classical function spaces. Brudnyi [B] has given a characterization of these spaces based on rearrangements of the local degree of polynomial approximation.

6.2. The spaces $A_q^\alpha(\text{rational})$ and $A_q^\alpha(\text{splines})$ are not the same when the approximation takes place in C or BMO since as we have already noted for spline approximation these spaces are not Besov spaces while for rational approximation, Peller has mentioned to us at this conference that he has used his results for the disc [P1] to show that $A_{1/\alpha}^\alpha(\text{rational in BMO}) = B_{1/\alpha}^\alpha(L_{1/\alpha})$. It has not yet been determined whether or not the approximation spaces for rationals and splines coincide when p=1.

6.3. It follows from Theorem 5.2 and the fact that Jackson and Bernstein inequalities are necessary for approximation spaces (Lemma 3.2) that the following inequalities hold for rational approximation:

$$r_n(f)_p \le c\, n^{-\alpha} \|f\|_{B^\alpha}, \quad n=0,1,\ldots$$

$$||R||_{B^\alpha} \leq c\, n^\alpha\, ||R||_p, \quad R \text{ a rational function of degree } n.$$

The first of these inequalities follows from the work of Petrushev [Pt]. There is at present not a simple direct proof for either of these important inequalities.

6.4. For complex rational approximation on the disc, Pekarskii [Pe] has shown that $A^\alpha_\sigma(\text{rational in } L_p) = B^\alpha_\sigma(L_\sigma)$, provided $1/\sigma = \alpha + 1/p$ and $1<p<\infty$.

REFERENCES

[B-L] J. Bergh - J. Löfström, Interpolation Spaces, Springer Ver., Grundlehren, vol. 223, Berlin, 1976.

[B-P] J. Bergh - J. Peetre, On the spaces V_p $(0<p<\infty)$, Bolletino V.M.I. (4)10(1074), 632-648.

[B] Yu. Brudnyi, Spline approximation and functions with bounded variation, Sov. Math. Doklady,

[Bu] H. Burchard, On the degree of convergence of piecewise polynomial approximation on optimal meshes II, TAMS, 234(1977) 531-557.

[B-S] P. Butzer - K. Scherer, Approximationsprozesse und interpolations-methoden, B.I. Hochschulskripten, 1968, Mannheim.

[P-S] J. Peetre - G. Sparr, Interpolation of normed Abelian groups, Ann. Mat. Pura Appl. 92(1972), 217-262.

[Pe] P. Pekarskii, Classes of analytic functions determined by best rational approximation in H_p, Math. Sbornik 127(169)(1985), 3-20. (Russian)

[Pe_1] P. Pekarskii, Estimates for the derivatives of rational function in $L_p[-1,1]$, Math. Zametki, 39(1986), 388-394.

[Pl] V. Peller, Description of Hankel operators of the class σ_p for p<1, investigation of the order of rational approximation, and other applications, Math USSR Sbornik 50(1985), 465-494.

[Pt] P. Petrushev, On the connection between spline and rational approximation, J. Approx. Th. (to appear).

[Pt_1] P. Petrushev, Direct and converse theorems for best spline approximation with free knots and Besov spaces, C.R. Acad. Bullg. Sci. 39(1986).

[P-P] P. Petrushev - V. Popov, Rational Approximation of Real Valued Functions, Cambridge Univ. Press., 1987.

Acknowledgements: Ronald A. DeVore was supported by NSF Grant DMS 8320562. This work wasdone while Vasil Popov was a Visiting Professor at the University of South Carolina.

ATOMIC DECOMPOSITIONS IN HARDY SPACES ON BOUNDED LIPSCHITZ DOMAINS

Klas Forsman
Department of Mathematics
University of Umeå
S-901 87 Umeå, Sweden

Abstract

The theory of Hardy spaces on R^n has been generalized to Hardy spaces of distributions f on certain closed subsets F of R^n. In this paper we present some new results for the case when F is bounded and the boundary is locally Lipschitzian.

Let f have its support contained in F. If a suitable maximal function of f belongs to L^p, then f belongs to the local Hardy space $h^p(F)$. Moreover, if f belongs to the standard Hardy space on R^n, then f has an atomic decomposition whose atoms are supported in F.

1. Introduction

This paper continues the work by Jonsson, Sjögren, and Wallin [7]. We present a theorem concerning the case when the closed subset F of R^n is bounded and the boundary locally is of class Lip 1. By using the technique of partition of unity we prove in Proposition 1, that if the distribution f has its support contained in F and $m_\sigma' f$ belongs to L^p, $0 < p \leq 1$, where $m_\sigma' f$ is a maximal function adapted to the local Hardy space $h^p(R^n)$, then f belongs to $h^p(F)$. The maximal function $m_\sigma f$, which is defined on F, is used in a corresponding theorem by Jonsson, Sjögren, and Wallin [7, p.150], leading to the same result about f. This theorem contains the additional requirement that f should belong to L^1_{loc}, but puts no Lipschitz condition on the boundary of F. Their paper also contains an analogous theorem [7, p.165] to our Theorem 1. Instead of our Lipschitz condition on the boundary they require that F should be convex. We use the result of

Proposition 1 to prove in our theorem that if f belongs to the standard Hardy space on R^n, $H^p(R^n)$, $0 < p \leq 1$, and the support of f is contained in F, then the atoms in the decomposition of f may be chosen in such a way that their support are contained in F. Thus a result concerning global H^p-theory is achieved by means of local h^p-technique.

2. Definitions and notation

Throughout the paper we use the following notation. F denotes a non-empty closed subset of the n-dimensional Euclidean space R^n and μ is a positive Borel measure with support F, $\operatorname{supp} \mu = F$, such that μ is finite on compact sets. The closed ball with center $x \in R^n$ and radius $r>0$ is denoted by $B(x,r)$. $c, c_1, c_2, \ldots$ stand for different positive constants.

F is a d-set, $0 < d \leq n$, means that there exists a measure μ with support F (a d-measure on F), which behaves like a d-dimensional Hausdorff measure, allowing us to treat Lebesgue classes $L^p(\mu)$ of functions on F in a natural way. d-sets were introduced and studied by A. Jonsson and H. Wallin in [8] and [10].

DEFINITION 2.1. A positive measure μ supported by F is a d-measure on F $(0<d\leq n)$ and F is a d-set, if, for some constants $c_1, c_2>0$, $x \in F$, and $0<r\leq 1$, we have

$$c_1 r^d \leq \mu(B(x,r)) \leq c_2 r^d \tag{2.1.1}$$

Clearly, R^n is a d-set with $d=n$.

DEFINITION 2.2. Consider a fixed closed subset F of R^n, all balls $B = B(x_0,r)$, all positive integers N, and all polynomials $P(x) = \sum_{|\beta|\leq N} a_\beta (x-x_0)^\beta$ of degree $\leq N$, $x_0 \in F$ and $0<r\leq 1$.

F preserves the Markov inequality if, for all $P(x)$ and B,

$$\max_{F\cap B} |\operatorname{grad} P| \leq \frac{c}{r} \max_{F\cap B} |P| \tag{2.2.1}$$

where $c = c(F,N)$.

From now on, we assume that μ is a d-measure on F and F a d-set preserving Markov's inequality. We refer to [9], Chapter II, for examples of such sets.

Our definitions of atoms are adapted to the set F and the measure μ:

DEFINITION 2.3. A local (p,q)-atom on F (with respect to μ) is a function $a \in L^q(\mu)$ with compact support such that for some ball $B = B(x_0,r)$, $x_0 \in F$, satisfying $\operatorname{supp} a \subset B$, we have

$$\left\{\frac{1}{\mu(B)} \int |a|^q d\mu\right\}^{1/q} < \mu(2B)^{-1/p}, \tag{2.3.1}$$

and the moment condition

$$\int aPd\mu = 0 \ , \quad \text{if} \quad r \leq 1 \ , \tag{2.3.2}$$

for all polynomials P of degree at most $[d(1/p-1)]$.

A local p-atom on F is a local (p,∞)-atom on F.

By $2B$ we mean the ball $B = B(x_0,2r)$. We assume that p and q are admissible exponents in the sense that $0 < p \leq 1$, $1 \leq q < \infty$ and $p<q$.

DEFINITION 2.4. The local Hardy space $h^{p,q}(F,\mu)$ on F (with respect to μ) is the space of all distributions f having a representation $\Sigma \lambda_j a_j$ where a_j are local (p,q)-atoms on F and $\Sigma|\lambda_j|^p < \infty$. We introduce

$$||f||_{h^{p,q}} = \inf (\Sigma|\lambda_j|^p)^{1/p}$$

which is a quasi-norm.

We speak about the local Hardy space $h^p(F)$ by which we mean

$$h^p(F) = h^p(F,\mu) = h^{p,\infty}(F,\mu).$$

The maximal function characterization of $H^p(\mathbb{R}^n)$ was given in [3] by Fefferman and Stein and analogously for $h^p(\mathbb{R}^n)$ by Goldberg in [4]. We follow Jonsson, Sjögren, and Wallin, using the class S_σ of specially well-behaving test-functions to define maximal functions suitable to our Hardy spaces.

DEFINITION 2.5. Let

$$S_\sigma = \{\Phi \in C_0^\infty(R^n);\ \text{supp}\,\Phi \subset B(0,1),\ |D^\beta\Phi| \le 1 \quad \text{for} \quad |\beta| \le \sigma\}$$

for $\sigma \in N$. If f is a distribution with support in F, we set for $x \in F$ the maximal function f for the space $h^p(F,\mu)$

$$m_\sigma f(x) = \sup_{\Phi \in S_\sigma} \sup_{0<t\le 1} \mu(B(x,t))^{-1} \left\langle f, \Phi\left(\frac{x-\cdot}{t}\right)\right\rangle.$$

In the case that $F=R^n$ and μ is the n-dimensional Lebesgue measure, we denote the maximal function by $m'_\sigma f(x)$.

Now we want to make the corresponding definitions for global atoms and for the global Hardy space on F. We shall say that F is a <u>global d-set</u> if there exists a measure $\mu \ge 0$, called a <u>global d-measure</u>, whose support is F and which satisfies (2.1.1) for arbitrary $r>0$ and $x \in F$. F <u>preserves the global Markov inequality</u> if, for any integer $N>0$, (2.2.1) holds, with a constant $c=c(F,n,N)$ for all polynomials P of degree at most N and all balls $B=B(x_0,r)$ with $x_0 \in F$ and $r>0$.

DEFINITION 2.6. <u>Global</u> (p,q)-<u>atoms</u> are defined like local (p,q)-atoms, except that the moment condition is always present, independently of the radius of the supporting ball.

DEFINITION 2.7. The <u>global Hardy space</u> $H^{p,q}(F,\mu)$ is defined by means of global (p,q)-atoms with ℓ^p-coefficients, like $h^{p,q}(F,\mu)$. We introduce the (global) <u>Hardy space on</u> F, $H^p(F) = H^p(F,\mu)$ by $H^p(F) = H^{p,\infty}(F,\mu)$.

DEFINITION 2.8. The <u>maximal function for</u> $H^{p,q}(F,\mu)$ is

$$M_\sigma f(x) = \sup_{\Phi \in S_\sigma} \sup_{t>0} \mu(B(x,t))^{-1} \left\langle f, \Phi\left(\frac{x-\cdot}{t}\right)\right\rangle..$$

In this paper we will study Hardy spaces on such bounded subsets F of R^n whose boundaries are locally Lipschitzian. We consider the points $x \in R^n$ as pairs (x',x_j) for $j=1,\dots,n$ with $x' \in R^{n-1}$ and $x_j \in R$.

DEFINITION 2.9. F:s *boundary* ∂F *is locally of class* Lip 1 means that for every point $x_0 \in \partial F$ there is one $j \in \{1,2,\dots,n\}$ and a function $g_j\colon R^{n-1} \to R$ satisfying, with $h' \in R^{n-1}$, and for some $M \in R$,

$$|g_j(x'+h') - g_j(x')| \le M|h'|$$

and such that in a neighbourhood of x_0, $x \in \partial F \Leftrightarrow x_j = g_j(x')$.

3. Main results

We will now mention some of the results by Jonsson, Sjögren, and Wallin [7], since our results are closely related to theirs. See also [9] for a more complete description of their theory.

For the case when F is a d-set preserving the Markov inequality, they prove a characterization by means of the maximal function $m_\sigma f$ of the local Hardy space $h^p(F)$ and $M_\sigma f$ of the global Hardy space $H^p(F)$.

We quote from [9, Chapter IV]:

THEOREM A (JONSSON, SJÖGREN, WALLIN). *Let* $F \subset R^n$ *be a d-set preserving Markov's inequality and let* μ *be a d-measure on* F. *Assume that* p *and* q *are admissible exponents. Then* (a) *and* (b) *hold:*

(a) *If* $f \in h^{p,q}(F,\mu)$ *then* $m_\sigma f \in L^p(\mu)$ *if* $\sigma > [d(1/p - 1)]$ *and, for some constant* $c = c(F,\mu,p,q,\sigma)$.

$$||m_\sigma f||_{L^p(\mu)} \le c||f||_{h^{p,q}(F,\mu)}$$

(b) *Conversely, if* $f \in L^1_{loc}(\mu)$, f *is identified with the distribution given by* $\langle f,\Phi\rangle = \int f\Phi d\mu$, *for* $\Phi \in C_0^\infty(R^n)$, *and* $m_\sigma f \in L^p(\mu)$ *for some positive integer* σ , *then* $f \in h^{p,\infty}(F,\mu)$ *and, for some constant* $c = c(F,\mu,p,\sigma)$.

$$||f||_{h^{p,\infty}(F,\mu)} \le c||m_\sigma f||_{L^p(\mu)}.$$

COROLLARY A (JONSSOn, SJÖGREN, WALLIN). *Let* F *and* μ *be as in Theorem A and let* p *and* q *be admissible exponents. Then the local Hardy space* $h^{p,q}(F,\mu)$ *is independent of* q *and* μ, *i.e. for fixed* p and F different q and μ give the same spaces with equivalent norms.

Further, as an application, Jonsson, Sjögren, and Wallin obtain a theorem concerning ordinary Hardy spaces in R^n. We quote from [7, p.165]:

THEOREM B (JONSSON, SJÖGREN, WALLIN). Let F be a closed convex bounded set with non-empty interior. If f belongs to the standard Hardy space $H^p(R^n)$, $0<p\leq 1$, and supp $f \subset F$, then f has an atomic decomposition whose atoms are supported in F.

In this paper, we improve Theorem B.

In Proposition 1 we prove (see below for the precise statement) that if f is a distribution, supported by F, satisfying $m_\sigma' f \in L^p(R^n)$, where $0<p\leq 1$, and F is the closure of a bounded, open and non-empty subset of R^n such that the boundary of F locally belongs to Lip 1, then f belongs to $h^p(F)$. Thus, the condition $f \in L^1_{loc}(\mu)$ of Theorem A(b) may be dropped in this case.

As a corollary, under the same assumptions about F , we get that a product of $f \in H^p(R^n)$ and $\varphi \in C_0^\infty(\Omega)$ belongs to $h^p(F)$ if F contains the intersection of the supports of F and φ. The product has a decomposition in local atoms with supports contained in F.

Finally, under the same assumptions about F, we prove in Theorem 1 that if $f \in H^p(R^n)$ has its support contained in F, then f has a decomposition in global atoms which also have their supports contained in F.

This is an amplification of Theorem B. In Theorem 1 we replace the convexity requirement on F of Theorem B by the condition that the boundary of F should belong to Lip 1 locally.

We will now give the precise statements of our results.

PROPOSITION 1. Let f be a distribution supported in F, where F is the closure of a bounded, open and non-empty subset of R^n with the boundary ∂F locally of class Lip 1. If $m_\sigma' f \in L^p(R^n)$, $0<p\leq 1$, then $f \in h^p(F)$ and f has a decomposition in local p-atoms which are supported in F. For some constant $c=c(F,\mu,p,\sigma)$.

$$||f||_{h^p(F,\mu)} \leq c\cdot||m_\sigma^{\prime}f||^1_{L^p(R^n)}.$$

Proposition 1 yields a localization corollary. Before stating the corollary, we mention that Goldberg [4] has shown that $h^p(R^n)$ is stable under multiplication by $\mathcal{S}$, where $\mathcal{S}$ is the Schwartz class of rapidly decreasing test functions. Let us also, like Goldberg, notice that every H^p-atom is an h^p-atom.

COROLLARY 1. Let F be as in Proposition 1, let f belong to the standard Hardy space $H^p(R^n)$, $0<p\leq 1$, and let $\varphi \in C_0^\infty(\Omega)$, where Ω is an open subset of R^n. If the intersection of the supports of f and φ is contained in F then the product φf belongs to the local Hardy space $h^p(F)$ and there exists an atomic decomposition of φf in local p-atoms which are supported in F.

THEOREM 1. Let F be the closure of a bounded, open, and non-empty subset of R^n with the boundary ∂F locally of class Lip 1. If f belongs to the standard Hardy space $H^p(R^n)$, $0<p\leq 1$, and $\operatorname{supp} f \subset F$, then f has an atomic decomposition whose atoms are supported in F.

In Section 4 we will give the proofs, using the method of partition of unity. We start locally by proving a proposition concerning the intersection of F with an n-dimensional cylinder S.

4. Proofs

We note that if $F= R^n$ and μ is the n-dimensional Lebesgue measure on R^n, then $h^p(F,\mu)$ gives the local Hardy space $h^p(R^n)$ introduced by Goldberg.

In order to prove Proposition 1, using the method of partition of unity, we need Proposition 2 concerning the situation locally.

In [7], Jonsson, Sjögren, and Wallin prove, that when F is a d-set, satisfying the Markov inequality, $0<p\leq 1$, and the distribution $f \in L^1_{loc}(\mu)$ satisfies $m_\sigma f \in L^p(\mu)$ for some σ, then $f \in h^p(F)$ and $||f||_{h^p} \leq c\cdot||m_\sigma f||_p$. (Theorem A).

We will use this result in the proof of Proposition 2 concerning the case when F:s boundary ∂F locally belongs to Lip 1. Such a set F is a d-set with $d=n$. Thus [7,Theorem 1.4] it preserves the Markov

Now let F be the closure of a bounded, open and non-empty subset of R^n whose boundary ∂F locally is of class Lip 1.

We denote the usual n-dimensional Lebesgue measure on R^n by m, and as μ we take the restriction of m to F.

Let $x_0 \in \partial F$. We choose x_0 to be the origin of a coordinate system.

Assume that in a neighbourhood of the origin the positive x_j-axis lies in the complement of F.

Let S be a closed circular cylinder, with its axis along the x_j-axis and bounded in the directions of the axis by (n-1)-dimensional balls in the following way:

the part of the cylinder lying below ∂F (in the negative x_j-direction) shall belong to the interior of F; (4.1)

the part of the cylinder lying above ∂F (in the positive x_j-direction) shall belong to the complement of F. (4.2)

Let $S' = S \cap F$ and denote $S \cap \partial F$ by $\partial F'$.

PROPOSITION 2. <u>If the distribution</u> g, <u>with</u> $\operatorname{supp} g \subset S'$, <u>satisfies</u> $m_\sigma^! g \in L^p(R^n)$, <u>for a fixed</u> p, $0<p\leq 1$, <u>then</u> g <u>belongs to</u> $h^p(F)$, <u>and</u> g <u>has an atomic decomposition where all the atoms have their supports contained in</u> F. <u>For some constant</u> $c = c(F,\mu,p,\sigma)$,

$$||g||_{h^p(F,\mu)} \leq c||m_\sigma^! g||_{L^p(R^n)}.$$

The proof of this proposition follows closely the proof of Theorem B by Jonsson, Sjögren, and Wallin. See [7, Theorem 5.3]. We use a translation, instead of a contraction, in order to get a distribution g_ε supported in a set S_ε' with a positive distance to ∂F.

<u>Proof of Proposition 2</u>. By the translation $x_j^{(\varepsilon)} = x_j - \varepsilon$ we move S' the distance ε in the negative x_j-direction. If $\varepsilon>0$ is chosen small enough, we get a closed set $S_\varepsilon' \subset F$ with $S_\varepsilon' \cap \partial F = \emptyset$. Via this translation g yields a distribution g_ε with support in S_ε'. I.e. if $y=(y',y_j)$ and φ is a test function we define

$$\left\langle g_\varepsilon, \varphi(y) \right\rangle = \left\langle g_\varepsilon, \varphi(y', y_j) \right\rangle = \left\langle g, \varphi(y', y_j+\varepsilon) \right\rangle .$$

By this definition, $m_\sigma' g_\varepsilon$ belongs to $L^p(R^n)$ too, since the Lebesgue measure is translation invariant.

Then take a C^∞ approximate identity $\eta_\delta(x) = \delta^{-n}\eta(\frac{x}{\delta})$, $\delta>0$, with $\int \eta dx = 1$ and $\operatorname{supp} \eta \subset B(0,1)$. If $\delta=\delta(\varepsilon)$ is small, $\tilde{g}_\varepsilon = g_\varepsilon * \eta_{\delta(\varepsilon)}$ is a C^∞-function with support in F. By [5, Theorem 1.6.3], $\tilde{g}_\varepsilon \to g_\varepsilon$ as $\delta(\varepsilon)\to 0$, in the distribution sense. Using the definition of g_ε we find that $\tilde{g}_\varepsilon \to g$ as $\varepsilon\to 0$, in the distribution sense. In Lemma 1 (see below) we prove that $m_\sigma \tilde{g}_\varepsilon$ belongs to $L^p(\mu)$, with its $L^p(\mu)$-norm independent of ε, so by Theorem A(b), $\tilde{g}_\varepsilon \in h^p(F)$, and

$$\|\tilde{g}_\varepsilon\|_{h^p(F,\mu)} \leq c_1 \|m_\sigma \tilde{g}_\varepsilon\|_{L^p(\mu)} \leq c\|m_\sigma' g\|_{L^p(R^n)}$$

where c and c_1 depend on F, μ, p and σ. Thus $\tilde{g}_\varepsilon$ has an atomic decomposition with local p-atoms.

It remains to let $\varepsilon\to 0$. We follow the method of [2. Lemma 4.2, p. 638] or [12, Lemma 5.13, p. 42].

For each $m\in Z$, let $(B_i^m)_{i=1}^\infty$ be an enumeration of the balls with radii $2\sqrt{n}2^m$ and centers at the lattice points whose coordinates are multiples of 2^m. Then any ball of radius r, $2^{m-1}<r\leq 2^m$, is contained in some B_i^m. Thus $\tilde{g}_\varepsilon \in h^p(F)$ can be written as

$$\tilde{g}_\varepsilon = \sum_{m\in Z} \sum_{i\geq 1} \lambda_i^m a_i^m$$

where a_i^m is a local (p,∞)-atom supported in B_i^m and

$$\left(\sum_{m,i} |\lambda_i^m|^p\right)^{1/p} \leq c_2 \|\tilde{g}_\varepsilon\|_{h^p}.$$

So if $\|\tilde{g}_{\varepsilon_k}\|_{h^p} \leq c_3$, we have

$$\tilde{g}_{\varepsilon_k} = \sum_m \sum_i \lambda_{i,k}^m a_{i,k}^m, \quad \sum_{m,i} |\lambda_{i,k}^m|^p \leq c_4. \tag{4.3}$$

By means of a diagonal process, we may find a sequence $k(j)\to\infty$ as $\varepsilon_{k(j)}\to 0$ such that each $a_{i,k(j)}^m$ converges weakly* in L^∞ to an atom

a_i^m and each $\lambda_{i,k(j)}^m$ converges to some λ_i^m and

$$\sum_{m,i} |\lambda_i^m|^p \le c_4.$$

To verify that the atomic sum converges in the distribution sense, we notice that the terms corresponding to cubes larger than some $\delta_0>0$ are finite in number, so their sum converges. We need thus only verify that the integrals of the remaining terms have a small sum when integrated against a test function Φ. This is proved in Lemma 2.

Hence, the atomic decomposition of $\tilde{g}_{\varepsilon_j}$ converges in the distribution sense to an atomic sum representing an $h^p(F)$ distribution which must be g, because $\tilde{g}_\varepsilon$ converges in the distribution sense to g as $\varepsilon\to 0$.

This gives the required decomposition of g, and we have the estimate

$$||g||_{h^p(F,\mu)} \le c\cdot ||m_\sigma' g||_{L^p(R^n)}, \quad c=c(F,\mu,p,\sigma),$$

which completes the proof.

LEMMA 1. <u>If</u> $\tilde{g}_\varepsilon = g_\varepsilon * \eta_{\delta(\varepsilon)}$, <u>where</u> g_ε <u>is defined in the proof of Proposition</u> 1, $m_\sigma' g_\varepsilon \in L^p(R^n)$, <u>and</u> $\eta_{\delta(\varepsilon)} = \delta^{-n}\eta(\frac{x}{\delta})$, $\delta=\delta(\varepsilon) > 0$, <u>with</u> $\int \eta dx = 1$, $\eta \in C_0^\infty(R^n)$, $\eta>0$, $\operatorname{supp}\eta \subset B(0,1)$ <u>and</u> $\eta(-x) = \eta(x)$, <u>then</u> $m_\sigma \tilde{g}_\varepsilon \in L^p(\mu)$ <u>with</u> $||m_\sigma \tilde{g}_\varepsilon||_{L^p(\mu)} \le c||m_\sigma' g||_{L^p(R^n)}$, $c=c(F,\mu,p,n,\sigma)$.

<u>Proof</u>. Let $\delta=\delta(\varepsilon)$, $\bar{x} \in F$, $0<t\le 1$ and $\Phi \in S_\delta$. Since we have

$$\left\langle \tilde{g}_\varepsilon, \Phi\left(\frac{\bar{x}-\cdot}{t}\right)\right\rangle = \left\langle g_\varepsilon * \eta_\delta, \Phi\left(\frac{\bar{x}-\cdot}{t}\right)\right\rangle = \left\langle g_\varepsilon, \Phi\left(\frac{\bar{x}-\cdot}{t}\right) * \eta_\delta\right\rangle ,$$

let us define the function Φ_δ by

$$\Phi_\delta\left(\frac{\bar{x}-z}{a}\right) = \Phi\left(\frac{\bar{x}-\cdot}{t}\right) * \eta_\delta(z) = \int \eta_\delta(y-z)\Phi\left(\frac{\bar{x}-y}{t}\right) d\mu(y)$$

where a is a constant which will be defined below.

Substituting $\frac{\bar{x}-z}{a} = x$ we get

$$\Phi_\delta(x) = \int \eta_\delta(\bar{x}-y-ax)\Phi\left(\frac{\bar{x}-y}{t}\right)d\mu(y) = \tag{4.4}$$

$$= \int \delta^{-n}\eta\left(\frac{\bar{x}-y-ax}{\delta}\right)\Phi\left(\frac{\bar{x}-y}{t}\right)d\mu(y).$$

Now, because $|ax| = |(ax-(\bar{x}-y))+(\bar{x}-y)| \leq \delta+t$, $|x| \leq \frac{\delta+t}{a}$, we have

$\operatorname{supp}\Phi_\delta \subset B\left(0, \frac{\delta+t}{a}\right)$ and $\Phi_\delta \in C_0^\infty(R^n)$.

We want to majorize $m_\sigma\tilde{g}_\varepsilon$ by some constant multiple of $m_\sigma g_\varepsilon$, and we start by estimating the derivatives of Φ_δ. The two cases $\delta \leq t$ and $t<\delta$ must be treated in somewhat different ways.

(I) $0<t\leq\delta$, $|\beta|\leq\sigma$:

$$D_x^\beta\Phi_\delta(x) = \int \delta^{-n}D_x^\beta\eta\left(\frac{\bar{x}-y-ax}{\delta}\right)\cdot\left(-\frac{a}{\delta}\right)^{|\beta|}\Phi\left(\frac{\bar{x}-y}{t}\right)d\mu(y)$$

(II) $\delta<t\leq 1$, $|\beta|\leq\sigma$:

Substituting $u=\bar{x}-y-ax$ in (4.4) we have

$$\Phi_\delta(x) = -\int \eta_\delta(u)\Phi\left(\frac{ax+u}{t}\right)d\mu(u)$$

and

$$D_x^\beta\Phi_\delta(x) = -\int \eta_\delta(u)\, D_x^\beta\, \Phi\left(\frac{ax+u}{t}\right)\cdot\left(\frac{a}{t}\right)^{|\beta|}d\bar{\mu}(u).$$

Thus, for $0<t\leq 1$, $|\beta|\leq\sigma$ and a fix constant c' depending only on η and σ, we have

$$|D_x^\beta\Phi_\delta(x)| \leq c'\cdot\min\left(\left(\frac{a}{\delta}\right)^\sigma, \left(\frac{a}{t}\right)^\sigma\right) \leq c'\cdot 2^\sigma \quad \text{if} \quad a=\delta+t.$$

With this choice of a, we have that $\Phi_\delta/(c'\cdot 2^\sigma) \in S_\sigma$.

But since a is larger than 1 if t is close to 1, we have to treat values of t between $t_0=1-\delta$ and 1 in a special way. Suppose that $1<a\leq 1+\delta<2$.

Let $\sum_{\nu=1}^{N}\chi_\nu$ be a partition of unity on the ball $B(0,2)$ such that

χ_ν is a C_0^∞-function, $\chi_\nu \geq 0$, $\sum_{\nu=1}^{N}\chi_\nu=1$, $\operatorname{supp}\chi_\nu \subset B(x_\nu,r_\nu)$ with

$r_\nu < \frac{1}{2}$, and

$$\Phi_\delta = \sum_{\nu=1}^{N} \chi_\nu \Phi_\delta .$$

We define the function Φ_ν by

$$c_\nu \Phi_\nu \left(\frac{\bar{x}-x_\nu-y}{t} \right) = \chi_\nu \Phi_\delta \left(\frac{\bar{x}-y}{a} \right).$$

It is easy to see that $\Phi_\nu \in S_\sigma$ if $c_\nu = c_\nu(\sigma,n,\chi_\nu)$ is suitably chosen. Since the Lebesgue measure is translation invariant, we get the estimate, where c'' and c''' are constants depending only on F,μ and n:

$$m_\sigma \tilde{g}_\varepsilon(\bar{x}) = \sup_{\Phi\in S_\sigma} \sup_{0<t\leq 1} \frac{1}{\mu(B(\bar{x},t))} \left\langle \tilde{g}_\varepsilon, \Phi\left(\frac{\bar{x}-\cdot}{t}\right)\right\rangle \leq$$

$$\leq \sup_{\Phi\in S_\sigma} \sup_{0<t\leq 1} \frac{c''}{m(B(\bar{x},t))} \left\langle g_\varepsilon, \Phi_\delta\left(\frac{\bar{x}-\cdot}{a}\right)\right\rangle =$$

$$= \sup_{\Phi\in S_\sigma} \sup_{0<t\leq 1} \frac{c''}{m(B(\bar{x},t))} \left\langle g_\varepsilon, \sum_{\nu=1}^{N} (\chi_\nu \Phi_\delta)\left(\frac{\bar{x}-\cdot}{a}\right)\right\rangle \leq$$

$$\leq \sum_{\nu=1}^{N} \sup_{\Phi\in S_\sigma} \sup_{0<t\leq 1} \frac{c'''}{m(B(\bar{x}-x_\nu,t))} \left\langle g_\varepsilon, c_\nu \Phi_\nu\left(\frac{\bar{x}-x_\nu-\cdot}{t}\right)\right\rangle \leq$$

$$\leq c''' \sum_{\nu=1}^{N} c_\nu m'_\sigma g_\varepsilon(\bar{x}-x_\nu) \in L^p(R^n)$$

because $m'_\sigma g_\varepsilon(\bar{x}-x_\nu)$ belongs to $L^p(R^n)$, with the same $L^p(R^n)$-norm as $m'_\sigma g(\bar{x})$, and

$$\left|\left| \sum_{\nu=1}^{N} c_\nu m'_\sigma g_\varepsilon(\bar{x}-x_\nu)\right|\right|_{L^p(R^n)} \leq c\cdot ||m'_\sigma g||_{L^p(R^n)},$$

where $c=c(F,\mu,p,n,\sigma)$.

Note that for $a\leq 1$ just one term is needed in the last sum of the estimate.
Lemma 1 is proved.

LEMMA 2. <u>Let</u> $\Phi \in C_0^\infty(R^n)$ <u>and let</u> $\{a^m_{i,k(j)}\}_{m<-M,\, i\geq 1}$ <u>for any</u> $j=1,\dots,\infty$ <u>denote a sequence of such local</u> p-<u>atoms, which have supporting balls</u>

of radii $< 2\sqrt{n}\cdot 2^{-M} \leq \delta_0 = \delta_0(\varepsilon_0)$ from the atomic decomposition of $\tilde{g}_\varepsilon$, $\varepsilon=\varepsilon(j)$, in the proof of Proposition 2. δ_0 is the same δ_0 as in that proof.

If M is large enough, then the sum

$$\sum_{m<-M} \sum_i \lambda^m_{i,k(j)} \int a^m_{i,k(j)}(x)\Phi(x)d\mu(x) < \varepsilon_0$$

where $\varepsilon_0>0$ is arbitrarily chosen.

Proof. Choose $\varepsilon_0>0$. Let $\Phi \in C_0^\infty(R^n)$ and let $a^m_{i,k(j)}$ be a local p-atom with a supporting ball $B^m_{i,k(j)} = B(x^m_{i,k(j)},r)$, where $2^{m-1} < r \leq 2^m$.

For $0<r\leq 1$ we get by (2.3.2), if $P_{\Phi^m_{i,k(j)}}(x-x^m_{i,k(j)})$ is the Taylor polynomial of degree s of Φ around $x^m_{i,k(j)}$,

$$\Big|\int_{B^m_{i,k(j)}} a^m_{i,k(j)}\Phi d\mu\Big| =$$

$$= \Big|\int_{B^m_{i,k(j)}} a^m_{i,k(j)}(x)\ [\Phi(x)-P_{\Phi^m_{i,k(j)}}(x-x^m_{i,k(j)})]d\mu(x)\Big| \leq$$

$$\leq \|a^m_{i,k(j)}\|_\infty\cdot c_1\cdot r^{s+1}\cdot\mu(B^m_{i,k(j)}) \leq$$

$$\leq \frac{1}{\mu(B^m_{i,k(j)})^{1/p}}\cdot c_1\cdot r^{(n/p)-n-1+\delta+1}\cdot c_2\cdot r^n \leq$$

$$\leq c_3\cdot r^{-n/p}\cdot c_1\cdot r^{(n/p)-n-1+\delta+1}\cdot c_2\cdot r^n = c_4\cdot r^\delta < \varepsilon \quad \text{for} \quad r<\delta_0,$$

$c_1=c_1(\Phi)$, $c_2=c_2(n)$, $c_3=c_3(n)$, $c_4=c_1c_2c_3$, because we have

$$\deg P_{\Phi^m_{i,k(j)}} = [n(\tfrac{1}{p}-1)] = s \quad \text{and thus} \quad s = n(\tfrac{1}{p}-1)-1+\delta,\ \delta>0.$$

Then we have

$$\sum_{m<-M}\sum_i \lambda^m_{i,k(j)} \int a^m_{i,k(j)}(x)\Phi(x)d\mu(x) \leq \sum_{m<-M}\sum_i \varepsilon|\lambda^m_{i,k(j)}|.$$

But since $|\lambda^m_{i,k(j)}| \leq |\lambda^m_{i,k(j)}|^p$ for $|\lambda^m_{i,k(j)}| \leq 1$ when $0<p\leq 1$

and $\sum_{m<-M} \sum_i |\lambda^m_{i,k(j)}|^p < c$ we have by (4.3)

$$\sum_{\substack{m<-M \\ |\lambda|\leq 1}} \sum_i \varepsilon|\lambda^m_{i,k(j)}| + \sum_{\substack{m<-M \\ |\lambda|>1}} \sum_i \varepsilon|\lambda^m_{i,k(j)}| \leq \varepsilon(c+c_5) < \varepsilon_0$$

if we choose $\varepsilon < \frac{\varepsilon_0}{c+c_5}$.

Lemma 2 is proved.

By partition of unity we are now able to prove Proposition 1. To the i:th supporting ball U_i, $1\leq i\leq m$, of the partition function φ_i we associate a cylinder S_i with the same radius as U_i. Using Proposition 1 we find that $\varphi_i f \in h^p(F)$.

<u>Proof of Proposition 1.</u> Since ∂F is compact and belongs to Lip 1 locally, we may choose a finite number m of balls $U_i(x_i,r_i)$, $x_i\in\partial F$, which cover ∂F in such a way that the intersection of two neighbouring balls has a non-empty intersection with the interior of F. We choose x_i to be the origin of a coordinate system such that in the ball U_i and in a neighbourhood of it, ∂F can be described by a function g_j like in Definition 2.9.

If we choose a suitable $U_0\subset F$, with a distance to ∂F which is greater than zero, but smaller than some ε, we have

$$F \subset \bigcup_{i=0}^{m} U_i .$$

Then there exists a partition of unity on F such that $\varphi_i \in C_0^\infty(U_i)$, $\varphi_i\geq 0$, $\sum_{i=0}^{m} \varphi_i = 1$, i.e. we have $f = \sum_{i=0}^{m} \varphi_i f$. Now let S_i be a closed, circular cylinder with its axis along the x_j-axis, bounded in the directions of the axis by $(n-1)$-dimensional balls of the same radius as the ball U_i and satisfying the conditions (4.1) and (4.2), so that $(U_i\cap F)\subset S_i$, $i=1,\ldots,m$.

Let $S_i' = S_i\cap F$ and denote $S_i\cap\partial F$ by $\partial F_i'$.

We want to use Proposition 2 to prove that $\varphi_i f \in h^p(F)$, $i=1,\ldots,m$. Clearly supp $\varphi_i f\subset F\cap U_i\subset S_i'$, $i=1,\ldots,m$. We need to prove that

$m'_\sigma(\varphi_i f) \in L^p(R^n)$. For $x \in U_i$, $0<t\leq 1$, and $\Psi \in S_\sigma$ we have

$$\frac{1}{m(B(x,t))}\left\langle \varphi_i f, \Psi(\tfrac{x-\cdot}{t})\right\rangle = \frac{1}{m(B(x,t))}\left\langle \varphi_i(y)f(y), \Psi(\tfrac{x-y}{t})\right\rangle =$$

$$\frac{1}{m(B(x,t))}\left\langle f(y), \varphi_i(y)\Psi(\tfrac{x-y}{t})\right\rangle = \frac{1}{m(B(x,t))}\left\langle f(y), \overline{\Psi}(\tfrac{x-y}{t})\right\rangle$$

where we define $\overline{\Psi}(\frac{x-y}{t}) = \varphi_i(y)\Psi(\frac{x-y}{t})$.

Now let $z = \frac{x-y}{t}$. Then $\overline{\Psi}(z) = \varphi_i(x-tz)\Psi(z)$. $\operatorname{supp}\overline{\Psi} \subset B(0,1)$ since $\operatorname{supp}\Psi(z) \subset B(0,1)$.

Further, we have, with β, β', β'' multiindices and $|\beta'|+|\beta''|=|\beta|\leq\sigma$,
$|D_z^\beta(\varphi_i(x-tz)\Psi(z)| = |\sum_{|\beta|\leq\sigma} c_{\beta'\beta''} D_z^{\beta'}\varphi_i(x-tz)D_z^{\beta''}\Psi(z)| \leq c_1$, $c_1=c_1(\sigma)$,

since $|D_z^{\beta''}\Psi(z)| \leq 1$ and, with $c_2=c_2(\sigma)$,

$|D_z^{\beta'}\varphi_i(x-tz)| = |t^{|\beta'|}(D^\beta\varphi_i)(x-tz)| \leq t^{|\beta'|}\cdot c_2$.

Thus $\frac{\overline{\Psi}}{c_1} \in S_\sigma$, because $0<t\leq 1$, and

$$\frac{1}{m(B(x,t))}\left\langle f(y), \overline{\Psi}(\tfrac{x-y}{t})\right\rangle = \frac{c_1}{m(B(x,t))}\left\langle f, \frac{\overline{\Psi}}{c_1}(\tfrac{x-\cdot}{t})\right\rangle \leq c_1\cdot m'_\sigma f(x),$$

i.e. $m'_\sigma(\varphi_i f)(x) \leq c_1\cdot m'_\sigma f(x)$ and we have $m'_\sigma(\varphi_i f)(x) \in L^p(R^n)$.

Then Proposition 2 proves that

$$\varphi_i f \in h^p(F) \text{ with } ||\varphi_i f||_{h^p(F,\mu)} \leq c\cdot||m'_\sigma\varphi_i f||_{L^p(R^n)},\quad c=c(F,\mu,p,\sigma),$$

and we have $\sum_{i=1}^m \varphi_i f \in h^p(F)$. It remains to prove that $\varphi_0 f \in h^p(F)$.

But this is a consequence of arguments analogous to those concerning the distribution g_ε in the proof of Proposition 2. Because $\operatorname{supp}\varphi_0 f$ has a positive distance to ∂F and $\widetilde{\varphi_0 f} = \varphi_0 f * \eta_{\delta(\varepsilon)}$ is a C^∞-function supported in F if $\delta(\varepsilon)$ is small.

Thus $\varphi_i f = \sum_\nu \lambda_\nu^i a_\nu^i$, a_ν^i local p-atoms with $\operatorname{supp} a_\nu^i \subset F$,
$i=0,1,\ldots,m$ (according to Proposition 2), and $f = \sum_{i=0}^m \varphi_i f = \sum_{i,\nu} \lambda_\nu^i a_\nu^i$,

$\sum_{i=0}^{m} \sum_{\nu} |\lambda_z^i|^p < \infty$, with the estimate $||f||_{h^p(F,\mu)} \leq c||m_\sigma' f||_{L^p(R^n))}$, $c=c(F,\mu,p,\sigma)$.

Proof of Corollary 1. We observe, that since φf is in $h^p(R^n)$, the maximal function $m_\sigma'(\varphi f)$ belongs to $L^p(R^n)$. The support of φf is contained in F, and the corollary follows immediately from Proposition 1.

Remark. φf in the corollary does not belong to the global Hardy space $H^p(F)$, since a multiplication by φ destroys global moment conditions.

Proof of Theorem 1. We make the same partition of unity on F as in the proof of Proposition 1 and get with the same notation

$$f = \sum_{i=0}^{m} \varphi_i f,$$

where, according to Corollary 1, $\varphi_i f = \sum_{\nu} \lambda_\nu^i a_\nu^i$, a_ν^i local p-atoms supported in F.

Now, in the decomposition

$$f = \sum_{i=0}^{m} \varphi_i f = \sum_{i,\nu} \lambda_\nu^i a_\nu^i$$

we want the atoms to be global.

Since F is bounded, we may argue like Jonsson, Sjögren, and Wallin in [7, Theorem 5.3, the middle part of the proof] to find an atomic decomposition of $\tilde{g}_\varepsilon$ in Proposition 2, which contains at most one atom whose support is not contained in a unit ball $B(y_i,1)$, $y_i \in F$.

Thus, for every i, $i=0,1,\ldots,m$ we may make the atomic decomposition in such a way that at most one atom a_ν^i has a support which is not contained in some ball $B(y_i,1)$, $y_i \in F$, i.e. we have got at most $(m+1)$ atoms which are not global. But the sum a of these atoms must have the global moment 0 since a is the difference between f, for which the sum of the global moments is 0, and the sum of the atoms with supports contained in unit balls, which also have the global moments 0.

Thus we may find a c such that $\frac{a}{c}$ is a standard p-atom for $H^p(R^n)$. And $f = c \cdot \frac{a}{c} + \sum_{\substack{i,\nu \\ \text{supp } a_\nu^i \\ \text{small}}} \lambda_\nu^i a_\nu^i$

is a decomposition of f in global atoms supported in F. Theorem 1 is proved.

REFERENCES

[1] R.R. Coifman, A real variable characterization of H^p, Studia Math. 51 (1974), 269-274.

[2] R.R. Coifman and G. Weiss, Extensions of Hardy spaces and their use in analysis. Bull. Amer. Math. Soc. 83 (1977), 569-645.

[3] C. Fefferman and E.M. Stein, H^p-spaces of several variables. Acta Math. 129 (1972), 137-193.

[4] D. Goldberg, A local version of real Hardy spaces. Duke Math.J. 46 (1979), 27-42.

[5] L. Hörmander, Linear partial differential operators, I, Springer-Verlag, Berlin (1976).

[6] D.S. Jerison and C.E. Kenig, Boundary behaviour of harmonic functions in non-tangentially accessible domains. Adv. in Math. 46 (1982), 80-145.

[7] A. Jonsson, P. Sjögren and H. Wallin, Hardy and Lipschitz spaces on subsets of R^n. Studia Math. 80.2 (1984), 141-166.

[8] A. Jonsson and H. Wallin, A Whitney extension theorem in L^p and Besov spaces. Ann. Inst. Fourier (Grenoble) 28 (1978), 139-142.

[9] A. Jonsson and H. Wallin, Functions spaces on subsets of R^n. Harwood Acad. Publ., New York (1984).

[10] A. Jonsson and H. Wallin, Local polynomial approximation and Lipschitz type conditions on general closed sets. Dept. of Math., Univ. of Umeå, No. 1 (1980).

[11] R.H. Latter, A decomposition of $H^p(R^n)$ in terms of atoms. Studia Math. 62 (1977), 92-101.

[12] P. Sjögren, Lectures on atomic H^p space theory in R^n. Dept. of Math., Univ. of Umeå. No. 5 (1981).

[13] E.M. Stein, Singular integrals and differentiability properties of functions. Princeton Univ. Press (1970).

[14] E.M. Stein and G. Weiss, On the theory of harmonic functions of several variables, The theory of H^p spaces. Acta Math. 103 (1960), 25-62.

[15] H. Wallin, Markov's inequality on subsets of R^n. In: Canadian Math. Soc., 2nd Edmonton Conf. on Approximation Theory (Edmonton 1982), 377-388, CMS Conf. Proc., Vol. 3. Providence, Rhode Island: Amer. math. Soc. 1984.

THE φ-TRANSFORM AND APPLICATIONS TO DISTRIBUTION SPACES

Michael Frazier
Department of Mathematics
University of New Mexico
Albuquerque, NM 87131

Björn Jawerth
Department of Mathematics
Washington University
St. Louis, MO 63130

1 Introduction

Our purpose here is to study a discrete transform S_φ, the "φ-transform," which maps a distribution f on $\mathbf{R}^n$ into a sequence $\{(S_\varphi f)_Q\}$ of complex numbers indexed by the dyadic cubes $Q \subseteq \mathbf{R}^n$. This transform is well-suited for the study of many of the familiar function spaces in analysis (e.g. Lebesgue, Hardy, and Sobolev spaces) in the sense that there is an explicit condition on the sequence $\{(S_\varphi f)_Q\}$ which characterizes each of these spaces. Inversion of the φ-transform then yields a representation (or "decomposition") of each element of these spaces. Furthermore, many classical operators, for example, Calderón-Zygmund operators, are almost diagonalized by the φ-transform.

The φ-transform has been more or less explicitly introduced in [11], which deals with the Besov spaces on $\mathbf{R}^n$ (see in particular Remark 2.3 in [11]). The present work is intended primarily as an introduction to [12], where a more thorough development of the ideas introduced here is given. However, the proofs of the results in §1-3 of this paper are complete without reference to [12], and we have included applications in §3 that do not appear in [12].

For the sake of perspective, we first make some simple remarks about Fourier series. For a function f on the torus with Fourier series $\sum_{n \in \mathbf{Z}} c_n(f)e^{inx}$, a fundamental fact is that $f \in L^2$ if and only if $\{c_n(f)\}_{n \in \mathbf{Z}} \in \ell^2$. However, virtually none of the standard spaces in analysis, other than L^2, has a satisfactory characterization in terms of Fourier coefficients. One should note the quantity of literature devoted to partial results (and related counterexamples) for various spaces. In particular, for $p \neq 2$, no condition depending only on the magnitudes of the Fourier coefficients can possibly characterize L^p—see [7], pp. 47-9 for a discussion of this question and its relation to Littlewood-Paley theory. Analogous problems exist for the Fourier transform in $\mathbf{R}^n$. Part of the difficulty is that the characters $\{e^{inx}\}_{n \in \mathbf{Z}}$ (or $\{e^{ix\cdot\xi}\}_{\xi \in \mathbf{R}^n}$) are not localized (in the x-variable) and hence their interaction is complicated.

For the φ-transform, a collection of functions $\{\psi_Q\}_{Q \text{ dyadic}}$ plays the role of the "universal" functions that the set $\{e^{inx}\}_{n \in \mathbf{Z}}$ plays for Fourier series. To define the ψ_Q's, select functions φ and ψ satisfying

$$\varphi, \psi \in S\,, \tag{1.1}$$

$$\operatorname{supp} \hat{\varphi}, \hat{\psi} \subseteq \{\xi \in \mathbf{R}^n : 1/2 \le |\xi| \le 2\}\,, \tag{1.2}$$

$$|\hat{\varphi}(\xi)|, |\hat{\psi}(\xi)| \ge c > 0 \quad \text{if} \quad 3/5 \le |\xi| \le 5/3\,, \tag{1.3}$$

and

$$\sum_{\nu \in \mathbf{Z}} \overline{\hat{\varphi}(2^\nu \xi)}\hat{\psi}(2^\nu \xi) = 1 \quad \text{if} \quad \xi \ne 0\,. \tag{1.4}$$

For $Q = Q_{\nu k} = \{(x_1, \ldots, x_n) \in \mathbf{R}^n : k_i \le 2^\nu x_i \le k_i + 1, i = 1, \ldots, n\}$, where $\nu \in \mathbf{Z}$ and $k \in \mathbf{Z}^n$, set

$$\psi_Q(x) = |Q|^{-1/2}\psi(2^\nu x - k)\,.$$

(Note that this normalization differs from the one in [11].) Then $\|\psi_Q\|_{L^2} = \|\psi\|_{L^2}$ for each dyadic cube $Q \subseteq \mathbf{R}^n$,

$$\operatorname{supp} \hat{\psi}_Q(\xi) \subseteq \{\xi : 2^{\nu - 1} \le |\xi| \le 2^{\nu + 1}\} \quad \text{if} \quad \ell(Q) = 2^{-\nu}\,, \tag{1.5}$$

and, if we write $x_Q = 2^{-\nu}k$ for $Q = Q_{\nu k}$, we have

$$|\partial^\gamma \psi_Q(x)| \le c_{\gamma, L}|Q|^{-1/2 - |\gamma|/n}(1 + 2^\nu|x - x_Q|)^{-L - |\gamma|} \tag{1.6}$$

for each $L \in \mathbf{Z}^+$ and each multi-index γ of length $|\gamma| \ge 0$. Thus the ψ_Q's are dually localized, near Q in the x-variable and on an appropriate annulus in ξ. Note that the ψ_Q's are obtained from one function ψ of a particularly simple form via dyadic translations and dilations.

The φ-transform is defined for appropriate distributions f on $\mathbf{R}^n$ by $(S_\varphi f)_Q = \langle f, \varphi_Q \rangle$, where $\varphi_Q(x) = |Q|^{-1/2}\varphi(2^\nu x - k)$ if $Q = Q_{\nu k}$. Inverting the φ-transform will yield the formal representation

$$f = \sum_Q (S_\varphi f)_Q \psi_Q = \sum_Q \langle f, \varphi_Q \rangle \psi_Q\,,$$

where from now on we adopt the convention that when Q appears as a summation index, the sum runs over all dyadic cubes in $\mathbf{R}^n$. Note that if we choose φ so that $\sum_{\nu \in \mathbf{Z}} |\hat{\varphi}(2^\nu \xi)|^2 = 1$ for $\xi \ne 0$, then we can take $\varphi = \psi$. We obtain $f = \sum_Q \langle f, \psi_Q \rangle \psi_Q$, as if the ψ_Q's formed a complete orthonormal basis in $\mathbf{R}^n$. However, the ψ_Q's are not an orthogonal family, although they are almost orthogonal in an appropriate sense. Namely, if Q and P are dyadic cubes with side lengths

$\ell(Q)$ and $\ell(P)$, respectively, such that $\ell(Q)/\ell(P) < 1/2$ or $\ell(Q)/\ell(P) > 2$, then $\langle\psi_Q, \psi_P\rangle = 0$ by (1.2); if $1/2 \leq \ell(Q)/\ell(P) \leq 2$ and $|x_Q - x_P|$ is large, then $|\langle\psi_Q, \psi_P\rangle|$ is small by (1.6) for $\gamma = 0$.

Many of the important spaces in analysis are characterized in terms of the magnitude of the φ-transform. For example, if $1 < p < +\infty$, we will see that a necessary and sufficient condition for f to belong to L^p is $\|(\sum_Q(|(S_\varphi f)_Q|\tilde{\chi}_Q)^2)^{1/2}\|_{L^p} < +\infty$, where $\tilde{\chi}_Q(x) = |Q|^{-1/2}\chi_Q(x)$ is the L^2-normalized characteristic function of Q. For $0 < p \leq 1$, this same condition characterizes the Hardy spaces $H^p(\mathbf{R}^n)$. The (homogeneous) Sobolev and potential spaces of order $\alpha \in \mathbf{R}$ are characterized by a similar expression, differing only by a factor of $|Q|^{-\alpha/n}$ in the sum. (There are related results for the usual (inhomogeneous) Sobolev and potential spaces.) With modifications, similar results hold for BMO, the space of functions of bounded mean oscillation. For the corresponding results for the Lipschitz spaces Λ_α, $0 < \alpha < 1$, or more generally the Besov spaces $\dot{B}_p^{\alpha q}$ or $B_p^{\alpha q}$, see [11].

The theory of the Triebel-Lizorkin spaces $\dot{F}_p^{\alpha q}$ (homogeneous) and $F_p^{\alpha q}$ (inhomogeneous) provides the proper context for the simultaneous consideration of the various spaces above. To define the $\dot{F}$-spaces, select φ satisfying (1.1-3), and define $\varphi_\nu(x) = 2^{\nu n}\varphi(2^\nu x)$ for $\nu \in \mathbf{Z}$. For $\alpha \in \mathbf{R}$, $0 < q \leq +\infty$, and $0 < p < +\infty$, $\dot{F}_p^{\alpha q}$ is the collection of all $f \in S'/P$ (tempered distributions modulo polynomials) such that

$$\|f\|_{\dot{F}_p^{\alpha q}} = \|(\sum_{\nu \in \mathbf{Z}} (2^{\nu\alpha}|\varphi_\nu * f|)^q)^{1/q}\|_{L^p} < +\infty\,,$$

where the ℓ^q-norm is replaced by the sup on v if $q = +\infty$. This definition is independent of the choice of φ. (For the inhomogeneous spaces $F_p^{\alpha q}$, see [12]. All of our results have analogues in the inhomogeneous case, which are stated in [12], but for simplicity we treat only the homogeneous case here.)

For $\alpha = 0$ and $q = 2$, the expression $(\sum_{\nu \in \mathbf{Z}} |\varphi_\nu * f|^2)^{1/2}$ is a discrete version of the classical Littlewood-Paley g-function. The theory of the Triebel-Lizorkin spaces can be viewed as a development and extension of the standard Littlewood-Paley theory, as presented, say, in [28], §4. We will see that the classical spaces which have a natural characterization via the φ-transform are exactly those which have a Littlewood-Paley characterization and hence fit into the $\dot{F}$-space (or Besov space) scales. For example, we obtain the Hardy spaces $H^p(= \dot{F}_p^{02})$, $0 < p < +\infty$ (in particular L^p, $1 < p < +\infty$) and the potential spaces $L_\alpha^p(= \dot{F}_p^{\alpha 2})$, $1 < p < +\infty$, $\alpha \in \mathbf{R}$ (in particular, the Sobolev spaces L_k^p, $k \in \mathbf{Z}^+$, $1 < p < +\infty$). (See e.g. [30], pp. 51-2 for these equivalences.) Our approach is motivated by techniques developed for the $\dot{F}$-spaces by Peetre, Triebel, and others, which in turn reflect the development of Littlewood-Paley theory by Zygmund, Calderón, Stein, Fefferman, and others. See [26], [30], and [11] for more specific historical background.

The φ-transform sets up a correspondence between the $\dot{F}_p^{\alpha q}$-spaces and certain sequence spaces $\dot{f}_p^{\alpha q}$. We define $\dot{f}_p^{\alpha q}$ to be the collection of all complex-valued sequences $s = \{s_Q\}_Q$ dyadic such that

$$\|s\|_{\dot{f}_p^{\alpha q}} = \|(\sum_Q (|Q|^{-\alpha/n}|s_Q|\tilde{\chi}_Q)^q)^{1/q}\|_{L^p} < +\infty \,, \tag{1.7}$$

for $\alpha \in \mathbf{R}$, $0 < q \leq +\infty$, and $0 < p < +\infty$. For φ and ψ selected satisfying (1.1-4), we define the inverse φ-transform formally as the operator T_ψ taking a sequence $s = \{s_Q\}_Q$ into $T_\psi s = \Sigma_Q s_Q \psi_Q$. Our basic result concerning the φ-transform can be stated as follows.

Theorem I. Suppose $\alpha \in \mathbf{R}$, $0 < p < +\infty$, $0 < q \leq +\infty$. The operators $S_\varphi : \dot{F}_p^{\alpha q} \to \dot{f}_p^{\alpha q}$ and $T_\psi : \dot{f}_p^{\alpha q} \to \dot{F}_p^{\alpha q}$ are bounded. Furthermore, $T_\psi \circ S_\varphi$ is the identity on $\dot{F}_p^{\alpha q}$. In particular, $\|f\|_{\dot{F}_p^{\alpha q}} \approx \|S_\varphi f\|_{\dot{f}_p^{\alpha q}}$ and $\dot{F}_p^{\alpha q}$ can be identified with a complemented subspace of $\dot{f}_p^{\alpha q}$. (Here $\approx$ means that the (quasi)-norms are equivalent.)

In other words, $\dot{F}_p^{\alpha q}$ is a retract of $\dot{f}_p^{\alpha q}$, since the following diagram commutes:

$$
\begin{array}{ccc}
 & \dot{f}_p^{\alpha q} & \\
S_\varphi \nearrow & & \searrow T_\psi \\
\dot{F}_p^{\alpha q} & \xrightarrow{\text{Id}} & \dot{F}_p^{\alpha q}
\end{array}
$$

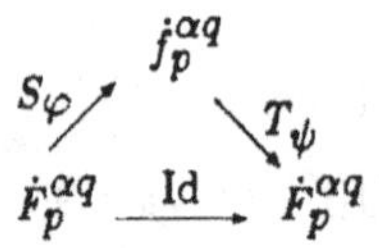

This diagram indicates that problems about $\dot{F}_p^{\alpha q}$ can be lifted to the associated sequence space $\dot{f}_p^{\alpha q}$, where many problems become more transparent. This transparency results partially from the discreteness of the sequence space norm, but also partially from the fact that the $\dot{f}_p^{\alpha q}$ (quasi)-norm depends only on "size" estimates, i.e. only on the magnitudes of the sequence elements. Cancellation aspects of the $\dot{F}_p^{\alpha q}$-spaces are "absorbed" into the functions $\{\psi_Q\}_Q$ dyadic, which have vanishing moments of all orders because of (1.5). In addition, there is a simple geometry implicit in the $\dot{f}_p^{\alpha q}$ (quasi)-norm, which can be exploited in applications.

Theorem I can also be regarded as a decomposition result. In this form it is convenient to state a related result simultaneously. For this purpose we define a *smooth molecule* for a cube Q to be a function m_Q satisfying

$$\int x^\gamma m_Q(x)dx = 0 \quad \text{if} \quad |\gamma| \leq N \,,$$

and

$$|\partial^\gamma m_Q(x)| \le |Q|^{-1/2 - |\gamma|/n}(1 + \ell(Q)^{-1}|x - x_Q|)^{-M - |\gamma|} \quad \text{if} \quad |\gamma| \le K,$$

where K, N, and M are sufficiently large integers, which depend on α, p, q, and n. In particular, we require $K \ge 0$, $K > \alpha$, and $N > -1 - \alpha + n(1/\min(1,p,q) - 1)$, where $N \le -1$ means that the vanishing moment condition is void. Also, $M > N + 10n(1/\min(1,p,q))$ will do here, although this number could be sharpened (cf. [11], [12]). By (1.5-6), each ψ_Q is a molecule for Q, up to a constant independent of Q. Finally, we define a *smooth atom* for Q to be any function a_Q satisfying

$$\operatorname{supp} a_Q \subseteq 3Q,$$

$$|\partial^\gamma a_Q(x)| \le |Q|^{-1/2 - |\gamma|/n} \quad \text{if} \quad |\gamma| \le K,$$

and

$$\int x^\gamma a_Q(x)\,dx = 0 \quad \text{if} \quad |\gamma| \le N,$$

where $3Q$ is the cube concentric with Q with side length three times the side length $\ell(Q)$ of Q. Then any smooth atom for Q is also a smooth molecule for Q.

Theorem II. Let $\alpha \in \mathbf{R}$, $0 < q \le +\infty$, and $0 < p < +\infty$.

A.) Each $f \in \dot{F}_p^{\alpha q}$ can be decomposed as follows:

i. $f = \sum_Q s_Q \psi_Q$, where $s_Q = (S_\varphi f)_Q = \langle f, \varphi_Q \rangle$,

and

ii. $f = \sum_Q s_Q a_Q$, where each a_Q is a smooth atom for Q.

In each case, we can obtain

$$\|\{s_Q\}\|_{\dot{f}_p^{\alpha q}} \le c\|f\|_{\dot{F}_p^{\alpha q}}. \tag{1.8}$$

B.) Conversely, if $f = \sum_Q s_Q m_Q$, where each m_Q is a smooth molecule for Q, then

$$\|f\|_{\dot{F}_p^{\alpha q}} \le c\|\{s_Q\}\|_{\dot{f}_p^{\alpha q}}. \tag{1.9}$$

We prove Theorem II (which includes Theorem I) in §2. As in [11], a key role is played by a discrete version of Calderón's reproducing formula. A new feature of crucial importance throughout the following is the use of the Fefferman-Stein vector-valued maximal inequality. For future reference, we state this result next. Let M be the Hardy-Littlewood maximal operator,

$$Mf(x) = \sup_{x \in Q} |Q|^{-1} \int_Q |f(y)|\,dy,$$

where the sup is taken over all cubes Q with sides parallel to the axes.

Theorem 1.1. (Fefferman-Stein, [9]) Suppose $1 < p < +\infty$, and $1 < q \leq +\infty$. Then

$$\|(\sum_{i=1}^{\infty} |Mf_i|^q)^{1/q}\|_{L^p} \leq c_{p,q}\|(\sum_{i=1}^{\infty} |f_i|^q)^{1/q}\|_{L^p}.$$

For our purposes this result is a powerful tool for "sweeping up" the approximately localised functions ψ_Q into the absolutely localized $\tilde{\chi}_Q$'s, as well as for other necessary "vacuum-cleaner" operations.

In §3 we consider several applications of our results. One is the restriction, or trace, problem for the $\dot{F}$-spaces. The key fact is that the trace of $\dot{F}_p^{\alpha q}(\mathbf{R}^n)$ on $\mathbf{R}^{n-1}$ does not depend on the parameter q. Relating the definition of the $\dot{f}_p^{\alpha q}$ (quasi)-norm to the geometry of the trace problem gives us a simple understanding of this fact. We obtain thereby the known results and a sharp new endpoint result. The second application is a simple proof of the real interpolation result that

$$(\dot{F}_{p_0}^{\alpha q}, \dot{F}_{p_1}^{\alpha q})_{\theta,p} = \dot{F}_p^{\alpha q}$$

if $\alpha \in \mathbf{R}$, $0 < q \leq +\infty$, $0 < p_0 < p_1 < +\infty$, $0 < \theta < 1$, and $1/p = (1-\theta)/p_0 + \theta/p_1$. This result is known for $q \geq 1$, see e.g. [4]. For $q = 2$ and $\alpha = 0$, we recover the Fefferman-Rivière-Sagher result ([8]) on H^p-interpolation. Our proof uses the commutative diagram above to reduce our computations to the simpler case of the associated sequence spaces. We also reconsider some known imbedding theorems from this perspective. Finally, we prove that $\dot{F}_p^{\alpha q}$ has the lower (Fourier) majorant property if $0 < p < 1$ or if $p = 1$ and $0 < q \leq 1$. In fact, we show that elements of a certain Besov space in which $\dot{F}_p^{\alpha q}$ is imbedded have lower majorants lying in $\dot{F}_p^{\alpha q}$.

Further results and extensions of the theory, which are carried out in detail in [12], are described in §4 briefly. For example, with modifications it is possible to extend these results to the case $p = +\infty$, so as to include the case of $BMO(= \dot{F}_\infty^{02})$ in a natural way. Also, our sequence space results can be extended to $p = 0$ to obtain interpolation results in the full range $0 \leq p \leq +\infty$. The results for $0 < p_0 < p_1 < +\infty$ follow from the case $p_0 = 0$, $p_1 = +\infty$ by reiteration. Using the sequence space corresponding to $p_0 = 0$ as one endpoint, one obtains the "standard" atomic decomposition results for H^p, $0 < p \leq 1$ ([5], [21], [31]) from the known equivalence between different descriptions of the real interpolation method.

We would like to note that simultaneously with our extension of our results from the Besov space case in [11] to the $\dot{F}$-space case presented here, Lemarié and Meyer ([22], [23]) extended these developments in another way. By a different construction, they obtained an expansion in $\mathbf{R}^1$

of the type $f = \Sigma_Q \langle f, \psi_Q \rangle \psi_Q$ with the ψ_Q's of a similar form to those above, but which are, in addition, orthonormal. They then showed that the set $\{\psi_Q\}_{Q \text{ dyadic}}$ forms an unconditional basis for many of the usual spaces, e.g. Hardy, Lebesgue, and Sobolev spaces, as well as BMO. They also obtained a certain tensor product type of extension to the case of $\mathbf{R}^n$, $n > 1$.

2 Proof of Theorems I and II

We present a proof of Theorem II (and hence I) in this section. For the sake of brevity, the proof we give is of a computational and perhaps unenlightening nature. We do note that a relatively straightforward approach succeeds. However, it is also possible to carry out the proof from a standpoint focused on the sequence spaces $\dot{f}_p^{\alpha q}$. This perspective is more revealing, shedding light in particular on the discretization process. It also yields sharper results in some cases. Further, it is useful in extending the theory to the case $p = +\infty$, and essential for $p = 0$. This approach and these extensions of the theory are carried out in [12].

We begin with the following key lemma in [11], which gives a discretized version of the Calderón reproducing formula. We note that the discretization is accomplished by a familiar sampling technique (cf. [6], p. 129, [26], pp. 237-8, and [30], p. 21).

Lemma 2.1. Suppose φ and ψ satisfy (1.1-4). If $f \in S'/P$, then

$$f(\cdot) = \sum_{\nu \in \mathbf{Z}} 2^{-\nu n} \sum_{k \in \mathbf{Z}^n} \tilde{\varphi}_\nu * f(2^{-\nu} k) \psi_\nu(\cdot - 2^{-\nu} k),$$

where $\tilde{\varphi}_\nu(x) = \overline{\varphi_\nu(-x)}$.

Proof. For $f \in S$, say, this follows formally by Fourier inversion. We refer to [11], Lemma 2.1, for details in the general case. □

We will also require Peetre's maximal function characterization of $\dot{F}_p^{\alpha q}$. Define, for $f \in S'/P$ and φ satisfying (1.1-3),

$$\varphi_\nu^{**} f(x) = \sup_{y \in \mathbf{R}^n} 2^{\nu\alpha} |\varphi_\nu * f(x-y)| / (1 + 2^\nu |y|)^L,$$

where we take $L > n/\min(p, q)$.

Theorem 2.2. (Peetre, [25]) Suppose $\alpha \in \mathbf{R}$, $0 < p < +\infty$, and $0 < q \leq +\infty$. Then

$$\|f\|_{\dot{F}_p^{\alpha q}} \approx \|(\sum_{\nu \in \mathbf{Z}} (\varphi_\nu^{**} f)^q)^{1/q}\|_{L^p}.$$

Proof of Theorem II A. For $Q = Q_{\nu k}$, we have $\psi_Q(x) = 2^{-\nu n/2}\psi_\nu(x - 2^{-\nu}k)$, and a simple computation gives $s_Q = \langle f, \varphi_Q \rangle = 2^{-\nu n/2}\tilde{\varphi}_\nu * f(2^{-\nu}k)$. Thus the representation $f = \sum_Q s_Q \psi_Q$ in Theorem II A i follows from Lemma 2.1.

For the "smooth atomic" decomposition in Theorem II A ii, we follow the procedure in [11], Theorem 2.6 ii. There we show that one can select $\theta, \varphi \in S$ such that φ satisfies (1.1-3), supp $\theta \subseteq \{x : |x| \leq 1\}$, $\int x^\gamma \theta(x)dx = 0$ if $|\gamma| \leq N$, and $\sum_{\nu \in \mathbf{Z}} \hat{\theta}_\nu(\xi)\hat{\tilde{\varphi}}_\nu(\xi) = 1$ if $\xi \neq 0$, where, as usual, $\theta_\nu(x) = 2^{\nu n}\theta(2^\nu x)$. Thus $f = \sum_{\nu \in \mathbf{Z}} \theta_\nu * \tilde{\varphi}_\nu * f$. If $\ell(Q) = 2^{-\nu}$, we set

$$s_Q = C|Q|^{1/2} \sup_{y \in Q} |\tilde{\varphi}_\nu * f(y)|$$

for C sufficiently large, and

$$a_Q(x) = \frac{1}{s_Q} \int_Q \theta_\nu(x-y)\tilde{\varphi}_\nu * f(y)dy \, .$$

Then a_Q is a smooth atom for Q and we have $f = \sum_Q s_Q a_Q$.

To obtain (1.8) in both cases, we apply Theorem 2.2 with φ replaced by $\tilde{\varphi}$ (which satisfies (1.1-3) also since $\hat{\tilde{\varphi}} = \tilde{\hat{\varphi}}$). We obtain

$$\sum_{\ell(Q) = 2^{-\nu}} |Q|^{-\alpha/n}|s_Q|\tilde{\chi}_Q(x) \leq c2^{\nu\alpha} \sup_{|y| \leq 2^{-\nu}\sqrt{n}} |\tilde{\varphi}_\nu * f(x-y)| \leq c\tilde{\varphi}_\nu^{**}f(x) \, ,$$

which yields (1.8). □

For Theorem II B, we use estimates (3.3-4) from [11]. Taking into account our change in normalization, these give the following.

Lemma 2.3. If m_Q is a smooth molecule for Q with $\ell(Q) = 2^{-\mu}$, then

i. $|\varphi_\nu * m_Q(x)| \leq c|Q|^{-1/2}2^{-(\nu-\mu)K}/(1 + 2^\mu|x - x_Q|)^\lambda$ if $\mu \leq \nu$,

and

ii. $|\varphi_\nu * m_Q(x)| \leq c|Q|^{-1/2}2^{-(\mu-\nu)(N+1+n)}/(1 + 2^\nu|x - x_Q|)^\lambda$ if $\nu \leq \mu$,

where λ is sufficiently large ($\lambda > n/\min(1, p, q)$).

We also need the following lemma.

Lemma 2.4. Let $0 < A \leq 1$, $\lambda > n/A$, and $\eta \geq 0$, $\eta \in \mathbf{Z}$. For any sequence $\{s_Q\}_Q$, we have

$$\sum_{\ell(Q) = 2^{-\mu}} |s_Q|/(1 + (2^\eta \ell(Q))^{-1}|x - x_Q|)^\lambda \leq c2^{\eta n/A}(M(\sum_{\ell(Q) = 2^{-\mu}} |s_Q|^A\chi_Q))^{1/A}(x) \, .$$

Proof. We may assume $x \in Q_{\mu 0}$. Let $A_0 = \{Q \text{ dyadic: } \ell(Q) = 2^{-\mu}, |x_Q| \le 2^{\eta - \mu}\}$, and, for $k = 1, 2, 3, \ldots$, let $A_k = \{Q \text{ dyadic: } \ell(Q) = 2^{-\mu}, 2^{k+\eta-\mu-1} < |x_Q| \le 2^{k+\eta-\mu}\}$. Then

$$\begin{aligned}\sum_{Q \in A_k} |s_Q|/(1 + (2^{\eta}\ell(Q))^{-1}|x - x_Q|)^{\lambda} &\le c2^{-k\lambda} \sum_{Q \in A_k} |s_Q| \le c2^{-k\lambda}(\sum_{Q \in A_k} |s_Q|^A)^{1/A} \\ &= c2^{-k\lambda + \mu n/A}(\int \sum_{Q \in A_k} |s_Q|^A \chi_Q)^{1/A} \\ &\le c2^{-k(\lambda - n/A)} 2^{\eta n/A}(M(\sum_{\ell(Q) = 2^{-\mu}} |s_Q|^A \chi_Q))^{1/A}.\end{aligned}$$

The result follows by summing on k. □

Proof of Theorem II B. Select A such that $n/\min(1, p, q) < n/A < N + 1 + n + \alpha$ and pick $\lambda > n/A$. If $\mu \le \nu$, we estimate $|\varphi_\nu * \sum_{\ell(Q) = 2^{-\mu}} s_Q m_Q|$ using Lemma 2.3i and Lemma 2.4 with $\eta = 0$, while if $\mu > \nu$, we use Lemma 2.3 ii and Lemma 2.4 with $\eta = \mu - \nu$. We obtain

$$\|f\|_{\dot F_p^{\alpha q}} = \|(\sum_{\nu \in \mathbf{Z}} (2^{\nu\alpha}|\varphi_\nu * \sum_Q s_Q m_Q|)^q)^{1/q}\|_{L^p}$$

$$\le c\|(\sum_{\nu \in \mathbf{Z}} (\sum_{\mu = -\infty}^{\nu} 2^{-(\nu - \mu)(K - \alpha)}(M(\sum_{\ell(Q) = 2^{-\mu}} |Q|^{-\alpha/n}|s_Q|\tilde\chi_Q)^A)^{1/A})^q)^{1/q}\|_{L^p}$$

$$+c\|(\sum_{\nu \in \mathbf{Z}} (\sum_{\mu = \nu + 1}^{\infty} 2^{-(\mu - \nu)(N + 1 + n + \alpha - n/A)}(M(\sum_{\ell(Q) = 2^{-\mu}} |Q|^{-\alpha/n}|s_Q|\tilde\chi_Q)^A)^{1/A})^q)^{1/q}\|_{L^p}.$$

Using $\|a * b\|_{\ell^q} \le \|a\|_{\ell^1}\|b\|_{\ell^q}$ if $q \ge 1$, or $|a + b|^q \le |a|^q + |b|^q$ and $\|a * b\|_{\ell^1} \le \|a\|_{\ell^1}\|b\|_{\ell^1}$ if $0 < q < 1$, and our assumptions $K - \alpha > 0$ and $N + 1 + n + \alpha - n/A > 0$, we obtain

$$\|f\|_{\dot F_p^{\alpha q}} \le c\|(\sum_{\nu \in \mathbf{Z}} (M(\sum_{\ell(Q) = 2^{-\nu}} |Q|^{-\alpha/n}|s_Q|\tilde\chi_Q)^A)^{q/A})^{A/q}\|_{L^{p/A}}^{1/A}.$$

Since $p/A > 1$ and $q/A > 1$, we can remove M from the last expression by Theorem 1.1, which yields (1.9). □

We note here for the record that Theorem II B implies that for $q < +\infty$, the convergence in Theorem II A is in $\dot F_p^{\alpha q}$ (quasi)-norm. Further, the convergence is unconditional, since the $\dot f_p^{\alpha q}$ (quasi)-norm depends only on the magnitudes of the sequence elements.

3 Applications

In this section we consider applications of Theorems I and II to "trace" problems, to real interpolation, to imbedding theorems, and to the construction of lower Fourier majorants. These applications have been selected for their simplicity and as examples of how the underlying geometry implicit in the $\dot{f}_p^{\alpha q}$ (quasi)-norm can be exploited in problems about $\dot{F}_p^{\alpha q}$. The following easy lemma will be of vital importance repeatedly.

Lemma 3.1. Let $\epsilon > 0$. Suppose that for each dyadic cube Q there is a set $E_Q \subseteq Q$ with $|E_Q|/|Q| > \epsilon$. Then

$$\|\{s_Q\}\|_{\dot{f}_p^{\alpha q}} \approx \|(\sum_Q (|Q|^{-\alpha/n}|s_Q|\tilde{\chi}_{E_Q})^q)^{1/q}\|_{L^p},$$

where $\tilde{\chi}_{E_Q} = |E_Q|^{-1/2}\chi_{E_Q}$.

Proof. Since $\chi_{E_Q} \leq \chi_Q$, one direction is trivial. For the other, note that for all $A > 0$, $\chi_Q \leq \epsilon^{-1/A}(M\chi_{E_Q}^A)^{1/A}$. Select A such that $p/A, q/A > 1$. Then by Theorem 1.1,

$$\begin{aligned}\|\{s_Q\}\|_{\dot{f}_p^{\alpha q}} &\leq \epsilon^{-1/A}\|(\sum_Q (M(|Q|^{-\alpha/n}|s_Q|\tilde{\chi}_{E_Q})^A)^{q/A})^{A/q}\|_{L^{p/A}}^{1/A} \\ &\leq c\epsilon^{-1/A}\|(\sum_Q (|Q|^{-\alpha/n}|s_Q|\tilde{\chi}_{E_Q})^q)^{1/q}\|_{L^p}. \qquad \square\end{aligned}$$

Lemma 3.1 allows us to pass to appropriate subsets of the dyadic cubes whenever this simplifies matters.

As an illustration, consider the problem of determining the space of restrictions to $\mathbf{R}^{n-1}$ (identified with $\{x \in \mathbf{R}^n : x_n = 0\}$) of elements of $\dot{F}_p^{\alpha q}(\mathbf{R}^n)$. These "traces" are defined originally on a subspace of sufficiently smooth functions in $\dot{F}_p^{\alpha q}$. If possible, then, the trace operator Tr is extended to all of $\dot{F}_p^{\alpha q}$ by continuity. (See e.g. [30], p. 131.) We discuss the trace problem only informally here since it is treated in detail in [12]. Our main interest is to illustrate the use of Lemma 3.1.

By Theorem II A ii, each $f \in \dot{F}_p^{\alpha q}$ can be written as $\sum_Q s_Q a_Q$ where each a_Q is a smooth atom for Q and (1.8) holds. By smoothness, the restriction of each a_Q to $\mathbf{R}^{n-1}$ is well-defined. Further, since supp $a_Q \subseteq 3Q$, this restriction is 0 for Q not belonging to

$$\mathcal{A} = \{Q \subseteq \mathbf{R}^n : Q \text{ is dyadic and } Q \cap \mathbf{R}^{n-1} \neq \phi\}.$$

Set $\tilde{s}_Q = s_Q$ if $Q \in \mathcal{A}$ and $\tilde{s}_Q = 0$ otherwise. Then $Trf = Tr(\sum_Q \tilde{s}_Q a_Q)$. For each $Q \in \mathcal{A}$, let

$$E_Q = \{x \in Q : \ell(Q)/2 \leq |x_n| < \ell(Q)\}.$$

By Lemma 3.1,

$$\|\{\tilde{s}_Q\}\|_{\dot{f}_p^{\alpha q}} \approx \|(\sum_Q (|Q|^{-\alpha/n}|\tilde{s}_Q|\tilde{\chi}_{E_Q})^q)^{1/q}\|_{L^p}.$$

But the E_Q's for $Q \in \mathcal{A}$ are disjoint, so q and $1/q$ cancel in the last expression. This explains the fact that $Tr\dot{F}_p^{\alpha q}$ depends only on α and p. In particular, we get $Tr\dot{F}_p^{\alpha q} = Tr\dot{F}_p^{\alpha p} = Tr\dot{B}_p^{\alpha p}$, since the Besov spaces $\dot{B}_p^{\alpha q}$ and the $\dot{F}_p^{\alpha q}$-spaces coincide on the diagonal $p = q$. The determination of the traces of the Besov spaces is relatively easy and was carried out in full detail in [11]. Hence, we obtain complete results for the trace problem for the $\dot{F}$-spaces as well, including a new endpoint result. See [12] for the explicit statements.

We now consider a problem of real interpolation for the $\dot{F}_p^{\alpha q}$-spaces. The functorial property of interpolation easily implies that the retract diagram in §1 persists after application of any interpolation method. Thus, we will work directly with the sequence spaces and then push down to the $\dot{F}$-spaces.

We first recall the definition of the K-functional:

$$K(t, x; X_0, X_1) = \inf_{x = x_0 + x_1} (\|x_0\|_{X_0} + t\|x_1\|_{X_1}) .$$

For a sequence $s = \{s_Q\}_{Q \text{ dyadic}}$, we let

$$G^{\alpha q}(s)(x) = (\sum_Q (|Q|^{-\alpha/n}|s_Q|\tilde{\chi}_Q)^q)^{1/q}(x) .$$

Theorem 3.2. Let $\alpha \in \mathbf{R}$, $0 < q \le +\infty$, and $0 < p_0 < p_1 < +\infty$. Let s be any sequence $s = \{s_Q\}$. Then

$$K(t, s; \dot{f}_{p_0}^{\alpha q}, \dot{f}_{p_1}^{\alpha q}) \approx K(t, G^{\alpha q}(s); L^{p_0}, L^{p_1}) .$$

Proof. The inequality

$$K(t, G^{\alpha q}(s); L^{p_0}, L^{p_1}) \le c_q K(t, s; \dot{f}_{p_0}^{\alpha q}, \dot{f}_{p_1}^{\alpha q}) \tag{3.1}$$

follows from the subadditivity $G^{\alpha q}(s_0 + s_1) \le c_q(G^{\alpha q}(s_0) + G^{\alpha q}(s_1))$ by a standard argument. To prove the converse estimate to (3.1), it follows by a well-known argument, presented below, that it suffices to show that there exists a splitting $s = s_0 + s_1$ such that

$$\|s_0\|_{\dot{f}_{p_0}^{\alpha q}}^{p_0} \le c \int_{\{x \,:\, G^{\alpha q}(s)(x) > t\}} |G^{\alpha q}(s)|^{p_0} , \quad \text{and} \quad \|s_1\|_{\dot{f}_{p_1}^{\alpha q}}^{p_1} \le c \int_{\{x \,:\, G^{\alpha q}(s)(x) \le t\}} |G^{\alpha q}(s)|^{p_1} . \tag{3.2}$$

Let $Q_t^+ = Q \cap \{x : G^{\alpha q}(s)(x) > t\}$ and $Q_t^- = Q \setminus Q_t^+$. Let $\mathcal{A}_t = \{Q \text{ dyadic} : |Q_t^+| > |Q|/2\}$ and $\mathcal{A}_t^c = \{Q \text{ dyadic} : |Q_t^-| \ge |Q|/2\}$. Let $s_Q^0 = s_Q$ if $Q \in \mathcal{A}_t$, $s_Q^0 = 0$ otherwise and $s_Q^1 = s_Q$ if $Q \in \mathcal{A}_t^c$, $s_Q^1 = 0$ otherwise. If $s_0 = \{s_Q^0\}$ and $s_1 = \{s_Q^1\}$, then $s = s_0 + s_1$. By Lemma 3.1, then,

$$\|s_0\|_{\dot{f}_{p_0}^{\alpha q}}^{p_0} = \int (\sum_Q (|Q|^{-\alpha/n}|s_Q^0|\tilde{\chi}_Q)^q)^{p_0/q}$$

$$\leq c\int(\sum_Q(|Q|^{-\alpha/n}|s_Q^0|\tilde{\chi}_{Q_t^+})^q)^{p_0/q}$$

$$\leq c\int_{\{x\,:\,G^{\alpha q}(s)(x)>t\}}|G^{\alpha q}(s)|^{p_0}\,.$$

Arguing similarly with s_1, (3.2) follows.

The argument showing the sufficiency of (3.2) runs as follows. Define the E_ϵ-functional for $\epsilon > 1$, $0 < t < +\infty$ by

$$E_\epsilon(t,a;\bar{A}) = \inf_{a=a_0+a_1}\max((\|a_0\|_{A_0}/t)^{1/\epsilon},(\|a_1\|_{A_1}/t)^{1/(\epsilon-1)})\,.$$

(See [19], §2.) If $K_\infty(t,a;\bar{A}) = \inf_{a=a_0+a_1}\max(\|a_0\|_{A_0}, t\|a_1\|_{A_1})$, then it is easy to verify that $E_\epsilon(t,a;\bar{A}) < s$ if and only if $K_\infty(s,a;\bar{A})/s^\epsilon < t$. In other words, $K_\infty(t,a;\bar{A})/t^\epsilon$ is the right continuous inverse of $E_\epsilon(t,a;\bar{A})$ as functions of t ([19], §2). Thus, it suffices to prove

$$E_\epsilon(t,s;\dot{f}_{p_0}^{\alpha q},\dot{f}_{p_1}^{\alpha q}) \leq cE_\epsilon(t,G^{\alpha q}(s);L^{p_0},L^{p_1})\,, \tag{3.3}$$

for some $\epsilon > 1$. If we take $\epsilon = p_1/(p_1-p_0)$, then it is known ([19], §4.2) that

$$E_\epsilon(t,f;L^{p_0},L^{p_1})^{\epsilon p_0} \approx \int_{\{x\,:\,|f(x)|>t\}}(|f|/t)^{p_0} + \int_{\{x\,:\,|f(x)|\leq t\}}(|f|/t)^{p_1}\,.$$

This yields the desired conclusion. □

Corollary 3.3. Let $\alpha \in \mathbf{R}$, $0 < q \leq +\infty$, $0 < p_0 < p_1 < +\infty$. Then

$$K(t,f;\dot{F}_{p_0}^{\alpha q},\dot{F}_{p_1}^{\alpha q}) \approx K(t,G^{\alpha q}(S_\varphi f);L^{p_0},L^{p_1})\,.$$

Proof. The retract diagram immediately gives

$$K(t,f;\dot{F}_{p_0}^{\alpha q},\dot{F}_{p_1}^{\alpha q}) \approx K(t,S_\varphi f;\dot{f}_{p_0}^{\alpha q},\dot{f}_{p_1}^{\alpha q})\,,$$

so the result follows from Theorem 3.2. □

Of course, the last two results and the standard characterization of $K(t,f;L^{p_0},L^{p_1})$ (see e.g. [3]), yield explicit characterizations of $K(t,s;\dot{f}_{p_0}^{\alpha q},\dot{f}_{p_1}^{\alpha q})$ and $K(t,f;\dot{F}_{p_0}^{\alpha q},\dot{F}_{p_1}^{\alpha q})$. In terms of interpolation spaces, we obtain the following two results.

Corollary 3.4. Let $\alpha \in \mathbf{R}$, $0 < q \leq +\infty$, $0 < p_0 < p_1 < +\infty$, $0 < \theta < 1$, and $1/p = (1-\theta)/p_0 + \theta/p_1$. Then

$$(\dot{f}_{p_0}^{\alpha q},\dot{f}_{p_1}^{\alpha q})_{\theta,p} = \dot{f}_p^{\alpha q}\,.$$

Proof. This follows immediately from Theorem 3.2 and the basic fact that $(L^{p_0},L^{p_1})_{\theta,p} = L^p$ under the same restrictions. □

Corollary 3.5. Under the same conditions on α, q, p_0, p_1, and θ as in Corollary 3.4,

$$(\dot{F}_{p_0}^{\alpha q}, \dot{F}_{p_1}^{\alpha q})_{\theta, p} = \dot{F}_p^{\alpha q}.$$

Proof. This is a consequence of Corollary 3.4, since $(\dot{F}_{p_0}^{\alpha q}, \dot{F}_{p_1}^{\alpha q})_{\theta, p}$ is a retract of $(\dot{f}_{p_0}^{\alpha q}, \dot{f}_{p_1}^{\alpha q})_{\theta, p}$ under S_φ and T_ψ. □

As noted earlier, for $\alpha = 0$ and $q = 2$ we obtain $(H^{p_0}, H^{p_1})_{\theta, p} = H^p$ if $0 < p_0 < p_1 < +\infty$ and $1/p = (1-\theta)/p_0 + \theta/p_1$, which is the key result in [8].

Corollary 3.5 provides the crucial step in proving an imbedding result (Theorem 3.8 iii) below that will be used in our proof of the lower majorant property. We write $X \to Y$ if $X \subseteq Y$ and $\|\cdot\|_Y \leq c\|\cdot\|_X$. We will need the following standard interpolation result for the Besov spaces.

Theorem 3.6. ([26], p. 64, or [30], p. 64) Suppose $\alpha, \alpha_0, \alpha_1 \in \mathbf{R}$, $\alpha_0 \neq \alpha_1$, $0 < \theta < 1$, $0 < q, q_0, q_1 \leq +\infty$, $0 < p < +\infty$, and $\alpha = (1-\theta)\alpha_0 + \theta\alpha_1$. Then

$$(\dot{B}_p^{\alpha_0 q_0}, \dot{B}_p^{\alpha_1 q_1})_{\theta, q} = \dot{B}_p^{\alpha q}.$$

It is convenient now to restate our results for the Besov spaces from [11], modified by our change in normalization. Let $\dot{b}_p^{\alpha q}$ be the set of all sequences $s = \{s_Q\}_{Q \text{ dyadic}}$ such that

$$\begin{aligned}\|s\|_{\dot{b}_p^{\alpha q}} &\equiv \Big(\sum_{\nu \in \mathbf{Z}} \Big\| \sum_{\ell(Q) = 2^{-\nu}} |Q|^{-\alpha/n} |s_Q| \tilde{\chi}_Q \Big\|_{L^p}^q\Big)^{1/q} \\ &= \Big(\sum_{\nu \in \mathbf{Z}} \Big(\sum_{\ell(Q) = 2^{-\nu}} (|Q|^{-\alpha/n - 1/2 + 1/p} |s_Q|)^p\Big)^{q/p}\Big)^{1/q} < +\infty.\end{aligned}$$

Theorem 3.7. ([11]) Suppose $\alpha \in \mathbf{R}$, $0 < p, q \leq +\infty$. The operators $S_\varphi : \dot{B}_p^{\alpha q} \to \dot{b}_p^{\alpha q}$ and $T_\psi : \dot{b}_p^{\alpha q} \to \dot{B}_p^{\alpha q}$ are bounded, and $T_\psi \circ S_\varphi$ is the identity on $\dot{B}_p^{\alpha q}$.

We pause now to reconsider some known imbedding theorems. It is clear from Theorem I and Theorem 3.7 that any imbedding between $\dot{F}$- and/or $\dot{B}$-spaces is equivalent to one between the corresponding $\dot{f}$ and/or $\dot{b}$-spaces. However, the latter imbedding may be simpler. Note for example in the proof of Theorem 3.8 i below that facts about functions of exponential type that are used in the traditional $\dot{B}$-space proof are replaced in the sequence space case by trivial relations between the L^{p_0} and L^{p_1} norms of χ_Q. We repeat the proof of Theorem 3.8 iii from [17] only to exhibit the use of Corollary 3.5, whose proof was not detailed in [17].

Theorem 3.8. (e.g. [17]) Suppose $0 < p_0 < p_1 \leq +\infty$, $\alpha_0, \alpha_1 \in \mathbf{R}$, $0 < q \leq +\infty$, and $\alpha_0 - n/p_0 = \alpha_1 - n/p_1$. Then

i) $\dot{B}_{p_0}^{\alpha_0 q} \to \dot{B}_{p_1}^{\alpha_1 q}$,

ii) if $p_0 \le q$, then $\dot{F}^{\alpha_0 q}_{p_0} \to \dot{B}^{\alpha_1 q}_{p_1}$,

and

iii) $\dot{F}^{\alpha_0 q}_{p_0} \to \dot{B}^{\alpha_1 p_0}_{p_1}$.

Proof. i) Replacing $-\alpha_1/n + 1/p_1$ in the second expression above for $\|s\|_{\dot{b}^{\alpha_1 q}_{p_1}}$ by $-\alpha_0/n + 1/p_0$, and applying $\ell^{p_0} \to \ell^{p_1}$, we obtain $\dot{b}^{\alpha_0 q}_{p_0} \to \dot{b}^{\alpha_1 q}_{p_1}$ immediately.

ii) For $p_0 \le q$, we have $\dot{F}^{\alpha_0 q}_{p_0} \to \dot{B}^{\alpha_0 q}_{p_0}$ trivially by Minkowski's integral inequality, so ii) follows from i).

iii) Pick $p' < p_0$, $p'' < p_1$ and $\theta \in (0,1)$ such that $1/p_0 = (1-\theta)/p' + \theta/p''$. Let $\alpha' = \alpha_0 + n(1/p_1 - 1/p')$ and $\alpha'' = \alpha_0 + n(1/p_1 - 1/p'')$. Then $(1-\theta)\alpha' + \theta\alpha'' = \alpha_1$. By ii), $\dot{F}^{\alpha_0 \infty}_{p'} \to \dot{B}^{\alpha' \infty}_{p_1}$ and $\dot{F}^{\alpha_0 \infty}_{p''} \to \dot{B}^{\alpha'' \infty}_{p_1}$. By Corollary 3.5, Theorem 3.6, and the imbedding $\ell^q \to \ell^\infty$, we obtain

$$\dot{F}^{\alpha_0 q}_{p_0} \to \dot{F}^{\alpha_0 \infty}_{p_0} = (\dot{F}^{\alpha_0 \infty}_{p'}, \dot{F}^{\alpha_0 \infty}_{p''})_{\theta p_0} \to (\dot{B}^{\alpha' \infty}_{p_1}, \dot{B}^{\alpha'' \infty}_{p_1})_{\theta p_0} = \dot{B}^{\alpha_1 p_0}_{p_1}. \qquad \square$$

A space $X \subseteq S'/P$ is said to have the *lower (Fourier) majorant property* if, for each $f \in X$, there exists $g \in X$ with $|\hat{f}(\xi)| \le \hat{g}(\xi)$ for all $\xi \in \mathbf{R}^n \setminus \{0\}$, and $\|g\|_X \le c\|f\|_X$, with c independent of f. We will obtain the following result.

Theorem 3.9. Suppose either $\alpha \in \mathbf{R}$, $0 < p < 1$, and $0 < q \le +\infty$, or $\alpha \in \mathbf{R}$, $p = 1$, and $0 < q \le 1$. Then $\dot{F}^{\alpha q}_p$ has the lower majorant property.

Note that by Theorem 3.9 iii, $\dot{F}^{\alpha q}_p \to \dot{B}^{\alpha - n(1/p - 1), p}_1$ if $\alpha \in \mathbf{R}$, $0 < p < 1$, and $0 < q \le +\infty$. By [11], Theorem 6.1, $\dot{B}^{\alpha - n(1/p - 1), p}_1$ has the lower majorant property. The following result sharpens this by showing that the lower majorants for $\dot{B}^{\alpha - n(1/p - 1), p}_1$ may be taken from the subspace $\dot{F}^{\alpha q}_p$.

Theorem 3.10. Suppose $\alpha \in \mathbf{R}$, $0 < p \le 1$, and $0 < q \le +\infty$. For each $f \in \dot{B}^{\alpha - n(1/p - 1), p}_1$, there exists $g \in \dot{F}^{\alpha q}_p$ such that $|\hat{f}(\xi)| \le \hat{g}(\xi)$ for all $\xi \in \mathbf{R}^n \setminus \{0\}$, and $\|g\|_{\dot{F}^{\alpha q}_p} \le c\|f\|_{\dot{B}^{\alpha - n(1/p - 1), p}_1}$, with c independent of f.

Proof. Select φ and ψ satisfying (1.1-4) and, in addition, $\hat{\psi}(\xi) \ge 0$ for all $\xi \in \mathbf{R}^n$. By Theorem 3.7, each $f \in \dot{B}^{\alpha - n(1/p - 1), p}_1$ can be written as $f = \sum_Q s_Q \psi_Q$, where

$$\Big(\sum_{\nu \in \mathbf{Z}} \Big(\sum_{\ell(Q) = 2^{-\nu}} |Q|^{-\alpha/n + 1/p - 1/2} |s_Q|\Big)^p\Big)^{1/p} \le c\|f\|_{\dot{B}^{\alpha - n(1/p - 1), p}_1}. \tag{3.4}$$

Let $g(x) = \sum_Q |s_Q| \psi_Q(x + x_Q) = \sum_Q |s_Q| |Q|^{-1/2} \psi(2^\nu x)$. Then $\hat{g}(\xi) = \sum_Q |s_Q| |\hat{\psi}_Q(\xi)| \ge |\hat{f}(\xi)|$. Letting $t_\nu = \sum_{\ell(Q) = 2^{-\nu}} |s_Q|$, then $g = \sum_{\nu \in \mathbf{Z}} t_\nu \psi_{Q_{\nu 0}}$. Let $E_{Q_{\nu 0}} = Q_{\nu 0} \setminus Q_{\nu + 1, 0}$.

Then $|E_{Q_{\nu 0}}| \geq |Q_{\nu 0}|/2$ for all $\nu \in \mathbf{Z}$, and the E_Q's for $\nu \in \mathbf{Z}$ are pairwise disjoint. Applying Theorem I, Lemma 3.1, and (3.4),

$$\begin{aligned}\|g\|_{\dot F_p^{\alpha q}} &\leq c\|(\sum_{\nu \in \mathbf{Z}}(|Q_{\nu 0}|^{-\alpha/n} t_\nu \tilde\chi_{Q_{\nu 0}})^q)^{1/q}\|_{L^p} \\ &\leq c\|(\sum_{\nu \in \mathbf{Z}}(|Q_{\nu 0}|^{-\alpha/n} t_\nu \tilde\chi_{E_{Q_{\nu 0}}})^q)^{1/q}\|_{L^p} \\ &\leq c(\sum_{\nu \in \mathbf{Z}}(|Q_{\nu 0}|^{-\alpha/n - 1/2 + 1/p} t_\nu)^p)^{1/p} \\ &\leq c\|f\|_{\dot B_1^{\alpha - n(1/p-1),p}} . \qquad \square\end{aligned}$$

Proof of Theorem 3.9. The result for $0 < p < 1$ follows immediately from Theorem 3.8 iii and Theorem 3.10. For $p = 1$ and $0 < q \leq 1$, the trivial imbedding $\dot F_1^{\alpha q} \to \dot F_1^{\alpha 1} = \dot B_1^{\alpha 1}$ and Theorem 3.10 yield the result. $\square$

Note that we have obtained the fact ([1], [2]) that for $0 < p < 1$, $H^p = \dot F_p^{02}$ has the lower majorant property. It is also known that H^1 has the lower majorant property ([1], [2]), but this seems to require a different proof.

4 Remarks on further results

We take this opportunity to discuss some of the work carried out in [12] which extends and generalizes the theory presented in §1-3 above.

We noted in §2 that it is possible to approach the proofs of Theorems I and II on the level of the sequence spaces $\dot f_p^{\alpha q}$. To do this, we define, for any sequence $s = \{s_Q\}_{Q \text{ dyadic}}$, any r with $0 < r \leq +\infty$, and an appropriate $\lambda > 0$, the maximal sequence s_r^*, by

$$(s_r^*)_Q = (\sum_{P : \ell(P) = \ell(Q)} |s_P|^r/(1 + \ell(Q)^{-1}|x_P - x_Q|)^\lambda)^{1/r} .$$

The case $r = +\infty$ corresponds to Peetre's φ_ν^{**} above; more generally, s_r^* is a discrete version of the Littlewood-Paley function g_λ^*. The following is an analogue of Theorem 2.2.

Theorem 4.1. Suppose $\alpha \in \mathbf{R}$, $0 < p < +\infty$, $0 < q \leq +\infty$, and $\lambda > n$. For any sequence $s = \{s_Q\}$,

$$\|s\|_{\dot f_p^{\alpha q}} \approx \|s^*_{\min(p,q)}\|_{\dot f_p^{\alpha q}} .$$

As an indication of the geometric meaning of Theorem 4.1, consider a collection $\{Q_i\}_{i=1}^\infty$ of disjoint dyadic cubes. Let $s_{Q_i} = |Q_i|^{1/2}$, $i = 1, 2, 3, \ldots$, and $s_Q = 0$ if $Q \notin \{Q_i\}_{i=1}^\infty$. Then

Theorem 4.1, with $\alpha = 0$, yields a slight refinement of the classical estimates for the Marcinkiewicz integral (cf. [9], [28]).

The key fact that enables us to pass back and forth from $\dot{f}_p^{\alpha q}$ to $\dot{F}_p^{\alpha q}$ is the fact that functions of exponential type are sufficiently smooth that their sups and infs are essentially comparable on cubes of appropriate side length (inversely related to the exponential type). We express this more precisely via the following result, which is a sequence space analogue of the classical result of Plancherel-Pólya (see e.g. [11]).

Lemma 4.2. Suppose $f \in S'$ and supp $\hat{f} \subseteq \{\xi \in \mathbf{R}^n : |\xi| \le 2\}$. Let $a_Q = \sup_{y \in Q} |f(y)|$. For $\gamma \in \mathbf{Z}$, $\gamma \ge 0$, let $b_{Q,\gamma} = \max\{\inf_{y \in \tilde{Q}} |f(y)| : \ell(\tilde{Q}) = 2^{-\gamma}\ell(Q), \tilde{Q} \subseteq Q\}$. Let $a = \{a_Q\}$ and $b = \{b_{Q,\gamma}\}$ be the associated sequences. If $0 < r \le +\infty$, $\ell(Q) = 1$, and γ is sufficiently large, then

$$(a_r^*)_Q \approx (b_r^*)_Q ,$$

with constants independent of f and Q.

To use this, we define, for $f \in S'$, φ satisfying (1.1-3), and Q dyadic with $\ell(Q) = 2^{-\nu}$,

$$\sup_Q(f) = |Q|^{1/2} \sup_{y \in Q} |\tilde{\varphi}_\nu * f(y)| ,$$

with $\tilde{\varphi}_\nu(x) = \overline{\varphi_\nu(-x)}$, as in §2. Also, for $\gamma \in \mathbf{Z}$, $\gamma \ge 0$, define

$$\inf_{Q,\gamma}(f) = |Q|^{1/2} \max\{ \inf_{y \in \tilde{Q}} |\tilde{\varphi}_\nu * f(y)| : \ell(\tilde{Q}) = 2^{-\gamma}\ell(Q), \tilde{Q} \subseteq Q\} .$$

Then the feasibility of our entire discretization process is to some extent explained by the following result.

Lemma 4.3. For γ sufficiently large and $f \in S'/\mathcal{P}$,

$$\|f\|_{\dot{F}_p^{\alpha q}} \approx \|\{\sup_Q(f)\}\|_{\dot{f}_p^{\alpha q}} \approx \|\{\inf_{Q,\gamma}(f)\}\|_{\dot{f}_p^{\alpha q}} .$$

In particular, the boundedness of $S_\varphi : \dot{F}_p^{\alpha q} \to \dot{f}_p^{\alpha q}$ follows immediately from $(S_\varphi f)_Q \le \sup_Q(f)$. Using (1.2), one can give a simple proof of the boundedness of $T_\psi : \dot{f}_p^{\alpha q} \to \dot{F}_p^{\alpha q}$ as well. Since $T_\psi \circ S_\varphi$ is the identity on $\dot{F}_p^{\alpha q}$ by Lemma 2.1, we obtain Theorem I.

The "molecular" result in Theorem II B is in fact a consequence of a more general result about linear operators on the sequence spaces $\dot{f}_p^{\alpha q}$. Consider a linear operator T defined formally on sequences $s = \{s_Q\}$ by $(Ts)_Q = \sum_P a_{QP} s_P$, where $\{a_{QP}\}$ is an infinite matrix doubly indexed by all dyadic cubes P and Q. Let $J = n/\min(1,p,q)$, and set

$$\omega_{QP}(\epsilon) = \left(\frac{\ell(Q)}{\ell(P)}\right)^{\alpha} \left(1 + \frac{|x_Q - x_P|}{\max(\ell(P),\ell(Q))}\right)^{-(J+\epsilon)} \min\left(\left(\frac{\ell(Q)}{\ell(P)}\right)^{\frac{n+\epsilon}{2}}, \left(\frac{\ell(P)}{\ell(Q)}\right)^{\frac{n+\epsilon}{2}+J-n}\right),$$

for $\epsilon > 0$. We say that an operator T as above is *almost diagonal on* $\dot{f}_p^{\alpha q}$ if there exists $\epsilon > 0$ such that

$$\sup_{P,Q} |a_{QP}|/\omega_{QP}(\epsilon) < +\infty .$$

For $\alpha = 0$ and $p, q \geq 1$, this condition reduces to

$$|a_{QP}| \leq c_\epsilon e^{-(n+\epsilon)d(Q,P)} ,$$

where d is a certain distance defined on the set of dyadic cubes, obtained from an analogue for $\mathbf{R}_+^{n+1}$ of the Poincaré metric on the upper half-plane. The following theorem is proved by methods related to those in the proof of Theorem II B above.

Theorem 4.4. An almost diagonal operator on $\dot{f}_p^{\alpha q}$ is bounded (on $\dot{f}_p^{\alpha q}$).

If $f = \sum_Q s_Q m_Q$, where each m_Q is a smooth molecule for Q, then by Theorem I we can write $m_Q = \sum_P \langle m_Q, \varphi_P \rangle \psi_P$. Letting T be the operator corresponding to the matrix $a_{PQ} = \langle m_Q, \varphi_P \rangle$, we have

$$f = \sum_P \sum_Q a_{PQ} s_Q \psi_P = \sum_P (Ts)_P \psi_P ,$$

where $s = \{s_Q\}$. The estimates in Lemma 2.3 above show that T is almost diagonal for $\dot{f}_p^{\alpha q}$. Applying Theorem I and Theorem 4.4,

$$\|f\|_{\dot{F}_p^{\alpha q}} \leq c\|Ts\|_{\dot{f}_p^{\alpha q}} \leq c\|s\|_{\dot{f}_p^{\alpha q}} .$$

Thus we recover Theorem II B.

If T is a linear operator on $\dot{F}_p^{\alpha q}$ and $f = \sum_P s_P \psi_P \in \dot{F}_p^{\alpha q}$, we can write

$$Tf = \sum_P s_P T\psi_P = \sum_P s_P \sum_Q \langle T\psi_P, \varphi_Q \rangle \psi_Q = \sum_Q \sum_P s_P \langle T\psi_P, \varphi_Q \rangle \psi_Q .$$

This formally identifies T with the linear operator on $\dot{f}_p^{\alpha q}$ with matrix $a_{QP} = \langle T\psi_P, \varphi_Q \rangle$. With this identification, many of the familiar operators is harmonic analysis, for example, classical Calderón-Zygmund operators and certain pseudo-differential operators, are almost diagonal. In this sense, the functions $\{\psi_Q\}_Q$ are "almost eigenfunctions" of these operators. In other words, our representing functions $\{\psi_Q\}_Q$ are chosen to "almost diagonalize" many standard operators. Our general perspective is that whereas exact diagonalization may restrict the choice of representing functions in such a way that simple norm characterizations of the type in Theorem I are impossible, an almost diagonalization is likely to be sufficient for most purposes, and yet may allow such a characterization.

There is an alternate approach to the "smooth atomic decomposition" in Theorem II A ii as well. One can obtain this result from the ψ_Q-decomposition in Theorem II A i essentially by slicing up the ψ_Q's and collecting terms appropriately. However, in this proof and in the proof in §2, the atoms a_Q depend on the function f. We ask when it is possible to obtain a representation as in Theorem II with a fixed, canonical set $\{a_Q\}$ (as for ψ_Q). Since the coefficients s_Q in the representation $f = \sum_Q s_Q a_Q$ for our proofs of the smooth atomic decomposition above do not depend linearly on f, we also ask when this decomposition can be obtained linearly. More generally, we say that a family $\{\sigma^Q\}_{Q \text{ dyadic}}$ of distributions (not necessarily of compact support) *represents* $\dot{F}_p^{\alpha q}$ if there exists a family $\{\tau^Q\}_{Q \text{ dyadic}}$ such that for all $f \in \dot{F}_p^{\alpha q}$, we have

i) $$f = \textstyle\sum_Q \langle f, \tau^Q\rangle \sigma^Q$$

with

$$\|\{\langle f, \tau^Q\rangle\}\|_{\dot{f}_p^{\alpha q}} \leq c\|f\|_{\dot{F}_p^{\alpha q}},$$

and

ii) for each sequence $\{s_Q\}_{Q \text{ dyadic}}$, we have

$$\|\sum_Q s_Q \sigma^Q\|_{\dot{F}_p^{\alpha q}} \leq c\|\{s_Q\}\|_{\dot{f}_p^{\alpha q}}.$$

If we instead start with a family $\{\tau^Q\}_Q$ and there exists a family $\{\sigma^Q\}_Q$ such that i) and ii) hold, we say that $\{\tau^Q\}_Q$ *norms* $\dot{F}_p^{\alpha q}$. In both cases, we may define the operator S_τ, for $f \in \dot{F}_p^{\alpha q}$, and T_σ, on sequences $s = \{s_Q\}_Q$, by

$$(S_\tau f)_Q = \langle f, \tau^Q\rangle, \text{ and } T_\sigma(s) = \sum_Q s_Q \sigma^Q.$$

The conditions i) and ii) say that the diagram

$$\begin{array}{ccc} & \dot{f}_p^{\alpha q} & \\ {}^{S_\tau}\nearrow & & \searrow^{T_\sigma} \\ \dot{F}_p^{\alpha q} & \xrightarrow{\text{Id}} & \dot{F}_p^{\alpha q} \end{array}$$

is commutative.

Theorem 4.5. Suppose $\alpha \in \mathbf{R}$, $0 < p < +\infty$, $0 < q \leq +\infty$, and $u \in \mathcal{S}$ satisfies

$$|\hat{u}(\xi)| \geq c > 0 \text{ if } 1/2 \leq |\xi| \leq 2 \tag{4.1}$$

and

$$\int u(x)x^{\gamma}dx = 0 \text{ if } |\gamma| \leq \max(K-1, N)\,. \tag{4.2}$$

Let $\{x^Q\}_{Q \text{ dyadic}}$ be any sequence of points such that $x^Q \in Q$ for each Q. For $\mu \in \mathbf{Z}$, let $u_\mu(x) = 2^{\mu n}u(2^\mu x)$, as usual, and let $u_\mu^Q(x) = |Q|^{-1/2}u_\mu(2^\nu(x - x^Q))$, if $\ell(Q) = 2^{-\nu}$. Then there exists $\mu_0 \in \mathbf{Z}$ such that for all $\mu \leq \mu_0$, $\{u_\mu^Q\}_Q$ both represents and norms $\dot{F}_p^{\alpha q}$.

The condition $u \in S$ in Theorem 4.5 can be relaxed. Thus, up to a dilation, a large class of functions represent and norm $\dot{F}_p^{\alpha q}$. In particular, there exists $u \in S$ satisfying (4.1-2) with supp $u \subseteq \{x : |x| \leq 1\}$ (see [11], p. 783). Then u_μ will satisfy supp $u_\mu \subseteq \{x : |x| \leq c\}$ for some $c > 0$. Except that c may not be 3, we thus obtain a linear, canonical version of the smooth atomic decomposition from the result that $\{u_\mu^Q\}_Q$ represents $\dot{F}_p^{\alpha q}$. The fact that $\{u_\mu^Q\}_Q$ norms $\dot{F}_p^{\alpha q}$ is a dual formulation of our decomposition result in which the determination of the coefficients in the expansion is localized.

The φ-transform and the *generalized φ-transform* in Theorem 4.5 are related to certain results in mathematical physics, for example in the theory of coherent states in quantum mechanics, and in signal analysis ([13], [14], [15], [16], [20], and [24]). Consider the group

$$G = \{(a,b) : a \in \mathbf{R}^n, b > 0\}\,,$$

with multiplication

$$(a_0, b_0)\cdot(a_1, b_1) = (b_0a_1 + a_0, b_0b_1)\,.$$

Then the map $U : G \to \mathcal{B}(L^2(\mathbf{R}^n))$ defined by

$$U(a,b)f(x) = b^{n/2}f(bx - a)$$

is a unitary representation of G. ($\mathcal{B}(L^2)$ is of course the space of bounded linear operators on L^2.) If we allow $b < 0$ as well and take $n = 1$, this representation is irreducible (see e.g. [14]). (For $b > 0$ only, U would be irreducible if we replace L^2 by H^2.) It follows then that for any non-zero $g \in L^2$, the span of the orbit of g under $U(G)$ is dense in L^2 ([15]). However, for $n > 1$ this representation is reducible; for example, the set of functions with Fourier transform supported in some proper angular sector (symmetric with respect to the origin if $b < 0$ is allowed) is invariant under $U(G)$. Nevertheless one can ask when the span of the orbit of $g \in L^2$ under $U(G)$ is dense in L^2. Our collection $\{\psi_Q\}$ (or $\{u_\mu^Q\}$, if we pick $x^Q = x_Q$) is the orbit of ψ (or u_μ) under the image by U of the discrete "semi-lattice" (not subgroup)

$$\{(k, 2^\nu) : k \in \mathbf{Z}^n\,,\ \nu \in \mathbf{Z}\}$$

contained in G. Hence in particular our results give conditions on the initial vector φ (or u_μ) in L^2 such that the discrete subset $\{\psi_Q\}$ (or $\{u_\mu^Q\}$) of the full orbit under $U(G)$ still has dense span in L^2, or more generally $\dot{F}_p^{\alpha q}$. Note that the condition (4.1) on u is related to the example above showing the reducibility of U on $L^2(\mathbf{R}^n)$ for $n > 1$. Also, (4.2) is related to the admissibility criterion in [14] (see in particular (2.25-27)) which is used to obtain a continuous reproducing formula similar to Calderón's reproducing formula, and hence related to our discrete (generalised) φ-transform above. The passage from u to u_μ in Theorem 4.5 indicates that an initial dilation may be needed before our fixed semi-lattice is sufficient. If we consider the minimal subgroup generated by our semi-lattice (namely $\{(2^\mu k, 2^\nu) : \mu, \nu \in \mathbf{Z}\ ,\ k \in \mathbf{Z}^n\}$), then Theorem 4.5 shows that the image under the action of this subgroup of any $u \in S$ satisfying (4.1-2) for $K = 1$ and $N = 0$, has dense span in L^2.

So far we have not discussed $\dot{F}_p^{\alpha q}$ with $p = +\infty$. The obvious definition of the norm, namely $\|(\sum_{\nu \in \mathbf{Z}} (2^{\nu\alpha}|\varphi_\nu * f|)^q)^{1/q}\|_{L^\infty}$, is not satisfactory—for example, Triebel remarks [30, p. 46] that this definition is not independent of the choice of φ satisfying (1.1-3). We obtain a satisfactory definition, however, by localizing appropriately. For $\alpha \in \mathbf{R}$, and $0 < q \le +\infty$, let $\dot{F}_\infty^{\alpha q}$ be the set of all $f \in S'/P$ such that

$$\|f\|_{\dot{F}_\infty^{\alpha q}} = \sup_{P \text{ dyadic}} \left(\frac{1}{|P|} \int_P \sum_{\nu = -\log_2 \ell(P)}^{\infty} (2^{\nu\alpha}|\varphi_\nu * f(x)|)^q dx \right)^{1/q} < +\infty\,,$$

where φ satisfies (1.1-3). We also define $\dot{f}_\infty^{\alpha q}$ to be the set of all sequences $s = \{s_Q\}_{Q \text{ dyadic}}$ such that

$$\|s\|_{\dot{f}_\infty^{\alpha q}} = \sup_{P \text{ dyadic}} \left(\frac{1}{|P|} \int_P \sum_{Q \subseteq P} (|Q|^{-\alpha/n}|s_Q|\tilde{\chi}_Q(x))^q dx \right)^{1/q} < +\infty\,.$$

(Note that this norm corresponds to a type of Carleson measure condition.) Then the analogue of Theorem I holds for $p = +\infty$.

Theorem 4.6. Let $\alpha \in \mathbf{R}$, and $0 < q \le +\infty$. Then $S_\varphi : \dot{F}_\infty^{\alpha q} \to \dot{f}_\infty^{\alpha q}$ and $T_\psi : \dot{f}_\infty^{\alpha q} \to \dot{F}_\infty^{\alpha q}$ are bounded operators. Also $T_\psi \circ S_\varphi$ is the identity on $\dot{F}_\infty^{\alpha q}$.

Using this, and working with the spaces $\dot{f}_\infty^{\alpha q}$, we obtain the following results, which indicate that our definition of $\dot{F}_\infty^{\alpha q}$ is appropriate.

Theorem 4.7. Suppose $\alpha \in \mathbf{R}$ and $0 < q \le +\infty$.

i) $\dot{F}_\infty^{\alpha q}$ is independent of the choice of φ satisfying (1.1-3).

ii) If $1 \le q < +\infty$, then $(\dot{F}_1^{\alpha q})^* = \dot{F}_\infty^{-\alpha q'}$, where $1/q + 1/q' = 1$.

iii) There exists an operator $A^{\alpha q}$ such that

$$\|A^{\alpha q} f\|_{L^p} \approx \|f\|_{\dot{F}_p^{\alpha q}}$$

for $\alpha \in \mathbf{R}$, $0 < p, q \leq +\infty$.

iv) If $0 < p_0 < +\infty$, $0 < \theta < 1$, and $1/p = (1-\theta)/p_0$, then

$$(\dot{F}_{p_0}^{\alpha q}, \dot{F}_{\infty}^{\alpha q})_{\theta, p} = \dot{F}_p^{\alpha q} .$$

We obtain as well a characterization of the K-functional for the couple $(\dot{F}_{p_0}^{\alpha q}, \dot{F}_{\infty}^{\alpha q})$, an almost diagonality condition for a matrix $\{a_{PQ}\}$ to be bounded on $\dot{f}_{\infty}^{\alpha q}$ (yielding a molecular estimate as in Theorem II B), as well as $p = +\infty$ versions of the smooth atomic decomposition and the generalized φ-transform above.

The proof of Theorem 4.7 relies on a discrete analogue of the local square function (see [10], [29], and [18]). This is natural, since the local square function developed in the theory of BMO. (Of course, by Theorem 4.7 ii and the $H^1 - BMO$ duality ([10]), $\dot{F}_{\infty}^{02} = (\dot{F}_1^{02})^* = (H^1)^* = BMO$; alternately §4 of [11] yields a direct proof of this fact. Similarly, Theorem 4.7 ii shows that the spaces $\dot{F}_{\infty}^{\alpha q}$ above coincide with those defined, at least for $q > 1$, by Triebel ([30], p. 50).)

It is also possible to extend the theory in the other direction; that is, to $p = 0$. We let $\dot{f}_0$ be the set of all sequences $s = \{s_Q\}$ such that

$$\|s\|_{\dot{f}_0} = |\bigcup_{s_Q \neq 0} Q| < +\infty .$$

If we define L^0 to be the set of all measurable functions f on $\mathbf{R}^n$ such that

$$\|f\|_{L^0} = |\{x \in \mathbf{R}^n : f(x) \neq 0\}| < +\infty ,$$

then $\|s\|_{\dot{f}_0} = \|(\Sigma_Q(|Q|^{-\alpha/n}|s_Q|\tilde{\chi}_Q)^q)^{1/q}\|_{L^0}$. Hence $\dot{f}_0$ is in some sense the simultaneous end-point space for all $\dot{f}_p^{\alpha q}$ as $p \to 0$ with α and q fixed. We obtain the following, which should be compared with the fact that $(L^0, L^\infty)_{\theta, 1/(1-\theta)}^{1/\theta} = L^p$ if $0 < \theta < 1$ and $p = \theta/(1-\theta)$ ([27]).

Theorem 4.8. Suppose $\alpha \in \mathbf{R}$, $0 < q \leq +\infty$, $0 < \theta < 1$, and $p = \theta/(1-\theta)$. Then

$$(\dot{f}_0, \dot{f}_{\infty}^{\alpha q})_{\theta, 1/(1-\theta)}^{1/\theta} = \dot{f}_p^{\alpha q} .$$

We can now obtain Corollary 3.4 above by reiteration from Theorem 4.8, thereby obtaining the $\dot{F}$-space results (Corollary 3.5 and Theorem 4.7 iv) as well.

Although one would define distribution space versions of the $p = 0$ sequence spaces, we do not do so because these spaces are no longer independent of the choice of φ.

The finite support condition in the definition of $\dot{f}_0$ is useful in connecting our "smooth atomic" decomposition of $\dot{F}_p^{\alpha q}$ (Theorem II A ii) with the traditional "non-smooth" atomic decomposition ([5], [21]) of H^p, $0 < p \le 1$. This approach will clarify the close connection between the latter decomposition and abstract interpolation theory. First, arguing on the sequence space level, we define an *atom for* $\dot{f}_p^{\alpha q}$, $0 < p \le 1$, $q \ge p$, and $\alpha \in \mathbf{R}$, to be any sequence $r = \{r_Q\}$ such that there exists a dyadic cube $\tilde{Q}$ so that $r_Q = 0$ if $Q \not\subseteq \tilde{Q}$, and

$$|\tilde{Q}|^{1/p}\|r\|_{\dot{f}_\infty^{\alpha q}} \le 1 .$$

Using Theorem 4.8 in conjunction with a known characterization of the real interpolation method and basic properties of the dyadic cubes, we obtain the following "atomic" decomposition of our sequence spaces.

Theorem 4.9. Suppose $\alpha \in \mathbf{R}$, $0 < p \le 1$, and $p \le q \le +\infty$. Then

$$\|s\|_{\dot{f}_p^{\alpha q}} \approx \inf\{(\sum_{\ell \in \mathbf{Z}^+} |\gamma_\ell|^p)^{1/p} : s = \sum_{\ell \in \mathbf{Z}^+} \gamma_\ell r_\ell \text{, where each } r_\ell \text{ is an atom for } \dot{f}_p^{\alpha q}\} .$$

To lift this result to the $\dot{F}$-spaces, we define an *atom for* $\dot{F}_p^{\alpha q}$ to be a function A satisfying

$$A = \sum_{Q \subseteq \tilde{Q}} r_Q a_Q$$

where $r = \{r_Q\}$ is an atom for $\dot{f}_p^{\alpha q}$ associated to $\tilde{Q}$, and each a_Q is a smooth atom for Q.

Theorem 4.10. Suppose $\alpha \in \mathbf{R}$, $0 < p \le 1$, and $p \le q \le +\infty$. Then

$$\|f\|_{\dot{F}_p^{\alpha q}} \approx \inf\{(\sum_{\ell \in \mathbf{Z}^+} |\gamma_\ell|^p)^{1/p} : f = \sum_{\ell \in \mathbf{Z}^+} \gamma_\ell A_\ell \text{, where each } A_\ell \text{ is an atom for } \dot{F}_p^{\alpha q}\} .$$

Starting instead with the ψ_Q decomposition, or the decomposition resulting from the generalized φ-transform, we obtain in each case a result of a similar kind to the one in Theorem 4.10.

When $\alpha = 0$ and $q = 2$, an atom A for $\dot{F}_p^{02} (0 < p \le 1)$ satisfies supp $A \subseteq 3\tilde{Q}$, $\|A\|_{\dot{F}_\infty^{02}} = \|A\|_{\mathrm{BMO}} \le |\tilde{Q}|^{-1/p}$, and $\int x^\gamma A(x)dx = 0$ if $|\gamma| \le N$, for some dyadic cube $\tilde{Q}$. Hence we recover the traditional atomic decomposition of H^p, $0 < p \le 1$, into "BMO-atoms" (in particular L^2-atoms). (See [5], [21], and [31].) Alternately, Theorem 4.10 and Theorem 4.7 iv can be used to give a proof of the equivalences $\dot{F}_p^{02} = H^p$, $0 < p < +\infty$, and $\dot{F}_\infty^{02} = BMO$.

By considering the analogues of Theorems I and II for the inhomogeneous spaces $F_p^{\alpha q}$, we can obtain simple proofs of various results ([30], p. 158) about pointwise multipliers. Of particular interest is the question of when the characteristic function χ of the upper half-plane $\mathbf{R}^n_+$ is a pointwise multiplier for $F_p^{\alpha q}(\mathbf{R}^n)$; i.e. when does

$$\|\chi f\|_{F_p^{\alpha q}} \leq c\|f\|_{F_p^{\alpha q}}$$

hold? Using the smooth atomic decomposition of $F_p^{\alpha q}$, the geometry of this problem is very similar to that of the trace problem discussed in §3. Because of this, we can reduce the case $q \geq p$ to the Besov space case $q = p$, which is easier. In addition, the simple geometry involved allows us to extend this result to the case where χ is the characteristic function of a domain in $\mathbf{R}^n$ of a certain type. In particular, Lipschitz domains as well as certain less smooth domains are included. Precise statements are given in [12].

In conclusion, we note that our basic results indicate that any problem regarding Lebesgue, Hardy, Sobolev, or, more generally, Triebel-Lizorkin spaces, implicitly involves the associated sequence spaces. Our applications have indicated that these sequence space norms reveal an elementary geometry that can be exploited. The extension of our results to other circumstances requires only a corresponding underlying geometry. Some possible extensions are discussed in [12].

References

[1] A. B. Alexandrov, A majorization property for the several variable Hardy-Stein-Weiss classes (in Russian), Vestnik Leningrad Univ. No. 13 (1982), 97-98.

[2] A. Baernstein II and E. Sawyer, Embedding and multiplier theorems for $H^p(\mathbf{R}^n)$, Mem. Amer. Math. Soc. 318 (1985), 1-82.

[3] J. Bergh and J. Löfström, *Interpolation Spaces: An Introduction*, Springer-Verlag, Berlin-Heidelberg-New York, 1976.

[4] H.-Q. Bui, Weighted Besov and Triebel spaces: Interpolation by the real method, Hiroshima Math. J. 12 (1982), 581-605.

[5] R. R. Coifman, A real variable characterization of H^p, Studia Math. 51 (1974), 269-274.

[6] H. Dym and H. P. McKean, *Fourier Series and Integrals*, Academic Press, Inc., New York, 1972.

[7] C. Fefferman, Harmonic analysis and H^p spaces, in *Studies in Harmonic Analysis*, J. M. Ash, editor, M.A.A. Studies in Mathematics, Vol. 13 (1976), 38-75.

[8] C. Fefferman, N. Rivière, and Y. Sagher, Interpolation between H^p-spaces, the real method, Trans. Amer. Math. Soc. 191 (1974), 75-81.

[9] C. Fefferman and E. M. Stein, Some maximal inequalities, Amer. J. Math. 93 (1971), 107-115.

[10] C. Fefferman and E. M. Stein, H^p spaces of several variables, Acta. Math. 129 (1972), 137-193.

[11] M. Frazier and B. Jawerth, Decomposition of Besov spaces, Indiana Univ. Math. J. 34 (1985), 777-799.

[12] M. Frazier and B. Jawerth, A discrete transform and decompositions of distribution spaces, to appear.

[13] P. Goupillaud, A. Grossmann, and J. Morlet, Cycle-octave and related transforms in seismic signal analysis, Geoexploration 23 (1984/85), 85-102.

[14] A. Grossman and J. Morlet, Decomposition of functions into wavelets of constant shape, and related transforms, to appear in *Mathematics and Physics, Lectures on Recent Results*, L. Streit, ed., World Scientific Publishing Co., Singapore.

[15] A. Grossman, J. Morlet, and T. Paul, Transforms associated to square integrable group representations I. General results, J. Math. Phys. 26 (1985), 2473-2479.

[16] A. Grossman and T. Paul, Wave functions on subgroups of the group of affine canonical transformations, (preprint).

[17] B. Jawerth, Some observations on Besov and Lizorkin-Triebel spaces, Math. Scand. 40 (1977), 94-104.

[18] B. Jawerth, The K-functional for H^1 and BMO, Proc. Amer. Math. Soc. 92 (1984), 67-71.

[19] B. Jawerth, R. Rochberg, and G. Weiss, Commutator and other second order estimates in real interpolation theory, to appear in Arkiv. för Mat.

[20] J. R. Klauder and B. S. Skagerstam, *Coherent States-Applications in Physics and Mathematical Physics*, World Scientific, 1984.

[21] R. Latter, A characterization of $H^p(\mathbf{R}^n)$ in terms of atoms, Studia Math. 62 (1978), 65-71.

[22] P. G. Lemarié and Y. Meyer, La transformation en ondelettes dans $\mathbf{R}^n$, manuscript, 1985.

[23] Y. Meyer, La transformation en ondelettes, manuscript, 1985.

[24] T. Paul, Affine coherent states and the radial Schrödinger equation I. Radial harmonic oscillator and hydrogen atom, (preprint).

[25] J. Peetre, On spaces of Triebel-Lizorkin type, Arkiv. för Mat. 13 (1975), 123-130.

[26] J. Peetre, *New Thoughts on Besov Spaces*, Duke Univ. Math. Series, Duke Univ., Durham, 1976.

[27] J. Peetre and G. Sparr, Interpolation of normed Abelian groups, Ann. Mat. Pura. Appl. 104 (1975), 187-207.

[28] E. M. Stein, *Singular Integrals and Differentiability Properties of Functions*, Princeton Univ. Press, Princeton, NJ, 1970.

[29] J.-O. Strömberg, Bounded mean oscillation with Orlicz norms and duality of Hardy spaces, Indiana Univ. Math. J. 28 (1979), 511-544.

[30] H. Triebel, *Theory of Function Spaces*, Monographs in Mathematics, Vol. 78, Birkhäuser Verlag, Basel, 1983.

[31] J. M. Wilson, On the atomic decomposition for Hardy spaces, Pacific J. Math. 116 (1985), 201-207.

1980 AMS Subject Classification: 46E35, 42B30

Key Words and Phrases: atomic decomposition, Hardy spaces, Sobolev spaces, Triebel-Lizorkin spaces, trace, interpolation

First author partially supported by NSF Grant DMS-8541317.
Second author partially supported by NSF Grants DMS-8403234 and 8604528.

Approximation by solutions of elliptic boundary value problems in various function spaces

U. Hamann and G. Wildenhain
Wilhelm-Pieck-Universität Rostock
Sektion Mathematik
Universitätsplatz 1
DDR-2500 Rostock

In the present paper an approximation problem is solved in a definitive way, first considered by H. Beckert [2] and further treated in papers by A. Göpfert [4], G. Anger [1], B.-W. Schulze and G. Wildenhain [9], G. Wanka [10], G. Wildenhain [11], [12], K. Beyer [3], U. Hamann [5], [6], [7], U. Hamann and G. Wildenhain [8].

(1) Let $\Omega \subset R^n$ be a bounded domain with a smooth boundary $\partial\Omega$, $U \subset \overline{U} \subset \Omega$ an open set and $V \subseteq \partial\Omega$ a subset of the boundary with inner points relative to $\partial\Omega$.

(2) $L = \sum_{|\alpha| \leq 2m} a_\alpha(x) \cdot D^\alpha$ is a properly elliptic differential operator with real coefficients in $C^\infty(R^n)$. On the boundary $\partial\Omega$ we consider a normal system $B = (B_1, \dots, B_m)$ of boundary operators with $\operatorname{ord} B_j \leq 2m-1$ and smooth coefficients. Further we suppose that L is covered on $\partial\Omega$ by the B_j.

Subject of our consideration are the sets

$$\mathfrak{M}_U(\Omega) := \{u \in C^\infty(\overline{\Omega}) : \operatorname{supp} Lu \subset U \ , \ Bu = 0\} \ ,$$

$$\mathfrak{N}_V(\Omega) := \{u \in C^\infty(\overline{\Omega}) : Lu = 0 \text{ in } \Omega \ , \ Bu|_{\partial\Omega \setminus V} = 0\} \ .$$

Let $\Gamma \subset \Omega$ be a (n-1)-dimensional, not necessary smooth surface, which splits up the domain Ω in an inner domain Ω_i and $\Omega_a = \Omega \setminus \overline{\Omega}_i$. We suppose $U \subset \overline{U} \subset \Omega_a$. Let $W^{m-1}(\Gamma)$ be the space of Whitney-Taylor fields of order m-1, i.e.

$$W^{m-1}(\Gamma) := \{g = (g_\alpha)_{|\alpha| \leq m-1} : g_\alpha \in C(\Gamma), \ \exists \ \varphi \in C^{m-1}(R^n) \text{ with } D^\alpha \varphi|_\Gamma = g_\alpha\}.$$

We want to prove the density of

$$\mathfrak{M}_U(\Gamma) := \{(D^\alpha u|_\Gamma)_{|\alpha|\le m-1} : u \in \mathfrak{M}_U(\Omega)\} \subset W^{m-1}(\Gamma) \quad \text{and}$$

$\mathfrak{N}_V(\Gamma) \subset W^{m-1}(\Gamma)$ ($\mathfrak{N}_V(\Gamma)$ is defined analogous) in $W^{m-1}(\Gamma)$.

(3) Let $W^{m-1}(\Gamma)$ be a Banach space with respect to the norm $\|g\| = \sum_{|\alpha|\le m-1} \|g_\alpha\|_{C(\Gamma)}$. This is a (weak) condition for Γ. Moreover we suppose the existence of absolutely continuous generalized harmonic measures with respect to the Dirichlet problem for Ω_i (see [9]) and that Γ fulfils the outer cone condition quasi everywhere. Quasi everywhere (q.e.) means "everywhere except a set of Wiener capacity zero".

(4) We suppose that the Dirichlet problem $Lu=0$ in Ω_i, $D^\alpha u|_\Gamma = g_\alpha$ ($|\alpha|\le m-1$; $g \in W^{m-1}(\Gamma)$) has an unique solution.

(5) We assume the existence of a global fundamental solution for the operator L in Ω.

(6) For the adjoint operator L^* the unique continuation property is assumed to hold in Ω_a in the following sense. If u is a solution of $L^*u=0$ in Ω_a, vanishing on a non-empty open subset $\omega \subset \Omega_a$, then u must be identically zero in Ω_a.

<u>Theorem 1</u>: Under the conditions (1)-(6) the density $\overline{\mathfrak{M}_U(\Gamma)} = \overline{\mathfrak{N}_V(\Gamma)} = W^{m-1}(\Gamma)$ holds.

P r o o f: For simplicity we suppose that the boundary value problem $Lu=0$ in Ω, $Bu=0$ only has the solution $u=0$. In the general case some technical modifications are to take into consideration.
Let $T \in (W^{m-1}(\Gamma))'$ be a continuous linear functional with $\langle T,u\rangle = 0$ for each $u \in \mathfrak{M}_U(\Gamma)$. We have to show $T=0$.

1. By (3) we have a representation $\langle T,u\rangle = \sum_{|\alpha|\le m-1} \int_\Gamma D^\alpha u(x)\, d\mu_\alpha(x)$ by measures, supported on Γ. Using the representation of $u \in \mathfrak{M}_U(\Omega)$ (i.e. $Lu=f \in C_0^\infty(U)$, $Bu=0$) by the Green function $G(x,y)$ of the boundary value problem (L,B) in Ω we get

$$D^\alpha u(x) = \int_U f(y) D_x^\alpha G(x,y)\,dy \quad \text{and} \quad \langle T,u\rangle = \int_U f(y) G^*\mu(y)\,dy = 0 \quad \text{for}$$

each $f \in C_0^\infty(U)$, where $G^*\mu(y) = \sum_{|\alpha|\le m-1} \int_\Gamma D_x^\alpha G(x,y)\, d\mu_\alpha(x)$ is a Green potential with respect to the tupel $\mu = (\mu_\alpha)$.

This means $G^*_{\mathcal{V}}\mu\equiv 0$ in U, and by (6) $G^*_{\mathcal{V}}\mu\equiv 0$ in Ω_a follows.

2. Using (5), classical estimates of the fundamental solution (see [9]) and standard results for Newton potentials with respect to general measures with compact support, we get $D^\beta G^*_{\mathcal{V}}\mu=0$ q.e. on Γ $(|\beta|\le m-1)$.

3. By our condition (4) the solution of the Dirichlet problem $L^*w=0$ in Ω_i, $D^\beta w|_\Gamma=g_\beta$ $(|\beta|\le m-1$, $g=(g_\beta)$ a given Whitney-Taylor field) can be represented by generalized harmonic measures (see [9]) in the form $w(z)=\sum_{|\beta|\le m-1}\int_\Gamma g_\beta(y)d\tau_z^\beta(y)$ $(z\in\Omega_i)$. Replacing g_β by $D^\beta G^*_{\mathcal{V}}\mu$, using 2., the Fubini-theorem and a generalized balayage principle (see [9]) we get $G^*_{\mathcal{V}}\mu(z)=0$ for all $z\in\Omega_i$.

4. Summarizing we have $G^*_{\mathcal{V}}\mu(z)=0$ q.e. in Ω such that $T=0$ easily follows.

5. $\overline{\mathcal{N}_V(\Gamma)}=W^{m-1}(\Gamma)$ can be proved by reduction to the previous case (see [11], [7]). □

In the case of the Laplace operator with Dirichlet conditions we get the assertion of Theorem 1 under the only supposition that quasi everywhere on Γ an outer cone condition is fulfilled.

Now we consider an other situation. The following results are given by the first author.

(3)' Let $\Gamma\in C^\infty$ a smooth open surface, $\overline{\Gamma}\subset\Omega$, $\dim\Gamma=n-1$ (for simplicity), $\Omega\setminus\overline{\Gamma}$ connected, $U\subset\overline{U}\subset\Omega\setminus\overline{\Gamma}$.

Let $F_j(\Gamma)$ $(j=1,\dots,2m)$ be suitable function spaces over Γ and $R_{2m-1}\mathcal{M}_U(\Omega):=\{R_{2m-1}u=(D_n^{j-1}u|_\Gamma)_{j=1,\dots,2m} : u\in\mathcal{M}_U(\Omega)\}$.
D_n is the normal derivation with respect to Γ. $R_{2m-1}\mathcal{N}_V(\Omega)$ is defined in a corresponding way.

(7) Let be exist an integer $s_o\ge 0$ such that $C^{s_o}(\overline{\Gamma})$ is continuously imbedded in $F_j(\Gamma)$ and $C^\infty(\overset{\circ}{\Gamma})$ is dense in $F_j(\Gamma)$ for $j=1,\dots,2m$.

Examples are $C^k(\overline{\Gamma})$, the Sobolev-Slobodeckij spaces $W_p^t(\Gamma)$ $(0\le t<\infty$, $1\le p<\infty)$ and the dual spaces $(W_p^t(\Gamma))'$ $(0\le t<\infty$, $1<p<\infty)$.

<u>Theorem 2</u>: Under the conditions (1),(2),(3)',(6) (with $\Omega\setminus\overline{\Gamma}$ instead of Ω_a), (7) the density

$$\overline{R_{2m-1}\mathcal{M}_U(\Omega)}=\overline{R_{2m-1}\mathcal{N}_V(\Omega)}=\prod_{j=1}^{2m}F_j(\Gamma)\quad\text{follows.}$$

Proof: We consider a

$$T \in \left(\prod_{j=1}^{2m} F_j(\Gamma)\right)' = \prod_{j=1}^{2m} (F_j(\Gamma))' \quad \text{with} \quad \langle T, R_{2m-1}u\rangle = 0 \qquad (*)$$

for each $u \in \mathfrak{M}_U(\Omega)$. In several steps we shall prove T=0.

1. Instead of the Green function in the proof of Theorem 1 here we use the general Green operator G of the boundary value problem as a continuous linear mapping from $\mathring{C}^{s+1-2m}(\Omega)$ into $C^s(\overline{\Omega})$, i.e. $G \in L(\mathring{C}^{s+1-2m}(\Omega), C^s(\overline{\Omega}))$ ($s \geq 2m-1$; $\mathring{C}^k(\Omega)$ is the closure of $C_0^\infty(\Omega)$ in $C^k(\overline{\Omega})$). The Green operator has the following properties:

- $LG\varphi = \varphi$ for all $\varphi \in C_0^\infty(\Omega)$ with $\varphi \perp N^*$
- $GL\varphi = \varphi + u_0$ for all $\varphi \in C_0^\infty(\Omega)$ with a suitable function $u_0 \in N$

(N and N^* are the kernels of the given resp. of the adjoint boundary value problem).

From (7) follows $\prod_{j=1}^{2m} (F_j(\Gamma))' \subset \prod_{j=1}^{2m} (C^{s-j+1}(\overline{\Gamma}))'$ for a suitable integer $s \geq 2m-1$. We have $R_{2m-1} \in L(C^s(\overline{\Omega}), \prod_{j=1}^{2m} C^{s-j+1}(\overline{\Gamma}))$. Considering T as an element of $\prod_{j=1}^{2m} (C^{s-j+1}(\overline{\Gamma}))'$ we get $G'R'_{2m-1}T \in (\mathring{C}^{s+1-2m}(\Omega))'$ (G' and R'_{2m-1} are dual operators). $G'R'_{2m-1}T$ corresponds to the Green potential in the first step of the proof of Theorem 1.

2. From (*) and from the properties of the Green operator we get $\langle G'R'_{2m-1}T, Lu\rangle = \langle T, R_{2m-1}u\rangle$ for all $u \in \mathring{C}^{s+1}(\Omega)$. It follows $L^*(G'R'_{2m-1}T) = 0$ on $\Omega \setminus \overline{\Gamma}$. Further we conclude $G'R'_{2m-1}T = w_0 \in N^*$ on U. The last equality corresponds to $G^*_\nu \mu \equiv 0$ in U in the proof of Theorem 1. There we have $w_0 \equiv 0$.

3. Because $L^*(G'R'_{2m-1}T - w_0) = 0$ in $\Omega \setminus \overline{\Gamma}$ and $G'R'_{2m-1}T - w_0 \equiv 0$ in $U \subset \Omega \setminus \overline{\Gamma}$, from (3)' and (6) we get $G'R'_{2m-1}T - w_0 \equiv 0$ in $\Omega \setminus \overline{\Gamma}$. Moreover, using elementary density properties, $G'R'_{2m-1}T - w_0 \in (\mathrm{Ker}(R_{s+1-2m}))^\perp$ follows.

4. From the last result, from a suitable continuation theorem and from the closed range theorem we conclude that the equation $R'_{s+1-2m}\Lambda = G'R'_{2m-1}T - w_0$ has an unique solution $\Lambda = (\lambda_j)_{j=1,\dots,s+2-2m} \in \prod_{j=1}^{s+2-2m} (C^{s+2-2m-j}(\overline{\Gamma}))'$, i.e. we have

$$\langle G'R'_{2m-1}T - w_0, h\rangle = \sum_{j=1}^{s+2-2m} \langle \lambda_j, D_n^{j-1} h|_\Gamma\rangle \quad \text{for each } h \in \mathring{C}^{s+1-2m}(\Omega).$$

Hereby we deduce

$$\sum_{j=1}^{s+2-2m} \langle \lambda_j, D_n^{j-1}(Lu)|_\Gamma \rangle = \langle G'R'_{2m-1}T - w_o, Lu \rangle = \langle T, R_{2m-1}u \rangle$$

for each $u \in \mathring{C}^{s+1}(\Omega)$.

5. Step by step, using once more an appropriate technical continuation theorem, we prove $\lambda_j = 0$ for $j=1,\dots,s+2-2m$, i.e. $\langle T, R_{2m-1}u \rangle = 0$ for each $u \in \mathring{C}^{s+1}(\Omega)$. Because $\prod_{j=1}^{2m} C^\infty(\bar{\Gamma})$ is dense in $\prod_{j=1}^{2m} F_j(\Gamma)$, from this $\langle T, f \rangle = 0$ for each $f = (f_j)_{j=1,\dots,2m} \in \prod_{j=1}^{2m} F_j(\Gamma)$ follows, i.e. $T = 0$. The density $\overline{R_{2m-1}\mathcal{N}_V(\Omega)} = \prod_{j=1}^{2m} F_j(\Gamma)$ can be reduced to the case above. □

The result of Theorem 2 is best possible in the sense that elements of $\prod_{j=1}^{l} F_j(\Gamma)$ for $l \geq 2m$ cannot be approximated by $R_l u$ ($u \in \mathcal{M}_U(\Omega)$ resp. $u \in \mathcal{N}_V(\Omega)$).

If we consider the situation of Theorem 1, namely $\Gamma \in C^\infty$, Γ closed, $\Omega = \Omega_i \cup \Omega_a \cup \Gamma$, Ω_i and Ω_a connected and $U \subset \bar{U} \subset \Omega_a$, then, roughly speaking, a result analogous Theorem 2 can be proved, replacing R_{2m-1} by R_{m-1}. Also this result is best possible in the sense that it fails for R_l with $l \geq m$.

Finaly we observe that the condition (6) in some sense is necessary for our density statements. So we get an analytical characterization of the unique continuation property.

References

1. Anger, G.: Eindeutigkeitssätze und Approximationssätze für Potentiale II. Math. Nachr. 50, 229-244 (1971).
2. Beckert, H.: Eine bemerkenswerte Eigenschaft der Lösungen des Dirichletschen Problems bei linearen elliptischen Differentialgleichungen. Math. Ann. 139, 255-264 (1960).
3. Beyer, K.: Approximation durch Lösungen elliptischer Randwertprobleme. Rostock. Math. Kolloq. 26, 27-34 (1984).
4. Göpfert, A.: Über L_2-Approximationssätze - eine Eigenschaft der Lösungen elliptischer Differentialgleichungen. Math. Nachr. 31, 1-24 (1966).

5. Hamann, U.: Zur Approximation in Sobolev-Räumen durch Lösungen elliptischer Randwertprobleme. Math. Nachr. 123, 73-82 (1985).
6. Hamann, U.: Approximation durch Normalableitungen von Lösungen elliptischer Randwertprobleme in beliebigen Sobolev-Räumen. Math. Nachr. (in print).
7. Hamann, U.: Approximation durch Lösungen allgemeiner elliptischer Randwertprobleme bei Gleichungen beliebiger Ordnung. Dissertation Wilhelm-Pieck-Universität Rostock 1986.
8. Hamann, U.; Wildenhain, G.: Approximation by solutions of elliptic equations. Z. Anal. Anw. 5(1), 59-69 (1986).
9. Schulze, B.-W.; Wildenhain G.: Methoden der Potentialtheorie für elliptische Differentialgleichungen beliebiger Ordnung. Berlin: Akademie-Verlag 1977, Basel-Stuttgart: Birkhäuser-Verlag 1977.
1o. Wanka, G.: Gleichmäßige Approximation durch Lösungen von Randwertproblemen elliptischer Differentialgleichungen zweiter Ordnung. Beiträge zur Analysis 17, 19-29 (1981).
11. Wildenhain, G.: Uniform approximation by solutions of general boundary value problems for elliptic equations of arbitrary order I,II. Z. Anal. Anw. 2(6), 511-521 (1983), Math. Nachr. 113, 225-235 (1983).
12. Wildenhain, G.: Approximation in Sobolev-Räumen durch Lösungen allgemeiner elliptischer Randwertprobleme bei Gleichungen beliebiger Ordnung. Rostock. Math. Kolloq. 22, 43-56 (1983).

INTERPOLATION OF SUBSPACES AND QUOTIENT SPACES BY THE COMPLEX METHOD

Eugenio Hernandez[1]
Department of Mathematics
Universidad Autonoma de Madrid
Madrid - 34
Spain

Richard Rochberg[2]
Department of Mathematics
Washington University
St. Louis, MO 63130
USA

Guido Weiss[3]
Department of Mathematics
Washington University
St. Louis, MO 63130
USA

§1. Introduction and Summary. We begin with an informal statement of our viewpoint and results. More precise formulations of the results are in the later sections. More detail on the background material is in [2], [3] and [14].

The theory of complex interpolation of Banach spaces is a method of constructing norms on a family of vector spaces so that the operator norms of an analytic family of maps into or out of the family of spaces will satisfy a maximum principle. More

---

[1]Supported in part by Spanish-American Grant CCB-8402-058
[2]Supported in part by NSF Grant DMS-8402191
[3]Supported in part by NSF Grant DMS-8200884

precisely, let Δ be the open unit disk in the complex plane. Suppose that for each point ζ in the boundary $\partial\Delta$ we are given a Banach space C_ζ. In this paper we will only consider finite dimensional spaces so $C_\zeta=(\mathbb{C}^n,N_\zeta)$ for some norming function N_ζ. The goal is to construct intermediate spaces $C_z=(\mathbb{C}^n,N_z)$, z in Δ, with several properties. First, if X is any fixed Banach space and T_z, $z\in\bar{\Delta}$, is an analytic family of linear maps of X into $\mathbb{C}^n$ which satisfies

$$(1.1) \qquad N_z(T_z(x)) \le \|x\|_X \quad \forall x\in X$$

for all z in $\partial\Delta$; then (1.1) persists for all z in Δ. In short, the operator norm of T satisfies the maximum principle. ((1.1) is the maximum principle for the case where the operator norm of T is one on the boundary. The general case where estimates on T vary from point to point in $\partial\Delta$ can be reduced to this by a function theoretic device presented in [2].)

Second, we want estimates for maps out of the spaces normed by the N_z. That is, if T_z, $z\in\bar{\Delta}$, is an analytic family of maps from $\mathbb{C}^n$ to a fixed Banach space Y such that

$$(1.2) \qquad \|T_z(v)\|_Y \le N_z(v) \quad \forall v\in\mathbb{C}^n,$$

for all z in $\partial\Delta$, then (1.2) also holds for all z in Δ.

Finally, as z in Δ converges to ζ in $\partial\Delta$, the functions N_z should converges to N_ζ.

Although conditions (1.1) and (1.2) push in opposite directions they can both be met. For u in $\mathbb{C}^n$ define

$$(1.3) \quad N_z(u) = \inf\{\textstyle\int_0^{2\pi}[N_\zeta(F(\zeta))]^2 P_z(\theta)d\theta)^{1/2}: F\in\mathcal{A}, F(z)=u\}.$$

<u>Here and throughout,</u> <u>$\zeta=e^{i\theta}$</u>, P_z is the Poisson kernel, and $\mathcal{A}$ is

the family of all $\mathbb{C}^n$ valued analytic functions on Δ which satisfy a certain growth condition near $\partial\Delta$. N_z satisfies all of the requirements and is the unique function that does so.

The construction of N_z settles the questions which were raised but there are further, more refined, questions related to (1.1) and (1.2). First consider (1.1). Instead of considering all analytic families of linear maps into $\mathbb{C}^n$ we could consider only those which have their range contained in certain subspaces. That is, suppose we are given, for each ζ in $\partial\Delta$, a subspace D_ζ contained in $\mathbb{C}^n$. We now want to construct a function N_z, $z \in \Delta$, so that whenever T_z is an analytic family of linear maps of a fixed Banach space X into $\mathbb{C}^n$ which satisfies (1.1) for all z in $\partial\Delta$ and which satisfies the additional condition

$$(1.4) \qquad T_\zeta(x) \in D_\zeta \quad \forall x \in X \quad \text{a.e. } d\theta$$

then (1.1) persists for all z in Δ. A fortiori the function N_z just described will have the required property. Now, however, there is a change. The functions defined by (1.3) are the largest functions which satisfy the maximum principle for all T_z. If we only want this maximum principle for T which also satisfy (1.4) then we can replace the function N_z given by (1.3) with the larger

$$(1.5) \qquad N_z(u) = \inf\{\left(\int_0^{2\pi}[N_\zeta(F(\zeta))]^2 P_z(\theta)d\theta\right)^{1/2} : F\in\mathcal{A},\ F(z)=u,\ F(\zeta)\in D_\zeta \text{ a.e. } d\theta\}.$$

(In contrast to the function defined by (1.3) this is not a norm on $\mathbb{C}^n$). We study the function N_z defined by (1.5) in detail in section 2. We will see that the study splits naturally into two parts. First, we must understand the <u>intermediate subspaces</u>, D_z, consisting of those vectors u in $\mathbb{C}^n$ for which the set $\{F\in\mathcal{A},$

$F(z)=u$, $F(\zeta)\in D_\zeta$ a.e. $d\theta\}$ is not empty (i.e. the subspace on which N_z is finite). This can be a complicated issue. For example, the dimension of the D_z can be strictly smaller than that of the D_ζ. If, for instance, n=2 and D_ζ is spanned by (1,0) on half the circle and spanned by (0,1) on the other half, then $\mathcal{A}$ contains only the zero function and hence the D_z are zero dimensional. To study the D_z it is enough to consider the case in which the N_ζ are all the ordinary Euclidean norm, $\| \|_{\mathbb{C}^n}$. Thus we set $\mathcal{A} = H^2_D = \{F\in H^2_{\mathbb{C}^n}: F(\zeta)\in D_\zeta$ a.e.$\}$. Here $H^2_{\mathbb{C}^n}$ denotes the Hardy space of $\mathbb{C}^n$-valued functions. In this case (1.5) becomes

$$(1.6)\quad N_z(u) = \inf\{(\int_0^{2\pi} \|F(\zeta)\|_{\mathbb{C}^n}^2 P_z(\theta)d\theta)^{1/2}: F\in H^2_D,\ F(z)=u\}.$$

An advantage of (1.6) is that we can (and will) use Hilbert space techniques extensively.

We will find that associated with the extremal problems (1.5) and (1.6) are <u>extremal functions</u>, That is given z in Δ and u in D_z there is $F_{z,u} \in \mathcal{A}$ such that $F_{z,u}(z) = u$ and

$$(1.7)\qquad N_z(u) = (\int_0^{2\pi} [N_\zeta(F_{z,u}(\zeta))]^2 P_z(\theta)\ d\theta)^{1/2};$$

moreover,

$$(1.8)\qquad N_\zeta(F_{z,u}(\zeta)) = N_z(u) \quad \text{a.e.}$$

The fact that these extremal functions have "constant modulus" on the boundary is similar to what happens in the classical Szego extremal problem. In fact, when n=1 (1.5) reduces to the classical Szego problem. Studies of the role of vector valued versions of Szego's theorem in vector valued Hardy space theory are in the book of Rosenblum and Rovnyak [15]. This work can be seen as part of that program.

Using the functions $F_{z,v}$ associated with (1.6) we will find an integer k and will construct a map $U(z)$ which is an analytic function of z, $z \in \Delta$, and, for fixed z, is a one to one linear map of $\mathbb{C}^k$ into $\mathbb{C}^n$ and which is onto D_z. That is

(1.9) $$U(z)\mathbb{C}^k = D_z.$$

If we regard $U(z)$ as an $n\times k$ matrix, then the coefficients of $U(z)$ belong to H^∞ and we have the isometric property

(1.10) $$\|U(\zeta)v\|_{\mathbb{C}^n} = \|v\|_{\mathbb{C}^k} \quad \forall v\in\mathbb{C}^k \quad \text{a.e. } d\theta$$

Moreover, if $u = U(0)v$, then $F_{o,u}(z) = U(z)v$.

It turns out that the same algebraic description of D_z, (1.9), is valid when each subspace D_ζ is endowed with a general Banach space norm N_ζ. This allows us to "pull back" the intermediate subspace construction to the <u>full range</u> case on $\mathbb{C}^k$. That is, we can reduce the study of (1.5) on $\mathbb{C}^n$ to the study of (1.3) on $\mathbb{C}^k$. More precisely, we can use the norms $M_\zeta(v) = N_\zeta(U(\zeta)v)$ on $\mathbb{C}^k$ to obtain an interpolation family $(\mathbb{C}^k, M_\zeta)$ on the boundary $\partial\Delta$. The intermediate norms M_z given by (1.3) on $\mathbb{C}^k$ then give the norms N_z for the intermediate subspaces D_z in $\mathbb{C}^n$ by

(1.11) $$N_z(u) = M_z(U(z)^{-1}u)$$

for each $u \in D_z$ and $z \in \Delta$.

Now consider (1.2). N_z given by (1.3) is the smallest function which will satisfy (1.2) for all choices of T. If, however, we also require that the $\ker(T_\zeta)$ contain subspaces D^ζ of $\mathbb{C}^n$, then we can replace N_z in (1.2) with N^z defined by

(1.12) $$N^z(u) = \inf\{\left(\int_0^{2\pi}[N^\zeta(F(\zeta))]^2 P_z(\theta)d\theta\right)^{1/2}: F\in\mathcal{A}, F(z)=u\}.$$

Here N^ζ is the quotient norm on $Q^\zeta = \mathbb{C}^n/D^\zeta$ (regarded as a seminorm on $\mathbb{C}^n$). We can, in fact, consider such N^z when N^ζ is any given family of seminorms. We study this construction in detail in section 3. The intermediate functional N^z is also a seminorm. We call this construction the <u>interpolation of families of quotient spaces</u>. The reason for the name is that the general seminorm N^ζ defines a norm on the quotient $\mathbb{C}^n/D^\zeta$ where $D^\zeta = \{v \in \mathbb{C}^n: N^\zeta(v) = 0\}$ is the kernel of the seminorm N^ζ. Thus we are interpolating the normed quotient spaces $(\mathbb{C}^n/D^\zeta, N^\zeta)$.

The interpolated quotient spaces give a natural realization of the (isometric) duals of the interpolated subspaces. This duality manifests itself in various aspects of the theory. For instance, we will see that the extremal functions exist for (1.12). Using the extremal functions we obtain an integer k, a norm N on $\mathbb{C}^{n-k}$, and a function $V(z)$, $z \in \Lambda$, which depends analytically on z and which for each z is a linear map of $\mathbb{C}^n$ into $\mathbb{C}^{n-k}$, so that

$$N^z(w) = N(V(z)w). \tag{1.13}$$

when $w \in \mathbb{C}^n$ and $z \in \Lambda$ (compare with equality (1.11)).

(1.11), (1.13), and the isometric duality between interpolation families of subspaces and of quotient spaces are our main conclusions. (We postpone the precise formulation of the duality until section 3 since it requires more notation.) Many aspects of the theory follow in routine ways from these results. In some of these cases we indicate the derivations but in other cases we are quite sketchy.

When T_ζ maps into D_ζ then the interpolation theorem, i.e. the maximum principle described by (1.1), can be improved by replacing N_z as defined by (1.3) with the N_z defined using (1.5). The quotient space interpolation theory lets us improve the maximum principle related to (1.2). For instance, if the analytic family T_ζ of operators in (1.2) satisfies $\ker(T_\zeta) \supset D^\zeta$, for ζ in $\partial\Delta$ then

(1.14) $$\|T_z(v)\|_Y \leq N^z(v) \quad \forall v \in \mathbb{C}^n$$

for $z \in \Delta$ and $v \in \mathbb{C}^n$ (which is a stronger conclusion than (1.2)).

One of the consequences of the construction induced by (1.3) is the Wiener-Masani theorem. This theorem asserts that a rather general positive definite matrix valued function $B(\zeta)$ on $\partial\Delta$ has the factorization

(1.15) $$B(\zeta) = A(\zeta)A^*(\zeta), \text{ for a.e. } \zeta \in \partial\Delta,$$

where $A(\zeta)$ are the boundary values of a n by n analytic matrix valued function $A(z)$, $z \in \Delta$, and $A^*(\zeta)$ are the adjoints of the $A(\zeta)$. We shall show that a similar factorization is true for positive <u>semi</u>-definite matrix valued functions $B(\zeta)$. However in that case the $A(z)$ will be rectangular matrices. We obtains a factorization (1.15) in the semi-definite case when the ranges D_ζ of the operators $B(\zeta)$ vary in a "smooth" way (that is, when $D = \{D_\zeta\}$ is an <u>analytic range</u>, a concept we define and study in §2). If the kernels of $B(\zeta)$ form an analytic range, the quotient space theory gives us the factorization

(1.16) $$B(\zeta) = A^*(\zeta)A(\zeta).$$

As we shall explain in §2, either (1.15) or (1.16) are true in the full range case (not necessarily with the same $A(\zeta)$).

The last section contains a few comments and questions, and mentions connections of this work to other topics (e.g. invariant subspaces, polynomial hulls).

We are grateful to our colleagues Bjorn Jawerth and Yves

Meyer who have gone through all this material and proposed many valuable suggestions.

§2. Interpolation of subspaces. Suppose $D = \{D_\zeta\}$ is a family of subspaces associated with the points $\zeta \in \partial\Delta$. The *intermediate subspaces* D_z, $z \in \Delta$, that are determined by D, consist of the values $F(z)$ of those functions F belonging to H^2_D:

$$D_z = \{F(z): F \in H^2_D\}. \tag{2.1}$$

In the firdt part of this section we study the case where the norm of $u \in D_z$ is given by (1.6). It is easy to see that there exists a unique extremal function $F_{z,u} \in H^2_D$ for which the infimum in (1.6) is attained, that is

$$N_z(u) = \left(\int_0^{2\pi} \|F_{z,u}(\zeta)\|^2_{\mathbb{C}^n} P_z(\theta)\, d\theta\right)^{1/2} \equiv \|F_{z,u}\|_z. \tag{2.2}$$

We let Π_z be the projection of H^2_D onto the orthogonal complement of $H_z = \{F \in H^2_D: F(z) = 0\}$, where we are using the Hilbert spaces structure induced by the norm $\|\ \|_z$ in (2.2). It is clear that Π_z is constant on the coset $F + H_z$ determined by the condition $F(z) = u$ and that $F_{z,u} = \Pi_z F$ for any F in this coset. These extremal functions have a very simple characterization:

Lemma (2.3). *$F \in H^2_D$ is an extremal function associated with $z \in \Delta$ and $u = F(z)$ if and only if $F \in H_z^\perp$.*

Corollary (2.4). *If $F_{z,u}$ is the extremal function associated with $z \in \Delta$ and $u \in D_z$ then it is also the extremal function associated with $z_1 \in \Delta$ and $u_1 = F_{z,u}(z_1) \in D_{z_1}$.*

Proof By (2.3) $F = F_{z,u}$ is extremal if and only if $\int_0^{2\pi} (F(\zeta), G(\zeta))\overline{(\zeta - z)} P_z(\theta) d\theta = 0$ for all $G \in H^2_D$. Since

$$P_z(\theta) = \frac{1}{2\pi} \frac{1 - |z|^2}{(1-\bar{\zeta}z)(1-\zeta\bar{z})}$$

this condition is equivalent to

$$\int_0^{2\pi}(F(\zeta),G(\zeta))\frac{d\theta}{\zeta-z} = 0 \tag{2.5}$$

for all $G \in H_D^2$. If $H \in H_D^2$ let $G(\zeta) = \dfrac{1-\zeta\bar{z}}{1-\zeta\bar{z}_1} H(\zeta)$. Then $G \in H_D^2$ and, therefore, satisfies (2.5). But this means that

$$\int_0^{2\pi}(F(\zeta),H(\zeta))\frac{d\theta}{\zeta-z_1} = 0$$

for all $H \in H_D^2$, which is equivalent to $F \perp H_{z_1}^{\perp}$; (2.4) now follows from (2.3).

<u>Corollary</u> (2.6). <u>If</u> $v \in D_z$ <u>then</u> $F = F_{z,v}$ <u>cannot assume the value</u> 0 <u>in</u> Δ <u>unless</u> $v = 0$. <u>Thus, an extremal function is either never zero or identically zero</u>.

<u>Proof</u>. If $F(z_1) = 0$ then, by (2.4), $F = F_{z_1,0}$ and, by (2.2), $\|F\|_{z_1} = N_{z_1}(0) = 0$. But this means $F(\zeta) = 0$ for a.e. $\zeta \in \partial\Delta$.

The extremal is given by an orthogonal projection; hence

<u>Lemma</u> (2.7). N_z <u>is a Hilbert space norm. The mapping</u> $v \to F_{z,v}$ <u>from</u> D_z <u>into</u> H_D^2 <u>is linear</u>.

Another consequence of (2.5) is the following constancy property of extremals (compare with (1.8)):

<u>Corollary</u> (2.8). $\|F_{z,v}(\zeta)\|_{\mathbb{C}^n} = N_z(v)$ <u>for a.e</u> $\zeta \in \partial\Delta$.

<u>Proof</u> Let $F = F_{z,v}$. By (2.4) we can assume $z = 0$. By (2.5)

$$0 = \int_0^{2\pi} (F(\zeta),G(\zeta))\frac{d\theta}{\zeta} = \int_0^{2\pi} (F(\zeta),\zeta G(\zeta))\, d\theta \tag{2.9}$$

for all $G \in H_D^2$. In particular, let $G(\zeta) = p(\zeta)F(\zeta)$, where p is any polynomial. Then (2.9) becomes $0 = \int_0^{2\pi}(F(\zeta),F(\zeta))\overline{\zeta p(\zeta)}d\theta$ for all polynomials $q(\zeta) = \zeta p(\zeta)$ of mean zero (since $F \in H^2_D$ we have $p(\zeta)F(\zeta) \in H^2_D$). This means that the $L^1(\partial\Delta)$ function $\|F(\zeta)\|^2_{\mathbb{C}^n}$ is conjugate analytic and real-valued. This implies it is constant and (2.8) follows.

Corollary (2.10). *Suppose* $\{v_1,v_2,\ldots,v_k\}$ *is an orthonormal basis of* D_z *and* $F_1,F_2,\ldots,F_k$ *are the associated extremal functions, then* $\{F_1(\zeta),F_2(\zeta),\ldots,F_k(\zeta)\}$ *is an orthonormal system in* $\mathbb{C}^n$ *for almost every* $\zeta \in \partial\Delta$.

This is an immediate consequence of lemma (2.7), corollary (2.8) and polarization. This same argument together with (2.4) allows us to extend (2.10) to $z \in \Delta$:

Corollary (2.11). *Suppose* $F_1,F_2,\ldots,F_k$ *are extremal functions associated with an orthonormal basis* $\{v_1,v_2,\ldots,v_k\}$ *of* D_{z_o}, *then* $\{F_1(z),F_2(z),\ldots,F_k(z)\}$ *is an orthonormal basis of* D_z *for all* $z \in \Delta$.

That the last orthonormal system forms a basis of D_z follows from the following observations. Let $e_1 = (1,0,\ldots,0)$, $e_2 = (0,1,\ldots,0),\ldots$, $e_k = (0,0,\ldots,1)$ be the canonical basis elements of $\mathbb{C}^k$, where k is the dimension of the intermediate subspace D_{z_o}. Since the dimension of each D_z cannot exceed n, we can choose z_o so that D_{z_o} has maximal dimension. An immediate consequence of the orthonormality in (2.11) is that $\dim D_z = k$ for all $z \in \Delta$. Hence we can assume that $z_o = 0$. Let $\{v_1,v_2,\ldots,v_k\}$ be an orthonormal basis of D_0 and $F_1,F_2,\ldots,F_k$ the corresponding extremal functions. We then define a linear transformation $U(z)$: $\mathbb{C}^k \to \mathbb{C}^n$ by letting $U(z)e_j = F_j(z)$ and then extending $U(z)$

linearly on $\mathbb{C}^k$. By (2.11) $U(z)\mathbb{C}^k = D_z$ for all $z \in \Delta$. Suppose $u = \Sigma_1^k u_j e_j \in \mathbb{C}^k$, then, using (2.7) and the definition of $U(z)$ we see that $U(z)u = \Sigma_1^k u_j F_j(z)$ is the extremal function associated with $v = \Sigma_1^k u_j v_j \in D_o$ and $0 \in \Delta$. By (2.8) we then have

$$\text{(2.12)} \qquad \|U(\zeta)u\|^2_{\mathbb{C}^n} = [N_o(v)]^2 = \Sigma_1^k |u_j|^2 = \|u\|^2_{\mathbb{C}^k}$$

for almost every $\zeta \in \partial\Delta$.

These facts give a rather complete description of the intermediate subspaces:

<u>Theorem</u> (2.13). <u>Suppose</u> $D = \{D_\zeta\}$ <u>is a family of subspaces associated with the points of</u> $\partial\Delta$, <u>then either</u> $D_z = \{0\}$ <u>for all</u> $z \in \Delta$ <u>or for some integer</u> k, $1 \leq k \leq n$, <u>there exists an</u> $n\times k$ <u>matrix</u> $U(z)$ <u>whose coefficients belong to</u> H^∞ <u>such that</u> $U(z)\mathbb{C}^k = D_z$ <u>for all</u> $z \in \Delta$. <u>Furthermore</u> $U(z)$ <u>satisfies the isometry property</u> (2.12) <u>and is a one-one operator on</u> $\mathbb{C}^k$ <u>into</u> $\mathbb{C}^n$.

The operator $U(z)$ also gives us a simple characterization of the space H^2_D:

<u>Theorem</u> (2.14). $H^2_D = UH^2_{\mathbb{C}^k} = \{F(z) = U(z)G(z): G \in H^2_{\mathbb{C}^k}\}$.

<u>Proof</u> If $G = \Sigma_1^k g_j e_j \in H^2_{\mathbb{C}^k}$, then $U(z)G(z) = \Sigma_1^k g_j(z)U(z)e_j = \Sigma_1^k g_j(z)F_j(z)$ clearly belongs to $H^2_{\mathbb{C}^n}$ (since $|F_j(z)| \leq 1$, $g_j \in H^2$) and $F_j(\zeta) \in D_\zeta$ for a.e. ζ. This shows that $UH^2_{\mathbb{C}^k} \subset H^2_D$.

Suppose there exists an $F \in H^2_D \cap (UH^2_{\mathbb{C}^k})^\perp$. We must first show that the vectors $F(\zeta), F_1(\zeta), \ldots, F_k(\zeta)$ are linearly dependent for a. e. $\zeta \in \partial\Delta$: Since D_z is spanned by $F_1(z), \ldots, F_k(z)$, this is certainly true if we replace ζ by $z \in \Delta$. Thus, all the $(k+1)$ minor determinants of the $n\times(k+1)$ matrix with columns $F(z)$, $F_1(z), \ldots, F_k(z)$ must be 0. Since these determinants are constructed by H^2 functions this fact about $(k+1)$ minor determinants must be true for a. e. $\zeta \in \partial\Delta$. Thus, we must have

$$F(\zeta) = \Sigma_1^k a_j(\zeta)F_j(\zeta)$$

with $a_j(\zeta) = (F(\zeta), F_j(\zeta))$. Since $\|F_j(\zeta)\|_{\mathbb{C}^n} = 1$ a.e. and $F(\zeta) \in L^2_{\mathbb{C}^n}(\partial\Delta)$ we must have $a_j \in L^2(\partial\Delta)$. We claim that $a_j(\zeta) = 0$ for a.e. $\zeta \in \partial\Delta$.

By (2.3) $F_j \in H_o^{\perp}$ (recall that we are using $z = 0$), $j = 1,2,\ldots,k$, and $z^m F(z) \in H_o$ when m is a positive integer. Thus, $\int_o^{2\pi} e^{im\theta} a_j(e^{i\theta})d\theta = \int_o^{2\pi}(e^{im\theta}F(e^{i\theta}), F_j(e^{i\theta}))d\theta = 0$, $j = 1,2,\ldots,k$. If $m \geqq 0$ we can make use of our assumption $F \in (UH^2_{\mathbb{C}}k)^{\perp}$ to conclude as well that

$$\int_o^{2\pi} e^{-im\theta} a_j(e^{i\theta})d\theta = \int_o^{2\pi}(F(e^{i\theta}), U(e^{i\theta})e^{im\theta}e_j)d\theta = 0,$$

$j = 1,2,\ldots,k$. Thus, for any integer m, $\int_o^{2\pi} a_j(e^{i\theta})e^{-im\theta} = 0$. Hence, $a_j(\zeta) = 0$ a.e. It follows that $F(\zeta) = 0$ a. e. and, consequently, $UH^2_{\mathbb{C}^k} \supset H^2_D$.

Other properties of intermediate spaces and their norms follow from these observations. An easy consequence of (2.4) and (2.8) is the constancy of the norm of an extremal function F:

(2.15) $$N_{z_1}(F(z_1)) = N_{z_2}(F(z_2))$$

for all $z_1, z_2 \in \Delta$.

We now turn our attention to the relation between the family of boundary subspaces $D = \{D_\zeta\}$ and the intermediate subspaces D_z. We noted in the first section that the dimension of D_z can be zero even if the D_ζ are non-trivial. If we want the interior spaces to be non-trivial and to converge to the boundary spaces D_ζ, the latter must vary analytically. We now make that precise using the results we have obtained.

If $D = \{D_\zeta\}$ let $U(\zeta)$ be the (non-tangential) boundary values of the operator $U(z)$ of theorem (2.13). Let κD be the family

$\{U(\zeta)\mathbb{C}^k\}$. Since $U(\zeta)\mathbb{C}^k$ is spanned by $F_1(\zeta),\ldots,\ F_k(\zeta)$ a.e., it follows that $U(\zeta)\mathbb{C}^k \subset D_\zeta$ for a.e. $\zeta \in \partial\Lambda$. Thus, $H^2_{\kappa D} \subset H^2_D$. On the other hand, if $F \in H^2_D$ then, by (2.14), $F = UG$ with $G \in H^2_{\mathbb{C}}k$. That is, writing $G = (g_1,\ldots,g_k)$ with each $g_j \in H^2$, we must have $F(\zeta) = U(\zeta)\Sigma_1^k g_j(\zeta)e_j \in U(\zeta)\mathbb{C}^k$. But this means that $F \in H^2_{\kappa D}$. Thus, $H^2_D \subset H^2_{\kappa D}$ and we have shown that

$$(2.16) \qquad H^2_D = H^2_{\kappa D}.$$

Equality (2.16) tells us that the intermediate subspaces D_z obtained from the family $D = \{D_\zeta\}$ are the same as those obtained from the family $\kappa D = \{U(\zeta)\mathbb{C}^k\}$; that is,

$$(2.17) \qquad D_z = (\kappa D)_z$$

for all $z \in \Lambda$. We can also say that the boundary spaces $U(\zeta)\mathbb{C}^k \in \kappa D$ are the non-tangential limits of the intermediate spaces D_z in the sense that we can find an orthonormal basis $\{F_1(z),\ldots,F_k(z)\}$ of D_z converging non-tangentially to the orthonormal basis $\{F_1(\zeta),\ldots,F_k(\zeta)\}$ of $U(\zeta)\mathbb{C}^k$ for a.e. $\zeta \in \partial\Lambda$.

These observations show how the boundary spaces must vary in order to have an <u>analytic range</u> $\{D_\zeta\}$ on $\partial\Lambda$. Suppose $T(z)$ is an $n\times k$ matrix valued function whose entries are in H^2. Then, the boundary values $T(\zeta)$, $\zeta \in \partial\Lambda$, exist a.e. in the nontangential sense. Let us assume also that the operator from $\mathbb{C}^k$ to $\mathbb{C}^n$ induced by $T(z)$ is one-one for each $z \in \Lambda$. By an argument similar to the one used in the proof of theorem (2.14), the spaces $D_\zeta = T(\zeta)\mathbb{C}^k$ must be k dimensional for a.e. $\zeta \in \partial\Lambda$ (as are each of the spaces $D_z = T(z)\mathbb{C}^k$). If $D = \{D_\zeta\}$ is a family of boundary spaces, each of the form $D_\zeta = T(\zeta)\mathbb{C}^k$, then we say that D is an analytic range and T is called a <u>range function</u> (In this case it also follows that

$D_z = T(z)$ for $z \in \Delta$). One of the consequences of theorem (2.13) is that $T(z)$ can be chosen with coefficients in H^∞ and with boundary values satisfying the isometry property (2.12) a.e..

Another characterization of analytic ranges can be given in terms of the family κD we associated with a family D. Given two families $D^j = \{D^j_\zeta\}$, $j = 1,2$, of subspaces we say that $D^1 \subset D^2$ if and only if $D^1_\zeta \subset D^2_\zeta$ for a.e. $\zeta \in \Lambda$. Suppose $D = \{D_\zeta\}$ is given, then, clearly, $\kappa D \subset D$ and it is not hard to check that κD is the largest analytic range contained in D (this follows from (2.14) and (2.16)); we shall call κD the <u>kernel</u> of D. Our construction shows that the intermediate spaces D_z are completely determined by the kernel κD. In fact, from these remarks we see that D is an analytic range if and only if $D = \kappa D$. <u>We assume this to be the case for the remainder of this section</u>.

We now return to the case where each $D_\zeta \in D$ has a general norm N_ζ. As in [2] some assumptions have to be made on the norms N_ζ. In order to carry out the construction of the intermediate subspaces we assume that for each $u \in \mathbb{C}^k$,

$$\text{(2.18)} \qquad \begin{aligned} &\text{(a)} \quad \zeta \to N_\zeta(T(\zeta)u) \text{ is measurable,} \\ &\text{(b)} \quad m_1(\zeta)\|u\|_{\mathbb{C}^k} \le N_\zeta(T(\zeta)u) \le m_2(\zeta)\|u\|_{\mathbb{C}^k} \text{ a. e.,} \end{aligned}$$

where m_j are positive functions on $\partial\Delta$ not depending on u, such that $\log m_j$ is integrable, $j = 1,2$.

The construction of the intermediate subspaces under the above assumptions involves the introduction of certain auxiliary spaces, as is done in the full range case (see [2]). We simplify the situation here by assuming that $T(z)$ and $T(z)^{-1}$ (the left inverse of $T(z)$) are uniformly bounded operators on Δ; thus, we can assume that m_1 and m_2 are constant functions. The definitions of the intermediate norms N_z, under these more restrictive assumptions, only involve the space H^2_D. That is, $N_z(u)$, for $u \in D_z = T(z)\mathbb{C}^k$, is given by (1.5) with $\mathcal{A} = H^2_D$.

These intermediate norms N_z are not, in general, Hilbert space norms and we cannot use an orthogonal projection to show that extremal functions exist. Nevertheless, extremal functions $F_{u,z}$ associated with each $u \in D_z$ and $z \in \Delta$ do exist (but they may fail to be unique). This can be shown by reducing the problem to the full range case where, in [2], it is completely worked out.

This reduction is obtained by introducing the norms M_ζ on $\mathbb{C}^k$ given by $M_\zeta(u) = N_\zeta(T(\zeta)u)$ for $u \in \mathbb{C}^k$. Then $C_\zeta \equiv (\mathbb{C}^k, M_\zeta)$ is a family of spaces associated with the points $\zeta \in \partial\Delta$ for which the construction developed in [2] can be applied by letting

$$M_z(w) = \inf\{(\int_0^{2\pi}[M_\zeta(F(\zeta))]^2 P_z(\theta)d\theta)^{1/2}: F \in H^2_{\mathbb{C}^k},\ F(z) = w\}$$

for each $z \in \Delta$. Under our assumptions on our range function we have the analog of (2.14) with $U(z)$ replaced by $T(z)$:

$$H^2_D = TH^2_{\mathbb{C}^k}. \tag{2.19}$$

From the definition of the norms M_ζ and (2.19) we immediately obtain

<u>Theorem</u> (2.20). <u>If</u> $v \in D_z$ <u>then</u> $N_z(v) = M_z(T(z)^{-1}v)$ <u>for each</u> $z \in \Delta$.

Many of the properties of the spaces D_z can be derived from those of the family $C_z = (\mathbb{C}^k, M_z)$ by applying theorem (2.20). For example, let us describe briefly the behaviour of linear operators on these intermediate subspaces. In [2] it is shown that an analytic family $\{B(z)\}$ of linear operators on the spaces C_z has operator norms $\|B(z)\|$ for which $\log\|B(z)\|$ is subharmonic in Δ. This result extends the Riesz - Thorin theorem and, in fact, the complex method of interpolation developed by A. P. Calderon. Let us show how this interpolation result extends to the subspaces.

Suppose A is a linear operator defined on the spaces D_z, $|z| \leq 1$, and let $\eta(z)$ be the norm of A acting on D_z. The operator $B(z) \equiv T(z)^{-1}AT(z)$ acts on C_z. If $u \in \mathbb{C}^k$, then, by (2.20),

$$M_z(T(z)^{-1}AT(z)u) = N_z(AT(z)u) \leq \eta(z)N_z(T(z)u) = \eta(z)M_z(u).$$

$\{B(z)\}$ is an analytic family on $\mathbb{C}^k$ and, thus, $\log \eta(z)$ is subharmonic since $\eta(z)$ is also the operator norm of $B(z)$. This argument extends to the case where A acts on the intermediate subspaces obtained from one family D into the intermediate subspaces derived from another family D'. Moreover, we can also assume that $A = A(z)$ is an analytic family. We leave the details of this to the reader.

We now turn to the extension of the Wiener-Masani theorem we discussed briefly in the introduction. This theorem asserts that if $\Omega(\zeta)$ is a $k\times k$ positive definite matrix for each $\zeta \in \partial\Delta$ which is measurable in ζ and such that $\log \|\Omega(z)\|$ and $\log \|\Omega(z)^{-1}\|$ are integrable ($\|\ \|$ denotes the operator norm), then there exists an <u>analytic</u> matrix valued function $A(z)$ on Δ having non-tangential boundary values $A(\zeta)$ such that

$$A^*(\zeta)A(\zeta) = \Omega(\zeta) \tag{2.21}$$

for almost every $\zeta \in \partial\Delta$. Moreover, $m_1(z) \leq \|A(z)\| \leq m_2(z)$ for all $z \in \Delta$, where $2\log[m_j(z)]$, $j = 1,2$, are the Poisson integrals of $\log \|\Omega(\zeta)^{-1}\|$ and $\log \|\Omega(\zeta)\|$.

This factorization, together with the tools developed here,

can be used to factor positive <u>semi</u>-definite matrices $\Omega(\zeta)$. Let $D = \{T(\zeta)\mathbb{C}^k\}$ be an analytic range. We first consider the factorization of the orthogonal projection $Q(\zeta)$ of $\mathbb{C}^n$ onto $D_\zeta = T(\zeta)\mathbb{C}^k$. Observe that $\Omega(\zeta) = T^*(\zeta)T(\zeta)$ is a positive definite $k\times k$ matrix for which the factorization (2.21) holds; that is, for almost every $\zeta \in \partial\Delta$

$$A^*(\zeta)A(\zeta) = T^*(\zeta)T(\zeta). \tag{2.22}$$

Let $S(\zeta) = T(\zeta)A(\zeta)^{-1}$. We claim that

$$Q(\zeta) = S(\zeta)S^*(\zeta). \tag{2.23}$$

To see this first observe that $S(\zeta)$ maps $\mathbb{C}^k$ into $\mathbb{C}^n$ and, thus, $S^*(\zeta)$ maps $\mathbb{C}^n$ into $\mathbb{C}^k$. Hence, $S(\zeta)S^*(\zeta)$ is a self adjoint operator on $\mathbb{C}^n$. It is also idempotent: using (2.22) and suppressing ζ we have $[SS^*]^2 = [TA^{-1}A^{*-1}T^*]^2 = TA^{-1}A^{*-1}A^*AA^{-1}A^{*-1}T^* = SS^*$. Moreover, it is clear that SS^* maps $\mathbb{C}^n$ into D_ζ. Finally, we show that this operator is onto D_ζ: if $v \in D_\zeta$ we can find $w \in \mathbb{C}^k$ such that $Tw = v$. Thus, using (2.22) again, $SS^*v = TA^{-1}A^{*-1}T^*Tw = TA^{-1}A^{*-1}A^*Aw = Tw = v$.

We should emphasize that (2.23) is a factorization with the analytic factor on the left. The factorization similar to (2.23) but with the analytic factor on the right shows up in the next section in the quotient space theory. This difference provides yet another sense in which the subspace theory and quotient space theories are dual to each other. (In the classical case of the Wiener-Masani theorem the two types of factorizations are closely related. For a strictly definite Ω we can obtain a factorization

$$\tilde{A}(\zeta)\tilde{A}^*(\zeta) = \Omega(\zeta) \tag{2.24}$$

with analytic $\tilde{A}(\zeta)$ by applying (2.21) to the positive definite matrix $\bar{\Omega}$.) Both types of factorization are considered in detail n [15].

Using our results we now extend (2.24) to semi-definite operators. First we make two observations concerning the factorization (2.23) of the projection $Q(\zeta)$. If the range function $T(z) = U(z)$ (the operator of theorem (2.13)), then it follows easily that

$$Q(\zeta) = U(\zeta)U^*(\zeta). \tag{2.25}$$

On the other hand, if $I_{\mathbb{C}}k$ denotes the identity operator on $\mathbb{C}^k$, then

$$I_{\mathbb{C}}k = U^*(\zeta)U(\zeta). \tag{2.26}$$

Theorem (2.27). *Suppose $\{\Omega(\zeta)\}$ is a family of positive semi definite operators on $\mathbb{C}^n$ with images $\{D_\zeta\}$ forming an analytic range. If the operator norms, and their bounds from below on D_ζ, are logarithmically integrable, then there exists an analytic $n\times k$ matrix valued function $C(z)$ on Λ such that*

$$\Omega(\zeta) = C(\zeta)C^*(\zeta)$$

for a.e. $\zeta \in \partial\Lambda$, where $C(\zeta)$ is the non-tangential limit of $C(z)$ as z approaches ζ.

Proof. Suppose $D = \{D_\zeta\} = \{T(\zeta)\mathbb{C}^k\}$, where $T(z)$ is a range function as described above, which in fact can be chosen to have the isometry property (2.12). $\Omega(\zeta)$ must be self adjoint and, therefore, a one to one map of D_ζ onto itself, since D_ζ must be the span of the k eigenvectors of $\Omega(\zeta)$ which have non zero eigenvalues. Since $T(\zeta)$ and $T^*(\zeta)$ define one to one maps of $\mathbb{C}^k$ onto D_ζ and D_ζ onto $\mathbb{C}^k$ respectively, it is easy to see that the $k\times k$ self adjoint matrix

$$B(\zeta) = T^*(\zeta)\Omega(\zeta)T(\zeta) \tag{2.28}$$

is one to one and positive semidefinite, and, therefore, positive definite for $\zeta \in \Lambda$. The integrability of $\log\|B(\zeta)\|$ and $\log\|B(\zeta)^{-1}\|$ follows from the isometry property of $T(\zeta)$ and so we can apply the Masani-Wiener theorem to obtain a factorization $B(\zeta) = \tilde{A}(\zeta)\tilde{A}^*(\zeta)$ as in (2.24). Now, using (2.25), (2.26) and the fact that $\Omega(\zeta)$ annihilates $D_\zeta^\perp$ we obtain that $\Omega(\zeta) = T(\zeta)B(\zeta)T^*(\zeta) = T(\zeta)\tilde{A}(\zeta)\tilde{A}^*(\zeta)T^*(\zeta)$ which is the desired factorization in (2.27) with $C(\zeta) = T(\zeta)\tilde{A}(\zeta)$.

An iteration theorem for the subspaces D_z also follows from theorem (2.20). That is, it can be shown that the construction of intermediate normed subspaces is stable under repetition of the construction on subdomains. The precise formulation follows the pattern in [2] and the proof is reduced to that case using theorem (2.20). We omit the details.

Finally let us consider the question of duality. In [2] it is shown that the family $C^* = \{C_\zeta^*\}$ of the duals to the spaces $\{C_\zeta\}$ ζ in $\partial\Lambda$, are the given boundary data for interpolation then

$$(C^*)_z = (C_z)^* \tag{2.29}$$

for all $z \in \Lambda$. That is, the dual of the intermediate space C_z, obtained from the family $\{C_\zeta\}$, is the intermediate space obtained from the boundary family of duals C^*. To make this precise we must explain what we mean by the "dual of the space $C = (\mathbb{C}^k, M)$", where M is a norm on $\mathbb{C}^k$. If $v = (v_1, v_2, \ldots, v_k)$, $w =$

$(w_1, w_2, \ldots, w_k) \in \mathbb{C}^k$ let $\langle v,w\rangle = \Sigma_1^k v_j w_j$. The <u>dual norm</u>, M^*, is then defined by letting

(2.30) $$M^*(w) = \sup\{|\langle v,w\rangle| : M(v) = 1\},$$

for each $w \in \mathbb{C}^k$. A simple computation shows that if, for all $v \in \mathbb{C}^k$, $m_1 \|v\|_{\mathbb{C}^k} \leqq M(v) \leqq m_2 \|v\|_{\mathbb{C}^k}$ then $(1/m_2)\|w\|_{\mathbb{C}^k} \leqq M^*(w) \leqq (1/m_1)\|w\|_{\mathbb{C}^k}$ for all $w \in \mathbb{C}^k$. Thus, if $\{C_\zeta\} = \{(\mathbb{C}^k, M_\zeta)\}$ is an interpolation family so is $\{C_\zeta^*\} = \{(\mathbb{C}^k, M_\zeta^*)\}$. This clarifies what we meant by (2.29).

Suppose we are given an analytic range $D = \{D_\zeta\} = \{T(\zeta)\mathbb{C}^k\}$ with each D_ζ endowed with a norm N_ζ so that (2.18) is satisfied. If $w \in D_z$ we then let

$$L_w(v) = \langle T(z)^{-1}v, T(z)^{-1}w\rangle$$

for all $v \in D_z$. Then L_w is a linear functional on D_z and (by a dimension argument) all linear functionals on D_z have this form. We define the <u>dual norms</u>, N_z^*, of N_z by letting

$$N_z^*(w) = \sup\{|\langle T(z)^{-1}v, T(z)^{-1}w\rangle| : N_z(v) = 1,\ v \in D_z\}$$

for $|z| \leqq 1$. By theorem (2.20) we see that

$$N_z^*(w) = M_z^*(T(z)^{-1}w).$$

Applying (2.20) and (2.29) we can then deduce that N_z^* is also the intermediate norm obtained from the norms N_ζ^*, $\zeta \in \partial\Delta$. We express this duality result by the equality

(2.31) $$(D^*)_z = (D_z)^*.$$

§3. <u>Interpolation of quotient spaces</u>. We now present the main features of the quotient spaces interpolation theory. As in the previous section we begin with the case amenable to Hilbert space techniques. We again obtain a linear map, this time <u>out of</u> $\mathbb{C}^n$, which can be used to reduce most questions to the situation of [2] (i.e. to considering a family of norms as opposed to semi-norms). That same linear map can also be used to analyze the case of general semi-norms (as opposed to the quotient norms for inner product spaces which we consider). That final extension is left to the reader.

There is another way of representing the duals of intermediate subspaces that is more natural than the result we have just obtained. This involves quotient spaces; some simple general observations will make it clear why this is the case. Let $D \subset \mathbb{C}^n$. Then $\bar{D}^{\perp} = \{w:(\bar{v},w) = 0$ for all $v \in D\} = \{\bar{u}:(v,u) = 0$ for all $v \in D\} = \overline{D^{\perp}}$. Let $D^+ = \bar{D}^{\perp} = \overline{D^{\perp}}$. We shall point out that there is a natural duality between $\mathbb{C}^n/D$ and D^+. Observe that if $[v] = v + D \in \mathbb{C}^n/D$ and $w \in D^+$ then $L_w([v]) = \langle v,w\rangle = (v,\bar{w})$ is well defined. That is, $\langle u,w\rangle = \langle v,w\rangle$ whenever $u \in [v]$. Consequently, L_w is a linear functional on the vector space $\mathbb{C}^n/D$ and, in fact, all linear functionals on this quotient space have this form. We denote this representation of the dual of $\mathbb{C}^n/D$ by

$$(\mathbb{C}^n/D)^* = D^+.$$

It is also easily checked that the linear functional L_w has norm $\|L_w\| = \|w\|_{\mathbb{C}^n}$ when $\mathbb{C}^n/D$ is endowed with the quotient norm $\|[v]\| = \inf\{\|u\|:u \in [v]\}$.

On the other hand, if we regard $\langle v,w\rangle = L_{[v]}(w)$ as the value of a linear functional on D^+ induced by the coset $[v]$, we also have a natural representation of the dual of D^+ by $\mathbb{C}^n/D$:

$$(D^+)^* = \mathbb{C}^n/D. \tag{3.2}$$

Moreover, $\|L_{[v]}\| = \|[v]\|$.

If we couple these facts with the result (see [6]) that $D = \{D_\zeta\}$ is an analytic range if and only if $D^+ \equiv \{D_\zeta{}^+\}$ is an analytic range, it is not surprising that the interpolation of subspaces is closely related, via duality, to the interpolation of quotient spaces.

We now present the main features of the quotient spaces interpolation theory. Let $D = \{D_\zeta\}$ be a family of subspaces, $\zeta \in \partial\Delta$, and $Q = \{\mathbb{C}^n/D_\zeta\}$ be the accompanying family of quotient spaces. Let $P_{D_\zeta^\perp}$ denote the projection of $\mathbb{C}^n$ onto $D_\zeta^\perp$ (we assume the matrix entries of this projection to be measurable). This projection induces the operator $P_{D^\perp}$ on $\mathbb{C}^n$-valued functions F: $((P_{D^\perp})F)(\zeta) = P_{D_\zeta^\perp}(F(\zeta))$; it is clear that $P_{D^\perp}$ is a projection of $L^2_{\mathbb{C}^n} = \{f = (f_1, f_2, \ldots, f_n): f_j \in L^2(\partial\Delta),\ j = 1,2,\ldots,n\}$ onto $L^2_{D^\perp} = \{f \in L^2_{\mathbb{C}^n}: f(\zeta) \in D_\zeta^\perp \text{ for a.e. } \zeta \in \partial\Delta\}$.

Each vector $v \in \mathbb{C}^n$ determines the coset $[v]_\zeta \in \mathbb{C}^n/D_\zeta$. The norm of this coset is $N^\zeta([v]_\zeta) = \inf\{\|u\|_{\mathbb{C}^n}: u-v \in D_\zeta\}$. We shall also write $N^\zeta(v)$ instead of $N^\zeta([v]_\zeta)$; N^ζ is a semi norm on $\mathbb{C}^n$ with kernel $\{v: N^\zeta(v) = 0\} = D_\zeta$. Observe that $N^\zeta(v) = \|P_{D_\zeta^\perp}(v)\|_{\mathbb{C}^n}$.

Let

$$\begin{aligned} |F|_z &= \left(\int_0^{2\pi} (N^\zeta(F(\zeta))^2 P_z(\theta)\, d\theta\right)^{1/2} \\ &= \left(\int_0^{2\pi} \|P_{D_\zeta^\perp}(F(\zeta))\|^2 P_z(\theta)\, d\theta\right)^{1/2} \end{aligned}$$

when F is a $\mathbb{C}^n$-valued function on $\partial\Delta$ and, for $z \in \Delta$, put

(3.3) $$N^z(v) = \inf\{|F|_z: F \in H^2_{\mathbb{C}^n},\ F(z) = v\}.$$

N^z is, then, a seminorm on $\mathbb{C}^n$, the <u>intermediate seminorm</u> induced by the boundary seminorms N^ζ, $\zeta \in \partial\Delta$.

In view of the development in §2 it is natural to ask whether an extremal function $F^{z,v}$ for which the infimum in (3.3) is attained does exist. A moment's reflection shows that this is a harder problem than the one resolved in the last section since the projection argument we gave there is not obviously available. We shall, however, present a version of it.

In order to do this we need to work in the quotient space $L^2_{\mathbb{C}^n}/L^2_D$ consisting of the cosets $[[f]] = \{g \in L^2_{\mathbb{C}^n}: f-g \in L^2_D\}$. We put

$$\|[[f]]\| = \inf\{(\int_0^{2\pi} \|g(\zeta)\|^2_{\mathbb{C}^n} \frac{d\theta}{2\pi})^{1/2}: g \in [[f]]\}$$

$$= (\int_0^{2\pi} \|(P_{D^\perp} f)(\zeta)\|^2_{\mathbb{C}^n} \frac{d\theta}{2\pi})^{1/2}$$

for the norm of the coset $[[f]]$. For simplicity we let $z = 0$; then (3.3) can be written

$$N^o(v) = \inf\{\|[[v]] + [[G]]\|: G \in H^o\}, \tag{3.4}$$

where $H^o = \{[[G]]: G \in \zeta H^2_{\mathbb{C}^n}\}$. Let M be the projection of $L^2_{\mathbb{C}^n}/L^2_D$ onto the orthogonal complement of H^o. Then

$$N^o(v) = \|M[[v]]\|. \tag{3.5}$$

The projection $P_{D^\perp}$ is constant on each coset $[[f]] \in L^2_{\mathbb{C}^n}/L^2_D$. Thus, $\overline{P_{D^\perp} M[[v]]} = F = F^{o,v}$ is a well defined function and, clearly,

$$\begin{aligned} N^o(v) &= (\int_0^{2\pi} \|P_{D^\perp}(M[[v]])(\zeta)\|^2_{\mathbb{C}^n} \frac{d\theta}{2\pi})^{1/2} \\ &= (\int_0^{2\pi} \|F^{o,v}(\zeta)\|^2_{\mathbb{C}^n} \frac{d\theta}{2\pi})^{1/2}. \end{aligned} \tag{3.6}$$

This function $F^{o,v}$ does satisfy some of the basic properties of an extremal function. We first show that it is analytic:

<u>Theorem</u> (3.7). $F^{o,v} \in H^2_{\mathbb{C}^n}$.

<u>Proof</u>. Let $w \in \mathbb{C}^n$ and m a positive integer. Since $\zeta^m w \in \zeta H^2_{\mathbb{C}^n}$, $M[[v]]$ is orthogonal to $[[\zeta^m w]]$ (with respect to the Hilbert space structure of $L^2_{\mathbb{C}^n}/L^2_D$). Since the inner product in $L^2_{\mathbb{C}^n}/L^2_D$ is given by the formula

$$([[f]],[[g]]) = \int_0^{2\pi} ((P_{D^\perp} f)(\zeta), (P_{D^\perp} g)(\zeta)) \frac{d\theta}{2\pi}$$

$$= \int_0^{2\pi} ((P_D\perp f)(\zeta), g(\zeta)) \frac{d\theta}{2\pi},$$

this means that, writing F for $F^{o,v}$,

$$o = ([[\zeta^m w]], M[[v]]) = \int_0^{2\pi} (e^{im\theta}w, P_D\perp(M[[v]])(e^{i\theta})) \frac{d\theta}{2\pi}$$
$$= \int_0^{2\pi} \langle w, F(e^{i\theta})\rangle e^{im\theta} \frac{d\theta}{2\pi}$$

for all positive integers m and vectors $w \in \mathbb{C}^n$. This shows that the $L^2_{\mathbb{C}^n}$ function F is of power series type and, thus, $F \in H^2_{\mathbb{C}^n}$. This proves the theorem.

It is clear that the values of $F = F^{o,v}$ belong to $D^+ = (D_\zeta^{\ +})$; that is, $F(\zeta) \in D_\zeta^{\ +}$ a.e.. Thus, $F \in H^2_{D^+}$. We also have

$$(3.8) \qquad (F,G) = \int_0^{2\pi} (F(\zeta), G(\zeta)) \frac{d\theta}{2\pi} = o$$

for all $G \in \zeta H^2_{D^+}$. To see this observe that, by the definition of M, $M[[v]] = [[v]] - [[g]]$, where $[[g]]$ belongs to the closure of H^o. Thus, $\overline{F} = P_D\perp v - P_D\perp g$ and

$$\int_0^{2\pi} (F(\zeta), G(\zeta)) d\theta = \int_0^{2\pi} \left\{ ((P_{D^+}\overline{v})(\zeta), G(\zeta)) - ((P_{D^+}\overline{g})(\zeta), G(\zeta)) \right\} d\theta.$$

Since $G \in H^2_{D^+}$, $P_{D^+}G = G$. Hence,

$$\int_0^{2\pi} ((P_{D^+}\overline{v})(\zeta), G(\zeta)) \, d\theta = \int_0^{2\pi} (\overline{v}, G(\zeta)) \, d\theta = o,$$

because G is holomorphic of mean o and $\overline{v}$ is a constant function. We also have $[[g]] = \lim_{m\to\infty} [[\zeta g_m]]$, where $g_m \in H^2_{\mathbb{C}^n}$. This means

$$\int_0^{2\pi} \| P_{D^+}(\overline{g(s)} - \overline{\zeta g_m(\zeta)}) \|^2_{\mathbb{C}^n} \, d\theta = \int_0^{2\pi} \| P_D\perp(g(\zeta) - \zeta g_m(\zeta)) \|^2_{\mathbb{C}^n} \, d\theta \to o$$

as $m \to \infty$. But,

$$\int_0^{2\pi} (P_{D^+}\overline{\zeta g_m(\zeta)}, G(\zeta)) \, d\theta = \int_0^{2\pi} (\overline{\zeta g_m(\zeta)}, G(\zeta)) \, d\theta = o,$$

since G is holomorphic and $\overline{\zeta g_m(\zeta)}$ is conjugate analytic of mean zero. It follows that

$$\int_0^{2\pi} ((P_{D^+}\overline{g})(\zeta), G(\zeta)) \, d\theta = o$$

and (3.8) follows. Thus, $F = F^{o,v}$ satisfies the hypotheses of lemma (2.3) (with D^+ replacing D) and we obtain

Theorem (3.9). $F^{o,v}$ *is an extremal function for the subspace*

<u>interpolation construction obtained from</u> D^+.

Now suppose $w \in (D^+)_o$. We claim that

(3.10) $$\langle v,w\rangle = \int_0^{2\pi} \langle F_{o,w}(\zeta), \overline{F^{o,v}(\zeta)}\rangle \frac{d\theta}{2\pi},$$

where $F_{o,w} \in H^2_{D^+}$ is the extremal function associated with $0 \in \Delta$ and $w \in (D^+)_o$. (The map of v to $F^{o,v}$ is conjugate linear. Therefore the conjugation in (3.10) is natural.)

To see this we use the equality $\overline{F^{o,v}} = P_{D^\perp}(v - g)$ we introduced in the proof of (3.8). Thus, using $P_{D^+}F_{o,w} = F_{o,w}$ and the mean value theorem,

$$\int_0^{2\pi} \langle \overline{F^{o,v}(\zeta)}, F_{o,w}(\zeta)\rangle \frac{d\theta}{2\pi} = \int_0^{2\pi} \Big\{\langle P_{D^\perp}v, F_{o,w}\rangle - \langle P_{D^\perp}g, F_{o,w}\rangle\Big\} d\theta$$

$$= \int_0^{2\pi} \langle v, F_{o,w}\rangle \frac{d\theta}{2\pi} - \int_0^{2\pi} \langle P_{D^\perp}g, F_{o,w}\rangle \frac{d\theta}{2\pi} = \langle v,w\rangle - \int_0^{2\pi} \langle P_{D^\perp}g, F_{o,w}\rangle \frac{d\theta}{2\pi}.$$

Again using $[[g]] = \lim_{m\to\infty} [[g_m]]$, we have

$$\int_0^{2\pi} \langle P_{D^\perp}g, F_{o,w}\rangle \frac{d\theta}{2\pi} = \lim_{m\to\infty} \int_0^{2\pi} (P_{D^\perp}\zeta g_m(\zeta), \overline{F_{o,w}(\zeta)}) \frac{d\theta}{2\pi}$$

$$= \lim_{m\to\infty} \int_0^{2\pi} (\zeta g_m(\zeta), \overline{F_{o,w}(\zeta)}) \frac{d\theta}{2\pi} = 0,$$

since $\zeta g_m(\zeta)$ is holomorphic of mean 0 and $\overline{F_{o,w}}$ is conjugate analytic. This establishes (3.10).

Two immediate consequences of (3.10) are

(3.11) $$|\langle v,w\rangle| \le N^o(v) N_o^+(w),$$

where $N_o^+(w) = (\int_0^{2\pi} \|F_{o,w}(\zeta)\|^2_{\mathbb{C}^n} \frac{d\theta}{2\pi})^{1/2}$ is the subspace norm of $w \in (D^+)_o$, and

(3.12) $$N^o(v)^2 = \langle w,v\rangle = N_o^+(w)^2,$$

which follows from (3.10) by choosing $w = F^{o,v}(0)$ (so that, by (3.9), $F^{o,v} = F_{o,w}$).

From this we can obtain the natural duality result promised at the beginning of this section. Let

$$D^o = \{v: N^o(v) = 0\}$$

be the kernel of the seminorm N^o. Then D^o is a subspace of $\mathbb{C}^n$ and N^o can also be considered to be a norm on $Q^o = \mathbb{C}^n/D^o$: $N^o([v]_o) = N^o(v)$. (3.10), (3.11) and (3.12) show that $L_w([v]_o) = \langle v,w\rangle = L_{[v]_o}(w)$ define either a linear functional L_w on Q^o for each w in $(D^+)_o$ or a linear functional $L_{[v]_o}$ on $(D^+)_o$ for each $[v]_o$ in Q^o. The norms of these linear functionals equal $N_o^+(w)$ and $N^o(v)$, respectively. By letting $\zeta - z$ play the role of ζ, making appropriate use of the Poisson kernel $P_z(\theta)$ and making the obvious modifications we have, therefore, shown all but the last statement of

Theorem (3.13). *Let N^z, $z \in \Delta$, be the seminorm (3.3) obtained from a family D of subspaces of $\mathbb{C}^n$. Let $D^z = \{v \in \mathbb{C}^n: N^z(v) = 0\}$ and $Q^z = \mathbb{C}^n/D^z$. Thus, N^z can also be considered to be the quotient norm $N^z([v]_z) \equiv N^z(v)$ on Q^z. Then*

$$[(D^+)_z]^* = Q^z \text{ and } [Q^z]^* = (D^+)_z.$$

More precisely, the pairing $\langle v,w\rangle$, $v \in \mathbb{C}^n$, $w \in (D^+)_z$ defines either a linear functional $L_{[v]_z}(w) = \langle v,w\rangle$ on $(D^+)_z$ or a linear functional $L_w([v]_z) = \langle v,w\rangle$ on Q^z. The linear functional norms $\|L_{[v]_z}\|$ and $\|L_w\|$ equal $N^z(v)$ and $N_z^+(w)$, respectively. Moreover, all linear functionals on $(D^+)_z$ and on Q^z have this form.

We have established all but the last statement of this theorem. Suppose L is a linear functional on $(D^+)_z$. Extending it to $\mathbb{C}^n$ we can then find $v \in \mathbb{C}^n$ such that $L(w) = \langle v,w\rangle$ for all $w \in (D^+)_z$. Thus, using (3.10), (3.11) and (3.12), we see that $L = L_{[v]}$. Similarly, to see that any linear functional on Q^z can be regarded as a linear functional L_w with $w \in (D^+)_z$ we first show that $(D^+)_z \subset (D^z)^+$. This inclusion is equivalent to

$$D^z \subset [(D^+)_z]^+. \tag{3.14}$$

But, if $v \in D^z$ then, by (3.11), $|(w,\bar{v})| = |\langle w,v\rangle| \leq N^z(v)N_z^+(w) = 0$; thus, $(w,\bar{v}) = 0$ for all $w \in (D^+)_z$. That is, $\bar{v} \in [(D^+)_z]^\perp$ and we obtain (3.14). On the other hand, the just established relation $[(D^+)_z]^* = Q^z$ tells us that $\dim (D^+)_z = \dim Q^z = n - \dim D^z = \dim (D^z)^+$. Thus,

$$(D^+)_z = (D^z)^+. \tag{3.15}$$

Suppose, then, that L is a linear functional on Q^z. We can regard L as a linear functional on $\mathbb{C}^n$ that is 0 on D^z. Thus, $L(v) = \langle v,w\rangle$ with $w \in (D^z)^+ = (D^+)_z$, by (3.15). This shows $L = L_w$, $w \in (D^+)_z$ (for this argument to be complete one must show each extremal function $F_{z,w}$, $w \in (D^+)_z$, equals $F^{z,v}$ for some $v \in \mathbb{C}^n$; a dimension argument establishes this fact). This completes the proof of theorem (3.13).

As was mentioned in the introduction, the duality relation between the subspace and quotient space interpolation theories is even more complete when D (and, thus, D^+) is an analytic range. We shall now show what we mean by this assertion. Suppose, then, that D is an analytic range. Let k be the dimension of the spaces D_z (thus, k is the dimension of almost all spaces D_ζ, $\zeta \in \partial\Delta$).

If $v \in D_z$, $F_{z,v}$ is a function $F \in H^2_{\mathbb{C}^n}$ with $F(z) = v$. Moreover, $\int_0^{2\pi} \|(P_{D^\perp}F)(\zeta)\|^2_{\mathbb{C}^n} P_z(\theta)\, d\theta = 0$ since $F_{z,v}(\zeta) \in D_\zeta$ for a. e. $\zeta \in \partial\Delta$. By the definition (3.3), therefore, $N^z(v) = 0$. Consequently, we have shown

$$D_z \subset D^z. \tag{3.16}$$

Our assumption that D is an analytic range implies that the

second space in (3.14) and the first space in (3.16) have dimension k. Hence, we obtain the equality

$$(3.17) \qquad D_z = D^z$$

for all $z \in \Delta$. Using this equality in (3.15) we obtain

$$(3.18) \qquad (D^+)_z = (D_z)^+$$

when D is an analytic range. Consequently, we obtain the following sharper version of theorem (3.13):

<u>Theorem</u> (3.19). <u>Suppose</u> D <u>is an analytic range then</u> $Q^z = \mathbb{C}^n/D_z$ <u>for all</u> $z \in \Delta$. <u>Thus,</u>

$$(3.20) \qquad [(D^+)_z]^* = \mathbb{C}^n/D_z \quad \text{and} \quad [\mathbb{C}^n/D_z]^* = (D^+)_z.$$

<u>That is, the pairing</u> $\langle v,w\rangle$, $v \in \mathbb{C}^n$, $w \in (D^+)_z$ <u>defines either a linear functional</u> $L_{[v]_z}(w) = \langle v,w\rangle$ <u>on</u> $(D^+)_z$ <u>or a linear functional</u> $L_w([v]_z) = \langle v,w\rangle$ <u>on</u> $\mathbb{C}^n/D_z$. <u>The linear functional norms</u> $\|L_{[v]_z}\|$ <u>and</u> $\|L_w\|$ <u>equal</u> $N^z(v)$ <u>and</u> $N_z^+(w)$, <u>respectively. Moreover, all linear functionals on</u> $(D^+)_z$ <u>and on</u> Q^z <u>have this form.</u>

If D is not an analytic range, then the inclusion relation (3.16) is strict: $D_z \subsetneq D^z$. To see this suppose $D = \{D_\zeta\}$ is not an analytic range and that the spaces D_ζ have constant dimension k. A consequence of theorem (2.13) and equality (2.16) is that the spaces D_z have constant dimension $j < k$. In this case the spaces $D_\zeta{}^+ \in D^+$ have constant dimension $n - k$ and, since D^+ cannot be an analytic range, the spaces $(D^+)_z$ have constant dimension $n - \ell < n - k$. From (3.14) we see that $\dim Q^z = n - \ell$. Consequently, $\dim D^z = \ell > k > j = \dim D_z$. We made the simplifying assumption that the spaces D_ζ have constant dimension; the argument we gave can easily be modified to the case $\kappa D \subsetneq D$. Thus, Theorem (3.19) gives us a characterization of analytic ranges.

We now obtain the map $V(z)\colon \mathbb{C}^n \longrightarrow X$ satisfying (1.13) from which we can obtain the intermediate seminorms N^z, $z \in \Delta$. In §2 we constructed the map U(z) that allowed us to "pull back" the subspace theory to the full range case (see theorems (2.13), (2.14) and (2.20)). The map $V(z)$ allows us to "push forward" the quotient space theory onto a full range situation. First we need a lemma.

Lemma (3.21). If H^{o} is closed then D is an analytic range.

Proof. If H^{o} is closed then $H^{o} \oplus (H^{o})^{\perp} = L^{2}_{\mathbb{C}}n/L^{2}_{D}$. In particular, we then have $[[v]] - M[[v]] \in H^{o}$. Thus, $M[[v]] = [[v + G]]$ for some $G \in \zeta H^{2}_{\mathbb{C}}n$. It follows that $F = F^{o,v}$ is of the form $\overline{P_{D}\perp(v + G)}$. But this means that $\overline{F^{o,v}} = v + G + f_{D}$, where $f_{D} \in L^{2}_{D}$. Suppose $v \in D^{o} = \{w: N^{o}(w) = 0\}$. Then $F^{o,v}(\zeta) = 0$ a. e. and, thus, we must have $v + G(\zeta) = -f_{D}(\zeta)$ a. e.. This means that $E \equiv v + G \in H^{2}_{D}$. Since $E(0) = v + G(0) = v$, we must have $v \in D_{o}$ and we conclude that $D^{o} \subset D_{o}$. The opposite inclusion has already been shown (see (3.16). This gives $D^{o} = D_{o}$. Using the notation introduced in the paragraph following theorem (3.19), this equality implies $\ell = j$. But, as we pointed out above, if D were not an analytic range we would then have $k < \ell = j < k$. This contradiction gives us Lemma (3.21).

Since this manuscript was prepared we have been able to establish the converse of this lemma: if D is an analytic range then H^{o} is closed. We will make use of this fact in finishing our analysis. We now suppose that D is an analytic range. Hence the space H^{o} is closed and, as shown in the proof of lemma (3.21), we have

$$\overline{F^{o,v}(\zeta)} = v + G(\zeta) + f_{D}(\zeta) \tag{3.22}$$

for a. e. $\zeta \in \partial\Delta$, where $G \in \zeta H^{2}_{\mathbb{C}}n$ and $f_{D} \in L^{2}_{D}$. Moreover, $\overline{F^{o,v}} \in P_{D}\perp(L^{2}_{\mathbb{C}}n)$. It follows that the mapping $v \longrightarrow \overline{F^{o,v}}$ is a linear map depending only on the cosets $[v]_o \in Q^o$. Thus, we can consider the conjugate analytic family of linear maps of Q^o into $\mathbb{C}^n$ defined by $S(z)[v]_o = \overline{F^{o,v}(z)}$. By theorem (3.9) $\overline{S(z)[v]_o}$ are extremal functions for the D^+ subspace interpolation construction. Since $S(z)[v]_o$ cannot be 0 for any $z \in \Lambda$ when $v \notin D_o$, by corollary (2.6), we must have

$$(3.23) \qquad S(z)Q^o = \{D_z\}^{\perp}.$$

The arguments used in §2 and identifying the dual of Q^o with itself give us

$$(3.24) \qquad \begin{matrix} \text{(i) } S(\zeta)S^*(\zeta) = P_{D^\perp} \\ \text{(ii) } S^*(\zeta)S(\zeta) = I_{Q^o} \end{matrix} \quad \text{for a. e. } \zeta \in \partial\Lambda.$$

By (3.22) $\overline{F^{o,v}} = K + f_D$, where $K = v + G$; or equivalently $S(\zeta)[v]_o = K(\zeta) = f_D(\zeta)$. Applying $S^*(\zeta)$ to both sides of this equality and using (3.23) (ii) (note that $S^*(\zeta)f_D(\zeta) = 0$ a. e.) we obtain $[v]_o = S^*(\zeta)K(\zeta)$ for a. e. $\zeta \in \partial\Lambda$. Since S^* and K are both analytic this last equality extends to $[v]_o = S^*(z)K(z)$ for all $z \in \Lambda$. Thus,

$$(3.25) \qquad [K(0)]_o = S^*(z)K(z), \ z \in \Lambda.$$

It is not hard to show that a necessary and sufficient condition for F to be an extremal function $F^{o,v}$ is that $F \in H^2_{D^+}$ and $\overline{F} = K + g_D$ with $K \in H^2_{\mathbb{C}^n}$, $K(0) = v$ and $g_D \in L^2_D$. Suppose, then, $F = F^{z,w}$. We have $\overline{F} = K + g_D$ with $K \in H^2_{\mathbb{C}^n}$, $K(z) = w$ and $g_D \in L^2_D$. Then $S^*(z)w = S^*(z)\overline{F^{z,w}(z)} = S^*(z)K(z)$. By (3.25) we also have $S^*(z)K(z) = [K(0)]_o$. But this implies

(3.26) $$F^{z,w}(\zeta) = F^{o,S^*(z)w}(\zeta)$$

for a. e. $\zeta \in \partial\Delta$. Thus, by the definition of N^o, N^z, (3.26) and (3.7),

$$N^o(S^*(z)w) = \left(\int_o^{2\pi} \|F^{o,S^*(z)w}(\zeta)\|^2_{\mathbb{C}^n} \frac{d\theta}{2\pi}\right)^{1/2}$$

$$= \left(\int_o^{2\pi} \|F^{z,w}(\zeta)\|^2_{\mathbb{C}^n} \frac{d\theta}{2\pi}\right)^{1/2} = \operatorname*{ess\,sup}_{\zeta \in \partial\Delta} \|F^{z,w}(\zeta)\|_{\mathbb{C}^n}$$

$$= \left(\int_o^{2\pi} \|F^{z,w}(\zeta)\|^2_{\mathbb{C}^n} P_z(\theta)\,d\theta\right)^{1/2} = N^z(w).$$

Let $V(z) = S^*(z)$ for $z \in \Delta$. We then have

Theorem (3.27). *There exists an analytic linear map* $V(z)$ *on* Δ *such that* $V(z): \mathbb{C}^n \longrightarrow Q^o$ *and* $N^z(w) = N^o(V(z)w)$ *for all* $z \in \Delta$ *and* $w \in \mathbb{C}^n$.

§4. Further results and comments. In this section we shall describe some further results, some open problems, and some relations between this work and other subjects. We will be quite sketchy.

(i) There is a close relationship between our constructions in section 2 and the theory of subspaces of $H^2_{\mathbb{C}^n}$ which are invariant under multiplication by χ, the coordinate function on the disk Δ. A closed subspace, S, of $H^2_{\mathbb{C}^n}$ is said to be *invariant* if $\chi S \subset S$. The construction of the intermediate subpaces D_z can be easily modified to obtain a theorem of P. Lax [10] characterizing the invariant subspaces of $H^2_{\mathbb{C}^n}$:

Theorem (4.1). *If* S *is an invariant subspace of* $H^2_{\mathbb{C}^n}$, *then there exists a non-negative integer* $k \leq n$ *and an* $n \times k$ *matrix valued function* $U(z): \mathbb{C}^k \longrightarrow \mathbb{C}^n$ *whose entries belong to* H^∞ *such that*

(a) $S = UH^2_{\mathbb{C}}k$ (*i. e. each* $F \in S$ *is of the form* $F(z) = U(z)G(z)$ *for some* $G \in H^2_{\mathbb{C}}k$);

(b) $\|U(\zeta)w\|_{\mathbb{C}^n} = \|w\|_{\mathbb{C}^k}$ *for a. e.* $\zeta \in \partial\Delta$ *whenever* $w \in \mathbb{C}^k$.

A proof of this theorem along the lines of our construction of intermediate subspaces is in [9]. To see the connection between theorem (4.1) and the construction given in §2 note that $S = H^2_D$ is an invariant subspace whenever $D = \{D_\zeta\}$ is a family of subspaces associated with the boundary of Δ. Theorems (2.13) and (2.14) give us Lax's theorem when $S = H^2_D$. These spaces, however, are not the general invariant subspaces; for example, $S = \chi H^2_{\mathbb{C}}n$ cannot be of this form since $D_o = \{F(0): F \in \chi H^2_{\mathbb{C}}n\} = \{0\}$, and this contradicts the fact that dim D_z is constant on Δ. The general invariant subspace has the form qH^2_D, where q is an inner function. This is shown in [9] by establishing the existence of extremal functions, as is done in §2, and then carrying out a more careful analysis of their structure.

(ii) Theorem (3.26) gives us a characterization of those families of subspaces D that are analytic ranges. This last concept was introduced by Helson and Lowdenslager in [6] where they mention that "hardly anything about analytic ranges is obvious." The special case where n = 2 and the subspaces are one dimensional illustrates this situation. Suppose, for example, that each D_ζ is spanned by the vector $(1,a(\zeta))$, where $a(\zeta)$ is a measurable function on $\partial\Delta$. The results of the previous sections imply that, in this case, the following are equivalent:

1) D is an analytic range;
2) D^+ is an analytic range ($D_\zeta{}^+$ is the span of $(-a(\zeta),1)$);
3) D_z is non trivial for some $z \in \Delta$ (or, all $z \in \Delta$);

4) Q^z is non trivial for some $z \in \Delta$ (or, all $z \in \Delta$).

A more subtle condition on the function $a(\zeta)$ is the following one involving the subclass N^+ of the Nevanlinna class N:

<u>Theorem</u> (4.2). <u>Suppose</u> D_ζ <u>is the span of</u> $(1,a(\zeta))$, $\zeta \in \partial\Delta$, <u>then</u> $D = \{D_\zeta\}$ <u>is an analytic range if and only if</u> $a = \bar{h}g$, <u>where</u> $g \in N^+$ <u>and</u> h <u>is an inner function.</u>

To see this observe that if D is an analytic range with D_ζ the span of $(1,a(\zeta))$, then $(1,a(\zeta)) = c(\zeta)(u_1(\zeta),u_2(\zeta))$ for a. e $\zeta \in \partial\Delta$, where

$$U(\zeta) = \begin{bmatrix} u_1(z) \\ u_2(z) \end{bmatrix}$$

is the 2×1 matrix in (2.13). In particular, $u_j \in H^\infty$, $j = 1,2$, and $|u_1(\zeta)|^2 + |u_2(\zeta)|^2 = 1$ a. e.. It follows that $u_1(\zeta)^{-1} = c(\zeta)$ and, thus, $a(\zeta) = u_2(\zeta)/u_1(\zeta)$ for a. e. $\zeta \in \partial\Delta$. Let $u_1 = hf$ with h an inner function and f an outer function, then

$$a(\zeta) = \overline{h(\zeta)}\left\{\frac{u_2(\zeta)}{f(\zeta)}\right\}$$

and we obtain the expression for a in theorem (4.2) with $g = u_2/f$.

If, on the other hand, D_ζ is the span of $(1,a(\zeta))$, with $a = \bar{h}g$, h inner and $g \in N^+$, then g can be written as a quotient f_1/f_2 with $f_j \in H^\infty$, $j = 1,2$. Thus,

$$(1,a(\zeta)) = \frac{1}{h(\zeta)f_2(\zeta)}(h(\zeta)f_2(\zeta),f_1(\zeta)).$$

But this means that D_ζ is the span of $(h(\zeta)f_2(\zeta),f_1(\zeta))$ and it follows that D is an analytic range.

This proof is quite simple given what we have done. It is an interesting exercise to go directly from, say, the condition that Q^0 is non-trivial to the conclusion of the theorem (without using the duality).

The condition in theorem (4.2) shows up several places. The

question "if $f \in H^\infty$, does there exist an inner function h such that $\bar{h}f \in H^\infty$?" is considered by Helson in [7], Chapter 9, where one can find conditions that insure that D and $D^\perp$ are both analytic ranges. [5] contains other references to work on fucntions f which satisfy the condition (see problem 14 on pg. 217).

(iii) Suppose we are given a positive definite n×n matrix valued function $\Omega(\zeta)$, $\zeta \in \partial\Delta$, Then it is possible to write $\Omega = A^*A$ on $\partial\Delta$ with A the boundary values of an analytic function and the values of A are invertible matrices (1.16). We can then extend Ω to Δ by $\Omega(z) = A^*(z)A(z)$. This gives us a solution of the Dirichlet problem for the differential equation

$$\frac{\partial}{\partial z}\left\{\Omega^{-1}\frac{\partial}{\partial \bar{z}}\Omega\right\} = 0 \tag{4.3}$$

(see (5.2) in [14]). In the previous sections we obtained factorizations for semidefinite Ω. We can again extend Ω to Δ. What is the replacement for (4.3)? (4.3) can't hold because Ω is not invertible. What can replace Ω^{-1}? That is, if $\Omega(z) = A^*(z)A(z)$ where A(z) is an analytic function which takes values which are <u>rectangular</u> matrices, what PDE must Ω satisfy?

(iv) Specifying a norm on $\mathbb{C}^n$ is equivalent to specifying the norming body, the closed unit ball. B is such a unit ball if and only if B is a closed, convex, bounded, circled set containing a neighborhood of the origin.

The interpolation construction of the intermediate norms can be described in terms of these unit balls. Suppose we are given the boundary data $\{B_\zeta: \zeta \in \partial\Delta\}$, where each B_ζ is a unit ball, then the unit ball B_z, $z \in \Lambda$, consists of those vectors $v \in \mathbb{C}^n$ for which there exist vector valued functions F such that $F(z) = v$ and $F(\zeta) \in B_\zeta$ a. e.. The results in §2 and §3 can also be described in this manner except that, in the subspace case, the unit balls need not contain a neighborhood of 0 and, in the seminorm case, the unit balls need not be bounded. (Actually, in the seminorm case we only have a sequence $\{F_n\}$ with $F_n(z) = (1-\epsilon_n)v$, $\epsilon_n \longrightarrow 0+$.)

This point of view, the study of analytic functions with boundary values in given target sets, is related to work of Alexander and Wermer [18] and of Slodkowski [16] on polynomially convex hulls. It is also related to the study of extremal problems in control theory as presented by Helton [8] and studied, for instance, by Hui [19]. In that context the norming functions N are "penalty functions."

(v) The extension of the theory presented here to the infinite dimensional case is still an open problem. The examples in Helson's book [7] suggest that a general theory might involve major technical difficulties and that new ideas are needed. Some natural examples, however, are infinite dimensional; a few are given in [13]. Another interesting example, involving the study of norms on analytically varying families of finite dimensional subspaces of an infinite dimensional space can be found in the work of Cowen and Douglas [4]. They consider a fixed operator acting on a Hilbert space and study the family of subspaces $D_\lambda = Ker(T-\lambda I)$ as a function of λ.

(vi) We came to these problems in order to extend the interpolation theory presented in [2]. Some of these topics, however, have been studied by others for different reasons. In particular, matrix factorizations such as (1.15) and (1.16) and their applications have been studied quite extensively. Many results and further refences are in [7], [8], [17], [12]. The case of semidefinite Ω has been less well studied; however, theorem (2.27) is related to [11].

References

[1] Calderon, A. P. , Intermediate Spaces and Interpolation, the Complex Method, Studia Math. 24 (1964) pp. 113-190.

[2] Coifman, R., Cwikel, M., Rochberg, R., and Weiss, Guido, The Complex Method for Interpolation of Operators Acting on Families of Banach Spaces, Lecture Notes in Mathematics 779, Springer-Verlag, Berlin, Heidelberg, New York (1980) pp. 123-153.

[3]________________, A Theory of Complex Interpolation for Families of Banach Spaces, Advances in Math. 33 (1982) pp. 203-229.

[4] Cowen, M., and Douglas, R., Complex Geometry and Operator Theory, Acta Math 114 (1978) pp. 187-261.

[5] Garnett, J., Bounded Analytic Functions, Academic Press, New York, 1981.

[6] Helson, H. and Lowdenslager, D., Prediction Theory and Fourier Series in Several Variables II, Acta Math. 106 (1961).

[7] Helson, H., Invariant Subspaces, Academic Press (1964), New York and London, pp. 1-130

[8] Helton, J. W., Operator Theory, Analytic Functions, Matrices and Electrical Engineering, CBMS Conference in Lincoln, Nebraska, Aug 13-23 (1985), to be published by the A.M.S..

[9] Hernandez, E., Topics in Complex Interpolation, Ph. D. Thesis, Washington University in St. Louis (1982).

[10] Lax, P., Translation Invariant Spaces, Acta Math. 101 (1959) pp. 163-178.

[11] Masani, P., and Wiener, N. On Bivariate Stationary Processes and the Factorization of Matrix Valued Functions, Theory of

Probability and its Application, Vol. IV (1959) pp. 301-308.

[12] Rochberg, R., Interpolation of Banach Spaces and Negatively Curved Vector Bundles, Pac. J. Math. 110 (1984) 355-376.

[13] Rochberg, R., and Weiss, Guido, Complex Interpolation of Subspaces of Banach Spaces, Suppl. Rendiconti Circ. Mat. Palermo, n. 1 (1981) pp. 179-186.

[14] ____________________, Analytic Families of Banach Spaces and Some of Their Uses, Recent Progress in Fourier Series, Peral and Rubio de Francia, editors, North-Holland (1985), Amsterdam, New York, Oxford, pp.173-202.

[15] Rosenblum, M., and Rovnyak, J., Hardy Classes and Operator Theory, Oxford University Press (1985), New York, Oxford.

[16] Slodkowski, Z, Polynomial hulls with convex sections and interpolation spaces, Proc. Amer. Math. Soc. to appear.

[17] Sz-Nagy, B., and Foias, C., Harmonic Analysis of Operators on Hilbert Space, North-Holland (1970), Amsterdam, New York, Oxford.

[18] Alexander, H., and Wermer, J, Polynomial hulls with convex fibers, Math. Ann. 271, (1985), 99-109.

[19] Hui, S. Existence, uniqueness, and continuity of solutions to optimization problems over spaces of analytic functions, Ph.D. Thesis, U. of Washington, 1986.

On interpolation of multi-linear operators

Svante Janson
Department of Mathematics
Uppsala University
Thunbergsvägen 3
S-752 38 UPPSALA, Sweden

1. Introduction.

It is well-known that the real method of interpolation does not behave as well as the complex method for multi-linear operators. For example (leaving quasi-Banach spaces, operators of more than two variables and other complications to later sections), suppose that (A_0,A_1), (B_0,B_1) and (C_0,C_1) are Banach couples, and that T is a bilinear operator (defined, say on $(A_0+A_1)\times(B_0+B_1)$, and mapping this product continuously into some Hausdorff vector space that contains (continuously) C_0 and C_1) such that T maps $A_0\times B_0$ continuously into C_0 and $A_1\times B_1$ into C_1. Then, for the complex method, T maps $[A_0,A_1]_\theta \times [B_0,B_1]_\theta$ into $[C_0,C_1]_\theta$ $(\theta\in(0,1))$, while for the real method we only have the less precise result that T maps $(A_0,A_1)_{\theta q_1}\times(B_0,B_1)_{\theta q_2}$ into $(C_0,C_1)_{\theta q}$ $(\theta\in(0,1),\ q_1,q_2,q\in[1,\infty])$, provided

$$1/q \le 1/q_1 + 1/q_2 - 1 . \tag{1}$$

(See e.g. Bergh and Löfström [1], Theorem 4.4.1 and Exercise 3.13.5, or Calderón [2], Lions and Peetre [6] and Zafran [8].)

In general, (1) cannot be improved, as is seen e.g. by the following example.

<u>Example.</u> Let $A_0 = B_0 = C_0 = \ell^1(\mathbb{Z}_+)$, $A_1 = B_1 = C_1 = \{(a_n)_0^\infty : \Sigma\, 2^n|a_n| < \infty\}$ and let T be the convolution of sequences. Then e.g. $(A_0,A_1)_{\theta q_1} = \{(a_n)_0^\infty : (2^{\theta n} a_n)\in \ell^{q_1}\}$, and T maps $(A_0,A_1)_{\theta q_1}\times(B_0,B_1)_{\theta q_2}$ into $(C_0,C_1)_{\theta q}$ iff $\ell^{q_1} * \ell^{q_2} \subset \ell^q$, i.e. iff (1) holds.

Nevertheless, in many applications the restriction (1) does not give the best possible result, and it becomes important to find conditions which enable us to relax it. The purpose of the present note is to point out the simple fact, which seems to be less well-known, that this can be done when "the spaces A and B can be varied independently of each other" in a sense that will be made precise below, and that then (1) may be replaced by

$$1/q \leq 1/q_1 + 1/q_2$$

(In special cases, this has been known for a long time. O'Neill [7] gives two early theorems of this type.)

The simplest case when "A and B can be varied independently" is when, in addition to the assumptions above, T also maps $A_1 \times B_0$ into D_0 and $A_0 \times B_1$ into D_1, where we assume that D_0 and D_1 belong to the same scale of spaces as C_0 and C_1; more precisely let us assume that $D_0 = (C_0,C_1)_{\alpha p_0}$ and $D_1 = (C_0,C_1)_{1-\alpha,p_1}$ with $0 < \alpha < 1$ and $p_0,p_1 \leq \infty$. It then follows from the result below that T maps $(A_0,A_1)_{\theta_1 q_1} \times (B_0,B_1)_{\theta_2 q_2}$ into $(C_0,C_1)_{\theta q}$ as soon as $\theta_1,\theta_2 \in (0,1)$ and $q_1,q_2 \in [1,\infty]$, provided $\theta = \alpha\theta_1 + (1-\alpha)\theta_2$ and (2) holds.

(The results below also cover more general situations where (θ_1,θ_2) only may range over a subset of the square $(0,1)^2$,

<u>Remark 1.</u> In the simple case just considered, one may also use an iterative approach as follows. By interpolating the <u>linear</u> operator $a \to T(a,b)$ with $b \in B_0$ or B_1 fixed, we obtain that T maps $(A_0,A_1)_{\theta q_1} \times B_j$ into $(C_0,C_1)_{\alpha\theta + (1-\alpha)j,q_1}$ $(j = 0$ or $1)$. A second interpolation then shows that T maps $(A_0,A_1)_{\theta q_1} \times (B_0,B_1)_{\theta q_2}$ into $C_{\theta q_2}$. Since we may do the interpolations in the opposite order, this shows that (1) may be replaced by $1/q \leq \max(1/q_1,1/q_2)$, a substantial improvement but still short of (2) (unless q_1 or $q_2 = \infty$).

2. Notation.

We will in the remainder of the paper be concerned with the following set-up, which is more general than the one considered in the introduction. A number of pertinent remarks are collected at the end of this section.

Let $\bar{A}_i = (A_{i0}, A_{i1})$, $1 \le i \le m$, and $\bar{B} = (B_0, B_1)$ be $m+1$ couples of complete quasi-normed Abelian groups (e.g. quasi-Banach spaces, see [1], Section 3.10) and let T be a fixed multi-additive operator of $\prod_{i=1}^{m} A_{i0} \cap A_{i1}$ into $B_0 + B_1$.

Let, for $0 < \theta < 1$ and $0 < q \le \infty$, $A_{i\theta q}$ and $B_{\theta q}$ be the real method interpolation spaces for the couples $\bar{A}_i$ and $\bar{B}$ respectively (equipped with any of the possible equivalent quasi-norms); when $\theta = 0$ or 1 we define $A_{i\theta q} = A_{i\theta}$ and $B_{\theta q} = B_\theta$ for all q.

For typographical convenience we will omit the letters A and B in the quasi-norms and write e.g. $\| \ \|_{i0}$ for the quasi-norm in A_{i0} and $\| \ \|_{\theta q}$ for some quasi-norm in $B_{\theta q}$.

Recall that any quasi-norm is equivalent to a p-norm for some p, $0 < p \le 1$. (A p-norm is a quasi-norm such that $\| \ \|^p$ is subadditive.) Hence we may assume that our spaces are p-normed for some $p > 0$.

We write $T: \prod_1^m A_{i\theta_i q_i} \to B_{\theta q}$ if the range of T is contained in $B_{\theta q}$ and T satisfies the norm inequality

$$\|T(a_1, \dots, a_m)\|_{\theta q} \le C \prod_1^m \|a_i\|_{i\theta_i q_i}, \quad a_i \in A_{0i} \cap A_{1i}, \tag{3}$$

for some $C < \infty$. (C denotes in this paper positive constants that may depend on T and the spaces and norms involved (and thus on the indices $\theta, q, \dots$).)

Finally, we fix real numbers $\alpha_0, \alpha_1, \dots, \alpha_m$ with $\alpha_1, \dots, \alpha_m$ non-zero, and define

$$\Omega = \{(\theta_1, \dots, \theta_m) \in [0,1]^m : 0 \le \alpha_0 + \sum_1^m \alpha_i \theta_i \le 1 \text{ and } T: \prod_1^m A_{i\theta_i q_i} \to B_{\theta q},$$

$$\text{with } \theta = \alpha_0 + \sum_1^m \alpha_i \theta_i, \text{ for some } q_1, \dots, q_m, \ q \in (0, \infty]\}.$$

Remark 2. The main reason for treating quasi-normed Abelian groups and not insisting upon vector spaces is the technical flexibility that it gives us (cf. the proof of Theorem 3 below), although the increased generality might also be useful in some applications.

Remark 3. The completeness of the spaces is not essential. Its main use is in Lemma 1 below, but if we instead take that result as a definition, the results extend to non-complete quasi-normed Abelian groups. (It is then really the case of interpolating two (quasi-)norms on the same space (group).)

Remark 4. The notation $T:\prod A_{i\theta_i q_i} \to B_{\theta q}$ is somewhat improper since T only is defined on the product of the intersections $A_{i0} \cap A_{i1}$. However, if every $q_i < \infty$, it means that T has a unique extension to a bounded multi-additive mapping of the entire product $\prod A_{i\theta_i q_i}$ into $B_{\theta q}$. If some q_i equals ∞, this is in general false (unless we replace $A_{i\theta_i\infty}$ by $(A_{i0},A_{i1})^0_{\theta_i\infty}$, the closure of $A_{i0} \cap A_{i1}$ in $A_{i\theta_i\infty}$), but we will in Remark 10 below see an important case where such an extension always is possible.

Remark 5. If we assume, as in the introduction, that T actually is defined on $\prod(A_{i0} + A_{i1})$ as a continuous multi-additive operator into some Hausdorff Abelian group that contains B_0 and B_1 continously, then the results of this paper may be applied to the restriction of T to $\prod_1^m A_{i0} \cap A_{i1}$. It is immediately seen that the extensions in Remarks 4 and 10 coincide with restrictions of the original operator T. Hence T itself maps the product $\prod A_{i\theta_i q_i}$ into $B_{\theta q}$.

Remark 6. The assumption that $\alpha_i (i \geq 1)$ be non-zero means that all couples $\bar{A}_i$ effectively take part in the interpolation. In general, our theorems fail if some α_i equals zero; but we may always treat such situations by keeping the corresponding variables) fixed and using our results on the remaining variables.

Remark 7. Perhaps it would be better to formulate the theory for general real scales of spaces $\{A_{i\theta q}\}$, $\{B_{\theta q}\}$ defined on some (maybe different) intervals, cf. [4].

3. Main results.

In order to avoid all worries about convergence and continuity when infinite sums are involved, we begin with a simple lemma which shows that on the intersection $A_{i0} \cap A_{i1}$, the interpolation norms $\| \ \|_{i\theta q}$ may be defined by a finite J-method.

Lemma 1. Let $0 < r < 1$ or $1 < r < \infty$. Then, for $a \in A_{i0} \cap A_{i1}$, $\|a\|_{i\theta q}$ is equivalent to

$$\inf\{\|\{r^{-\theta n} J(r^n, a_n, \bar{A}_i)\}_{-N}^{N}\|_{\ell^q} : N < \infty \text{ and } \sum_{-N}^{N} a_n = a\}. \tag{4}$$

Proof. It suffices to consider $r > 1$. It is well-known that the norm $\|a\|_{i\theta q}$ may be defined by (3.1) if we allow infinite decompositions $a = \sum_{-\infty}^{\infty} a_n$ (converging in the sum norm). Given such a decomposition and $N > 0$, set $a'_N = \sum_{N}^{\infty} a_n$ and $a'_{-N} = \sum_{-\infty}^{-N} a_n$. If A_{i0} and A_{i1} are normed, then

$$\|a'_N\|_{i0} \le \|a\|_{i0} + \sum_{-\infty}^{N-1} \|a_n\|_{i0} \le \|a\|_{i0} + \sum_{-\infty}^{N-1} r^{n\theta} \|\{r^{-n\theta} J(r^n, a_n)\}\|_{\ell^\infty} \le$$

$$\le r^{N\theta}(r^{-N\theta}\|a\|_{i0} + C\|\{r^{-n\theta} J(r^n, a_n)\}\|_{\ell^q})$$

and

$$\|a'_N\|_{i1} \le \sum_{N}^{\infty} \|a_n\|_{i1} \le \sum_{N}^{\infty} r^{n(\theta-1)} \|\{r^{-n\theta} J(r^n, a_n)\}\|_{\ell^\infty} \le$$

$$\le Cr^{N(\theta-1)} \|\{r^{-n\theta} J(r^n, a_n)\}\|_{\ell^q} .$$

Together with similar estimates for a'_{-N}, this shows that we can use the decomposition $a = a'_{-N} + \sum_{-N+1}^{N-1} a_n + a'_N$ in (3), with N large enough; we omit the details. The same argument applies also when the spaces are not normed, since we may assume that they are p-normed for some $p > 0$ and take p:th powers in the calculations above. □

Lemma 2. The following two conditions are equivalent, for every $(\theta_1,\ldots,\theta_m) \in [0,1]^m$.

(i) $(\theta_1,\ldots,\theta_m) \in \Omega$

(ii) $\theta = \alpha_0 + \Sigma\,\alpha_i\,\theta_i \in [0,1]$, and

$$\|T(a_1,\ldots,a_m)\|_{\theta\infty} \le C \prod_{i=1}^{m} \|a_i\|_{i0}^{1-\theta_i}\,\|a_i\|_{i1}^{\theta_i} \; . \qquad (5)$$

These conditions imply

(iii) $K(t, T(a_1,\ldots,a_m),\bar{B}) \le Ct^{\alpha_0} \prod_{i=1}^{m} t^{\alpha_i\theta_i}\,\|a_i\|_{i0}^{1-\theta_i}\,\|a_i\|_{i1}^{\theta_i}$.

When $0 < \alpha_0 + \Sigma\,\alpha_i\,\theta_i < 1$, all three conditions are equivalent.

Proof. If $(\theta_1,\ldots,\theta_m) \in \Omega$, then (ii) follows because

$\|a_i\|_{i\theta_i q_i} \le C\|a_i\|_{i0}^{1-\theta_i}\,\|a_i\|_{i1}^{\theta_i}$, and $\|\;\|_{\theta\infty} \le C\,\|\;\|_{\theta q}$.

Conversely, if (ii) holds and $B_{\theta\infty}$ is p-normed, then (using Lemma 1) $T: \prod_1^m A_{i\,\theta_i\,p} \to B_{\theta\infty}$, because $\|a_i\|_{i0}^{1-\theta_i}\,\|a_i\|_{i1}^{\theta_i} \le t^{-\theta_i} J(t,a_i)$ for every $t > 0$.

The implication (ii) $\Rightarrow$ (iii) when $\theta = 0$ or 1 and the equivalence of (ii) and (iii) when $0 < \theta < 1$ are immediate. □

Remark 8. All three conditions are equivalent also when $\theta = 0$ or 1. provided the couple $\bar{B}$ is Gagliardo complete. We will, however, not need this.

Remark 9. The proof shows that we may always take $q_1 = \ldots = q_m = p$ and $q = \infty$ in the definition of Ω, provided $B_{\theta\infty}$ is p-normed. In particular, in the realm of normed spaces, $(\theta_1,\ldots,\theta_m) \in \Omega$ iff $T: \prod A_{i\theta_i 1} \to B_{\theta\infty}$.

Theorem 1. Ω is convex.

Proof. By Lemma 2. □

If we do not care about the indices q_i and q, Theorem 1 is the only interpolation result that we need. However, we can now be much more precise.

<u>Theorem 2.</u> If $(\theta_1,\dots,\theta_m)$ belongs to the interior of Ω, $\theta = \alpha_0 + \sum_1^m \alpha_i \theta_i$ and $1/q \le \sum_1^m 1/q_i$, then $T: \prod_1^m A_{i\theta_i q_i} \to B_{\theta q}$

<u>Proof.</u> Let $a_i \in A_{i0} \cap A_{i1}$, $i = 1,\dots,m$. We use Lemma 1 with $r = 2^{\alpha_i}$; thus we may write each a_i as a finite sum $a_i = \sum_j a_{ij}$ such that if we set

$$x_{ij} = 2^{-\theta_i \alpha_i j} J(2^{\alpha_i j}, a_{ij}, \bar{A}_i),$$

then

$$\|\{x_{ij}\}_j\|_{\ell^{q_i}} \le C\|a_i\|_{i\,\theta_i\,q_i}. \tag{6}$$

For $J = (j_1,\dots,j_m) \in \mathbf{Z}^m$, let $b_J = T(a_{1j_1},\dots,a_{mj_m})$. Then, for any $(\lambda_1,\dots,\lambda_m) \in \Omega$, Lemma 2 yields

$$\begin{aligned} t^{-\theta}K(t,b_J,\bar{B}) &\le Ct^{-\theta}t^{\alpha_0}\prod_1^m t^{\alpha_i\lambda_i}\|a_{ij_i}\|_{i0}^{1-\lambda_i}\ \|a_{ij_i}\|_{i1}^{\lambda_i} \le \\ &\le Ct^{\alpha_0-\theta}\prod_1^m t^{\alpha_i\lambda_i}\, 2^{-\lambda_i\alpha_i j_i}\, J(2^{\alpha_i j_i}, a_{ij_i}, \bar{A}_i) = \\ &= C\prod_1^m t^{\alpha_i(\lambda_i-\theta_i)}\, 2^{(\theta_i-\lambda_i)\alpha_i j_i}\, x_{ij_i}. \end{aligned} \tag{7}$$

(C may depend on $(\lambda_1,\dots,\lambda_m)$.) Taking $t = 2^n$, $n \in \mathbf{Z}$, we obtain

$$2^{-n\theta}K(2^n,b_J) \le C\prod_1^m 2^{(\theta_i-\lambda_i)\alpha_i(j_i-n)}\, x_{ij_i}. \tag{8}$$

If $\delta > 0$ is small enough, we may here choose $\lambda_i = \theta_i \pm \delta/\alpha_i$ with arbitrary signs. Hence we obtain

$$2^{-n\theta}K(2^n,b_J) \le C\, 2^{-\delta\Sigma|j_i-n|}\prod_1^m x_{ij_i}. \tag{9}$$

We subsitute $J + n = (j_1+n,\dots,j_m+n)$ for J and obtain by Hölder's inequality and (6)

$$\begin{aligned} \|\{2^{-n\theta}K(2^n,b_{J+n})\}_n\|_{\ell^q} &\le C\,2^{-\delta|J|}\,\|\{\prod_1^m x_{ij_i+n}\}_n\|_{\ell^q} \le \\ &\le C\,2^{-\delta|J|}\prod_1^m\|\{x_{ij_i+n}\}_n\|_{\ell^{q_i}} \le C\,2^{-\delta|J|}\prod_1^m \|a_i\|_{i\theta_i q_i}. \end{aligned} \tag{10}$$

Finally we sum over J and obtain by Minkowski's inequality (first taking an appropriate power if $q < 1$ or B is not normed) the sought inequality

$$\| 2^{-n\theta} K(2^n, T(a_1,\ldots,a_m)) \|_{\ell^q} \leq C \prod_1^m \|a_i\|_{i\theta_i q_i} . \tag{11}$$

Remark 10. If the conditions of Theorem 2 are satisfied, then T can be extended to $\prod A_{i\theta_i q_i}$ also when some (or all) $q_i = \infty$. Viz., if ε is so small that $(\theta_1 \pm \varepsilon, \ldots, \theta_m \pm \varepsilon) \in \operatorname{int}(\Omega)$, then T may be uniquely extended to a bounded multi-additive mapping of $\prod(A_{i,\theta_i-\varepsilon,1} + A_{i,\theta_i+\varepsilon,1})$ into $B_0 + B_1$, and it is not difficult to show that the restriction of this extension to $\prod A_{i\theta_i q_i}$ maps into $B_{\theta q}$.

4. The boundary of Ω.

Theorem 2 gives a satisfactory result when $\bar\theta = (\theta_1,\ldots,\theta_m)$ lies in the interior of Ω, but what can be said when $\bar\theta$ lies on the boundary? Remark 9 above gives a choice of $q_1,\ldots,q_m,q$ that always works, and it is in general the best result when $\bar\theta$ is an extreme point of Ω. If $\bar\theta$ is not an extreme point, however, then a better result exists (assuming some non-degeneracy conditions on the geometry). In fact, the following theorem is just the Lions-Peetre-Zafran result stated in the introduction disquised in our terminology. The extension to quasi-Banach spaces was made by Karadzov [5]. (The theorem is sharper than his, however, since our condition sometimes allows a larger p than the result in [5], cf. Remark 12 below.)

Let ℓ_i, $i=1,\ldots,m$, denote the coordinate mappings $\theta \to \theta_i$, and ℓ_0 the linear form $\sum_1^m \alpha_i \theta_i$ (i.e. $\ell_0 = \sum \alpha_i \ell_i$).

Theorem 3. Suppose that $(\theta_1,\ldots,\theta_m)$ lies in the interior of a line segment $L \subset \Omega$ such that the $n+1$ forms $\ell_0,\ldots,\ell_m$ all are non-constant on L. Let $\theta = \alpha_0 + \Sigma\alpha_i \theta_i$. Suppose that p, $q_1,\ldots,q_m$, q are such that

$$B_{\theta q} \text{ is p-normed, } p \leq q_i \leq \infty, \text{ and } 0 \leq 1/q \leq \sum_1^m 1/q_i - (m-1)/p. \tag{12}$$

(This implies $p \leq q \leq \infty$.) Then $T: \prod A_{i\theta_i q_i} \to B_{\theta q}$.

Proof. We may replace all quasi-norms by their p:th powers (and q, q_i by q/p, q_i/p). Ω remains the same (cf. Lemma 2) and the power theorem ([1], Theorem 3.11.6) shows that we have reduced the theorem to the case $p = 1$. We will thus assume that $1 \leq q_i \leq \infty$, that $B_{\theta q}$ is normed and that

$$1/q = \sum_1^m 1/q_i - (m-1). \tag{13}$$

We use the notation of the proof of Theorem 2. Since (8) holds for $(\lambda_1,\ldots,\lambda_m) \in L$, there exist $\beta = (\beta_1,\ldots,\beta_m)$ and $\gamma = \Sigma\, \beta_i$ with $\beta_1,\ldots,\beta_m$, γ all non-zero $(\beta_i = (\lambda_i - \theta_i)\alpha_i$, for a suitable $(\lambda_1,\ldots,\lambda_m) \in L)$, such that

$$2^{-n\theta} K(2^n, b_J) \leq C\, 2^{-|\Sigma \beta_i j_i - \gamma n|} \prod_1^m x_{ij_i} . \tag{14}$$

We treat first the case when both B_0 and B_1 are normed. Then $K(2^n,\cdot)$ is a norm, and by (14) and (6) it suffices to show that

$$\| \{\sum_J 2^{-|\beta\cdot J - \gamma n|} \prod_1^m x_{ij_i}\}_n \|_{\ell^q} \leq C \prod_1^m \| (x_{ij})_j \|_{\ell^{q_i}} \tag{15}$$

for all (finite) sequences of positive numbers $\{x_{ij}\}$, provided $1 \leq q_i \leq \infty$ and (13) holds.

In fact, we prove (15) for all sequences of complex numbers $\{x_{ij}\}$. By the multi-linear interpolation theorem for the <u>complex</u> method (cf. the introduction), applied to the mapping $((x_{1j}),\ldots,(x_{mj})) \to (\sum_J 2^{-|\beta\cdot J - \gamma n|} \prod_1^m x_{ij_i})$, it suffices to consider the cases $q_1 = \ldots = q_m = q = 1$ and $q_2 = \ldots = q_m = 1$, $q_1 = q = \infty$ (the cases when some other $q_i = \infty$ and the remaining ones equal 1 are the same by symmetry). Both cases are easy and are left to the reader.

The argument with powers above now shows that the theorem holds provided both B_0 and B_1 are p-normed.

We return to the general case (with $p = 1$), where we only assume that $B_{\theta q}$ is normed. ($q \geq 1$ is fixed in the remainder of the proof.) There exists $r > 0$ such that B_0 and B_1 are r-normed and thus, by what we already have proved, $T: \prod A_{i\theta_i q_i} \to B_{\theta q}$ provided $q_1 = q$

and $q_2 = \ldots = q_m = r$. Since $B_{\theta q}$ is assumed to be normed, it immediately follows that we may as well take $q_1 = q$ and $q_2 = \ldots = q_m = 1$ (if $a \in A_{i\theta_i 1}$, then $a = \Sigma a_n$ with $\Sigma \|a_n\|_{i\theta_i r} \le C\|a\|_{i\theta_i 1}$, cf. Lemma 1). Now consider the set

$$E = \{(1/q_1,\ldots,1/q_m) \in [0,1]^m : T: \prod A_{i\theta_i q_i} \to B_{\theta q}\} \quad .$$

We have shown that $(1/q, 1,\ldots,1) \in E$, and by symmetry also $(1, 1/q,\ldots,1),\ldots,(1,1,\ldots,1/q) \in E$. Furthermore, by applying the already proved case of the theorem to the couples $(A_{i\theta_i\infty}, A_{i\theta_i 1})$ and $(B_{\theta q}, B_{\theta q})$ (extending T to $\prod A_{i\theta_i 1}$), and using reiteration, it follows that E is convex. (It is in this step sometimes necessary to keep some of the variables fixed and interpolate only the others (as in Remark 6), but that is a trivial complication.)

Consequently (13) implies that $(1/q_1,\ldots,1/q_m) \in E$ and the proof is completed. □

<u>Remark 11.</u> Suppose that (for a fixed θ) $B_{\theta q}$ is normed for $q \ge 1$, and q-normed for $q < 1$. (E.g., let $\bar{B}$ be normed.) The theorem then applies with $p = 1$ for $q_1,\ldots,q_m$, $q \ge 1$. For $q \le 1$ we obtain, with $p = q$, only the case $q_1 = \ldots = q_m = q$. We can trivially extend the latter case to $q_1,\ldots,q_m \le q \le 1$, and the first case to

$$1/q \le \sum_1^m 1/\max(q_i,1) - (m-1) = 1 - \sum_1^m (1-1/q_i)_+ \ ,$$

where $0 < q_i \le \infty$. In general, these results are the best possible as is seen by the example in the introduction.

<u>Remark 12.</u> If B_0 is p_0-normed and B_1 is p_1-normed, then $B_{\theta q}$ is $\min(p,q)$-normed, with $1/p = (1-\theta)/p_0 + \theta/p_1$. (This follows by the power theorem.) This gives the result in [5].

When $\bar{\theta}$ belongs to a facet of Ω of higher dimension than 1, better results can be obtained. For simplicity we consider only the normed case. Note that the special cases $d = 0,1$ and n are given by Remark 9, Theorem 3 and Theorem 2, respectively; the theorem may be interpreted as saying that the more degrees of freedom that we have in the interpolation, the better result we get.

Theorem 4. Suppose that M is a d-dimensional subspace of $\mathbb{R}^m$ such that any d of the $m+1$ forms $\ell_0,\dots,\ell_d$ are linearly independent on M, and suppose that $\bar{\theta} = (\theta_1,\dots,\theta_m) \in \Omega$ is such that $\bar{\theta} + M \cap U \subset \Omega$ for some neighbourhood of the origin U. Let $\theta = \alpha_0 + \Sigma\alpha_i\theta_i$. If B_0 and B_1 are normed and $q_1,\dots,q_m$, $q \in [1,\infty]$ are such that

$$1/q \le \Sigma\, 1/q_i - (m-d), \tag{16}$$

then $T: \prod A_{i\theta_i q_i} \to B_{\theta q}$.

Proof. This is similar to Theorem 3, so we will be brief. Let u_i denote α_i times the orthogonal projection of the i:th basis vector e_i on M, and let $u_0 = \sum_1^m u_i$ be the projection of $(\alpha_1,\dots,\alpha_m)$. By assumption, any d of $u_0,u_1,\dots,u_m$ are independent.

It follows from (8), choosing suitable $(\lambda_1,\dots,\lambda_m)$, that for some $\delta > 0$

$$2^{-n\theta}K(2^n,b_J) \le C\, 2^{-\delta|\Sigma j_i u_i - nu_0|} \prod x_{ij_i}, \tag{17}$$

and it suffices to show that the multi-linear mapping

$$((x_{1j}),\dots,(x_{mj})) \to \sum_J 2^{-\delta|\Sigma j_i u_i - nu_0|} \prod_1^m x_{ij_i} \quad \text{maps } \prod \ell^{q_i} \text{ into } \ell^q.$$

Again we use complex interpolation and reduce this to the cases $q_1 = q_2 = \dots = q_{d-1} = \infty$, $q_d = \dots = q_m = q = 1$ and $q_1 = q_2 = \dots = q_d = \infty$, $q_{d+1} = \dots = q_m = 1$, $q = \infty$. These cases both follow from the fact that if $v_1,\dots,v_d$ are independent, then for any vector v

$$\sum_{j_1,\dots,j_d} 2^{-\delta|\sum_1^d j_i v_i + v|} \le C. \tag{18}$$

5. An application: convolution.

We illustrate the use of Theorem 2 by considering convolutions of Lorentz spaces. The results of this section are essentially due to O'Neill [7], see also Hunt [3], although they only consider Banach spaces (i.e. $p, q \ge 1$).

We consider functions on $\mathbb{R}$, or more generally functions on a unimodular (e.g. Abelian) locally compact group (with a Haar measure).

Let $\bar{A}_1 = \bar{A}_2 = \bar{B} = (L^{\infty}, L^1)$, and $T(f,g) = f * g$. Then $A_{i\theta q} = B_{\theta q} = L^{1/\theta, q}$ $(0 < \theta < 1,\ 0 < q \leq \infty)$. We choose $\alpha_0 = -1$, $\alpha_1 = \alpha_2 = 1$. Since $L^{\infty} * L^1 \subset L^{\infty}$, $L^1 * L^{\infty} \subset L^{\infty}$ and $L^1 * L^1 \subset L^1$, Ω contains $(1,1)$, $(1,0)$ and $(1,1)$, and thus (by Theorem 1) the triangle $\{(\theta_1,\theta_2) \in [0,1]^2 : \theta_1 + \theta_2 \geq 1\}$. Theorem 2 gives the following result: If $1 < p_1, p_2 < \infty$ and $1/p = 1/p_1 + 1/p_2 - 1 > 0$, then

$$L^{p_1 q_1} * L^{p_2 q_2} \subset L^{pq} \tag{19}$$

holds, provided

$$1/q \leq 1/q_1 + 1/q_2 \,. \tag{20}$$

In particular, if $1 < p_1, p_2 < \infty$ and $1/p = 1/p_1 + 1/p_2 - 1 > 0$, then $L^{p_1} * L^{p_2} \subset L^{pq}$ with $1/q = 1/p_1 + 1/p_2$. Note that, in this case, $q < 1$ and that L^{pq} is not a Banach space.

Simple examples show that (e.g. on $\mathbb{R}$), (20) is also necessary for (19) to hold.

Remark 13. It is easy to treat the boundary cases directly. In fact, it is not difficult to show that, (19) holds in the following 5 cases, but not for any other $p_1, p_2, p, q_1, q_2, q \in (0,\infty]$.

(i) $1 < p_1 < \infty$, $1 < p_2 < \infty$, $0 < 1/p = 1/p_1 + 1/p_2 - 1$, $1/q \leq 1/q_1 + 1/q_2$ (the case treated above).

(ii) $1 < p_1 < \infty$, $1 < p_2 < \infty$, $1/p_1 + 1/p_2 = 1$, $p = \infty$, $1/q_1 + 1/q_2 \geq 1$.

(iii) $p_1 = 1$, $p_2 = p = \infty$, $q_1 \leq 1$ (or $p_2 = 1$, $p_1 = p = \infty$, $q_2 \leq 1$)

(iv) $p_1 = 1$, $1 < p_2 = p < \infty$, $q_1 \leq 1$, $q_2 \leq q < \infty$ (or conversely)

(v) $p_1 = p_2 = p = 1$, $q_1 \leq 1$, $q_2 \leq 1$, $1 \leq q$.

References

1. J. Bergh and J. Löfström, Interpolation spaces. Grundlehren Math. Wiss. 223, Springer-Verlag, Berlin-Heidelberg-New York 1976.
2. A.P. Calderón, Intermediate spaces and interpolation, the complex method. Studia Math. 24 (1964), 113-190.
3. R.A. Hunt, On L(p,q) spaces. Enseign. Math. 12 (1966), 249-276.

4. S. Janson, P. Nilsson and J. Peetre, Notes on Wolff's note on interpolation spaces. Proc. London Math. Soc. 48 (1984), 283-299.

5. G.E. Karadzov, An interpolation method of "means" for quasi-normed spaces. Dokl. Akad. Nauk SSSR 209:1 (1973), 33-36 (Russian). Translation in Soviet Math. Dokl. 14 (1973), 331-335.

6. J.L. Lions and J. Peetre, Sur une classe d'espaces d'interpolation. Inst. Hautes Études Sci. Publ. Math. 19 (1964), 5-68.

7. R. O'Neill, Convolution operators and L(p,q) spaces. Duke Math. J. 30 (1963), 129-142.

8. M. Zafran, A multilinear interpolation theorem. Studia Math. 62 (1978), 107-124.

MARKOV'S INEQUALITY AND LOCAL POLYNOMIAL APPROXIMATION

Alf Jonsson
Department of Mathematics
University of Umeå
S-901 87 Umeå, Sweden

0. Introduction

Let P be a polynomial in one variable of degree at most k. The classical Markov's inequality states that

$$\max_B |dP/dx| \leq cr^{-1}\max_B |P|,$$

where B is a closed interval of length $2r$, and the constant c is given by $c=k^2$. This can be generalized in many directions, see e.g. [1], [3], and [6] (cf. the comments in the third remark after Proposition 3 in this paper). In [3] (see also [4], [5], and [7]), we studied the following generalization to closed subsets F of $\mathbb{R}^n$, motivated by the need in the theory of function spaces on subsets of $\mathbb{R}^n$. Let $B(x,r)$ denote the closed ball with center x and radius r, and let P be a polynomial in n variables of degree at most k. The set F preserves Markov's inequality if there is a constant c, depending on k, n, and F, only, such that

$$\max_{B\cap F} |\text{grad } P| \leq cr^{-1}\max_{B\cap F} |P| \tag{1}$$

for every ball $B=B(x,r)$, $x \in F$, $r\leq 1$. In [3], sets F having this property are studied. In particular, they are characterized geometrically, and an L^p-version of (1) is given.

In this note we make some remarks on the results in [3]. We are motivated by the study of function spaces on manifolds with singularities in [2], where the observations made here are needed. Our approach differs from the one sketched above in the respect that we do not assume that (1) holds for all balls $B(x,r)$, $x \in F$, $0<r\leq 1$, but only for balls with centers in a subset S of F (Definition 1 below). The properties of the concept then obtained, are similar to those for sets

preserving Markov's inequality given in [3]. An essential difference is, however, that the concept given in Definition 1 depends in an essential way on k, the degree of the polynomials under consideration, see Example 1. The main results in Section 1 are Proposition 2 and Proposition 3, giving an L^p-version and, for $k=1$, a geometric characterization of sets preserving Markov's inequality at a subset.

The results are, in Section 2, applied to the theory of function spaces. In Theorems 1 and 2 we give constructive characterizations of Besov and Lipschitz spaces of functions defined on certain subsets of $\mathbb{R}^n$. The basic tool is then a lemma on polynomial approximation (Lemma 1), which requires the preceeding results in a crucial way.

1. Sets preserving Markov's inequality at a subset

In this section, we shall study the concept introduced in the following definition. The closed ball with center x_0 and radius r is denoted by $B(x_0,r)$.

DEFINITION 1. Let $S \subset E \subset \mathbb{R}^n$, and let k be a positive integer. The set E preserves Markov's inequality for polynomials of degree $\leq k$ at S if the following condition holds: For all polynomials P of degree at most k and all balls $B=B(x_0,r)$, $x_0 \in S$, $0<r\leq 1$, we have

$$\sup_{E\cap B}|\operatorname{grad} P| \leq cr^{-1}\sup_{E\cap B}|P|, \tag{2}$$

with a constant c depending on E, S, n, and k, only.

To get a shorter notation, we shall say that (E,S) has the property M_k if E preserves Markov's inequality for polynomials of degree $\leq k$ at S. Below, and in the rest of the paper, c denotes a positive constant, not necessarily the same each time it appears. We use the usual multi-index notation, see e.g. [3, p.xiii].

Suppose that (2) holds for a certain ball B and all polynomials P of degree at most k. Let $t>0$ and let $\sum_{|j|\leq k} a_j(x-x_0)^j$ be the Taylor expansion of P around x_0. A repeated application of (2) shows that

$$|a_j| = |D^jP(x_0)/j!| \leq cr^{-|j|}\sup_{E\cap B}|P|$$

from which it follows that

$$\sup_{tB}|P| \leq c \sup_{E\cap B}|P|, \tag{3}$$

where $tB=B(x_0,tr)$ and c depends on t, k, and the constant c in (2). This shows that the condition $0<r\leq 1$ in Definition 1 may be replaced by $0<r\leq r_0$, where r_0 is any positive number. We remark that if (3) holds, then (2) holds. This follows from

$$\sup_{E\cap B}|\text{grad } P| \leq \sup_{B}|\text{grad } P| \leq cr^{-1}\sup_{B}|P| \leq cr^{-1}\sup_{E\cap B}|P|.$$

Here the second inequality follows from the fact that $E=\mathbb{R}^n$ preserves Markov's inequality, cf. the second remark after Proposition 3 below. This means that the condition (2) in Definition 1 can be replaced by the condition that (3) holds for $t=1$.

If E is closed and $E=S$ and (E,S) has the property M_1, then (E,S) has the property M_k for any k, see [3, p.37]. This means that E preserves Markov's inequality in the sense defined in the introduction. The analogue of this does not hold in general if $S\subset E$, as is seen by the following example.

Example 1. Let $E\subset\mathbb{R}^2$ be the set $\{(x_1,x_2);x_1x_2=0\}$ and let $S=\{(0,0)\}$. Then (E,S) has the property M_1 (cf. Proposition 3 below), but not the property M_2, which is seen by taking $P=x_1x_2$.

If (E,S) has the property M_k, then (2) is satisfied for every ball $B=B(x_0,r)$, $x_0\in\bar{S}$, $0<r\leq 1$, where $\bar{S}$ is the closure of S. To see this, take a sequence $x_n\in S$, $n=1,2,3,\ldots$, such that $x_n\to x_0$, and balls $B_n=B(x_n,r_n)$ such that the balls B_n are precisely included in $B(x_0,r)$. Then, by (2) and (3),

$$\sup_{B_n}|\text{grad } P| \leq c \sup_{E\cap B_n}|\text{grad } P| \leq c\, r_n^{-1} \sup_{E\cap B_n}|P| \leq cr_n^{-1} \sup_{E\cap B}|P|,$$

and (2) follows by letting n tend to infinity. Similarly, if (E,S) has the property M_k, then (2) holds for every ball $B=B(x_0,r)$, $x_0\in S$, $0<r\leq 1$ if we replace E in (2) by $\bar{E}$, since

$$\sup_{B}|\text{grad } P| \leq c \sup_{E\cap B}|\text{grad } P| \leq cr^{-1} \sup_{E\cap B}|P|.$$

These and similar remarks give the following proposition, which shows that it would not be an essential limitation to assume in Definition 1 that E and S are closed.

PROPOSITION 1. *Let* $S \subset E \subset \mathbb{R}^n$, *and let* k *be a positive integer. Then* (E,S) *has the property* $M_k \Leftrightarrow (E, \bar{S}\cap E)$ *has the property* $M_k \Leftrightarrow (\bar{E},S)$ *has the property* $M_k \Leftrightarrow (\bar{E},\bar{S})$ *has the property* M_k.

To say that (E,S) has the property M_k is not the same thing as to say that $(E,\{x\})$ has the property M_k for every $x \in S$, even if S is compact. We give an example which shows this.

Example 2. Define a continuous function f of one variable by $f(1/n)=2^{-n}/n$, $n\geq n_0$ (n and n_0 are positive integers), $f(x)=0$ if $x\geq 0$ and x is not in an open interval $I_n=(1/n-2^{-n},1/n+2^{-n})$, $n\geq n_0$, f is linear in the intervals $(1/n-2^{-n}, 1/n)$ and $(1/n,1/n+2^{-n})$, and $f(x)=x$, $x<0$. Here n_0 is so big that the intervals I_n are disjoint. Consider now the set $E\subset \mathbb{R}^2$ given by the curve $E=\{(x,y);y=f(x),x\in\mathbb{R}\}$, and let $S=\{(1/n, 2^{-n}/n), n\geq n_0\}\cup\{0,0\}$. Then (E,S) has not the property M_1 (consider the polynomial $P(x,y)=y$ in the balls with centers $(1/n, 2^{-n})$ and radii 2^{-n}), but from Proposition 3 below, it follows that $(E,\{x\})$ has the property M_1 for every $x \in S$. To see this easily, recall that the condition $0<r\leq 1$ in Definition 1 and thus also in Proposition 3 may be replaced by $0<r\leq r_0$.

The next two propositions in this section are generalizations of propositions given in [3], Chapter II, Section 2. The proofs given here are essentially the same as in [3]. In the first proposition, μ denotes a positive Borel measure with $\mu(CE)=0$, finite on bounded sets, satisfying the doubling condition

$$0 < \mu(B(x,2r)) \leq c\mu(B(x,r)), \quad x \in E, \quad 0<r\leq 1, \tag{4}$$

where c is a positive constant. It has recently been proved, in [8], that any closed set carries a measure satisfying this doubling condition.

PROPOSITION 2. *Let* $1 \leq q \leq \infty$, $S \subset E \subset \mathbb{R}^n$, k *a positive integer, let* μ *be as above, and assume that* (E,S) *has the property* M_k. *Then*

$$\left\{ \frac{1}{\mu(B)} \int_B |P(x)|^q d\mu(x) \right\}^{1/q} \sim \sup_{E \cap B} |P| \tag{5}$$

<u>for all balls</u> $B=B(x_0,r)$, $x_0 \in S$, $0<r\leq 1$, <u>and all polynomials</u> P <u>of degree at most</u> k. <u>Here and in the rest of the paper</u> $a \sim b$ <u>means that</u> $c_1 b \leq a \leq c_2 b$ <u>for some positive constants</u> c_1 <u>and</u> c_2. <u>Here these constants depend on</u> k <u>and the constants</u> c <u>in</u> (2) <u>and</u> (4) <u>only</u>.

<u>Proof.</u> The right-hand inequality involved in (5) follows immediately from (3). Let A be the right-hand side of the relation (5). According to (3), used with $t=2$, there is a point $y \in E \cap B(x_0,r/2)$ with $|P(y)| \geq c\ A$. From (2) and (3) it follows that $|\text{grad } P| \leq c\ A/r$ in B, which means that $|P| \geq c\ A$ in a ball $B(y,c_0 r)$, where $1/2>c_0>0$. Using this one obtains that the left-hand side of (5) is larger than or equal to $c\ A\{\mu(B(y,c_0 r))/\mu(B)\}^{1/q} \geq c\ A$, where the inequality follows from the doubling condition upon writing, with $2^{-i} \leq c_0 < 2^{-i+1}$, $\mu(B(y,c_0 r)) \geq \mu(B(y,2^{-i}r)) \geq c\mu(B(y,2r)) \geq c\mu(B(x_0,r))$. This proves the left-hand inequality involved in (5).

<u>Remark.</u> Let $B=B(x_0,r)$, $x_0 \in S$, $0 < r \leq 1$ be a fixed ball, and assume that (2) holds for this B. This implies that (5) holds for B, if we furthermore assume that μ restricted to B satisfies the doubling condition on $E \cap B$ (but not in general). This follows since we may then choose y from $E \cap B$.

We shall need a slightly more general version of (5). Let $c \geq 1$, and let G be such that

$$B = B(x_0,r) \subset G \subset B(x_0,cr) = B_1, \tag{6}$$

where $x_0 \in S$ and $0<cr\leq 1$. Then we have

$$\left(\frac{1}{\mu(G)} \int_G |P(x)|^q d\mu(x) \right)^{1/q} \sim \sup_{E \cap G} |P|. \tag{7}$$

This follows from (3), (4), and (5), upon writing

$$\sup_{E \cap G} |P| \leq \sup_{E \cap B_1} |P| \leq c \sup_{E \cap B} |P| \leq c \left\{ \frac{1}{\mu(B)} \int_B |P(x)|^q d\mu(x) \right\}^{1/q}$$

and using that $\mu(B) \geq c\mu(B_1) \geq c\mu(E \cap G)$.

PROPOSITION 3. *Let* $S \subset E \subset \mathbb{R}^n$. *Then* (E,S) *has the property* M_1 *if and only if the following geometric condition holds: There is an* $\varepsilon>0$ *so that none of the sets* $E \cap B(x_0,r)$, $x_0 \in S$, $0<r\leq 1$, *is contained in any band of the type* $\{x; |b\cdot(x-x_0)|<\varepsilon r\}$ *where* $|b|=1$.

Proof. Suppose first that the geometric condition does not hold. Then, for any $\varepsilon>0$, we can find x_0, r, and b, as in the proposition such that $E \cap B(x_0,r) \subset \{x; |b\cdot(x-x_0)|<\varepsilon r\}$. If we check (2) on the polynomial $P(x) = b\cdot(x-x_0)$, we see that (2) does not hold.

Suppose next that the condition holds for a certain $\varepsilon>0$, and consider a ball $B(x_0,r)$, $x_0 \in S$, $0<r\leq 1$. Let P be a first-degree polynomial, $P(x)=b\cdot(x-x_0)+a$, where $|b|=1$. There exists an $x_1 \in E \cap B(x_0,r)$ with $|b\cdot(x_1-x_0)| \geq \varepsilon r$. Thus $|P(x_1)-P(x_0)| \geq \varepsilon r$, so at x_0 or x_1 we have $|P| \geq \varepsilon r/2$, and it follows that (2) holds.

Remark. The proof of the proposition gives the following more exact result. Consider a fixed ball $B(x_0,r)$, $x_0 \in E$. Let (i) denote the statement that (2) holds with a certain constant c, and (ii) the statement that the set $E \cap B(x_0,r)$ is not contained in any band of the type $\{x; |b\cdot(x-x_0)|<\varepsilon r\}$ where $|b|=1$. Then (i) implies that (ii) holds with $\varepsilon=1/(2c)$ and conversely (ii) implies that (i) holds with $c=2/\varepsilon$.

Remark. If E is closed and $S=E$, then, by the remarks preceeding Example 1, the geometric condition in Proposition 3 characterizes sets E preserving Markov's inequality.

Remark. Consider the following question. Let $E \subset \mathbb{R}^n$ be bounded. For which sets E is it true that

$$\sup_E |\operatorname{grad} P| \leq c \sup_E |P| \tag{8}$$

for all polynomials P of degree at most k, where c does not depend on P? In a way, this corresponds to the requirement that (2) holds for a fixed ball B, and the question can be answered as in the following proposition. The problem to obtain an estimate on the constant c in (8) is studied e.g. in [1] and [6].

PROPOSITION 4. *Let* E *be a bounded subset of* $\mathbb{R}^n$ *and let* $k\geq 1$. *Then there is a constant* c *such that* (8) *holds for all polynomials of* degree at most k, if and only if E is not a subset of some algebraic variety $\{x; P_0(x)=0,\ P_0$ a nonzero polynomial of degree at most $k\}$.

Proof. Take $x_0 \in E$ and $B = B(x_0,r)$ so that $B \supset E$. By the discussion after Definition 1, we may replace the condition (8) in the proposition by

$$\sup_B |P| \leq c \sup_E |P|,$$

and we prove the proposition with (8) replaced by this condition. The only-if part is then obvious, and we shall prove the converse part.

Assume that c does not exist, and choose polynomials $\tilde{P}_i$ of degree at most k so that $||\tilde{P}_i||_B = c_i ||\tilde{P}_i||_E$, where $c_i \to \infty$, $i \to \infty$, and $||\cdot||_E$ denotes the sup-norm over E. Considering the polynomials $P_i = \tilde{P}_i / ||\tilde{P}_i||_B$ we get polynomials P_i with $||P_i||_B = 1$ and $||P_i||_E \to 0$, $i \to \infty$.

In the Taylor expansion $P_i(x) = \Sigma\, a_k^i (x-x_0)^k$, we have, by Markov's inequality for the ball, $|a_k^i| < c$, where c does not depend on i. Thus we can choose a subsequence of the polynomials $\{P_i\}$ so that all $\{a_k^i\}$, $i=1,2,\ldots$, converge to a number a_k. The subsequence converges uniformly to P_0 given by $P_0(x) = \Sigma\, a_k (x-x_0)^k$; consequently $||P_0||_B = 1$ and $||P_0||_E = 0$. Thus $E \subset \{x; P_0(x)=0\}$ and P_0 is nonzero, and the if-part of the theorem follows.

The following example is our main motivation for studying sets preserving Markov's inequality at a subset. For examples in the case when E is closed and $S=E$, i.e. the case when E preserves Markov's inequality, we refer to [3, p. 39].

Example 3. Let E be the surface of a tetrahedron in $\mathbb{R}^3$, and let S be the union of the edges and corners of E. Then it is clear from Proposition 3 that (E,S) has the property M_1. More generally, it is clear that for a large class of $(n-1)$-dimensional manifolds in $\mathbb{R}^n$ with singularities on a subset S, we have that (E,S) has the property M_1.

2. Applications

The result in the previous section will be used to prove theorems on function spaces defined by means of local polynomial approximation. Then we need the following lemma. In the lemma, $\{\pi_j\}_{|j|\leq k}$ denotes an orthonormal base in the subspace V of $L^2(\mu,G)$, which consists of polynomials of degree at most k. This base may e.g. be the base obtained by orthogonalizing the polynomials c_j in part a) of the lemma by means of the Gram-Schmidt orthogonalization process. $L^2(\mu,G)$ denotes the L^2-space obtained by restricting μ to G.

LEMMA 1. *Let* $S \subset E \subset \mathbb{R}^n$, $1 \leq p \leq \infty$, k *a positive integer, let* μ *be a measure on* E *satisfying the doubling condition* (4), *and assume that* E *preserves Markov's inequality for polynomials of degree at most* k *at* S. *Let* G *satisfy* (6) *for some* $x_0 \in S$ *(thus* G *may e.g. be equal to a ball* $B(x_0,r)$, $x_0 \in S$, $0<r\leq 1$). *Then holds:*

a) *The polynomials* $\tau_j=(x-x_0)^j$, $|j|\leq k$, *are linearly independent in* $L^2(\mu,G)$.

b) *Let* L *be the projection of* $L^1(\mu,G)$ *onto* V *given by* $Lf = \sum_j \pi_j \int_G f\pi_j d\mu$, *and let* P *be a polynomial of degree at most* k. Then

$$\int_G |f-Lf|^p d\mu \leq c \int_G |f-P|^p d\mu,$$

(usual modification if $p=\infty$), *for* $f \in L^1(\mu,G)$, *where* c *does not depend on* f,P, $\{\pi_j\}$, or G.

Proof. Suppose that $\tau = \sum_j \lambda_j\tau_j$ equals zero in $L^2(\mu,G)$. From (7) we get that $\sup_{E\cap G}|\tau| = 0$, and thus $\sup_{E\cap B}|\tau| = 0$, where B is the ball in (6). Now, $\lambda_j = c\, D^j\tau(x_0)$, and thus a repeated use of Markov's inequality (2) gives $\lambda_j=0$. This proves a).

Using that L is a projection and Hölder's inequality we get

$$\int_G |Lf-P|^p d\mu = \int_G |L(f-P)|^p d\mu = \int_G |\sum_j \pi_j \int_G (f-P)\pi_j d\mu|^p d\mu \le$$

$$\le c \sum_j \int_G |\pi_j|^p (\int_G |f-P||\pi_j| d\mu)^p d\mu \le c \sum_j (\sup_{E\cap G} |\pi_j|)^{2p} \mu(G)^p \int_G |f-P|^p d\mu.$$

From (7) and the normalization of π_j we get $\sup_{E\cap G} |\pi_j| \le c\mu(G)^{-1/2}$, and statement b) in the lemma follows.

The lemma gives a method to construct a polynomial approximation Lf to a given function f. If $k=1$, $p=\infty$, and $f \in C(E)$ (i.e. f is a continuous function defined on E), then this may be achieved in a different way, using Lagrange interpolation.

PROPOSITION 5. <u>Let</u> $S \subset E \subset \mathbb{R}^n$, E <u>closed, let</u> $f \in C(E)$, <u>and assume that</u> E <u>preserves Markov's inequality at</u> S <u>for polynomials of degree at most</u> 1. <u>Then holds:</u>

a) <u>Let</u> $B=B(x_0,r)$, $x_0 \in S$, $0<r\le 1$. <u>Then it is possible to choose</u> $n+1$ <u>points</u> $x^0, x^1, \ldots, x^n$ <u>from</u> $E\cap B$ <u>such that, with</u> $X=\{x^0,\ldots,x^n\}$, <u>for all polynomials</u> P <u>of degree at most</u> 1 <u>we have</u>

$$|\text{grad } P| \le c\, r^{-1} \max_{B\cap X} |P|,$$

<u>where the constant</u> c <u>depends on</u> n <u>and the constant</u> c <u>in</u> (2).

b) <u>Let</u> G <u>satisfy</u> (6), <u>let</u> Lf <u>be the polynomial of degree at most one interpolating</u> f <u>at</u> $\{x^0,\ldots,x^n\}$, <u>where the points</u> x^i <u>are chosen as above with the ball</u> B <u>equal to the ball</u> B <u>in</u> (6), <u>and let</u> P <u>be any polynomial of degree at most one. Then</u>

$$\sup_{G\cap E} |f-Lf| \le c \sup_{G\cap E} |f-P|,$$

<u>where the constant</u> c <u>depends on</u> n <u>and the constants</u> c <u>in</u> (2) <u>and</u> (6).

<u>Proof</u>. We first prove a). Take ε as in Proposition 3, so that no band $|b\cdot(x-x_0)| < \varepsilon r$, $|b|=1$, contains $B\cap E$. Let $x^0=x_0$. Choose $x^1 \in B\cap E$ so far away from x^0 as possible, and let $v_1=x^1-x^0$. Choose $x^2 \in B\cap E$ as far away from the line through x^0 and x^1 as possible, and let $v_2=x^2-x_p^2$, where x_p^2 is the projection of x^2 onto the

line. Let next $x^3 \in B\cap E$ be as far as possible from the two-dimensional affine subspace containing x^0, x^1, and x^2, and let $v_3=x^3-x_p^3$. In this way we choose $x^1,\dots,x^n$ and mutually orthogonal vectors $v_1,\dots,v_n$. The point x^i may be written $x^i=x^0+\sum_{m=1}^{i}\lambda_m v_m$, where $|\lambda_m|\leq 1$ and $\lambda_i=1$. To see that $|\lambda_m|\leq 1$, note that by our choice of x^m, the distance from x^i to the affine subspace spanned by $x^0,x^1,\dots,x^{m-1}$ is less than or equal to $|v_m|$ for $i\geq m$. On the other hand, the vector $v=\sum_{s=m}^{i}\lambda_s v_s$ is orthogonal to the subspace, so this distance is $|v|$ and thus at least $|\lambda_m||v_m|$, so $|\lambda_m|\leq 1$.

Now, let D be any band $|b\cdot(x-x_0)|<\delta r$ containing X, and let λD denote the band $|b\cdot(x-x_0)|<\gamma\delta r$. We claim that the band 3^iD contains all vectors $x^0+\sum_{s=1}^{i}\varkappa_s v_s$ with $|\varkappa_s|\leq 1$. This is clear for $i=1$. Assume that the claim has been proved for $i-1$. As we saw above we have $x^i=x^0+\sum_{m=1}^{i-1}\lambda_m v_m+v_i$ where $|\lambda_m|\leq 1$. Thus

$|b\cdot(x_0+v_i-x_0)|=|b\cdot v_i|=|b\cdot(x^i-x^0)-b\cdot(x^0+\sum_{m=1}^{i-1}\lambda_m v_m-x_0)|\leq \delta r+3^{i-1}\delta r$,

so $|b\cdot(x^0+\sum_{s=1}^{i}\varkappa_s v_s-x_0)|\leq 3^{i-1}\delta r+3^{i-1}\delta r+\delta r\leq 3^i\delta r$.

Thus, since $|v_s|\geq\varepsilon r$, we have shown that the band 3^nD contains an n-dimensional ball with center x_0 and radius εr. This means that we must have $3^n\delta r\geq\varepsilon r$ so $\delta\geq\varepsilon 3^{-n}$. Using again Proposition 3, in the more precise version given in the remark after it, we get part a) of the proposition.

Next we prove b). From part a) and (3) it follows that $\max_{B_1}|P|\leq c\max_X|P|$, for any polynomial P of degree at most one, where B_1 is the ball in (6). Thus, since $LP=P$ and $L(f-P)(x)=f(x)-P(x)$, $x\in X$, we get

$$\sup_{G\cap E}|Lf-P|\leq\max_{B_1}|L(f-P)|\leq c\max_X|L(f-P)|\leq c\sup_{G\cap E}|f-P|.$$

We now give an application of Lemma 1 to the theory of Besov spaces on subsets of R^n. It may be seen as a completing of the theory given in Chapter V and Chapter VI in [3]. In [2], the lemma is applied in a similar way when treating Besov spaces on manifolds with singularities; then the results from Section 1 in this paper are needed in a more genuine way. We first give some definitions, referring to [3] for the background.

A closed set $F \subset \mathbb{R}^n$ is a d-set, $0<d\leq n$, (cf. [3, p.28-34]) if there exists a positive Borel measure μ with support F, called a d-measure, satisfying, for some constants $c_1, c_2 > 0$,

$$c_1 r^d \leq \mu(B(x,r)) \leq c_2 r^d, \quad x \in F, \quad 0<r\leq 1. \tag{9}$$

Here $0<r\leq 1$ may be replaced by $0<r\leq r_0$. Any two d-measures on F are equivalent, and below μ denotes a fixed d-measure on F.
Examples: The closure of Lipschitz domains in $\mathbb{R}^n$ are d-sets with d=n and their boundaries are d-sets with $d=n-1$, see [3, p.30]. Let now F be a d-set preserving Markov's inequality, as defined in the introduction. Then Besov spaces can be defined on F in the following way, cf. Chapter V, § 2.4 in [3].

Let π denote a division of $\mathbb{R}^n$ into equally big cubes Q with sides r, halfopen in the form $\{x: a_i < x_i \leq a_i+r, i=1,2,\ldots,n\}$, obtained by intersecting $\mathbb{R}^n$ with hyperplanes orthogonal to the axes. We call such a division a net π with mesh r. Let $P_k(\pi)$ be the set of all functions $m(\pi)$, such that $m(\pi)$ coincides in each cube Q in π with a polynomial of degree $\leq k$. Then $f \in B_\alpha^{p,q}(F)$ if and only if $f \in L^p(\mu)$ and to every net π with mesh $2^{-\nu}$, $\nu=0,1,2,\ldots$, there is a function $n(\pi) \in P_{[\alpha]}(\pi)$ such that for some sequence $(a_\nu)_0^\infty$ with $(\Sigma a_\nu^q)^{1/q} < \infty$,

$$||f-n(\pi)||_{p,\mu} \leq 2^{-\nu\alpha} a_\nu. \tag{10}$$

The $B_\alpha^{p,q}(F)$-norm of f is $||f||_{p,\mu} + \inf(\Sigma a_\nu^q)^{1/q}$, where the infimum is taken over all possible sequences (a_ν).

The next theorem shows that the functions $n(\pi)$ in (10) may be chosen in a constructive way. Consider a function f in $B_\alpha^{p,q}(F)$, a net π with mesh $2^{-\nu}$ and a cube Q in π intersecting F. Let $2Q$ be the cube obtained by expanding the cube Q with a factor 2 around its center. Then $2Q$ satisfies the condition (6) for some $x_0 \in Q$, with G in (6) equal to $2Q$ and $r=2^{-\nu-1}$. Note that μ satisfies the doubling condition (4) for $0<r\leq r_0$, which also means that (7) holds under the more general condition $0<cr\leq r_0$. We now

apply Lemma 1, with $k=[\alpha]$, $G=2Q$, $E=S$, and E in the lemma equal to F. Thus we associate to f and Q the polynomial

$$P_Q=Lf=\sum_j \pi_j \int_{2Q} f\pi_j d\mu.$$

Define $m(\pi)$ on each $Q\in\pi$ by $m(\pi)=P_Q$ if Q intersects F and $m(\pi)=0$ otherwise.

THEOREM 1. <u>Let</u> F <u>be a d-set preserving Markov's inequality</u>, $0<d\leq n$, <u>let</u> $f\in B_\alpha^{p,q}(F)$, $\alpha>0$, $1\leq p,q\leq\infty$, <u>and construct for every net</u> π <u>with mesh</u> $2^{-\nu}$, $\nu=0,1,\ldots$, $m(\pi)\in P_{[\alpha]}(\pi)$ <u>as above. Then</u> $m(\pi)$ <u>satisfies</u>

$$||f-m(\pi)||_{p,\mu}\leq 2^{-\nu\alpha}b_\nu,$$

<u>for some sequence</u> (b_ν) <u>with</u> $(\Sigma\, b_\nu^q)^{1/q}\leq c||f||_{B_\alpha^{p,q}(F)}$, <u>where</u> c <u>does not depend on</u> f.

<u>Proof</u>. To each net π, associate $n(\pi)$ as in (10). For each cube Q in a fixed net π with mesh $2^{-\nu}$, let π_{2Q} be the net with mesh $2^{-\nu+1}$ containing $2Q$. There are a finite number of different nets π_{2Q}, only. To f and Q, associate P_Q as above. Then, by Lemma 1,

$$\int_Q |f-P_Q|^p d\mu \leq c\int_{2Q} |f-n(\pi_{2Q})|^p d\mu,$$

and it follows, for $\nu\geq 1$, that $||f-m(\pi)||_{p,\mu}\leq c2^{-\nu\alpha}a_{\nu-1}$. For $\nu=0$ we obtain instead by means of a direct calculation $||f-m(\pi)||_{p,\mu}\leq c||f||_{p,\mu}$. This proves the theorem.

<u>Remark</u>. The theorem means that $||f||_{p,\mu}+(\Sigma\, b_\nu^q)^{1/q}$ is an equivalent norm in $B_\alpha^{p,q}(F)$, where $b_\nu=\sup||f-m(\pi)||_{p,\mu}$, the supremum taken over all nets π with mesh $2^{-\nu}$.

In [3] it is proved that every $f\in B_\beta^{p,q}(F)$, $\beta=\alpha-(n-d)/p>0$, can be extended to a function defined on $\mathbb{R}^n$ belonging to $B_\alpha^{p,q}(\mathbb{R}^n)$. The extension is obtained by means of an operator which depends on a certain family $\{f^{(j)}\}$ and, if β is an integer, on certain approximations $f_\nu^{(j)}$ of $f^{(j)}$, see [3, p.157]. The functions $f^{(j)}$ are uniquely determined by f if F preserves Markov's inequality (see [3, p.126]), but the functions $f_\nu^{(j)}$ are not. This leads to an extension operator which is not linear in the integer case. Now, the functions $m(\pi)$ in

Theorem 1 are obtained in a linear way from f, and the functions $f_\nu^{(j)}$ may be obtained in a linear way from the functions $m(\pi)$ (cf. the proof of Theorem 4 and Theorem 5 in Chapter V in [3]). This gives a linear extension operator in the integer case also. Thus, from the extension theorem in [3] we obtain the following corollary to Theorem 1.

COROLLARY 1. Let F, α, p, and q, be as in Theorem 1, and let $\beta=\alpha-(n-d)/p>0$. Then there is continuous linear extension operator from $B_\beta^{p,q}(F)$ to $B_\alpha^{p,q}(\mathbb{R}^n)$.

In [3] it is proved that the restriction operator from $B_\alpha^{p,q}(\mathbb{R}^n)$ to $B_\alpha^{p,q}(F)$ is linear and continuous, too.

Lemma 1 may also be applied to the Lipschitz spaces $\Lambda_\alpha(F)$. Let F be a closed set preserving Markov's inequality and $\alpha>0$. Then these spaces may be defined as follows (cf. [3, p. 72]). A function f belongs to $\Lambda_\alpha(F)$, $\alpha>0$, if and only if $|f|\leq M$ on F and, for every cube Q with center in F and side of length $\delta\leq 1$, there exists a polynomial P_Q of degree at most $[\alpha]$ such that

$$|f(x)-P_Q(x)| \leq M\delta^\alpha, \quad \text{for} \quad x \in Q\cap F.$$

The norm of f in $\Lambda_\alpha(F)$ is equal to the infimum of all possible constants M. Since, by [8], F carries a measure μ satisfying the doubling condition (4), we can use Lemma 1 in a similar way as in the application to Besov spaces given above, to get a constructive characterization of $\Lambda_\alpha(F)$. Together with an extension theorem for $\Lambda_\alpha(F)$ ([3, p.58]) this gives the following result.

THEOREM 2. Let $F \subset R^n$ be a closed set preserving Markov's inequality and $\alpha>0$. Then there exists a continuous linear extension operator from $\Lambda_\alpha(F)$ to $\Lambda_\alpha(\mathbb{R}^n)$.

This theorem has also been proved by P. Wingren (personal communication). He uses a different method which does not rely upon the results in [8].

REFERENCES

[1] Ju.A. Brudnyi and M.I. Gansbury, On an extremal problem for polynomials of n variables, Izviestia AN USSR, 37 (1973), 344-355.

[2] A. Jonsson, Besov spaces on manifolds with singularities, under preparation.

[3] A. Jonsson and H. Wallin, Function spaces on subsets of $\mathbb{R}^n$, Mathematical Reports 2, Part 1, Harwood Academic Publ. (1984).

[4] A. Jonsson and H. Wallin, Local polynomial approximation and Lipschitz functions on closed sets, Constructive function theory, Varna (1981), 368-375, Sofia (1984).

[5] A. Jonsson, P. Sjögren, and H. Wallin, Hardy and Lipschitz spaces on subsets of $\mathbb{R}^n$, Studia Mathematica 80 (1984), 141-166.

[6] W. Plésniak, Again on Markov's inequality, Constructive Theory of Functions, Varna (1984), 679-683, Sofia (1984).

[7] H. Wallin, Markov's inequality on subsets of $\mathbb{R}^n$, Canad. Math. Soc., Conf. Proc., 3 (1983), 377-388.

[8] A.L. Volberg and S.V. Konyagin, There is a homogeneous measure on any compact subset in $\mathbb{R}^n$, Dokl. Akad. Nauk, 278 (1984) No.4. English Transl.: Soviet Math. Dokl., 30 (1984), 453-456.

TWO WEIGHTS WEAK TYPE INEQUALITY
FOR THE MAXIMAL FUNCTION IN THE ZYGMUND CLASS

M. Krbec
Mathematical Institute
Czech. Acad. Sci.
Žitná 25, 115 67 Prague
Czechoslovakia

1. Preliminaries

For $f \in L_{1,loc}(\mathbb{R}^n)$ let

$$Mf(x) = \sup_{x \in Q} \frac{1}{|Q|} \int_Q |f(y)|\, dy$$

be *the standard maximal operator*, ρ and σ weights, i.e. measurable a.e. positive functions in $\mathbb{R}^n$. The goal of this note is to present a necessary and sufficient condition in order that the weak type inequality

$$\rho(\{Mf > \lambda\}) \leq \frac{C(\rho,\sigma)}{\lambda} \int |f(x)|\left(1 + \log^+ \frac{|f(x)|}{\lambda}\right)\sigma(x)\, dx \tag{1.1}$$

holds for all functions $f \in L(1 + \log^+ L)$.

The inequality (1.1) is a limiting case of the well known two weights weak type inequality ($1 < p < \infty$)

$$\rho(\{Mf > \lambda\}) \leq K\lambda^{-p} \int |f(x)|^p \sigma(x)\, dx \tag{1.2}$$

which holds iff $(\rho,\sigma) \in A_p$, i.e.

$$\left(\frac{1}{|Q|}\int_Q \rho\right)^{1/p}\left[\frac{1}{|Q|}\int_Q \left(\frac{1}{\sigma}\right)^{1/(p-1)}\right]^{(p-1)/p} \leq C \; ; \tag{A_p}$$

see the fundamental paper [4].

As to the strong type inequalities in Orlicz spaces, the characterization of all weights for which the maximal operator is bounded in a reflexive Orlicz space was given by Kerman and Torchinsky [2]. The related problem for Riesz potentials and fractional maximal functions was solved by Kokilashvili and the author in [3]. It is well known that the differentiability properties of functions deteriorate as we get "sufficiently close" to L_1 ; namely the maximal operator maps the class $L(1 + \log^+ L)$ into $L_{1,loc}$ and not into the Zygmund class (see [6], [5], [1]...). It turns out that under a reasonable limiting A_p condition the same occurs even in the case of the weak type inequalities.

Let $1 < p < \infty$ and $p' = p/(p-1)$. It is easy to check that $(\rho,\sigma) \in A_p$ iff

$$\sup_Q \left[\int_Q \left(\frac{\rho(Q)}{|Q|\sigma(x)}\right)^{p'} \frac{\sigma(x)}{\rho(Q)} dx\right]^{1/p'} = A_p(\rho,\sigma) < \infty \tag{1.3}$$

where $\rho(Q) = \int_Q \rho(x)\,dx$. This suggest *the limiting condition*

there exists $\eta > 0$ such that

$$\sup_Q \int_Q \exp\left(\frac{\eta\rho(Q)}{|Q|\sigma(x)}\right) \frac{\sigma(x)}{\rho(Q)} dx = A_{\log}(\rho,\sigma) < \infty . \tag{$A_{\log}$}$$

2. The result

Theorem. *The inequality* (1.1) *holds for each function* $f \in L(1 + \log^+ L)$ *iff* $(\rho,\sigma) \in A_{\log}$.

P r o o f . Let (1.1) be true and, for brevity, let us write $\Phi(t) = |t|(1 + \log^+|t|)$. The Taylor expansion of the function exp gives for any $\eta > 0$ and a positive integer j_0

$$S_\eta = \int_Q \exp\left(\frac{\eta\rho(Q_0)}{|Q|\sigma(x)}\right) \frac{\sigma(x)}{\rho(Q)} dx = \sum_{j=0}^{\infty} \frac{1}{j!} \int_Q \left(\frac{\eta\rho(Q)}{|Q|\sigma(x)}\right)^j \frac{\sigma(x)}{\rho(Q)} dx$$

$$= \sum_{j=0}^{j_0-1} \ldots + \sum_{j=j_0}^{\infty} \frac{\eta^j}{j!} \left(A_{j'}(\rho,\sigma)\right)^j ,$$

where $j' = j/(j-1)$.

The inequality

$$\rho(\{Mf > \lambda\}) \leq C(\rho,\sigma) \int \Phi(f(x)/\lambda)\sigma(x)\,dx$$

implies in a standard manner that

$$\rho(\{Mf > \lambda\}) \leq C(\rho,\sigma) \int_{|f(x)| > \lambda/2} \Phi(f(x)/\lambda)\sigma(x)\,dx . \tag{2.1}$$

Indeed, let $f = f^* + f_*$ where $f_*(x) = 0$ if $|f(x)| > \lambda/2$ and $f_*(x) = f(x)$ if $|f(x)| \leq \lambda/2$. Then $\{Mf > \lambda\} \subset \{Mf^* > \lambda/2\} \cup \{Mf_* > \lambda/2\}$ and (2.1) follows from the fact that $\{Mf_* > \lambda/2\} = \emptyset$.

As $t(1 + \log^+ t) \leq C(p-1)^{-1} t^p$ for $t > \frac{1}{2}$ and p near 1 ,

$$\rho(\{Mf > \lambda\}) \leq \frac{C\cdot C(\rho,\sigma)}{p-1} \int_{|f(x)| > \lambda/2} |f(x)/\lambda|^p \sigma(x)\,dx \leq$$

$$\leq \frac{C\cdot C(\rho,\sigma)}{(p-1)\lambda^p}\int |f(x)|^p\,\sigma(x)\,dx\ .$$

The best constant on the right hand side of the last inequality is less or equal to $C\cdot C(\rho,\sigma)/(p-1)$ and according to the Muckenhoupt pioneering theorem [4] the relation between $A_p(\rho,\sigma)$ from (1.3) and this best constant gives

$$\{A_p(\rho,\sigma)\}^p \leq C\cdot C(\rho,\sigma)/(p-1)\ ,\quad p \ \text{near}\ 1\ .$$

Let $\tilde{C}$ stand for $\{C\cdot C(\rho,\sigma)\}^{1/p}$; then

$$A_p(\rho,\sigma)(p-1) \leq \tilde{C}\ .$$

In particular,

$$A_{j'}(\rho,\sigma) \leq \tilde{C}/(j'-1) = \tilde{C}(j-1) < \tilde{C}j$$

and thus for j_0 sufficiently large

$$\sum_{j=j_0}^{\infty}\frac{\eta^j}{j!}\{A_{j'}(\rho,\sigma)\}^j \leq \sum_{j=j_0}^{\infty}\frac{\eta^j}{j!}\tilde{C}^j j^j\ ,$$

the right hand side series being convergent for $0 < \eta < (\tilde{C}e)^{-1}$.

Conversely, let $(A_{\log})$ hold. Using the Young inequality we get for any cube Q , $\eta > 0$, and $f \in L(1+\log^+L)$

$$\frac{1}{|Q|}\int_Q |f(x)|\,dx \leq CA_{\log}(\rho,\sigma) + C(\eta)\int_Q \Phi\{f(x)\}\,\frac{\sigma(x)}{\rho(Q)}\,dx\ .$$

Denoting by $I_Q(f)$ the last integral and using $(A_{\log})$ we can rewrite it as

$$\frac{1}{|Q|}\int_Q |f(x)|\,dx \leq CA_{\log}(\rho,\sigma) + C(\eta)I_Q(f)\ .$$

Therefore we can choose a sequence of cubes Q_j in such a way that

$$I_{Q_j}(f) > 1$$

and

$$\rho(\{Mf > CA_{\log}(\rho,\sigma) + C(\eta)\}) \leq \sum \rho(Q_j) \leq \sum \int_{Q_j}\Phi\{f(x)\}\sigma(x)\,dx$$
$$\leq C_n \int \Phi\{f(x)\}\sigma(x)\,dx\ .$$

If now $\lambda > 0$ is arbitrary, then

$$\rho(\{Mf > \lambda\}) = \rho\Big(\{M\{(CA_{\log}(\rho,\sigma) + C(\eta))f/\lambda\} > CA_{\log}(\rho,\sigma) + C(\eta)\}\Big)$$
$$\leq C_n \int \Phi\{(CA_{\log}(\rho,\sigma) + C(\eta))f(x)/\lambda\}\sigma(x)\,dx$$

$$\leq C_{n,\eta}(\rho,\sigma) \int \Phi\{f(x)/\lambda\}\sigma(x)\; dx \; ,$$

q.e.d.

Remark. The study of differentiability properties of functions from Zygmund classes is interesting in the connection with boundary value problems. Very likely there is no reasonable (preferably integral) characterization of trace spaces for Orlicz-Sobolev spaces generated by the Young function of the above type.

References

[1] M. de GUZMÁN: *Differentiation of Integrals in* $\mathbb{R}^n$. Springer-Verlag, Berlin-Heidelberg-New York 1975.

[2] R. KERMAN and A. TORCHINSKY: *Integral inequalities with weights for the Hardy maximal function.* Studia Math. 71 (1982), 277-284.

[3] V. M. KOKILASHIVILI and M. KRBEC: *Weighted inequalities for Riesz potentials and fractional maximal functions in Orlicz spaces* (Russian). Dokl. Akad. Nauk SSSR 283 (1985), 280-283. (English transl. in Soviet Math. Dokl. 32 (1985), 70-73.

[4] B. MUCKENHOUPT: *Weighted norm inequalities for the Hardy maximal function.* Trans. Amer. Math. Soc. 165 (1972), 207-226.

[5] C. SADOSKY: *Interpolation of Operators and Singular Integrals. An Introduction to Harmonic Analysis.* M. Dekker, Inc., New York and Basel 1979.

[6] A. ZYGMUND: *Trigonometric Series, Vol. 1.* Cambridge Univ. Press 1959.

BANACH ENVELOPES OF SOME INTERPOLATION QUASI-BANACH SPACES

Mieczysław Mastyło
Institute of Mathematics
A. Mickiewicz University
Matejki 48/49
60-769 Poznań, Poland

1. Introduction

Let (X, τ) be a topological vector space. The _Mackey topology_ of it is the strongest locally convex topology $\mu = \mu(X)$ on X which produces the same topological dual space as the original topology τ of X. If (X, τ) is metrizable, then μ coincides with the strongest locally convex topology on X which is weaker than τ. Moreover, if $\mathcal{B}$ is a base of neighbourhoods of zero for τ, then the family $\{\mathrm{conv}\, U : U \in \mathcal{B}\}$ is a base of neighbourhoods of zero for μ (see [15]). So if $(X, \tau)^*$ separates points of X, then the Mackey topology μ of X is metrizable. If X is a quasi-normed space with a separating dual, then the topology $\mu(X)$ is normed and the completion $\hat{X}$ of (X, μ) is a Banach space (which is called the _Banach envelope_ of X). In this case the topology μ of X is generated by the Minkowski functional (denoted by $\|\cdot\|_X^{\wedge}$) of the convex hull of the unit ball $B = \{x \in X : \|x\|_X \leq 1\}$, which is called the _Mackey norm_ on X. It is easy to see that

$$\|x\|_X^{\wedge} = \inf\{\sum_{i=1}^{n} \|x_i\|_X : \quad x = \sum_{i=1}^{n} x_i\}.$$

By $X^{\wedge}$ we denote $(X, \|\cdot\|_X^{\wedge})$.

Let A_0 and A_1 be two quasi-normed spaces which are continuously embedded in some Hausdorff topological vector space $\mathcal{A}$. An _admissible operator_ for the pair $\vec{A} = (A_0, A_1)$ is a linear operator on $A_0 + A_1$ whose restriction to A_i is a bounded operator on A_i, $i = 0, 1$. An _intermediate_ quasi-normed space A between A_0 and A_1 $(A_0 \cap A_1 \hookrightarrow A \hookrightarrow A_0 + A_1)$ is called an _interpolation space_ between A_0 and A_1 if each admissible operator T for $\vec{A}$ is bounded on A and

$$\|T\|_A \leq c \max\{\|T\|_{A_0}, \|T\|_{A_1}\}$$

for some positive constant c independent of T. If $c = 1$, then A is called an _exact interpolation space_ between $\vec{A}$.

Remark 1. If $\vec{A} = (A_0, A_1)$ is a pair of normed spaces and A is a quasi-normed space exact interpolation between $\vec{A}$, then A with the

Mackey norm is also exact interpolation between $\vec{A}$.

The main purpose of this paper is to investigate the Banach envelopes of some interpolation quasi-normed spaces.

Conventions. Two (quasi-) normed spaces, A and B, are considered as equal and we write $A = B$ if their (quasi-) norms are equivalent. The symbol $A \hookrightarrow B$ means that inclusion map is continuous. The equivalence notion $f \sim g$ means that $c_1 f(t) \leq g(t) \leq c_2 f(t)$ for some positive constants c_1, c_2 and all $t > 0$.

2. The Mackey norm on $\vec{A}_E$

Let $\vec{A} = (A_0, A_1)$ be a pair of quasi-normed spaces. The Peetre K-functional for $\vec{A}$ is defined by

$$K(t, a; \vec{A}) = \inf \{ \|a_0\|_{A_0} + t\|a_1\|_{A_1} : a = a_0 + a_1,\ a_0 \in A_0,\ a_1 \in A_1 \}$$

for each $a \in A_0 + A_1$ and $t > 0$.

Denote by L^∞ $[L^\infty_{1/s}]$ the Banach space of all measurable classes of real valued functions f on $\mathbb{R}_+$ such that $|f(t)|$ $[|f(t)|/t]$ is essentially bounded. Put $\vec{L}^\infty = (L^\infty, L^\infty_{1/s})$. For any quasi-normed lattice E intermediate between $\vec{L}^\infty$, the real interpolation space (or K-space) $\vec{A}_E$ is the set of all $a \in A_0 + A_1$ such that

$$\|a\|_{\vec{A}_E} = \|K(\cdot, a; \vec{A})\|_E < \infty .$$

For the measurable function $f: \mathbb{R}_+ \to \mathbb{R}$, we put

$$\hat{f}(t) = \inf \{ g(t) : g(t) \geq |f(t)| \text{ a. e., } g: \mathbb{R}_+ \to \mathbb{R}_+ \text{ is concave} \},\ t > 0.$$

For $f \in L^\infty + L^\infty_{1/s}$ we have (see [4])

$$\hat{f}(t) = K(t, f; \vec{L}^\infty),\quad t > 0. \tag{1}$$

Proposition 1. Let E be a quasi-normed lattice intermediate between $\vec{L}^\infty$ and let $F = \vec{L}^\infty_E$. If $\vec{A}$ is a Banach pair, then there exists a constant $\gamma = \gamma(\vec{A}) < 14$ such that the inequalities

$$\|a\|_{\vec{A}_{F^\wedge}} \leq \|a\|_{\vec{A}_{\hat{F}}} \leq \gamma \|a\|_{\vec{A}_{F^\wedge}}$$

hold for $a \in \vec{A}_E$.

Proof. Obviously, we have $\vec{A}_E = \vec{A}_F$ with equality of norms. Since $\|f\|_{\hat{F}} \leq \|f\|_F$, so $\|a\|_{\vec{A}_{F^\wedge}} \leq \|a\|_{\vec{A}_F}$. Hence we obtain the left inequality. On the other hand, let $a \in \vec{A}_E$, then $a \in \vec{A}_{F^\wedge}$. Let $\varepsilon > 0$, then we can find a decomposition

$$K(\cdot, a ; \vec{A}) = f_1 + \dots + f_n$$

such that $f_i \in E$, $i = 1, \dots, n$, and

$$\sum_{i=1}^{n} \| f_i \|_F < \| K(\cdot, a ; \vec{A}) \|_{F^\wedge} + \varepsilon = \| a \|_{\vec{A}_{F^\wedge}} + \varepsilon . \qquad (2)$$

We obtain

$$K(t, a ; \vec{A}) \leq \hat{f}_1(t) + \dots + \hat{f}_n(t)$$

for all $t > 0$, whence by a result of K-divisibility (see [3], [6]) there exists a constant $\gamma < 14$ such that $a = a_1 + \dots + a_n$ and

$$K(\cdot, a_i ; \vec{A}) \leq \gamma \hat{f}_i , \quad a_i \in A_0 + A_1 , \quad i = 1, \dots , n .$$

Since $f_i \in E$, $i = 1, \dots, n$, so $\| f_i \|_F = \| \hat{f}_i \|_E$ (by (1)). From the last inequalities we obtain $a_i \in \vec{A}_F$. Consequently we get

$$\sum_{i=1}^{n} \| a_i \|_{\vec{A}_E} \leq \gamma \sum_{i=1}^{n} \| \hat{f}_i \|_E = \gamma \sum_{i=1}^{n} \| f_i \|_F < \gamma (\| a \|_{\vec{A}_{F^\wedge}} + \varepsilon)$$

(by (2)). Hence

$$\| a \|^{\wedge}_{\vec{A}_E} \leq \gamma \| a \|_{\vec{A}_{F^\wedge}} .$$

This completes the proof.

Let X be a normed subspace of a Hausdorff vector space $\mathfrak{X}$. We shall say that X has the completion in $\mathfrak{X}$ if $\hat{X}$ is a subspace of $\mathfrak{X}$ and $\hat{X}$ is continuously embedded in $\mathfrak{X}$. The problem of existence of a completion of X in $\mathfrak{X}$ is solved by the following

<u>Theorem 1</u> [1]. If a normed space X is a subspace of a Hausdorff vector space $\mathfrak{X}$, then a completion of X in $\mathfrak{X}$ exists if and only if the following conditions hold:

(a) X is continuously embedded in $\mathfrak{X}$,

(b) every Cauchy sequence in X converges in $\mathfrak{X}$,

(c) when a Cauchy sequence in X converges to 0 in $\mathfrak{X}$, it also converges to 0 in X.

<u>Remark 2</u>. In the special case $\mathfrak{X} = L^0(\Omega, \Sigma, \nu)$ (the topological vector space of all equivalence classes of ν measurable real valued functions on a measure space Ω with a σ-finite measure ν, equipped with the topology of convergence in measure on ν-finite sets), a normed

(order) ideal X in $\mathfrak{X}$ has a completion in $\mathfrak{X}$ if and only if the condition (c) of Theorem 1 holds (conditions (a) and (b) are fulfilled, see [14, Proposition 2.7.2, p. 77]).

It is easy to see that the normed ideal X in $L^0(\Omega, \Sigma, \nu)$ satisfies condition (c) of Theorem 1, if the norm of X is <u>semi-continuous</u>, i.e., if $0 \leq x_n \uparrow x$ ν-a.e., $x \in X$, imply $\|x_n\|_X \uparrow \|x\|_X$. (Semi-continuity of $\|\cdot\|_X$ is equivalent to the condition: if $x, x_n \in X$, $x_n \to x$ in $L^0(\Omega, \Sigma, \nu)$, then $\|x\|_X \leq \liminf_{n\to\infty} \|x_n\|_X$).

<u>Proposition 2</u>. Let X be a normed ideal in $L^0(\Omega, \Sigma, \nu)$ and let $Y \subset X$ be an ideal subspace dense in X. If the norm of X is semi-continuous on Y, then it is semi-continuous.

<u>Proof</u>. Let $0 \leq x_n \uparrow x$, ν-a.e., $x \in X$, $(y_m) \subset Y$ and $\|y_m - x\|_X \to 0$. Then $z_n^m = \min(x_n, |y_m|)$ are in Y and $0 \leq z_n^m \uparrow z^m = \min(x, |y_m|)$ ν-a.e. as $n \to \infty$. Since $\| |y_m| - x \|_X \to 0$, so $\|z^m - x\|_X \to 0$ as $m \to \infty$. By the assumption $\|z^m\|_X = \lim_{n\to\infty} \|z_n^m\|_X$ for each $m \in \mathbb{N}$. Moreover, $\|z_n^m\|_X \leq \|x_n\|_X \leq \|x\|_X$ and $\|x\|_X = \lim_{m\to\infty} \|z^m\|_X$. Hence we obtain $\|x\|_X = \lim_{n\to\infty} \|x_n\|_X$, and the proof is complete.

Applying the Hahn-Banach-Kantorovič theorem (see [5] or the method of proof of Lemma 4.3 in [4]), we get

<u>Proposition 3</u>. Let E be an exact interpolation quasi-normed space between $\vec{L}^\infty$. Then $E = \vec{L}^\infty_E$ with equality of norms.

In what follows, let E be an exact interpolation quasi-normed space between $\vec{L}^\infty$.

<u>Theorem 2</u>. If $L^\infty \cap L^\infty_{1/s}$ is dense subspace in E, then there exists the Banach envelope of E in $L^0(\mathbb{R}_+, dt/t)$. It is of the following form:

$$\hat{E} = \Big\{ x \in L^0(\mathbb{R}_+, dt/t) : x = \sum_{n=1}^\infty x_n \text{ (convergence in } L^0(\mathbb{R}_+, dt/t)), \quad x_n \in E, \ \sum_{n=1}^\infty \|x_n\|\hat{}_E < \infty \Big\}$$

with the norm

$$(*) \quad \|x\|_{\hat{E}} = \inf\Big\{ \sum_{n=1}^\infty \|x_n\|\hat{}_E : x = \sum_{n=1}^\infty x_n, \ \sum_{n=1}^\infty \|x_n\|\hat{}_E < \infty \Big\}.$$

<u>Proof</u>. First, we show that the Mackey norm is semi-continuous. By Proposition 2, we need only to show that it is semi-continuous on $L^\infty \cap L^\infty_{1/s}$, since $L^\infty \cap L^\infty_{1/s}$ is dense in $(E, \|\cdot\|\hat{}_E)$. Observe that $K(t, \cdot; \vec{L}^\infty)$ is semi-continuous norm on $L^\infty + L^\infty_{1/s}$ for each $t > 0$. So if $0 \leq x_n \uparrow x$ a.e., $x \in L^\infty \cap L^\infty_{1/s}$, then $\tilde{x}_n, \tilde{x} \in L^\infty \cap L^\infty_{1/s}$ and $0 \leq \tilde{x}_n(t) \uparrow \tilde{x}(t)$ for each

$t>0$ by (1). It is not hard to show that for each $\lambda>1$ there exists $n_o \in \mathbb{N}$ such that

$$\hat{x}(t) \leq \lambda \hat{x}_n(t) \quad \text{for every } n>n_o \text{ and } t>0 . \tag{3}$$

Fix $\lambda>1$. Then combining (3), Remark 1 and Proposition 3 one obtains the following

$$\|x\|^{\wedge}_{E} = \|\hat{x}\|^{\wedge}_{E} \leq \lambda \|\hat{x}_n\|^{\wedge}_{E} = \lambda \|x_n\|^{\wedge}_{E} \quad \text{for every } n>n_o.$$

Hence $\|x\|^{\wedge}_{E} \leq \lim_{n\to\infty} \|x_n\|^{\wedge}_{E}$ by the arbitrariness of $\lambda>1$. The reverse inequality obviously holds and, consequently, $\|x_n\|^{\wedge}_{E} \to \|x\|^{\wedge}_{E}$. Hence there exists the Banach envelope of E in $L^0(\mathbb{R}_+, dt/t)$ (by Remark 2). Since $\hat{E}$ with the norm (*) is a Banach lattice and condition (c) of Theorem 1 implies $\|x\|^{\wedge}_{E} = \|x\|_{\hat{E}}$ for each $x\in E$, so $\hat{E}$ is the Banach envelope of E (by density of E in $\hat{E}$). The proof is complete.

<u>Theorem</u> 3. Let $\vec{A}$ be a Banach pair in $\mathcal{A}$. If there exists the Banach envelope of E in $L^0(\mathbb{R}_+, dt/t)$, then $\widehat{\vec{A}_E} = \vec{A}_{\hat{E}}$.

<u>Proof</u>. Since $\|f\|^{\wedge}_{E} = \|f\|_{\hat{E}}$ for $f\in E$, so

$$\|a\|_{\vec{A}_{\hat{E}}} \sim \|a\|^{\wedge}_{\vec{A}_E} \tag{4}$$

for $a\in \vec{A}_E$ by Proposition 1. By (4) and Theorem 1 there exists the Banach envelope of $\vec{A}_E$ in $\mathcal{A}$. Since $\vec{A}_{\hat{E}}$ is a Banach space, we need only to show that $\vec{A}_E$ is a dense subspace in $\vec{A}_{\hat{E}}$ (by (4)). Let $a\in\vec{A}_{\hat{E}}$, then

$$K(t, a; \vec{A}) = \sum_{i=1}^{\infty} f_i(t) \quad \text{a.e.},$$

where $f_i \in E$, $i\in\mathbb{N}$ and

$$\sum_{i=1}^{\infty} \|f_i\|^{\wedge}_{E} = \sum_{i=1}^{\infty} \|\hat{f}_i\|^{\wedge}_{E} < \infty . \tag{5}$$

Hence $\sum_{i=1}^{\infty} \hat{f}_i(1) < \infty$ and

$$K(t, a; \vec{A}) \leq \sum_{i=1}^{\infty} \hat{f}_i(t) \tag{6}$$

for all $t>0$. From the fundamental result about K-divisibility [3, 6] there exists a constant $\gamma>0$ depending only on $\vec{A}$ and elements $a_i \in A_o + A_1$ such that $a = \sum_i a_i$ (convergence in $A_o + A_1$) and such that for all $t>0$ and all $i\in\mathbb{N}$

$$K(t, a_i; \vec{A}) \leq \gamma \hat{f}_i(t). \tag{7}$$

By (7) we have $(a_i) \subset \vec{A}_E$. Moreover

$$K(\cdot, a - \sum_{i=1}^{n} a_i ; \vec{A}) \leq \sum_{i=n+1}^{\infty} K(\cdot, a_i ; \bar{A}) \leq \gamma \sum_{i=n+1}^{\infty} \hat{f}_i$$

and

$$\| a - \sum_{i=1}^{n} a_i \|_{\vec{A}_{\hat{E}}} \leq \gamma \sum_{i=n+1}^{\infty} \| \hat{f}_i \|_{\hat{E}} \to 0 \quad \text{as} \quad n \to \infty ,$$

by (5), (6) and (7). Since $\sum_{i=1}^{n} a_i \in \vec{A}_E$, the proof is complete.

3. Banach envelope of special quasi-Banach spaces

Using Peetre's method (cf. [13]), we give sufficient conditions on E for characterizing the Banach envelope of the interpolation space $\vec{A}_E$. We need some definitions.

Let $\mathcal{P}$ denote the set of all functions $\phi: \mathbb{R}_+ \to \mathbb{R}_+$ such that $\phi(s) \leq \max(1, s/t)\phi(t)$ for all $s, t > 0$. A function ϕ belongs to $\mathcal{P}^{+-}$ if $\min(1, 1/t) \cdot s_\phi(t) \to 0$ as $t \to 0, \infty$, where $s_\phi(t) = \sup_{u>0} (\phi(ut)/\phi(u))$. Let $\vec{A}$ be a Banach pair. By $\Lambda_\phi(\vec{A})$ ($\phi \in \mathcal{P}$) we denote the space of all $a \in A_0 + A_1$ which can be represented in the form

$$a = \int_{\mathbb{R}_+} u(t) \frac{dt}{t} \quad (\text{convergence in } A_0 + A_1) , \tag{8}$$

where u is strongly measurable with values in $A_0 \cap A_1$, and

$$J(t, u(t); \vec{A}) = \max(\|u(t)\|_{A_0}, t\|u(t)\|_{A_1}) \in L^1_{1/\phi}(\mathbb{R}_+, dt/t).$$

We equip $\Lambda_\phi(\vec{A})$ with the norm

$$\|a\|_\phi = \inf_u \|J(\cdot, u(\cdot); \vec{A})\|_{L^1_{1/\phi}} ,$$

where the infimum is taken over all representations of a in the form (8). It is known that $\Lambda_\phi(\vec{A})$ is an exact interpolation space between $\vec{A}$ ([7], [8]).

If $\phi \in \mathcal{P}^{+-}$ and $0 < p \leq \infty$, then the real interpolation space $\vec{A}_E$ with $E = L^p_{1/\phi}(\mathbb{R}_+, dt/t)$ is denoted by $\vec{A}_{\phi,p}$.

A quasi-normed space X intermediate for the Banach pair $\vec{A}$ belongs to the set $C_J(\phi, \vec{A})$ (cf. [8]), where $\phi \in \mathcal{P}$, if

$$\|a\|_X \leq c\, J(t, a ; \vec{A})/\phi(t)$$

for some $c > 0$ and all $a \in A_0 \cap A_1$.

Proposition 4 (cf. [2]). A Banach space X belongs to $C_J(\phi, \vec{A})$ if and only if $\Lambda_\phi(\vec{A}) \hookrightarrow X \hookrightarrow A_0 + A_1$.

Proposition 5. Let $\vec{A}$ be a Banach pair in $\mathcal{A}$ and let X be a quasi-

-Banach space in $C_J(\phi, \vec{A})$ with $X \subset \Lambda_\phi(\vec{A})$. If there exists the Banach envelope of X in $\mathcal{A}$, then it is equal to $\Lambda_\phi(\vec{A})$.

Proof. Since $X \in C_J(\phi, \vec{A})$, we have

$$\|a\|_X \leq c_1 J(t, a; \vec{A})/\phi(t), \quad a \in A_0 \cap A_1, \quad t > 0.$$

Now the same inequality holds also with $\|a\|_X$ replaced by $\|\cdot\|_{\hat{X}}$. Consequently

$$\|a\|_{\hat{X}} \leq c_1 J(t, a; \vec{A})/\phi(t), \quad a \in A_0 \cap A_1, \quad t > 0$$

(by the existence of the Banach envelope of X in $\mathcal{A}$). Using Proposition 4, we get $\Lambda_\phi(\vec{A}) \hookrightarrow \hat{X}$. On the other hand, since $X \subset \Lambda_\phi(\vec{A})$, then the closed graph theorem implies that the inclusion mapping is continuous. Thus, there is $c_2 > 0$ such that

$$\|a\|_\phi \leq c_2 \|a\|_{\hat{X}} \quad \text{for} \quad a \in X,$$

whence $\hat{X} \hookrightarrow \Lambda_\phi(\vec{A})$ by density of X in $\hat{X}$. Hence the proof is complete.

If E is a quasi-Banach space exact interpolation between $\vec{L}^\infty$, then by Proposition 5 and Theorem 2 we obtain

Corollary 1. Let $\vec{A}$ be a Banach pair in $\mathcal{A}$. If there exists the Banach envelope of $\vec{A}_E$ in $\mathcal{A}$ and $\vec{A}_E \hookrightarrow \Lambda_\psi(\vec{A})$, where $\psi(t) = \|\min(1, s/t)\|_E^{-1}$, then $\widehat{\vec{A}_E} = \Lambda_\psi(\vec{A})$. In particular, if $L^\infty \cap L^\infty_{1/s}$ is dense in E and $E \subset \Lambda_\psi(\vec{L}^\infty)$, then $\hat{E} = \Lambda_\psi(\vec{L}^\infty)$.

Corollary 2. Let $\vec{A}$ be a Banach pair in $\mathcal{A}$ and let $\phi \in \mathcal{P}^{+-}$, $0 < p < 1$. Then $\widehat{\vec{A}_{\phi,p}} = \vec{A}_{\phi,1}$.

Proof. Let $E = \vec{L}^\infty_{\phi,p}$, $0 < p < 1$. First, we show that $\phi \sim \psi$ (for the definition of ψ, see Corollary 1). In fact, fix some $t > 0$. Since the function $x_t(s) = \min(1, s/t)$ is concave on $\mathbb{R}_+$, we have $K(s, x_t; \vec{L}^\infty) = x_t(s)$ for $s > 0$ by (1). Hence

$$\psi(t)^{-p} = \int_0^\infty (\phi(s)^{-1}\min(1, s/t))^p \frac{ds}{s} = \int_0^\infty (\phi(t/u)\min(1, 1/u))^p \frac{du}{u}$$
$$\leq \int_0^\infty (\phi(t)^{-1}\min(1, 1/u) s_\phi(u))^p \frac{du}{u} = c(\phi)\, \phi(t)^{-p},$$

where $c(\phi) = \int_0^\infty (\min(1, 1/u) s_\phi(u))^p \frac{du}{u} < \infty$ (see [8]). Of course,

$$\psi(t)^{-p} \geq \int_0^t \left(\frac{s}{t\,\phi(t)}\right)^p \frac{ds}{s} \geq p^{-1} \phi(t)^{-p},$$

so $\phi \sim \psi$. Since $L^\infty \cap L^\infty_{1/s}$ is dense in E (see [8], [10]) and $\vec{A}_{\phi,p} \hookrightarrow \vec{A}_{\phi,1} = \Lambda_\phi(\vec{A})$, hence, by Theorem 2 and 3, there exists the Banach envelope of $\vec{A}$ in $\mathcal{A}$. Using Corollary 1 one obtains the required equality.

Remark 3. If $\phi(t) = t^{\theta}$, $0 < \theta < 1$ and $0 < p < 1$, then $\widehat{\vec{A}}_{\theta,p} = \vec{A}_{\theta,1}$ (see Peetre [13]).

Since $(\vec{A}_{\phi,p})^* = (\widehat{\vec{A}}_{\phi,p})^*$ with equality of norms, so by well-known results (see [3], [7] and [10]) and Corollary 2 we obtain

Corollary 3. Let $\vec{A}$ be a Banach pair such that $A_0 \cap A_1$ is dense in A_0 and A_1, $\phi \in \mathcal{P}^{+-}$, $0 < p < 1$. Then $(\vec{A}_{\phi,p})^* = (A_0^*, A_1^*)_{\phi_*,\infty}$, where $\phi_*(t) = 1/\phi(1/t)$ for $t > 0$.

Example. Let $\phi \in \mathcal{P}$ and $q \in (0, \infty)$. The Lorentz sequence space $\ell(\phi,q)$ consists of all bounded sequences of scalars $x = (x_n)$ such that

$$\|x\|_{\phi,q} = \left(\sum_{n=1}^{\infty} (\phi(n)\, x_n^*)^q\, n^{-1} \right)^{1/q} < \infty ,$$

where $x^* = (x_n^*)$ is the non-increasing rearrangement of $x = (x_n)$ defined by $x_n^* = \inf\{ s > 0 \colon \operatorname{card}\{k \colon |x_k| \geq s\} < n\}$. For $\phi(t) = t^{1/p}$, $1 \leq p < \infty$ we obtain the classical Lorentz sequence space $\ell(p, q)$. If $\phi \in \mathcal{P}^{+-}$, then $(\ell^{\infty}, \ell^1)_{\phi_*,q} = \ell(\phi, q)$ (see [11], Proposition 8 and Theorem 5). Hence for $0 < q < 1$ we get

$$\widehat{\ell(\phi, q)} = \ell(\phi, 1),$$

by Corollary 2. For an estimation of the Mackey norm of the Lorentz space $L(\phi, q)$, $0 < q < 1$, we refer to [9] (see also [12]).

References

1. Aronszajn, N., Gagliardo, E.: Interpolation spaces and interpolation methods. Ann. Mat. Pura Appl. 68, 51-118 (1965).

2. Bergh, J., Löfström, J.: Interpolation spaces. An introduction. Berlin - Heidelberg - New York: Springer 1976.

3. Brudnyĭ, Ju. A., Krugljak, N. Ja.: Real interpolation functor. Dokl. Akad. Nauk SSSR 256, 14-17 (1981); Soviet Math. Dokl. 23, 5-8 (1981).

4. Brudnyĭ, Ju. A., Krugljak, N. Ja.: Real interpolation functors. Book manuscript. Jaroslavl' 1981 [Russian].

5. Bukhvalov, A. W.: Theorems on interpolation of sublinear operators in spaces with mixed norms. In: Qualitative and approximate methods for the investigation of operator equations, Jaroslavl' 1984 [Russian].

6. Cwikel, M.: K-divisibility of the K-functional and Calderón couples. Ark. Mat. 22, 39-62 (1984).

7. Cwikel, M., Peetre, J.: Abstract K and J spaces. J. Math. Pures et Appl. 60, 1-50 (1981).

8. Gustavsson, J.: A function parameter in connection with interpolation of Banach spaces. Math. Scand. 42, 289-305 (1978).

9. Haaker, A.: On the conjugate space of a Lorentz space. Research Report, Lund 1970.

10. Janson, S.: Minimal and maximal methods of interpolation. J.Functional Analysis 44, 50-73 (1981).

11. Merucci, C.: Interpolation réelle avec parametre fonctionnel des espaces $L^{p,q}$, Université de Nantes, Sem. D'Anal. exposé 17, 350--373 (1980/81).

12. Nawrocki, M., Ortyński, A.: The Mackey topology and complemented subspaces of Lorentz sequence spaces d(w, p) for $0<p<1$. Trans. Amer. Math. Soc. 287, 713-722 (1985).

13. Peetre, J.: Remark on the dual of an interpolation space. Math. Scand. 34, 124-128 (1974).

14. Rolewicz, S.: Metric linear spaces. PWN-Polish Scientific Publishers, Warszawa, D. Reidel Publishing Company, Dordrecht Boston Lancaster 1984.

15. Shapiro, J. H.: Mackey topologies, reproducing kernels, and diagonal maps on the Hardy and Bergman spaces. Duke Math. J. 43, 187--202 (1976).

A CONSTRUCTION OF EIGENVECTORS FOR THE CANONICAL ISOMETRY

Tomasz Mazur
Department of Mathematics
Technical University
Malczewskiego 29
26-600 Radom, Poland

1. Introduction

In the theory of several complex variables the methods of functional analysis play an important rôle. Hilbert space methods in the study of biholomorphic mappings were applied and developed by S.Bergman [1,2]. In this context the following quantities appear:

1° the subspace $L^2H(D)$ of $L^2(D)$ consisting of all holomorphic functions on a domain $D \subset \mathbb{C}^N$ with scalar product

$$<f,g> = \int_D f(z)\overline{g(z)}dm(z) ,$$

2° the evaluation functional $\kappa_z^*: L^2H(D) \to \mathbb{C}$

$$\kappa_z^*(f) = f(z) ,$$

which is represented in terms of the scalar product by $\kappa_z \in L^2H(D)$

3° the Bergman function of a domain D defined as follows

$$K_D(w,z) = <\kappa_z,\kappa_w> \quad (z,w \in D),$$

and

4° the operator of canonical isometry $U_\phi: L^2H(\tilde{D})$ induced by a biholomorphic mapping $\phi: D \to \tilde{D}$ given by

$$(U_\phi f)(z) = f(\phi(z))\phi'(z) \qquad (z \in D) .$$

Here ϕ' denotes the complex Jacobian of ϕ.

Using the mapping U_ϕ one has a possibility to carry problems from a domain to its biholomorphic image. If $D = \tilde{D}$ then U_ϕ is unitary operator.

In [5] the following spectral property of U_ϕ is established:

Theorem 1. Let D be a domain in $\mathbb{C}^N$ such that $L^2H(D) \neq \{0\}$. If the biholomorphic automorphism ϕ of D has a fixed point then U_ϕ has a pure point spectrum.

To prove this theorem we used the properties of evaluation functional. Here we will apply a more geometrical method based on representative coordinates to construct infinitely many eigenvectors of U_ϕ in certain bounded domains. The method of representative coordinates turns out to be very practical, because a constructive way of describing the Bergman function is known. It is based on the alternating projections (see [8]).

2. Representative coordinates

We recall the notion of covariant representative coordinates [2,6].

Let D be a bounded domain in $\mathbb{C}^N$ and $t \in D$. Consider the functions μ_k $(k = 1,2,\dots N)$ defined as follows

$$\mu_k(z) = \frac{\partial \log K_D(z,t)}{\partial \bar{t}_k} - \frac{\partial \log K_D(t,t)}{\partial \bar{t}_k} = \frac{\partial}{\partial \bar{t}_k} \log \frac{K_D(z,t)}{K_D(t,t)}$$

Notice that

$$\frac{\partial \mu_k(z)}{\partial z_j} = \frac{\partial^2 \log K_D(z,t)}{\partial z_j \partial \bar{t}_k} \qquad (k,j = 1,2,\dots N)$$

for $z = t$ are the coefficients of the Bergman metric tensor [2]. From the positivity of this tensor on bounded domains [2,7] it follows that

$$\det\left(\frac{\partial \mu_k(u)}{\partial z_j}\right) \neq 0 \quad \text{for} \quad z = t$$

and there exists a neighbourhood of t in D in which μ_k $(k=1,2,\dots N)$ form a holomorphic coordinate system. If $\psi: D \to \tilde{D}$ is a biholomorpic mapping and $\tilde{z} = \psi(z)$, $\tilde{t} = \psi(t)$ then (see [6])

$$\tilde{\mu}(\tilde{z}) = T \cdot \mu(z)$$

Here $T = \overline{\left[\frac{\partial \tilde{t}_j}{\partial t_k}\right]}^{-1}$ $(j,k = 1,2,\dots N)$ is the conjugate inverse matrix to the Jacobian matrix of ψ and "." denotes multiplication of matrices.

In the literature (for example [3]) one often meets the contravariant representative coordinates. These coordinates were introduced by S. Bergman as a solution of certain extremal problem [2,6].

The covariant representative coordinates μ_i $(i = 1,2,\ldots N)$ are related to contravariant representative coordinates ν^j $(j = 1,2,\ldots N)$ by a nondegenerate linear transformation (see [6]).

$$\mu = G \cdot \nu, \quad \mu = (\mu_1,\ldots,\mu_N), \quad \nu = (\nu^1,\ldots,\nu^N)$$

where G denotes the matrix of the coefficients of the Bergman metric tensor.

3. Eigenvectors of the canonical isometry

Consider a biholomorphic automorphism ϕ of a bounded domain D with a fixed point $t \in D$.

From the well-known rule of transformation for the Bergman function under biholomorphic mappings [2,7]

$$K_D(z,t) = K_D(\phi(z),\phi(t))\phi'(z)\overline{\phi'(t)}$$

it follows that $K_D(\cdot,t) = \kappa_t$ is an eigenvector of U_ϕ:

$$U_\phi\kappa_t(z) = U_\phi K_D(z,t) = K_D(\phi(z),t)\phi'(z) =$$

$$= \frac{K_D(\phi(z),\phi(t))\phi'(z)\overline{\phi'(t)}}{\overline{\phi'(t)}} = \frac{1}{\overline{\phi'(t)}} K_D(z,t) = \frac{1}{\overline{\phi'(t)}} \kappa_t(z) .$$

Consider a neighbourhood of t in D which is mapped biholomorphically by covariant representative coordinates onto some neighbourhood of $0 \in \mathbb{C}^N$ and for every z in it

$$\mu(\tilde{z}) = T \cdot \mu(z), \qquad \tilde{z} = \phi(z) .$$

Lemma 1. There is a basis in $\mathbb{C}^N$ consisting of eigenvectors of T.

This follows from the fact that T is an isometry with respect to the image of the Bergman metric for D under representative coordinates.

Denote by $\mathcal{F}_w$ the space of all germs of holomorphic functions defined at $w \in \mathbb{C}^N$.

Consider the linear isomorhisms

$$V_o^T\colon \mathcal{F}_o \to \mathcal{F}_o, \quad V_o^T(h)_o = (h(T)\det T)_o),$$

$$V_t^\phi\colon \mathcal{F}_t \to \mathcal{F}_t, \quad V_t(t)_o = (f(\phi)\phi')_o$$

and

$$W: \mathcal{F}_o \to \mathcal{F}_t, \qquad W(h)_o = (h(\mu)\mu')_t \ .$$

Notice that

$$V_t^{\phi} = W \circ V_o^T \circ W^{-1}.$$

Theorem 2. If $v = (v_1,\ldots,v_N)$ is an eigenvector of the transpose matrix T^t to T then the germ definied by holomorphic function

$$(*) \qquad h(z) = v_1z_1 +\ldots+ v_Nz_N, \qquad z = (z_1,\ldots,z_N)$$

is an eigenvector of operator V_O^T.

Proof: We are looking for a solution of the equation

$$V_o^T(H)_o = l(h)_o$$

among the germs defined by functions of the form

$$h(z) = a_1z_1 +\ldots+ a_Nz_N, \qquad a = (a_1,\ldots,a_N) \in \mathbb{C}^N.$$

The above euqation may be written in the matrix form

$$a\cdot T\cdot z\cdot \det T = (a\cdot T)\cdot z\cdot \det T = la\cdot z \ .$$

Hence

$$a(T - \frac{1}{\det T} I) = 0.$$

The solutions of this system are the eigenvectors of T^t.

Lemma 2. If $(h_1)_o$ and $(h_2)_o$ are eigenvectors of V_o^T then $(h_1h_2)_o$ is an eigenvector of V_o^T too.

Proof: Let l_1 och l_2 be suitable eigenvalues. Then

$$V_o^T(h_1h_2)_o = (h_1(T))_o(h_2(T))_o\cdot \det T = \frac{l_1l_2}{\det T}(h_1)_o(h_2)_o = \frac{l_1l_2}{\det T}(h_1)_o(h_2)_o.$$

Remark. Let $m_1,m_2,\ldots,m_N$ be natural numbers or 0. The germs

$$(h)_o = (h_1^{m_1}h_2^{m_2}\ldots h_N^{m_N})_o,$$

where h_i $(i = 1,2,\ldots N)$ are functions of the form (*), are the eigenvectors of V_o^T. All of these are linearly independent.

Lemma 3. If $(h)_o \in \mathcal{F}_o$ is an eigenvector of V_o^T then $W(h)_o$ is an eigenvector of V_t^ϕ.

Proof:

$$(f)_t := (h(\mu)\mu')_t = W(h)_o \in \mathcal{F}_t$$

$$V_t^\phi(f)_t = W \circ V_o^T \circ W^{-1}(f)_t = W \circ V_o^T(h)_o =$$

$$= W\, l(h)_o = lW(h)_o = l(f)_t$$

Theorem 3. Let $D \subset \mathbb{C}^N$ be a bounded domain and $t \in D$ a fixed point of the biholomorphic automorphism ϕ of D. Let $v^i = (v_1^i, v_2^i, \ldots, v_N^i)$, $i = 1,2,\ldots,N$, be an eigenvector of the inverse, conjugate and transposed matrix to the complex Jacobian matrix of ϕ at $t \in D$. Suppose that the covariant representative coordinates map biholomorphically D onto bounded domain. Then the functions

$$(**) \qquad f_i(z) = \sum_{k=1}^{N} v_k^i \mu_k(z) \det\left(\frac{\partial \mu_n(z)}{\partial z_j}\right) =$$

$$= \sum_{k=1}^{N} v_k^i \frac{\partial}{\partial \bar{t}_k} \log\frac{K_D(z,t)}{K_D(t,t)} \det\left[\frac{\partial^2 \log K_D(z,t)}{\partial z_j \partial \bar{t}_n}\right] \qquad (j,n = 1,2,\ldots,N)$$

generate infinitely many linearly independent eigenvectors of the canonical isometry U_ϕ.

Proof: From lemma 3 it follows that in some neighbourhood of t in D we have

$$f_i(\phi)\phi' = lf_i \qquad (i = 1,2,\ldots,N)$$

for some $l \in \mathbb{C}$. By assumption the functions (**) to $L^2H(D)$. From the uniqueness theorem if follows that the above equality takes place on D. Because the operator W is an isomorphism, the products of the functions (**) are linearly independet eigenvectors of U_ϕ.

Remark. The statement of Theorem 3 is also true in a bounded domain $D \subset \mathbb{C}^N$ such that the representative coordinates are holomorphic functions on its closure.

4. Example

A bounded domain $D \subset \mathbb{C}^N$ is called completely N-circular (with respect to $0 \in \mathbb{C}^N$) if for every $z \in D$ and every $\lambda_i \in C$ such that $|\lambda_i| \leq 1$ $i = 1,2,\ldots,N$ the point $w = (\lambda_1 z_1, \lambda_2 z_2, \ldots, \lambda_N z_N)$ belongs to D.

It is a theorem due to Welke (see [2]) that in such domains the representative coordinates are proportional to the Cartesian coordinates

$$\mu_i(z) = c_i z_i \qquad (i = 1,2,\dots,N).$$

Consider an automorphism ϕ of completely N-circular domain D such that $\phi(0) = 0$. In view of a classical theorem by Cartan ϕ can be written as $\phi = (\phi_1,\phi_2,\dots,\phi_N)$ where

$$\phi_i(z) = a_{i1}z_1 +\dots+ a_{iN}z_N \quad (a_{ij} \in \mathbb{C},\ i,j = 1,2,\dots,N).$$

In this case the matrix T has a simple form

$$T = (a_{ij}) \qquad (i,j = 1,2,\dots,N).$$

The functions

$$(***) \quad f_i(z) = c_1\dots c_N \sum_{k=1}^{N} v_k^i c_k z_k = A \sum_{k=1}^{N} B_k z_k \qquad (A,B_k \in \mathbb{C})$$

and their products are linearly independent eigenvectors of U_ϕ.

From H. Cartan's theorem that in a complete N-circular domain D the set of all homogeneous polynomials of different degrees is a basis in $L^2H(D)$, it follows that the family of all vectors obtained here is linearly dense in $L^2H(D)$.

Problem. Assume that $D \subset \mathbb{C}^N$ is as in Theorem 3. Assume that the automorphishm $\phi: D \to D$ has a fixed point. Is the set of all eigenvectors of U_ϕ constructed in this paper linearly dense in $L^2H(D)$?

References.

1. Bergman, S.: Über die Kernfunktion eines Bereiches und ihr Verhalten am Rande. J.Reine Angew. Math. 169, 1-43 (1933) and 172 89-123 (1934).
2. Bergman, S.: The kernel function and conformal mapping, 2nd ed. (Math. Surveys 5.) Providence: Am. Math. Soc. 1970.
3. Fuks, B.: Special chapters of the theory of analytic functions of several complex variables. Moscow: Goź. Izdat. Fiz.-Mat. Lit. 1963 [Russian]. English translation: Providence, Ann. Math. Soc. 1966.
4. Mazur, T., Skwarczyński, M.: Spectral properties of automorphisms of the unit disc. Demonstratio Math. 4, 1069-1072 (1984).
5. Mazur, T. Skwarczyński, M.: Spectral properties of holomorphic automorphism with fixed point. Glasgow Math. J 28,25.30 (1986).

6. Skwarczyński, M.: The Bergman function and semiconformal mappings. Thesis. Warsaw University 1969.

7. Skwarczyński, M.: Biholomorphic invariants related to the Bergman functions. Dissertationes Math. 173 (1980).

8. Skwarczyński, M.: Alternating projections in complex analysis. In: Proc. 2nd Internat. Conf. on Complex Analysis and its Applications, Varna, May 2-9, 1983.

SMOOTHNESS OF SCHMIDT FUNCTIONS OF SMOOTH HANKEL OPERATORS

Vladimir V. Peller
Leningrad Branch
Steklov Institute of Mathematics
Academy of Sciences of the USSR
Fontanka 27
191011 Leningrad, USSR

§1. Introduction

In this paper we investigate smoothness properties of Schmidt functions of Hankel operators with symbols satisfying certain smoothness conditions.

Recall that for an operator T acting from a Hilbert space H_1 to a Hilbert space H_2 a non-zero vector f is called a *Schmidt vector* if $T^*Tf=s^2f$ for some $s>0$. In this case f is also called a *Schmidt* s-*vector*. If f is a Schmidt vector, the pair $\{f,1/s\cdot Tf\}$ is called a *Schmidt pair*. Note that $\{f,g\}$ is a Schmidt pair for T if and only if $\{g,f\}$ is a Schmidt pair for T*. If T is an operator on a function space, Schmidt vectors are called *Schmidt functions*.

Recall that each compact operator $T:H_1\to H_2$ admits the so-called Schmidt expansion: $Tx=\sum_{n\ge0}s_n(x,f_n)g_n$, $x\in H_1$, where $\{f_n\}$ is an orthonormal system in H_1, $\{g_n\}$ is an orthonormal system in H_2, and $\{s_n\}$ is the sequence of singular numbers (s-numbers) of T. Clearly, in this case the f_n are Schmidt functions, the $\{f_n,g_n\}$ are Schmidt pairs, and each Schmidt s-function is a linear combination of those f_n for which $s_n=s$.

The problem of the investigation of smoothness properties of Schmidt functions of Hankel operators has arisen in discussions with E.M.Dyn'kin in connection with the problem of computation of best rational approximation with given degree. The matter is that the best approximation of a function on the unit circle by rational functions of degree n in the norm of BMO (this norm is specified below) can be expressed via a theorem of Adamyan, Arov, and Krein [1] in terms of Schmidt pairs of the corresponding Hankel operator with $s=s_n$. So, to approximately compute the best rational approximation we can replace Schmidt pairs by their truncated Fourier expansions. To be sure that we have a good approximation we have to know that Schmidt pairs possess certain smoothness.

Now we recall some definitions. Let ϕ be a bounded function on the unit circle T. The Hankel operator H_ϕ acting from the Hardy class

H^2 to $H^2_- \overset{\text{def}}{=} L^2 \ominus H^2$ is defined by

$$H_\phi f = \mathbb{P}_- \phi f,$$

where $\mathbb{P}_-$ is the orthogonal projection from L^2 onto H^2_-. The norm of H_ϕ is given by Nehari's theorem (see e.g. [2], [4], [7]):

$$\|H_\phi\| = \inf\{ \|\phi - h\| : h \in H^\infty \}.$$

Together with Fefferman's theorem [3] this implies that $\|H_\phi\|$ is equivalent to the norm of $\mathbb{P}_-\phi$ in BMO. So we can supply BMO with a norm for which $\|H_\phi\| = \|\mathbb{P}_-\phi\|_{BMO}$.

By Hartman's theorem (see e.g. [2], [4], [7]) H_ϕ is compact if and only if ϕ can be represented as $\phi=\phi_1+\phi_2$ where $\phi_1 \in H^\infty$ and $\phi_2 \in C(\mathbb{T})$. In view of Sarason's theorem [8] H_ϕ is compact if and only if $\mathbb{P}_-\phi \in VMO$.

We also need the notion of Toeplitz operator. For $\phi \in L^\infty$ the Toeplitz operator T_ϕ is defined on H^2 by

$$T_\phi f = \mathbb{P}_+ \phi f, \quad f \in H^2,$$

where $\mathbb{P}_+$ is the orthogonal projection from L^2 onto H^2.

In § 2 we describe those function classes X for which we prove in § 3 that for the Hankel operators with symbols in X all their Schmidt functions also belong to X. We consider many concrete examples of such spaces (Hölder classes, Besov classes, Bessel potentials, Carleman classes etc.). The main tools of the paper are the results of [7] on hereditary properties of the operator of best approximation by analytic functions. In § 4 we examine properties of Schmidt functions of compact Hankel operators and of Hankel operators of the Schatten - von Neumann class S_p.

Note that in general every non-zero function in H^2 can be a Schmidt function of a bounded Hankel operator. Indeed, let $f \in H^2$ and g be a function in H^2_- with the same modulus, i.e. $|f|=|g|$ a.e. on $\mathbb{T}$. Put $\phi=g/f$. Then $\|H_\phi\| \le 1$ and $H_\phi f=g$. Therefore $\|H_\phi f\|_2 = \|g\|_2 = \|f\|_2$. It follows that $H^*_\phi H_\phi f = f$, i.e. f is a Schmidt function.

§2. Best uniform approximation by analytic functions

In this section we specify classes of function spaces for which we shall prove in §3 the main result of the paper. Throughout the paper X denotes a linear space of functions on $\mathbb{T}$ which contains the set of trigonometric polynomials and is contained in VMO.

Recall (see e.g. [7]) that for $f\in VMO$ there exists a unique function g_0 in $VMO_A \overset{def}{=} \mathbb{P}_+VMO$ which minimizes the distance

$$\|f-g_0\|_{L^\infty} = \inf\{\|f-g\|_{L^\infty(\mathbb{T})} : g\in VMO_A\}.$$

The function g_0 is called *the best uniform approximation of* f *by analytic functions*. The best approximation operator $\mathcal{A}$, which is non-linear, is defined on VMO by $\mathcal{A}f=g_0$.

In [7] it is proved that three classes of function spaces X are $\mathcal{A}$-invariant: $\mathcal{A}X\subset X$.

I. The so-called $\mathcal{R}$-spaces. We do not enter into detail here. We only mention that the Besov spaces $B_p^{1/p}$ (see the definition in §3) and the space VMO are $\mathcal{R}$-spaces and so they are $\mathcal{A}$-invariant (the case $p<1$ is treated in [6]).

II. The spaces satisfying the following system of axioms.

(A1) If $f\in X$ then $\bar{f}\in X$ and $\mathbb{P}_+f\in X$.

(A2) X is a Banach algebra with respect to the pointwise multiplication.

(A3) The set of trigonometric polynomials is dense in X.

(A4) Each multiplicative linear functional on X coincides with a point evaluation $f\to f(\zeta)$ for $\zeta\in\mathbb{T}$.

Note that (A1) - (A4) imply that $X\subset C(\mathbb{T})$ and the maximal ideal space of X coincides with $\mathbb{T}$.

III. The spaces satisfying the following system of axioms.

(B1) If $f\in X$ then $\bar{f}\in X$, $zf\in X$ and $\mathbb{P}_+f\in X$.

(B2) There exists a space Y such that Y satisfies (A1) - (A4), $X\subset Y$, and $T_{\bar{f}}X_A\subset X_A$ for any $f\in Y_A$. (Here $X_A = \mathbb{P}_+X$, $Y_A = \mathbb{P}_+Y$.)

(B3) If $f\in X$, $f\geq 0$, and $\inf\{f(t): t\in\mathbb{T}\}>0$ then $f^{-1}\in X$.

Note that X is not assumed to be a normed space.

It is proved in [7] that any function space satisfying one of the conditions I, II, III is A-invariant: $AX \subset X$.

Now we specify the class of function spaces X for which we shall establish in §3 the main result. This class consists of function spaces which satisfy the following three conditions:

(α) If $f \in X$ then $\bar{f} \in X$ and $\mathbb{P}_+ f \in X$.

(β) $AX \subset X$.

(γ) If $f \in X$ and ϕ is a function analytic in a neighbourhood of $f(\mathbb{T})$ then $\phi \circ f \in X$.

Note that (γ) implies both (A4) and (B3).

§3. Schmidt functions of smooth Hankel operators

In this section we establish the main result of the paper for the spaces satisfying (α), (β), and (γ) and give various examples of such spaces.

Lemma 1. *Suppose that* X *satisfies* (α), (β), *and* (γ) *and* $\phi \in X$. *Then* $\phi - A\phi$ *has the form* $cz^n\bar{h}/h$, *where* $c \in \mathbb{C}$, n *is a negative integer and* h *is an outer function invertible in* X.

Proof. From (β) we have $A\phi \in X$. It is well known (see e.g. [7], Theorem 1.7) that $\phi - A\phi$ has constant modulus on $\mathbb{T}$. Then $\phi - A\phi = cu$, where $c \in \mathbb{C}$ and u is a unimodular function in X. It follows from Theorem 3.15 of [7] that $u = z^n e^{i\phi}$ where n is an integer and ϕ is a real function in X. Now

$$\operatorname{dist}_{L^\infty}(u, H^\infty) = \|H_u\| = 1$$

since $A\phi$ is the best approximation of ϕ by analytic functions. Then it follows from Nehari's theorem that T_u is not left-invertible (see e.g. [2], [7]). But $T_{\exp(i\phi)}$ is invertible (see [7]) and so $n<0$.

Now put $h = \exp\frac{1}{2}(\tilde{\phi} - i\phi)$ where $\tilde{\phi}$ is the harmonic conjugate of ϕ. Then h is an outer function and it follows from (α) and (γ) that $h \in X$ and $h^{-1} \in X$.■

Lemma 2. *Suppose that* X *satisfies* (α), (β), *and* (γ), $\phi \in X$, $f \in H^2$, *and* $\|H_\phi f\|_2 = \|H_\phi\| \cdot \|f\|_2$. *Then* $f \in X$.

Proof. Let $v = \phi - A\phi$. Then $H_\phi = H_v$ and by Nehari's theorem $\|v\|_{L^\infty} = \|H_\phi\|$. By Lemma 1 $v = c\bar{z}^n\bar{h}/h$, where $n>0$, $c \in \mathbb{C}$, h is invertible in X.

Clearly

$$\{ g\in H^2 : \|H_v g\|_2 = \|H_v\| \cdot \|g\|_2 \} = \text{Ker } T_u .$$

It is easy to see that if g has the form $g=qh$ where q is a polynomial of degree at most $n-1$ then $g\in\text{Ker } T_u$. On the other hand $T_u = cT_{\bar{z}^n}T_{\bar{h}/h}$ and it is well known that $T_{\bar{h}/h}$ is invertible (see e.g. [7]). Clearly, $\dim \text{Ker } T_{\bar{z}^n} = n$ and so $\dim \text{Ker } T_u = n$. It follows that

$$\text{Ker } T_u = \{ qh : \deg q \le n-1 \} \subset X . \blacksquare$$

In the proof of the main result we shall use an idea of Adamyan, Arov, and Krein from their proof of their theorem mentioned in §1.

Theorem 1. *Suppose that* X *is a function space satisfying* (α), (β), *and* (γ). *Let* $\phi\in X$. *Then all Schmidt functions of* H_ϕ *also belong to* X.

Proof. Suppose that f is a Schmidt function which corresponds to a singular number s. If $s=s_0(H_\phi)$, the result follows from Lemma 2 since in this case $\|H_\phi f\|_2 = \|H_\phi\| \cdot \|f\|_2$.

Consider the subspace

$$E = \{ k\in H^2 : H_\phi^* H_\phi k = s^2 k \} .$$

Let $d = \dim E$. Choose now $n>0$ so that $s = s_n(H_\phi)$ and

$$s_{n-1}(H_\phi) > s_n(H_\phi) = \cdots = s_{n+d-1}(H_\phi) > s_{n+d}(H_\phi) .$$

We have $f\in E$. Put $g = s^{-1}H_\phi f$. It is proved by Adamyan, Arov, and Krein (see [7], Lemma 1.1) that $|g| = |f|$ a.e. Clearly, (f,g) is a Schmidt pair for H_ϕ.

Put $\phi_s = g/f$. Then ϕ_s is a unimodular function and $\|H_{\phi_s} f\|_2 = \|g\|_2 = \|f\|_2$. So f is a Schmidt function of H_{ϕ_s}. Since $H_{\phi_s} = H_{\mathbb{P}_-\phi_s}$, it remains to prove that $\mathbb{P}_-\phi_s\in X$ and the result will follow from Lemma 2.

It is proved by Adamyan, Arov, and Krein that H_ϕ and $H_{s\phi_s}$ coincide on E (see [7], Theorem 1.6) It is easy to see that if two

Hankel operators coincide on a function k, they coincide on $z^m k$ for $m \geq 0$. Therefore H_ϕ and $H_{s\phi_s}$ coincide on

$$L \stackrel{\text{def}}{=} \text{span}\{\ z^m E : \ m \geq 0\ \}.$$

L is invariant under multiplication by z. Hence by Beurling's theorem (see [4]) L has the form $L = \theta H^2$ where θ is an inner function. It is shown by Adamyan, Arov, and Krein (see [7], Theorem 1.6) that in our case $\dim (H^2 \ominus \theta H^2) \leq n$.

So $H_\phi \theta\psi = H_{s\phi_s} \theta\psi$ for any $\psi \in H^2$ which implies that $H_{\phi\theta} = H_{s\phi_s\theta}$. The latter means that $\mathbb{P}_-\phi\theta = \mathbb{P}_- s\phi_s\theta$. Therefore $\phi\theta = s\phi_s\theta + h$, where $h \in H^\infty$. It follows that $\mathbb{P}_-\phi_s = s^{-1}\mathbb{P}_-\phi - s^{-1}\mathbb{P}_-\bar{\theta}h$.

By (α) we have $\mathbb{P}_-\phi \in X$. It remains to prove that $\mathbb{P}_-\bar{\theta}h \in X$.

Since $\dim (H^2 \ominus \theta H^2) \leq n$, it follows that θ is a finite Blaschke product of degree at most n (see [4], Lecture II). It is easy to verify that in this case $\mathbb{P}_-\bar{\theta}h$ is a rational function with at most n poles. So $\mathbb{P}_-\phi_s \in X$. ∎

Remark 1. It is easy to see that under the assumptions of Theorem 1 for any $f \in X$ we have $H_\phi f \in X$. This implies that if (f,g) is a Schmidt pair for H_ϕ then not only $f \in X$ but also $g \in X$.

Remark 2. Suppose that X satisfies the assumptions of Theorem 1. Then for any invertible function $f \in X_A$ there exists $\phi \in X$ such that f is a Schmidt function of H_ϕ. Indeed, let $\phi = \overline{zf}/f$. Then $\phi \in X$ and $\|H_\phi f\|_2 = \|H_\phi\|_2 \cdot \|f\|_2$.

Examples. Here we present some examples of function spaces satisfying (α), (β), and (γ). The corresponding proofs are given in [7] where many other examples are also considered.

1. The Hölder - Zygmund classes. The classes Λ_α for $0<\alpha\leq 1$ are defined by

$$\Lambda_\alpha = \{\ f \in C(\mathbb{T}) \ : \ |f(\zeta_1) - f(\zeta_2)| \leq c_f|\zeta_1 - \zeta_2|^\alpha,\ \zeta_1,\ \zeta_2 \in \mathbb{T}\},\ \alpha<1,$$

$$\Lambda_1 = \{\ f \in C(\mathbb{T}) \ : \ |f(e^{i(x+t)}) + f(e^{i(x-t)}) - 2f(e^{ix})| \leq c_f|t|,\ x,\ t \in \mathbb{R}\}.$$

For $\alpha>1$

$$\Lambda_\alpha = \{\ f\in C^n(\mathbf{T})\ :\ f^{(n)}\in\Lambda_{\alpha-n}\},\ n<\alpha.$$

Denote by λ_α, $\alpha>0$, the closure of the set of polynomials in Λ_α.

The classes λ_α and Λ_α, $0<\alpha<\infty$, satisfy (α), (β), and (γ) (the λ_α satisfy (A1) - (A4) while the Λ_α satisfy (B1) - (B3)) (see [7]).

2. Besov classes. Let $s>0$, $1\le p,q\le\infty$. The Besov class B^s_{pq} is defined by

$$B^s_{pq} = \{\ f\in L^p\ :\ \int_{-\pi}^{\pi} \|\Delta^n_t f\|_{L^p}\cdot|t|^{-(1+sq)}dt < \infty\ \},\quad q<\infty,$$

$$B^s_{p\infty} = \{\ f\in L^p\ :\ \sup_{t\neq 0} \|\Delta^n_t f\|_{L^p}\cdot|t|^{-s} < \infty\ \},$$

where n is a non-negative integer, $n>s$, and $(\Delta_t f)(e^{ix}) = f(e^{i(x+t)}) - f(e^{ix})$, $\Delta^n_t = \Delta_t\Delta^{n-1}_t$. We also use the notation $B^s_p = B^s_{pp}$.

The spaces B^s_{pq} satisfy the assumptions of Theorem 1 for $s > 1/p$ or for $s = 1/p$ and $q=1$ (see [7]).

It is also possible to consider the case of $0<p\le\infty$, $0<q\le\infty$.

3. Sobolev spaces (Bessel potential spaces). Let $1<p<\infty$ and $s>0$. The space L^s_p is defined as the space of functions whose fractional derivatives of order s belong to L^p. For $s>1/p$ the spaces L^s_p satisfy the assumptions of Theorem 1 (see [7]).

4. Let Z be one of the following spaces

$$VMO,\ BMO,\ C_A+\overline{C}_A,\ H^\infty+\overline{H}^\infty,\ H^1+\overline{H}^1$$

$(C_A \stackrel{\text{def}}{=} C(\mathbf{T})\cap H^2)$.

Then for any positive integer n the space $X = \{\ f\ :\ f^{(n)}\in Z\ \}$ satisfies the assumptions of Theorem 1 (see [7]). If $Z = L^1 + \mathbb{P}_+L^1$ then for $n\ge 2$ the space $X = \{\ f\ :\ f^{(n)}\in Z\ \}$ satisfies the assumptions of Theorem 1 (see [7]).

Therefore if $n\ge 1$ and $(\mathbb{P}_-\phi)^{(n)}$ belongs to $C(\mathbf{T})$, L^∞ or L^1 then any Schmidt function f of H_ϕ satisfies the condition $f^{(n)}\in C(\mathbf{T})$, L^∞ or L^1 respectively. If $\phi^{(n)}$ belongs to $C(\mathbf{T})$ or L^∞

then any Schmidt function f of H_ϕ satisfies $f^{(n)} \in VMO$ or $f^{(n)} \in BMO$.

5. The spaces $F\ell^1(w)$. Let $w = \{w_n\}_{n \ge 0}$ be a sequence of positive numbers. The space $F\ell^1(w)$ consists of the functions f on $\mathbb{T}$ satisfying

$$\sum_{n=-\infty}^{\infty} |\hat{f}(n)| \cdot w_{|n|} < \infty,$$

where $\hat{f}(n)$ is the n-th Fourier coefficient of f. If w satisfies the condition

$$w_{|j+k|} \le w_{|j|} w_{|k|}$$

for any j, k then $F\ell^1(w)$ satisfies the assumptions of Theorem 1.

Note that in [7] the spaces $F\ell^p(w)$ with $1 < p \le \infty$ are also considered.

6. Carleman classes. Let $\{M_n\}_{n \ge 0}$ be an increasing logarithmically convex sequence. The Carleman class $C\{M_n\}$ consists of the functions f on $\mathbb{T}$ satisfying

$$\|f^{(n)}\|_{L^\infty} \le C_f Q_f^n \cdot n! \cdot M_n$$

for some C_f, Q_f. If $M_{n+1}/M_n \le CQ^{Q^n}$ for some $C, Q > 0$ then $C\{M_n\}$ satisfies the assumptions of Theorem 1 (see [7]).

§4. Schmidt functions of compact Hankel operators

In this section we consider properties of Schmidt functions of compact Hankel operators and of Hankel operators of the Schatten - von Neumann class S_p, $0 < p < \infty$ ($T \in S_p$ iff $\sum_{n \ge 0} s_n(T) < \infty$). Recall that H_ϕ belongs to S_p if and only if $\mathbb{P}_- \phi \in B_p^{1/p}$ (see [5] for $1 \le p < \infty$ and [6], [9] for $0 < p < 1$). Here we find sharp conditions on Schmidt functions of such operators.

Theorem 2. *Suppose that* H_ϕ *is compact (or* $H_\phi \in S_p$, $0 < p < \infty$*) and f is a Schmidt function of* H_ϕ*. Then f has the form* $f = qe^g$, *where q is a polynomial and g is a function in* VMO *(or in* $B_p^{1/p}$*).*

The spaces VMO and $B_p^{1/p}$ satisfy (α) and (β). But VMO and $B_p^{1/p}$ with $p>1$ do not satisfy (γ). The proof of Theorem 2 repeats that of Theorem 1. The only difference is that the condition $g \in VMO$ ($g \in B_p^{1/p}$, $p>1$) does not imply that $\exp g \in VMO$ ($\exp g \in B_p^{1/p}$).

Corollary. *Under the hypotheses of Theorem 2* $f \in \bigcap_{q<\infty} H^q$.

Proof. The result follows from Theorem 2 and Zygmund's theorem (see e.g. [7], Corollary 1.6) which implies that $\exp g \in \bigcap_{q<\infty} H^q$ for $g \in VMO$. ∎

Remark. Any function f of the form $f = q \cdot \exp g$ where q is a polynomial and g is an analytic function in VMO (in $B_p^{1/p}$) can be a Schmidt function of a compact Hankel operator (of a Hankel operator of class S_p). Indeed, let $n > \deg q$, $h = \exp g$ and $\phi = \bar{z}^n \bar{h}/h$. As we have already mentioned in the proof of the above corollary $f \in H^2$. We have $\|H_\phi f\|_2 = \|H_\phi\| \cdot \|f\|_2$ and $\phi = \bar{z}^n \exp(2i \cdot \operatorname{Im} g)$. Now $\operatorname{Im} g \in VMO$ (or $B_p^{1/p}$). It is easy to see that $\exp(2i \cdot \operatorname{Im} g) \in VMO$ (or $B_p^{1/p}$) which implies that $\phi \in VMO$ (or $B_p^{1/p}$).

Therefore it is not true for compact Hankel operators (and for Hankel operators of class S_p with $p>1$) that their Schmidt functions always belong to VMO or $B_p^{1/p}$ respectively.

But for $p \leq 1$ the spaces $B_p^{1/p}$ do satisfy the properties (α), (β), and (γ) (see [7] for $p=1$ and [6] for $p<1$). Therefore we obtain the following result.

Theorem 3. *Let* $0 < p \leq 1$. *Suppose that* H_ϕ *is a Hankel operator of class* S_p. *Then the Schmidt functions of* H_ϕ *blong to* $B_p^{1/p}$.

REFERENCES

1. V.M.Adamyan, D.Z.Arov, and M.G.Krein, *Analytic properties of the Schmidt pairs of a Hankel operator and the generalized Schur - Takagi problem*, Matem. Sbornik, 86 (1971), 34-75.
Translation: Math. USSR Sbornik, 15 (1971), 31-73.

2. R.G.Douglas, *Banach algebra techniques in operator theory*, Acad. Press, N.-Y., London, 1972.

3. C.Fefferman, *Characterizations of bounded mean oscillation*, Bull. Amer. Math. Soc., 77 (1971), 587-588.

4. N.K.Nikol'skii, *Lectures on the shift operator*, Nauka, Moscow, 1980.
Translation of the enlarged version: *Treatise on the shift operator*, Springer-Verlag, Berlin, Heidelberg, N.-Y., 1986.

5. V.V.Peller, *Hankel operators of class* S_p *and their applications (rational approximation, Gaussian processes, and the problem of majorizing operators)*, Matem. Sbornik, 113 (1980), 538-581.
Translation: Math. USSR Sbornik, 41 (1982), 443-479.

6. V.V.Peller, *A description of Hankel operators of class* S_p *for* $p>0$, *an investigation of the rate of rational approximation, and other applications*, Matem. Sbornik, 122 (1983), 481-510.
Translation: Math. USSR Sbornik, 50 (1985), 465-494.

7. V.V.Peller and S.V.Khrushchev, *Hankel operators, best approximations, and stationary Gaussian processes*, Uspekhi Matem. Nauk, 37 (1982), 53-124.
Translation:Russian Math. Surv., 37 (1982), 61-144.

8. D.Sarason, *Functions of vanishing mean oscillation*, Trans. Amer. Math. Soc., 207 (1975), 391-405.

9. S.Semmes, *Trace ideal criteria for Hankel operators and applications to Besov spaces*, Int. Equat. and Op. Theory, 7 (1984), 241-281.

REAL INTERPOLATION BETWEEN SOME OPERATOR IDEALS*

Lars Erik Persson
Dept. of Mathematics
Luleå University
S-95187 Luleå, SWEDEN

0. Introduction

We assume that the reader is somewhat acquainted with the real interpolation theory. See [10], [1] or [18] and compare also with our section 2 where we discuss some basic concepts and results from [9].

Let A_o and A_1 denote quasi-Banach spaces. We write $T \in L(A_o,A_1)$ if T is a bounded linear operator from A_o to A_1. We can obtain more precise information of the operator T by considering some associated sequence of s-numbers, $s = (s_n(T))_1^\infty$, in the sense of Pietsch [15]. Examples of s-numbers are the Weyl numbers, the Kolmogorov numbers, the Gelfand numbers and the approximation numbers $\alpha = (\alpha_n(T))_1^\infty$, where

$$\alpha_n(T) = \inf\{\|T-T_n\| : \text{rank } T_n < n;\ T_n \in L(A_o,A_1)\} .$$

We say that $T \in S^{(s)}_{\varphi,q}$, $\varphi = \varphi(t) \geq 0$, $0 < q \leq \infty$, if

$$\|T\|_{S^{(s)}_{\varphi,q}} = \left(\sum_1^\infty (s_n(T)\varphi(n))^q \frac{1}{n}\right)^{1/q} < \infty ,$$

(with the usual interpretation when $q = \infty$). If $\varphi(t) = t^{1/p}$, then we have the class $S^{(s)}_{p,q}$, which for the case $p = q$ coincide with the usual extension $S^{(s)}_p$ of the Schatten-Hilbert class $S_p(H)$ (H = Hilbert space) to the quasi-Banach case.

If we impose suitable restrictions on $\varphi(t)$, e.g. that $\varphi(t)$ is a "parameter function", then the classes $S^{(s)}_{\varphi,q}$ and $S^{(e)}_{\varphi,q}$ ($e = (e_n(T))_1^\infty$ is the sequence of entropy numbers) are complete operator ideals. For properties and definitions of s-numbers, entropy numbers and operator ideals we refer to [15] or [5].

By interpolation between operator ideals we can obtain new operator ideals. The

---

* This research was partly done under NFR contract F-FU 8685-100.

first result in this direction is due to Triebel [17]. Moreover, Peetre-Sparr [11] proved that

$$(S_{q_o}^{(\alpha)}, S_{q_1}^{(\alpha)})_{\theta,q_\theta} = S_{q_\theta}^{(\alpha)} , \tag{0.1}$$

where $0 < q_o , q_1 \leq \infty$, $0 < \theta < 1$ and $\frac{1}{q_\theta} = \frac{1-\theta}{q_o} + \frac{\theta}{q_1}$. This result was obtained by analysing the connection between the K-functional and the approximation E-functional. By using other methods (0.1) has been generalized by König [5], [6] and Cobos [4] (the parameter function case). In section 3 of this paper we shall see that these generalizations, and even more general results of this kind, can also be obtained by using some useful estimates by Nilsson [9], concerning the close connection between K- and E-functionals.

In this paper we also state and prove the corresponding embedding theorems obtained by interpolation between general operator ideals of the types $S_{\varphi,q}^{(s)}$ or $S_{\varphi,q}^{(e)}$. In particular our results extend some results by König [6], Pietsch [16] and Cobos [3]. We also confirm and generalize the estimate of the K-functional formulated by König [6]. Finally we remark that our proofs are rather elementary and easy to carry over to more general situations. In section 5 we give some examples of such generalizations.

This paper is organized in the following way. In section 1 we give some necessary definitions and lemmas. Section 2 is used to discuss some basic definitions and results from the paper [9] concerning real interpolation and approximation theory. The results obtained in section 2 are used in section 3 to prove our results on interpolation between operator ideals of the type $S_{\varphi,q}^{(\alpha)}$. We also give the corresponding estimate of the K-functional. In section 4 we prove the corresponding embedding theorems concerning interpolation between general operator ideals of the type $S_{\varphi,q}^{(s)}$ or between entropy operator ideals of the type $S_{\varphi,q}^{(e)}$. In section 5 we give some concluding remarks, raise some questions and state some generalizations of our results.

Conventions. The equivalence symbol $\varphi(t) \approx \psi(t)$ means that $c_1\varphi(t) \leq \psi(t) \leq c_2\varphi(t)$ for some positive constants c_1 and c_2 and every $t \geq 0$. Two quasi-normed spaces, A and B , are considered as equal and we write $A = B$ if their quasi-norms are equivalent. The inclusion $A \subset B$ means that we have continuous embedding. We use the notation $\alpha = (\alpha_\nu)_I$ for a sequence with I as its index set ($I = N$, $I = Z$ or $I = Z_+$). Moreover, C denotes a positive constant, not necessarily the same in different appearances.

1. Some definitions and lemmas

Let $\varphi = (\varphi_\nu)_Z$. By $\ell^q(\varphi)$ we denote the space of all sequences $g = (g_\nu)_Z$ such that

$$\|g\|_{\ell^q(\varphi)} = (\sum_{-\infty}^{\infty} |g_\nu \varphi_\nu|^q)^{1/q} < \infty .$$

If Z is replaced by N then $\Sigma_{-\infty}^{\infty}$ is replaced by Σ_1^{∞} in this definition.

By $Q(a,b)$, $-\infty < a < b < \infty$, we denote the class of non-negative functions on $]0,\infty[$ such that, for some $\varepsilon > 0$, $\lambda(t)t^{-a-\varepsilon}$ is increasing and $\lambda(t)t^{-b+\varepsilon}$ is decreasing.

Here and in the sequel we assume that θ, q, q_o, q_1 and q_θ are parameters satisfying $0 < \theta < 1$, $0 < q,q_o,q_1 \leq \infty$ and $\frac{1}{q_\theta} = \frac{1-\theta}{q_o} + \frac{\theta}{q_1}$. Moreover, $\varphi_o(t)$ and $\varphi_1(t)$ always denote functions satisfying $\varphi_i(t) \in Q(0,c)$, $i = 0,1$, for some $c > 0$.

<u>Lemma 1.1</u> <u>Let</u> $\lambda(t) \in Q(0,1), \varphi(t) = \varphi_o(t)/\varphi_1(t)$, $\psi_i = (\varphi_i(2^\nu))_Z$, $i = 0,1$, <u>and</u> $\mu = (\varphi_o(2^\nu)/\lambda(\varphi(2^\nu)))_Z$. <u>Then</u>

a) $$(\ell^{q_o}(\psi_o),\ell^{q_1}(\psi_1))_{\lambda,q_\theta} = \ell^{q_\theta}(\mu)$$

<u>if</u> $q_o = q_1 < \infty$ <u>or if</u> $\lambda(t) = t^\theta$.

b) $$(\ell^{q_o}(\psi_o),\ell^{q_1}(\psi_1))_{\lambda,q} = \ell^q(\mu)$$

<u>if</u> $\varphi(t) \in Q(0,c)$ <u>or</u> $\varphi(t) \in Q(-c,0)$ <u>for some</u> $c > 0$.

For the case $\lambda(t) = t^\theta$ Lemma 1.1 a) is a special case of the usual diagonal relation (see [1, p. 119]). For the general case Lemma 1.1 is a special case of the results obtained in [14]. The lemma holds as well if we replace Z by N or Z_+ .

In this paper we shall work with the parameter function class $Q(a,b)$. In order to be able to compare our results with some similar results in the literature we shall prove a lemma which can be of independent interest. First we need some more definitions.

Let B be the class of all non-negative and continuous functions on $]0,\infty[$ such that $\varphi(1) = 1$ and

$$\bar{\varphi}(t) = \sup_{s>0} \frac{\varphi(ts)}{\varphi(s)} < \infty \text{ , for every } t > 0 .$$

The Boyd indices $\alpha_{\bar{\varphi}}$ and $\beta_{\bar{\varphi}}$ of the sub-multiplicative function $\bar{\varphi}$ are defined by

$$\alpha_{\bar{\varphi}} = \inf_{t>1} \frac{\log \bar{\varphi}(t)}{\log t} = \lim_{t\to\infty} \frac{\log \bar{\varphi}(t)}{\log t} ,$$

$$\beta_{\bar{\varphi}} = \sup_{0<t<1} \frac{\log \bar{\varphi}(t)}{\log t} = \lim_{t\to 0+} \frac{\log \bar{\varphi}(t)}{\log t} .$$

It is well-known that $-\infty < \beta_{\bar{\varphi}} \le \alpha_{\bar{\varphi}} < \infty$.

The class B_ψ consists of all continuously differentiable functions ψ on $]0,\infty[$ satisfying

$$0 < \inf_{t>0} \frac{t\psi'(t)}{\psi(t)} \le \sup_{t>0} \frac{t\psi'(t)}{\psi(t)} < 1 .$$

Lemma 1.2

a) Let $\lambda(t) \in Q(a,b)$. Then there exists a function $\varphi(t)$ such that $\varphi(t) \in B$, $\varphi(t) \approx \lambda(t)$ and $a < \beta_{\bar{\varphi}} \le \alpha_{\bar{\varphi}} < b$.

b) Let $\varphi(t) \in B$ and $a < \beta_{\bar{\varphi}} \le \alpha_{\bar{\varphi}} < b$. Then there exists a function $\lambda(t)$ such that $\lambda(t) \approx \varphi(t)$ and $\lambda(t) \in Q(a,b)$.

Remark 1.1 According to Lemma 1.2 we say that the class $Q(0,1)$ is essentially equivalent to the sub-class of B consisting of all $\varphi(t)$ satisfying $0 < \beta_{\bar{\varphi}} \le \alpha_{\bar{\varphi}} < 1$. It is proved in [13] that the class $Q(0,1)$ is also essentially equivalent to the class B_ψ (and also to the class $\mathcal{P}^{+-}$).

Proof of Lemma 1.2 Assume that $\lambda(t) \in Q(a,b)$. Then there exists an ε , $0 < \varepsilon < (b-a)/2$, such that, for every $s , t > 0$,

$$\min(t^{a+\varepsilon}, t^{b-\varepsilon}) \le \frac{\lambda(st)}{\lambda(s)} \le \max(t^{a+\varepsilon}, t^{b-\varepsilon}) . \tag{1.1}$$

We put $\varphi(t) = \lambda(t)/\lambda(1)$. Obviously $\varphi(t) \in B$, $\varphi(t) \approx \lambda(t)$ and, by (1.1),

$$a + \varepsilon \le \frac{\log \bar{\varphi}(t)}{\log t} = \frac{\log \bar{\lambda}(t)}{\log t} \le b - \varepsilon ,$$

for every $t > 0$. We conclude that $a < \alpha_{\bar{\varphi}} \le \beta_{\bar{\varphi}} < b$ and the proof of a) is complete.

Let $a < \beta_{\bar{\varphi}} \le \alpha_{\bar{\varphi}} < b$ and choose $\varepsilon > 0$ so that $a + \varepsilon \le \beta_{\bar{\varphi}} \le \alpha_{\bar{\varphi}} \le b - \varepsilon$. We choose ε_o so that $0 < \varepsilon_o < \varepsilon$ and $\varepsilon_1 = \varepsilon - \varepsilon_o \le (b-a)/2$. According to the definition of the lower Boyed index $\beta_{\bar{\varphi}}$ we have

$$\frac{\log \bar{\varphi}(t)}{\log t} \ge \beta_{\bar{\varphi}} - \varepsilon_o \ge a + \varepsilon_1 \quad \text{if} \quad 0 < t \le \delta(\varepsilon_o) \le 1 .$$

Therefore

$$\bar{\varphi}(t) \leq t^{a+\varepsilon_1} \text{ , if } 0 < t \leq \delta(\varepsilon_o) \leq 1 \text{ ,}$$

so that

$$\bar{\varphi}(t) \leq Ct^{a+\varepsilon_1} \quad \text{if} \quad 0 < t \leq 1 \text{ .} \tag{1.2}$$

In a similar way we can use the definition of the upper Boyd index $\alpha_{\bar{\varphi}}$ and obtain that

$$\bar{\varphi}(t) \leq Ct^{b-\varepsilon_1} \quad \text{if} \quad t > 1 \text{ .} \tag{1.3}$$

We combine (1.2) with (1.3) and find that

$$\varphi(st) \leq C \max(t^{a+\varepsilon_1}, t^{b-\varepsilon_1})\varphi(s) \quad \text{for} \quad s > 0 \text{ , } t > 0 \text{ .}$$

Therefore, according to Proposition 1.2 in [13], $\varphi(t)$ is equivalent to some function $\lambda(t)$ such that $\lambda(t)t^{-a-\varepsilon_1}$ is non-decreasing and $\lambda(t)t^{-b+\varepsilon_1}$ is non-increasing. We conclude that $\lambda(t) \in Q(a,b)$ and the proof is complete.

2. Some real interpolation and approximation spaces

Let F be a quasi-normed Z-lattice of sequences. This means that if $\alpha = (\alpha_\nu)_Z \in F$ and $|\beta_\nu| \leq |\alpha_\nu|$, for $\nu \in Z$, then $\beta = (\beta_\nu)_Z \in F$ and $\|\beta\|_F \leq \|\alpha\|_F$. In particular we see that the spaces $\ell^q(\psi)$, $\psi = (\psi_\nu)_Z$, are quasi-normed Z-lattices.

Let $\bar{G} = (G_o, G_1)$ be a compatible quasi-Banach pair of Abelian groups, i.e. G_o and G_1 are subgroups of some Abelian group G satisfying the separation axiom.

The approximation E-functional is defined as

$$E(t,g) = E(t,g,\bar{G}) = \inf_{g_o \in (G_o)_t} \|g-g_o\|_{G_1} \text{ ,}$$

where $(G_o)_t = \{g_o \in G_o : \|g_o\|_{G_o} \leq t\}$. The approximation space $\bar{G}_{F:E}$ is the collection of all $g \in G_o + G_1$ such that $E(t,g,\bar{G}) < \infty$ for every $t > 0$ and

$$\|g\|_{\bar{G}_{F:E}} = \|(E(2^\nu,g,\bar{G}))_Z\|_F < \infty \text{ .}$$

The K-functional is defined for every $g \in G_o + G_1$ and every $t > 0$ as

$$K(t,g) = K(t,g,\bar{G}) = \inf_{g_o+g_1=g} (\|g_o\|_{G_o} + t\|g_1\|_{G_1}) \text{ .}$$

By the K-space $\bar{G}_{F:K}$ we mean the set of all $g \in G_o + G_1$ satisfying

$$\|g\|_{\bar{G}_{F:K}} = \|(K(2^\nu,g,\bar{G}))_\nu\|_F < \infty .$$

In particular for the case when $F = \ell^q(\psi)$, $\psi = ((\lambda(2^\nu))^{-1})_Z$, $\lambda \in Q(0,1)$, we find that $\bar{G}_{F:K}$ coincide with the parameter function space $\bar{G}_{\lambda,q}$ usually defined by using the equivalent quasi-norm

$$\|g\|_{\bar{G}_{\lambda,q}} = \left(\int_0^\infty \left(\frac{K(t,g)}{\lambda(t)}\right)^q \frac{dt}{t}\right)^{1/q} .$$

For the case $\lambda(t) = t^\theta$, $0 < \theta < 1$, we have the usual parameter space $\bar{G}_{\theta,q}$.

We shall now formulate a useful estimate by Nilsson [9, p. 320] but first we assume that F has the following additional properties:

$$\|(\alpha_{\nu-1})_Z\|_F \le C\|(\alpha_\nu)_Z\|_F , \quad \sup_{\|\alpha\|_F\le 1} (\sup_{\nu\ge k}|\alpha_\nu|) < c_k , \tag{2.1}$$

where $c_k \to 0$ as $k \to \infty$ and, for every $k \in Z$,

$$(\chi_{]0,2^k[}(2^\nu))_Z \in F \quad \text{and} \quad \ell^\infty((\max(1,2^{-\nu}))_Z) \subset F . \tag{2.2}$$

Here χ_I denotes the characteristic function of the interval I .

<u>Theorem 2.1 (Nilsson)</u> <u>Let</u> $\bar{F} = (F_0,F_1)$ <u>be a compatible pair of quasi-Banach</u> Z-<u>lattices where</u> F_0 <u>and</u> F_1 <u>satisfy</u> (2.1). <u>Then, for every</u> $t > 0$,

$$K(t,g,\bar{G}_{F_0:E},\bar{G}_{F_1:E}) \approx K(t,(E(2^\nu,g,\bar{G}))_Z,\bar{F}) . \tag{2.3}$$

<u>Remark 2.1</u> The estimate (2.3) implies that the following reiteration formula holds

$$(\bar{G}_{F_0:E},\bar{G}_{F_1:E})_{F_2:K} = \bar{G}_{(F_0,F_1)_{F_2:K}:E} , \tag{2.4}$$

if F_2 is any quasi-Banach lattice satisfying (2.2).

<u>Remark 2.2</u> If, in addition to the assumptions in Theorem 2.1, we can find partitions $Z = A_k \cup B_k$, for every $k \in Z$, such that

$$\|\chi_{A_k} g\|_{F_0} \le C2^k\|g\|_{F_1} ,$$

and (2.5)

$$\|\chi_{B_k} g\|_{F_1} \le C2^{-k}\|g\|_{F_0} ,$$

then (2.3) can be written as the following estimate of Holmstedt's type:

$$K(2^k,g,\bar{G}_{F_0:E},\bar{G}_{F_1:E}) \approx$$

$$\|(\chi_{A_k}E(2^\nu,g,\bar{G}))_Z\|_{F_0} + 2^k\|(\chi_{B_k}E(2^\nu,g,\bar{G}))_Z\|_{F_1} .$$

See [9, p. 322].

3. Interpolation between $S^{(\alpha)}_{\varphi,q}$ operator ideals

We shall now consider the normed Abelian pair $\bar{G} = (G_0,G_1)$ of operators, where $G_1 = L(A_0,A_1)$ and

$$G_0 = L_0(A_0,A_1) = \{T \in L(A_0,A_1):\text{rank } T < \infty\}$$

with the norms defined by

$$\|T\|_{G_1} = \sup_{\|g\|_{A_0}\leq 1} \|Tg\|_{A_1} \quad \text{and} \quad \|T\|_{G_0} = \text{rank } T ,$$

respectively. We put $\bar{L} = (L_0(A_0,A_1),L_1(A_0,A_1))$ and note that

$$E(2^\nu,T,\bar{L}) = \begin{cases} \alpha_1(T) & , \nu = -1, -2, -3, \dots , \\ \alpha_{2^\nu+1}(T) & , \nu = 0, 1, 2, \dots . \end{cases} \tag{3.1}$$

If F is a quasi-Banach Z-lattice of sequences we say that $T \in S^{(\alpha)}_F$ if

$$\|T\|_{S^{(\alpha)}_F} = \|(E(2^\nu,T,\bar{L}))_Z\|_F < \infty .$$

<u>Theorem 3.1</u> <u>Let</u> $\lambda(t) \in Q(0,1)$, $\varphi(t) = \varphi_0(t)/\varphi_1(t)$ <u>and</u> $\rho(t) = \varphi_0(t)/\lambda(\varphi(t))$. <u>Then</u>

a) $$(S^{(\alpha)}_{\varphi_0,q_0},S^{(\alpha)}_{\varphi_1,q_1})_{\lambda,q_\theta} = S^{(\alpha)}_{\rho,q_\theta} \tag{3.2}$$

<u>if</u> $q_0 = q_1$ <u>or if</u> $\lambda(t) = t^\theta$, $0 < \theta < 1$.

b) $$(S^{(\alpha)}_{\varphi_0,q_0},S^{(\alpha)}_{\varphi_1,q_1})_{\lambda,q} = S^{(\alpha)}_{\rho,q} , \tag{3.3}$$

<u>if</u> $\varphi(t) \in Q(0,c)$ <u>or</u> $\varphi(t) \in Q(0,c)$ <u>for some</u> $c > 0$.

<u>Remark 3.1</u> For the case when $\varphi_i(t) = t^{1/p_i}$, $i = 0,1$, and $\lambda(t) = t^\theta$ (3.2) and (3.3) coincide with the estimates by Peetre-Sparr [11, p. 257] and König [5, p. 121] respectively. According to Lemma 1.2 we see that Theorem 3.1 is also more general than a recent result by Cobos [4, Th 5.3].

We shall also state the expected estimate of the K-functional corresponding to Theorem 3.1.

Theorem 3.2 *Let* $\varphi(t) = \varphi_o(t)/\varphi_1(t)$. *If* $\varphi(t) \in Q(0,c)$ *for some* $c > 0$ *and if* $t \geq \varphi(1)$, *then*

$$K(t,T,S^{(\alpha)}_{\varphi_o,q_o},S^{(\alpha)}_{\varphi_1,q_1}) \approx$$

$$\approx \left(\sum_{1}^{h(t)} (\alpha_n(T)\varphi_o(n))^{q_o}\frac{1}{n}\right)^{1/q_o} + t\left(\sum_{h(t)+1}^{\infty} (\alpha_n(T)\varphi_1(n))^{q_1}\frac{1}{n}\right)^{1/q_1} ,$$

where $h(t) = [\varphi^{-1}(t)]$ *and* φ^{-1} *is the inverse of* φ .

Remark 3.2 For the case $\varphi_i(t) = t^{1/q_i}$, $i = 0,1$, this estimate has been formulated by König [6, p. 37]. However, the proof in [6] is not correct. See also [7, p. 133].

Proof of Theorem 3.1 For our operator case the formula (2.4) reads

$$(S^{(\alpha)}_{F_o},S^{(\alpha)}_{F_1})_{F_2:K} = S^{(\alpha)}_{(F_o,F_1)_{F_2:K}} . \tag{3.4}$$

We put $\mu = (\rho(2^\nu))_Z$, $F_i = \ell^{q_i}(\psi_i)$, where $\psi_i = (\varphi_i(2^\nu))_Z$, $i = 0,1$, and $F_2 = \ell^q(\psi_2)$, where $\psi_2 = ((\lambda(2^\nu))^{-1})_Z$. Let $\varphi(t) \in Q(0,c)$ or $\varphi(t) \in Q(-c,0)$ for some $c > 0$. Then, by Lemma 1.1 b),

$$(F_o,F_1)_{F_2:K} = (\ell^{q_o}(\psi_o),\ell^{q_1}(\psi_1))_{\lambda,q} = \ell^q(\mu) . \tag{3.5}$$

Let $0 < q < \infty$ and $F = \ell^q(\mu)$. By using (3.1) we obtain

$$\|T\|^q_{S^{(\alpha)}_F} = \sum_{-\infty}^{-1} (\alpha_1(T)\rho(2^\nu))^q + \sum_{0}^{\infty}(\alpha_{2^\nu+1}(T)\rho(2^\nu))^q . \tag{3.6}$$

If follows from our assumptions that also $\rho(t) \in Q(0,c)$. Accordingly, we can find constants c_o and c_1 such that, for $2^\nu \leq t \leq 2^{\nu+1}$, $\nu \in Z$,

$$0 < c_o \leq \frac{\rho(t)}{\rho(2^\nu)} \leq c_1 < \infty . \tag{3.7}$$

Therefore

$$\sum_{-\infty}^{-1} (\alpha_1(T)\rho(2^\nu))^q \approx (\alpha_1(T)\rho(1))^q , \tag{3.8}$$

and, for $\nu = 0, 1, 2, \ldots$ and some positive constants c_2 and c_3 ,

$$c_2(\alpha_{2^{\nu+1}+1}(T)\rho(2^{\nu+1}))^q \leq \sum_{2^\nu+1}^{2^{\nu+1}} (\alpha_n(T)\rho(n))^q \frac{1}{n} \leq c_3(\alpha_{2^\nu+1}(T)\rho(2^\nu))^q . \tag{3.9}$$

We combine (3.6) - (3.9) and obtain

$$\|T\|_{S_F^{(\alpha)}} \approx \|T\|_{S_{\rho,q}^{(\alpha)}} . \tag{3.10}$$

According to (3.7) we see that (3.10) holds also when $q = \infty$. In the same way we find that

$$\|T\|_{S_{F_i}^{(\alpha)}} \approx \|T\|_{S_{\varphi_i,q_i}^{(\alpha)}} , \quad i = 0,1 . \tag{3.11}$$

We can now complete the proof of Theorem 3.1 b) by using (3.4), (3.5), (3.10), (3.11) and get

$$(S_{\varphi_o,q_o}^{(\alpha)}, S_{\varphi_1,q_1}^{(\alpha)})_{\lambda,q} = (S_{F_o}^{(\alpha)}, S_{F_1}^{(\alpha)})_{\lambda,q} = (S_{F_o}^{(\alpha)}, S_{F_1}^{(\alpha)})_{F_2:K} =$$

$$= S_{(F_o,F_1)_{F_2:K}}^{(\alpha)} = S_F^{(\alpha)} = S_{\rho,q}^{(\alpha)} .$$

The proof of Theorem 3.1 a) follows in the same way. (We only need to use Lemma 1.1 a) instead of Lemma 1.1 b)).

Proof of Theorem 3.2 For $k \in Z$ we put $a_k = [\log(\varphi^{-1}(2^k))]$, $A_k = \{\nu \in Z: -\infty < \nu \leq a_k\}$ and $B_k = \{\nu \in Z: \nu \geq a_k\}$. Since $\varphi(t) \in Q(0,c)$ we have

$$\frac{\varphi(2^{\nu+1})}{\varphi(2^\nu)} \geq 1 + \delta , \tag{3.12}$$

for some $\delta > 0$ and every $\nu \in Z$. We put $F_i = \ell^{q_i}(\psi_i)$, where $\psi_i = (\varphi_i(2^\nu))_Z$, $i = 0,1$, and $g = (g_\nu)_Z$. Let $0 < q_o < q_1 < \infty$. By using (3.12) and Hölder's inequality with $p = q_1/(q_1-q_o)$ we obtain

$$\|\chi_{A_k} g\|_{F_o} = \Big(\sum_{\nu\in A_k} (g_\nu\varphi_o(2^\nu))^{q_o}\Big)^{1/q_o} \leq$$

$$\leq \Big(\sum_{\nu\in A_k} (\varphi(2^\nu))^{pq_o}\Big)^{1/pq_o} \cdot \Big(\sum_{\nu\in A_k} (g_\nu\varphi_1(2^\nu))^{q_1}\Big)^{1/q_1} \leq c2^k\|g\|_{F_1} . \tag{3.13}$$

According to well-known inclusions between ℓ^q-spaces and (3.12) we find

$$\|\chi_{B_k} g\|_{F_1} = \Big(\sum_{\nu\in B_k} (g_\nu \varphi_1(2^\nu))^{q_1}\Big)^{1/q_1} \le$$

$$\le \sup_{\nu\in B_k} \frac{1}{\varphi(2^\nu)} \Big(\sum_{\nu\in B_k} (g_\nu \varphi_o(2^\nu))^{q_1}\Big)^{1/q_1} \le C2^{-k}\|g\|_{F_o} . \tag{3.14}$$

It is easy to justify our arguments and see that the estimates (3.13) - (3.14) hold also if $q_1 = \infty$. In a similar way we find that (3.13) and (3.14) hold also for the remaining cases $q_o > q_1$ and $q_o = q_1$. Therefore (2.5) holds. Let $q_o, q_1 < \infty$. By using remark 2.2, (3.1) and (3.11) we obtain

$$K(2^k, T, S^{(\alpha)}_{\varphi_o,q_o}, S^{(\alpha)}_{\varphi_1,q_1}) \approx K(2^k, T, S^{(\alpha)}_{F_o}, S^{(\alpha)}_{F_1}) \approx$$

$$\approx \|(\chi_{A_k} E(2^\nu, T, \bar{L}))_Z\|_{F_o} + 2^k \|(\chi_{B_k} E(2^\nu, T, \bar{L}))_Z\|_{F_1} \approx$$

$$\approx \Big(\sum_{-\infty}^{-1} (a_1(T)\varphi_o(2^\nu))^{q_o} + \sum_{0}^{a_k} (\alpha_{2^\nu+1}(T)\varphi_o(2^\nu))^{q_o}\Big)^{1/q_o} + 2^k \Big(\sum_{a_k}^{\infty} (\alpha_{2^\nu+1}(T)\varphi_1(2^\nu))^{q_1}\Big)^{1/q_1} .$$

Since $\varphi_i(t) \in Q(0,c)$, $i = 0,1$, we see that the estimates (3.7) - (3.9) hold with $\rho(t)$ and q replaced by $\varphi_i(t)$ and q_i, $i = 0,1$, respectively. Therefore we can make some complementary calculations and obtain that

$$K(2^k, T, S^{(\alpha)}_{\varphi_o,q_o}, S^{(\alpha)}_{\varphi_1,q_1}) \approx$$

$$\approx \Big(\sum_{1}^{a'_k} (\alpha_n(T)\varphi_o(n))^{q_o} \frac{1}{n}\Big)^{1/q_o} + 2^k \Big(\sum_{a'_k+1}^{\infty} (\alpha_n(T)\varphi_1(n))^{q_1}\Big)^{1/q_1} ,$$

where $a'_k = [\varphi^{-1}(2^k)]$, $k \ge \log \varphi(1)$. This estimate implies the estimate in Theorem 3.2. For the cases $q_o = \infty$ or $q_1 = \infty$ we only need to make some obvious modifications of the proof above. The proof is complete.

Finally we remark that if $0 < t \le \varphi(1)$, then we obviously have

$$K(t, T, S^{(\alpha)}_{\varphi_o,q_o}, S^{(\alpha)}_{\varphi_1,q_1}) \approx t\Big(\sum_{1}^{\infty}(\alpha_n(T)\varphi_1(n))^{q_1}\Big)^{1/q_1} .$$

4. Interpolation between $S^{(s)}_{p,\varphi}$ and $S^{(e)}_{p,\varphi}$ operator ideals

We shall first generalize some results by König [6] and Pietsch [16]. Our proof is different and based upon Lemma 1.1.

Theorem 4.1 *Let* $\lambda(t) \in Q(0,1), \varphi(t) = \varphi_o(t)/\varphi_1(t)$ *and* $\rho(t) = \varphi_o(t)/\lambda(\varphi(t))$. *Assume that* $s = (s_n(T))_N$ *has the additivity property. Then*

a) $$(S^{(s)}_{\varphi_o,q_o}, S^{(s)}_{\varphi_1,q_1})_{\lambda,q_\theta} \subset S^{(s)}_{\rho,q_\theta} \tag{4.1}$$

if $q_o = q_1$ *or if* $\lambda(t) = t^\theta$, $0 < \theta < 1$.

b) $$(S^{(s)}_{\varphi_o,q_o}, S^{(s)}_{\varphi_1,q_1})_{\lambda,q} \subset S^{(s)}_{\rho,q} \tag{4.2}$$

if $\varphi(t) \in Q(0,c)$ *or* $\varphi(t) \in Q(-c,0)$ *for some* $c > 0$.

Remark 4.1 For the case $\lambda(t) = t^\theta$, $\varphi_i(t) = t^{1/p_i}$, $i = 0,1$, $0 < p_o < p_1 \leq \infty$, the inclusion (4.2) has been proved by Pietsch [16, p. 162]. See also [5] and [6].

Remark 4.2 The inclusions (4.1) and (4.2) can in fact be strict at least for the Weyl numbers and the Gelfand numbers (see [16, p. 163]).

The n:th outer entropy number $e_n(T)$, $n \in N$, of an operator $T \in L(A,B)$ is defined to be the infimum of all $\delta \geq 0$ such that there are $b_1, b_2, \ldots, b_p \in B$ with $p \leq 2^{n-1}$ and

$$T(U_A) \subset \bigcup_1^p \{b_\nu + \delta U_B\} ,$$

where U_A, U_B are the closed unit balls of A and B , respectively.

Theorem 4.2 *Theorem 4.1 holds also if we replace the sequence* $s = (s_\nu(T))_N$ *by the sequence* $e = (e_\nu(T))_N$ *of entropy numbers.*

Remark 4.3 For the case $\lambda(t) = t^\theta$, $\varphi_i(t) = t^{1/p_i}$, $i = 0, 1, p_o \neq p_1$, the inclusion in Theorem 4.2 a) can also be found in [2, p. 129]. According to Lemma 1.2 we see that Theorem 4.2 b) also contains a recent result by Cobos [3, Th 2.2].

Proof of Theorem 4.1 Let $T = T_o + T_1, T_i \in S^{(s)}_{\varphi_i,q_i}$, $i = 0,1$, be any decomposition of the operator $T \in (S^{(s)}_{\varphi_o,q_o}, S^{(s)}_{\varphi_1,q_1})_{\lambda,q}$. The additivity property of our sequence of s-numbers implies that $s_1(T) \leq s_1(T_o) + s_1(T_1)$ and, for every $\nu \in N$,

$$s_{2^{\nu+1}}(T) \leq s_{2^{\nu+1}-1}(T) \leq s_{2^\nu}(T_o) + s_{2^\nu}(T_1) . \tag{4.3}$$

We put $\psi_i = (\varphi_i(2^\nu))_{Z_+}$, $i = 0,1$, $\mu = (\rho(2^\nu))_{Z_+}$, $s' = (s_{2^\nu}(T))_{Z_+}$ and note that

$$\|T_i\|_{S^{(s)}_{\varphi_i,q_i}} \approx \left(\sum_0^\infty (s_{2^\nu}(T_i)\varphi_i(2^\nu))^{q_i}\right)^{1/q_i}, \quad i = 0,1, \tag{4.4}$$

and

$$\|T\|_{S^{(s)}_{\rho,q}} \approx \left(\sum_0^\infty (s_{2^\nu}(T)\rho(2^\nu))^q\right)^{1/q} = \|s'\|_{\ell^q(\mu)} . \tag{4.5}$$

Now we assume that $\varphi(t) \in Q(0,c)$ or $\varphi(t) \in Q(-c,0)$. Then, by Lemma 1.1 b), we have

$$\|s'\|_{\ell^q(\mu)} \approx \|s'\|_{(\ell^{q_o}(\psi_o),\ell^{q_1}(\psi_1))_{\lambda,q}} . \tag{4.6}$$

According to the growth conditions of $\varphi_i(t)$, $i = 0,1$, (4.3) and (4.4) we obtain

$$K(t,s',\ell^{q_o}(\psi_o),\ell^{q_1}(\psi_1)) = \inf_{s'=s'_o+s'_1} (\|s'_o\|_{\ell^{q_o}(\psi_o)} + t\|s'_1\|_{\ell^{q_1}(\psi_1)}) \leq$$

$$\leq C\left(\left(\sum_0^\infty (s_{2^\nu}(T_o)\varphi_o(2^\nu))^{q_o}\frac{1}{n}\right)^{1/q_o} + t\left(\sum_0^\infty (s_{2^\nu}(T_1)\varphi_1(2^\nu))^{q_1}\frac{1}{n}\right)^{1/q_1}\right) \leq$$

$$\leq C(\|T_o\|_{S^{(s)}_{\varphi_o,q_o}} + t\|T_1\|_{S^{(s)}_{\varphi_1,q_1}}) .$$

We conclude that

$$K(t,s',\ell^{q_o}(\psi_o),\ell^{q_1}(\psi_1)) \leq CK(t,T,S^{(s)}_{\varphi_o,q_o},S^{(s)}_{\varphi_1,q_1}) .$$

Hence, according to (4.5) - (4.6), we find that

$$\|T\|_{S^{(s)}_{\rho,q}} \leq C\|s'\|_{\ell^q(\mu)} \leq C\|s'\|_{(\ell^{q_o}(\psi_o),\ell^{q_1}(\psi_1))_{\lambda,q}} \leq$$

$$\leq C\|T\|_{(S^{(s)}_{\varphi_o,q_o},S^{(s)}_{\varphi_1,q_1})_{\lambda,q}} .$$

Therefore (4.2) holds. The proof of the inclusion (4.1) is similar. (We only need to use Lemma 1.1 a) instead of Lemma 1.1 b)).

Proof of Theorem 4.2 It is well-known that the entropy numbers are additive and that the map $T \to (e_\nu(T))_Z$ is at least a pseudo-s-function (see [15, p. 168]). Therefore the proof of Theorem 4.2 can be carried out exactly as that of Theorem 4.1.

5. Some generalizations and open questions

By analysing our proofs of the theorems 3.1, 4.1 and 4.2 we find that we in fact can formulate the following more general version of these theorems.

Theorem 5.1 Let F_i , $i = 0, 1, 2$, satisfy the conditions in Theorem 2.1 and Remark 2.1. Then

$$(S^{(\alpha)}_{F_o}, S^{(\alpha)}_{F_1})_{F_2:K} = S^{(\alpha)}_{(F_o,F_1)_{F_2:K}} ,$$

and

$$(S^{(\beta)}_{F_o}, S^{(\beta)}_{F_1})_{F_2:K} \subset S^{(\beta)}_{(F_o,F_1)_{F_2:K}} ,$$

if β is any sequence of additive s-numbers or if β is the sequence of entropy numbers.

The space $S^{(\beta)}_F$, $\beta = (\beta_n(T))_N$, consists of all T satisfying $\|T\|_{S^{(\beta)}_F} < \infty$, where $\|T\|_{S^{(\beta)}_F} = \|(\beta'_\nu)_Z\|_F$, $\beta'_\nu = \beta_1(T)$, $\nu = -1, -2, -3 \ldots$ and $\beta'_\nu = \beta_{2^\nu+1}$, $\nu = 0, 1, 2, \ldots$.

We can use the same arguments as in the proof of Theorem 3.2 and obtain the following more general estimate of the K-functional.

Theorem 5.2 Let $\bar{G} = (G_o, G_1)$ be a compatible pair of quasi-normed Abelian groups. If $\varphi(t) \in Q(0,c)$ for some $c > 0$, then

$$K(t,g,\bar{G}_{\varphi_o,q_o:E}, \bar{G}_{\varphi_1,q_1:E}) \approx$$

$$\approx \left(\int_0^{\varphi^{-1}(t)} (\varphi_o(u)E(u,g,\bar{G}))^{q_o} \frac{du}{u} \right)^{1/q_o} + t \left(\int_{\varphi^{-1}(t)}^{\infty} (\varphi_1(u)E(u,g,\bar{G}))^{q_1} \frac{du}{u} \right)^{1/q_1} , \tag{5.1}$$

where $\varphi^{-1}(t)$ is the inverse of the function $\varphi(t) = \varphi_o(t)/\varphi_1(t)$.

Remark 5.1 We have the following limiting case of the estimate (5.1): If $\varphi_o(t) \in Q(0,c)$, then

$$K(t,g,\bar{G}_{\varphi_o,q_o:E}, G_1) \approx \left(\int_0^{\varphi_o^{-1}(t)} (\varphi_o(t)E(u,g,\bar{G}))^{q_o} \frac{du}{u} \right)^{1/q_o} . \tag{5.2}$$

Compare with [9, p. 321]. In particular we can argue as in the proof of Theorem 3.2 to obtain that

$$K(t,T,S^{(\alpha)}_{\varphi_o,q_o},L(A_o,A_1)) \approx \left(\sum_{1}^{[\varphi_o^{-1}(t)]} (a_n(T)\varphi_o(n))^{q_o} \frac{1}{n}\right)^{1/q_o} ,$$

whenever $t \geq \varphi_o(1)$.

Example 5.1 Let $g^*(t)$ be the usual non-increasing rearrangement of a function $g(x)$ defined on a measure space (Ω,μ) . By $\Lambda^q(\mu)$ we denote the r.i-space of all functions $g(x)$ satisfying

$$||g||_{\Lambda^q(\mu)} = (\int_0^\infty (g^*(t)\mu(t))^q \frac{dt}{t})^{1/q} < \infty .$$

The relation $E(t,g,L^0,L^\infty) = f^*(t)$ is well-known (see [11]). Therefore we have the following special cases of (5.1) and (5.2).

$$K(t,g,\Lambda^{q_o}(\varphi_o),\Lambda^{q_1}(\varphi_1)) \approx$$

$$\approx \left(\int_0^{\varphi^{-1}(t)} (f^*(u)\varphi_o(u))^{q_o} \frac{du}{u}\right)^{1/q_o} + t\left(\int_{\varphi^{-1}(t)}^{\infty} (f^*(u)\varphi_1(u))^{q_1} \frac{du}{u}\right)^{1/q_1} , \tag{5.3}$$

and

$$K(t,g,\Lambda^{q_o}(\varphi_o),L^\infty) \approx \left(\int_0^{\varphi_o^{-1}(t)} (f^*(u)\varphi_o(u))^{q_o} \frac{du}{u}\right)^{1/q_o} \tag{5.4}$$

According to Lemma 1.2 we see that (5.3) contains a result by Merucci [8, p. 46]. For the case $\varphi_o(t) = t^{1/q_o}$ (5.4) is the usual estimate by Kree (see [1, p. 109]).

The general descriptions of sequence spaces obtained in [12] and [14] and the methods used in this paper imply that we can obtain exact descriptions of the spaces $(S^{(\alpha)}_{\varphi_o,q_o},S^{(\alpha)}_{\varphi_1,q_1})_{\lambda,q}$ also for the most troublesome case when both $q \neq q_\theta$ and φ_o is not separated from φ_1 by some a priore separation condition, e.g. of the type $\varphi_o/\varphi_1 \in Q(0,c)$. However, these descriptions will be rather complicated and not of the simple type $S^{(\alpha)}_{\rho,q}$.

Question 1 Let $\varphi(t) = \varphi_o(t)/\varphi_1(t) \approx (\log(1+t))^c$, $c \in R$. Prove or disprove that

$$(S^{(\alpha)}_{\varphi_o,q_o},S^{(\alpha)}_{\varphi_1,q_1})_{\lambda,q} = S^{(\alpha)}_{\rho,q}$$

if $\lambda(t) \in Q(0,1)$, $\rho(t) = \varphi_o(t)/\lambda(\varphi(t))$ and $q \neq q_\theta$.

The underlying estimate of the K-functional corresponding to Theorem 4.1 ought to be

$$K(t,T,S^{(s)}_{\varphi_o,q_o},S^{(s)}_{\varphi_1,q_1}) \geq$$

$$\geq C\Big(\Big(\sum_{1}^{h(t)} (s_n(T)\varphi_o(n))^{q_o}\frac{1}{n}\Big)^{1/q_o} + t\Big(\sum_{h(t)}^{\infty} (s_n(T)\varphi_1(n))^{q_1}\frac{1}{n}\Big)^{1/q_1}\Big), \tag{5.5}$$

where $\varphi(t) = \varphi_o(t)/\varphi_1(t)$, $\varphi(t) \in Q(0,c)$, $h(t) = [\varphi^{-1}(t)]$ and $t \geq \varphi(1)$.

For the case $\varphi_i(t) = t^{1/q_i}$, $i = 0,1$, (5.5) has been formulated by König [6, p. 37], but the proof in [6] is not correct.

Question 2 Is (5.5) true for all sequences of s-numbers or for the sequence of entropy numbers?

Question 3 Is it possible to replace the inclusions in Theorem 4.1 (or some special case of this theorem) by equalities for some sequence of s-numbers other than the approximation number sequence? Is the inclusion in Theorem 4.1 strict in some other case than that presented by Pietsch [16, p. 163]?

References

1. Bergh, J. and Löfström, J.: Interpolation spaces. An introduction. Grundlehren der Mathematischen Wissenschaften 223, Springer, Berlin-Heidelberg-New York (1976).

2. Carl, B. and Kühn, T.: Entropy and eigenvalues of certain integral operators. Math. Ann. 268, 127-136 (1984).

3. Cobos, F.: Entropy and Lorentz-Marcinkiewicz operator ideal. To appear in Ark. Mat.

4. Cobos, F.: On the Lorentz-Marcinkiewicz operator ideal. To appear in Math. Nachr.

5. König, H.: Eigenvalue distribution of compact operators. Birkhäuser Verlag, Basel-Boston-Stuttgart (1986).

6. König, H.: Interpolation of operator ideals with an application to eigenvalue distribution problems. Math. Ann. 233, 35-48 (1978).

7. König, H.: Type constants and (q,2)-summing norms defined by n vectors. Israel J. Math., vol 37, 130-138 (1980).

8. Merucci, C.: Interpolation réelle avec fonction paramètre; Dualite, reiteration et applications. Thesis, University of Nantes (1983).

9. Nilsson, P.: Reiteration theorems for real interpolation and approximation spaces. Ann. Mat. Pura Appl., vol 32, 291-330 (1982).

10. Peetre, J.: A theory of interpolation of normed spaces. Lecture notes, Brasilia (1963). [Notas de Matematica 39, 1-86 (1968)].

11. Peetre, J. and Sparr, G.: Interpolation of normed Abelian groups. Ann. Mat. Pura Appl. 92, 217-262 (1972).

12. Persson, L-E.: Descriptions of some interpolation spaces in off-diagonal cases. Lecture Notes in Mathematics 1070, 213-231, Springer, Berlin (1984).

13. Persson, L-E.: Interpolation with a parameter function. Technical report 3, Luleå (1985).

14. Persson, L-E.: Real interpolation between cross-sectional L^p-spaces in quasi-Banach bundles. Technical report 1, Luleå (1986).

15. Pietsch, A.: Operator Ideals. North-Holland publishing company, Amsterdam-New York-Oxford (1980).

16. Pietsch, A.: Weyl numbers and eigenvalues of operators in Banach spaces. Math. Ann. 247, 149-168 (1980).

17. Triebel, H.: Über die Venteilung der Approximationszahlen kompakter Operatoren in Sobolev-Besov-Räumen, Inv. Math. 4, 275-293 (1967).

18. Triebel, H.: Interpolation theory, Function spaces, Differential operators, North-Holland publishing Company, Amsterdam-New York-Oxford (1978).

DIRECT AND CONVERSE THEOREMS FOR SPLINE AND RATIONAL APPROXIMATION AND BESOV SPACES

Pencho P. Petrushev
Institute of Mathematics
Bulgarian Academy of Sciences
1090 Sofia, P.O. Box 373
Bulgaria

1. Introduction. A basic problem in approximation theory is to find complete direct and converse theorems for approximation by polynomials, splines, rational functions, etc. These are theorems which describe the intrinsic properties of classes of functions with a certain order of approximation. In our opinion the most natural way to obtain such theorems for linear and non-linear approximations is to prove pairs of adjusted inequalities of Jackson and Bernstein type and then to characterize the approximations considered by the K-functional of J. Peetre generated by the corresponding function spaces. This idea was realized by many mathematiciens, see [6] for the case of linear approximation and [10] for non-linear approximation. We shall use it to obtain direct and converse theorems for spline approximation in L_p and uniform metrics. Afterwards, using the relations between rational and spline approximations, we shall characterize the rational approximation in L_p $(1 < p < \infty)$ metric.

At first, we shall illustrate the main idea in one general situation. Let X_0, X_1 be two quasi-normed linear spaces and $X_1 \subset X_0$ and denote by $\|\cdot\|_{X_i}$ the quasi-norm in X_i $(i = 0,1)$. (More precisely we can suppose that X_i is Abelian group and the quasi-norm $\|\cdot\|_{X_i}$ has the following properties: $\|f\|_{X_i} \geq 0$, $\|0\|_{X_i} = 0$, $\|-f\|_{X_i} = \|f\|_{X_i}$ and $\|f+g\|_{X_i} \leq C(\|f\|_{X_i} + \|g\|_{X_i})$ for $f,g \in X_i$, where $C = \text{constant} \geq 1$; it is possible $\|f\|_{X_i} = 0$ for some $f \neq 0$. Also, we shall suppose that for $f,g \in X_1$

$$(1) \quad \|f+g\|_{X_1}^{\lambda} \leq \|f\|_{X_1}^{\lambda} + \|g\|_{X_1}^{\lambda}, \text{ where } 0 < \lambda \leq 1.$$

In the sequel the letter C will make us double service: i) C will denote the space of continuous functions and the uniform metric, ii) C will denote absolute constants (any constant) and also constants depending on some parameters which can be indicated in brackets.

Let $\{S_n\}_{n=1}^{\infty}$ be a family of subsets of X_1, which we shall consider as a tool for approximation. Suppose that this family satisfies the following conditions: i) $S_{n-1} \subset S_n$, ii) $S_{n-1} \pm S_n \subset S_{2n}$, i.e. if $f \in S_{n-1}$ and $g \in S_n$, then $f \pm g \in S_{2n}$, iii) $\|f\|_{X_1} = 0$ for each $f \in S_1$. Denote by $S_n(f)_{X_0}$ the best approximation of $f \in X_0$ by means of the elements of $S_n : S_n(f)_{X_0} = \inf\{\|f-S\|_{X_0} : s \in S_n\}$.

The following assertion gives direct and converse theorems in terms of K-functional:

$$K(f,t;X_0,X_1) = \inf_{g \in X_1} (\|f-g\|_{X_0} + t\|g\|_{X_1}), \quad f \in X_0.$$

<u>Proposition 1.1.</u> Under the previous assumptions suppose that the following inequalities of Jackson and Bernstein type hold: For each $f \in X_1$

(2) $S_n(f)_{X_0} \leq Cn^{-\alpha}\|f\|_{X_1}$ for $n = 1,2,\ldots$

and for each $S \in S_n$, $n \geq 1$,

(3) $\|S\|_{X_1} \leq Cn^{\alpha}\|S\|_{X_0}$,

where α is a positive constant. Then for each $f \in X_0$ and $n = 1,2,\ldots$ we have

(4) $S_n(f)_{X_0} \leq C.K(f,n^{-\alpha}; X_0, X_1)$ (direct inequality) and

(5) $K(f,n^{-\alpha};X_0,X_1) \leq Cn^{-\alpha}\Big(\sum_{\nu=1}^{n} \frac{1}{\nu}(\nu^{\alpha}S_{\nu}(f)_{X_0}^{\lambda}\Big)^{\frac{1}{\lambda}}$ (converse inequality),

where λ is from (1). Namely, (4) follows from (2) and (5) - from (3).

<u>Proof</u>. Estimate (4) follows from (2) immediately. Indeed, for each $g \in X_1$ we have

$$S_n(f)_{X_0} \leq C(\|f-g\|_{X_0} + S_n(g)_{X_0}) \leq C(\|f-g\|_{X_0} + n^{-\alpha}\|g\|_{X_1}),$$

which implies (4).

With no loss of generality we shall suppose that there exist $S_{2^\nu} \in S_{2^\nu}$, $\nu = 0,1,\ldots$ such that $\|f-S_{2^\nu}\|_{X_0} = S_{2^\nu}(f)_{X_0}$ and hence

(6) $\|S_{2^\nu}-S_{2^{\nu-1}}\|_{X_0} \leq C(\|f-S_{2^\nu}\|_{X_0} + \|f-S_{2^{\nu-1}}\|_{X_0}) \leq C.S_{2^{\nu-1}}(f)_{X_0}$.

By standard arguments, using (1), (3) and (6) we get

$$K(f,2^{-m\alpha};X_0,X_1) \le \|f-S_{2^m}\|_{X_0} + 2^{-m\alpha}\|S_{2^m}\|_{X_1}$$

$$\le S_{2^m}(f)_{X_0} + 2^{-m\alpha}\Big(\sum_{\nu=1}^{m}\|S_{2^\nu}-S_{2^{\nu-1}}\|_{X_1}^{\lambda}\Big)^{1/\lambda}$$

$$\le S_{2^m}(f)_{X_0} + C.2^{-m\alpha}\Big(\sum_{\nu=1}^{m}(2^{(\nu+1)\alpha}\|S_{2^\nu}-S_{2^{\nu-1}}\|_{X_0})^{\lambda}\Big)^{1/\lambda}$$

$$\le C.2^{-m\alpha}\Big(\sum_{\nu=0}^{m}(2^{\nu\alpha}S_{2^\nu}(f)_{X_0})^{\lambda}\Big)^{1/\lambda}$$

for each $m \geq 0$, which implies (5). □

Corollary 1.1. Suppose that the assumptions of Proposition 1.1. are satisfied and let ω be non-negative and non-decreasing function on $[0,\infty)$ such that $\omega(2.t) \le 2^{\beta}\omega(t)$ for $t \geq 0$, $\beta \geq 0$. Let $f \in X_0$. Then

$S_n(f)_{X_0} = O(n^{-\gamma}\omega(n^{-1}))$, $0 < \beta+\gamma < \alpha$, iff

$K(f,t;X_0,X_1) = O(t^{\gamma/\alpha}\omega(t^{1/\alpha}))$.

In particular

$S_n(f)_{X_0} = O(n^{-\gamma})$, $0 < \gamma < \alpha$, iff

$K(f,t;X_0,X_1) = O(t^{\gamma/\alpha})$.

Estimates (4) and (5) can be used successfully in more general situations than that in Corollary 1.1, but for orders of approximation not better than $O(n^{-\alpha})$.

Denote

$$S_q^{\gamma}(X_0) = \{f \in X_0 : \|f\|_{S_q^{\gamma}(X_0)} = \Big(\sum_{\nu=0}^{\infty}(2^{\nu\gamma}S_{2^\nu}(f)_{X_0})^{q}\Big)^{1/q} < \infty\},$$

When $0 < q < \infty$ and

$$S_{\infty}^{\gamma}(X_0) = \{f \in X_0 : \|f\|_{S_\infty^{\gamma}(X_0)} = \sup_n n^{\gamma}S_n(f)_{X_0} < \infty\}.$$

As usual, we shall denote by $(X_0,X_1)_{\theta,q}$ the real interpolation space between X_0 and X_1:

$$(X_0,X_1)_{\theta,q} = \{f \in X_0 : \|f\|_{(X_0,X_1)_{\theta,q}} = \Big(\sum_{\nu=0}^{\infty}(2^{\nu\theta}K(f,\tfrac{1}{2^\nu};X_0,X_1))^{q}\Big)^{1/q} < \infty\},$$

when $0 < q < \infty$ and

$$(X_0,X_1)_{\theta,\infty} = \{f \in X_0 : \|f\|_{(X_0,X_1)_{\theta,\infty}} = \sup_t t^{-\theta}K(f,t;X_0,X_1) < \infty\},$$

see [1].

Corollary 1.2. Under the assumptions of Proposition 1.1 we have $S_q^{\gamma}(X_0) = (X_0,X_1)_{\gamma/\alpha,q}$ provided $0 < \gamma < \alpha$, $0 < q \leqslant \infty$, with equivalent quasi-norms.

Corollaries 1.1 and 1.2 are immediate cosequences of Proposition 1.

In order to prove Corollary 1.1 one have only to use that $\omega(\delta.t) \leqslant (2(\delta+1))^{\beta}\omega(t)$ for $\delta,t \geqslant 0$, since $\omega(2t) \leqslant 2^{\beta}\omega(t)$. In the proof of Corollary 1.2 one has to apply Hardy's inequality:

$$(7) \quad \sum_{\nu=0}^{\infty} (2^{-\nu\rho} \sum_{\mu=0}^{\nu} a_{\mu})^p \leqslant C(p,\rho) \sum_{\mu=0}^{\infty} (2^{-\mu\rho}.a_{\mu})^p,$$

where $0 < p < \infty$, $\rho > 0$, $a_{\mu} \geqslant 0$. We omit the details.

2. Direct and converse theorems for spline approximation in L_p and uniform metric.

Denote by $\tilde{S}(k,n)$ the set of all piece wise polynomial functions of degree k-1 with n-1 free knots in (0,1), i.e. $s \in \tilde{S}(k,n)$ if there exist points (knots) $0 = x_0 < x_1 < \dots < x_n = 1$ and polynomials Q_i, degree Q $\leqslant k-1$, such that $s(x) = Q_i(x)$ for $x \in (x_{i-1},x_i)$, $i = 1,2,\dots,n$. We suppose that $S(x_i) = S(x_i-0)$ or $S(x_i) = S(x_i+0)$ at each knot x_i. Then $S(k,n) = \tilde{S}(k,n) \cap C^{k-2}$ is the set of all splines of degree k-1 with n+1 knots in [0,1].

Denote by $S_n^k(f)_p$, $S_n^k(f)_c$, $\tilde{S}_n^k(f)_p$ and $\tilde{S}_n^k(f)_c$ the best approximations of f in L_p and uniform metric by the elements of S(k,n) and $\tilde{S}(k,n)$, respectively. For instance

$$S_n^k(f)_p = \inf\{\|f-s\|_p : s \in S(k,n)\}.$$

It is well known that there is no substantial difference between piece wise polynomials and splines as tools for approximation in L_p and uniform metric: If $f \in L_p(0,1)$, when $0 < p < \infty$ and $f \in C_{[0,1]}$ when $p = \infty$, then for every $k \geqslant 2$ and $n \geqslant 1$ we have

$$(8) \quad \tilde{S}_m^k(f)_p \leqslant S_m^k(f)_p \leqslant \tilde{S}_n^k(f)_p,$$

where $n = (m-1)k+1$.

Besov spaces appear in a natural way in spline approximation. Denote

$$B_{p,q;k}^{\alpha} = \{f \in L_p(0,1) : \|f\|_{B_{p,q;k}^{\alpha}} = (\int_0^{1/k} (t^{-\alpha}\omega_k(f,t)_p)^q \frac{dt}{t})^{1/q} < \infty\},$$

where $0 < \alpha,p,q < \infty$, $k \geqslant 1$ and $\omega_k(f,t)_p$ is the usual k-th modulus of smoothness of f in L_p:

$$\omega_k(f,t)_p = \sup_{0\le h\le t} \|\Delta_h^k f(.)\|_{L_p(0,1-kh)}.$$

In our case we shall always have $p=q=\sigma$ and we shall denote briefly

$$B^\alpha_{\sigma;k} = B^\alpha_{\sigma,\sigma;k}, \|f\|_{B^\alpha_{\sigma;k}} = \|f\|_{B^\alpha_{\sigma,\sigma;k}}.$$

If $0<\sigma<1$ this is quasi-Banach space modulo P_{k-1}, where P_{k-1} is the set of all algebraic polynomials of degree $k-1$.

Apart from the quasi-norm $\|.\|_{B^\alpha_{\sigma;k}}$ we shall need the following equivalent quasi-norm

$$\|f\|^{(1)}_{B^\alpha_{\sigma;k}} = (\int_0^{1/k} (t^{-\alpha}\|\Delta_t^k f(.)\|_{L_\sigma(0,1-kt)})^\sigma \frac{dt}{t})^{1/\sigma}.$$

The equivalence of quasi-norms $\|.\|_{B^\alpha_{\sigma,k}}$ and $\|.\|^{(1)}_{B^\alpha_{\sigma,k}}$ follows immediately from the following inequality: If $f\in L_\sigma(0,1)$, $0<\sigma<\infty$, $k\ge 1$ and $0<t\le 1/k$, then

$$\omega_k(f,t)_\sigma \le C(t^{-1}\int_0^t\int_0^{1-kt} |\Delta_h^k f(x)|^\sigma dx dh)^{1/\sigma},$$

see [7].

Our first aim is to characterize the spline approximation in L_p $(0<p<\infty)$ metric. To this end we shall follow the plan from Section 1.

Theorem 2.1. (Jackson type inequality). Let $f\in B^\alpha_{\sigma;k}$, where $\alpha>0$, $\sigma=(\alpha+1/p)^{-1}$, $0<p<\infty$ and $k\ge 1$. Then $f\in L_p(0,1)$ and

(9) $\quad S_n^k(f)_p \le Cn^{-\alpha}\|f\|_{B^\alpha_{\sigma;k}}$ for $n=1,2,\dots,$

where $C=C(\alpha,p,k)$.

Theorem 2.2 (Bernstein type inequality). If $s\in\tilde{S}(k,n)$ (and of course, if $s\in S(k,n)$), $k,n\ge 1$, $0<p<\infty$, $\alpha>0$ and $\sigma=(\alpha+1/p)^{-1}$, then

(10) $\quad \|s\|_{B^\alpha_{\sigma;k}} \le Cn^\alpha\|s\|_p,$

where $C=C(\alpha,p,k)$. However, the estimate (10) does not hold in the case $p=\infty$.

According to Proposition 1.1 Theorems 2.1 and 2.2 imply the following direct and converse theorems:

Theorem 2.3 (direct theorem). Let $f\in L_p(0,1)$, $0<p<\infty$, $\sigma>0$, $\sigma=(\alpha+1/p)^{-1}$ and $k\ge 1$. Then

$S_n^k(f)_p \le CK(f,n^{-\alpha}; L_p, B^\alpha_{\sigma;k})$ for $n=1,2,\dots,$ where $C=C(\alpha,p,k)$.

Theorem 2.4 (converse theorem). Let $f\in L_p(0,1)$, $0<p<\infty$, $\alpha>0$, $\sigma=(\alpha+1/p)^{-1}$ and $k\ge 1$. Then

(11) $K(f,n^{-\alpha};L_p,B^{\alpha}_{\sigma;k}) \leqslant Cn^{-\alpha}(\sum_{\nu=1}^{n} \frac{1}{\nu}(\nu^{\alpha}\tilde{S}^{k}_{\nu}(f)_p)^{\lambda})^{1/\lambda}$

for $n = 1,2,\ldots$, where $\lambda = \min\{\sigma,1\}$, $C = C(\alpha,p,k)$.

As a consequence of Theorems 2.3 and 2.4 we obtain as in Section 1:

Corollary 2.1. Let $f\in L_p(0,1)$, $0 < p < \infty$, $k \geqslant 1$ and let ω be non-decreasing and non-negative function on $[0,\infty)$ such that $\omega(2t) \leqslant 2^{\beta}\omega(t)$ for $t \geqslant 0$ $(\beta \geqslant 0)$. Then we have $S^k_n(f)_p = O(n^{-\gamma}\omega(n^{-1}))$, $0 < \beta + \gamma < \alpha$ iff $K(f,t;\ L_p,\ B^{\alpha}_{\sigma;k}) = O(t^{\gamma/\alpha}\omega(t^{1/\alpha}))$.

In particular

$S^k_n(f)_p = O(n^{-\gamma})$, $0 < \gamma < \alpha$ iff $K(f,t;\ L_p,\ B^{\alpha}_{\sigma:k}) = O(t^{\gamma/\alpha})$.

Denote

$$S^{\gamma}_{q,k}(L_p) = \{f\in L_p(0,1):\|f\|_{S^{\gamma}_{q,k}(L_p)} = (\sum_{\nu=0}^{\infty}(2^{\nu\gamma}S^k_{2^{\nu}}(f)_p)^q)^{1/q} < \infty\}$$

when $0 < q < \infty$ and

$$S^{\gamma}_{\infty,k}(L_p) = \{f\in L_p(0,1):\|f\|_{S^{\gamma}_{\infty,k}(L_p)} = \sup_n n^{\gamma}S^k_n(f)_p < \infty\},$$

where $L_p(0,1) = C[0,1]$ when $p = \infty$.

Corollary 2.2. Let $0 < p < \infty$, $k \geqslant 1$, $0 < q \leqslant \infty$, $0 < \gamma < \alpha$ and $\sigma = (\alpha+1/p)^{-1}$. Then we have

(12) $S^{\gamma}_{q,k}(L_p) = (L_p,B^{\alpha}_{\sigma;k})_{\gamma/\alpha,q}$ with equivalent quasi-norms.

In particular, if $0 < p < \infty$, $k \geqslant 1$, $\gamma > 0$ and $\sigma_1 = (\gamma+1/p)^{-1}$, then

(13) $S^{\gamma}_{\sigma_1,k}(L_p) = B^{\gamma}_{\sigma_1;k}$ with equivalent quasi-norms.

In order to prove Theorem 2.1 we need the following embedding reuslt:

Theorem 2.5 (P. Oswald [8]). If $f\in B^{\alpha}_{\sigma,p;k}$ where $\alpha > 0$, $\sigma = (\alpha+1/p)^{-1}$, $0 < p < \infty$, $k > 1$, then $f\in L_p(0,1)$ and

$$E_{k-1}(f)_p \leqslant C\|f\|_{B^{\alpha}_{\sigma,p;k}},$$

where $E_{k-1}(f)_p$ denotes the best approximation of f in $L_p(0,1)$ by means of all algebraic polynomials of degree at most k-1.

Clearly, if $f\in B^{\alpha}_{\sigma,\sigma;k}$, $\sigma < p$, then $f\in B^{\alpha}_{\sigma,p;k}$ and $\|f\|_{B^{\alpha}_{\sigma,p;k}} \leqslant \|f\|_{B^{\alpha}_{\sigma,\sigma;k}}$. Consequently, if $f\in B^{\alpha}_{\sigma;k} = B^{\alpha}_{\sigma,\sigma;k}$, where $\alpha > 0$, $\sigma = (\alpha+1/p)^{-1}$, $0 < p < \infty$, $k \geqslant 1$, then $f\in L_p(0,1)$ and

(14) $E_{k-1}(f)_p \leq C\|f\|_{B^\alpha_{\sigma;k}}$, $C = C(\alpha,p,k)$.

Proof of Theorem 2.1. In this proof we shall use the quasi-norm $\|f\|^{(1)}_{B^\alpha_{\sigma;k}}$ instead of $\|f\|_{B^\alpha_{\sigma;k}}$. By the assumptions of Theorem 2.1 and by Theorem 2.5, see (14), it follows that $f\in L_p(0,1)$ and

(15) $E_{k-1}(f)_p \leq C\|f\|^{(1)}_{B^\alpha_{\sigma;k}}$.

Let $\Delta = (u,v)$ be an arbitrary subinterval of $(0,1)$. Denote briefly

(16) $\|f\|_{B(\Delta)} = \|f\|_{B(u,v)} = \left(\int_0^{(v-u)/k} (t^{-\alpha}\|\Delta_t^k f(.)\|_{L_\sigma(u,v-kt)})^\sigma dt/t\right)^{1/\sigma}.$

By simple change of variables using (15) we get

(17) $E_{k-1}(f,\Delta)_p \leq C\|f\|_{B(u,v)}$,

where $E_{k-1}(f,\Delta)_p$ is the best approximation of f in $L_p(\Delta)$ by means of all algebraic polynomials of degree k-1.

Now let $n \geq 1$. We define by induction points $x_0,x_1,x_2,\ldots$ Set $x_o = 0$. Let $x_0,x_1,\ldots,x_{i-1}$ be already defined so that $0 = x_0<x_1<\ldots<x_{i-1}<1$. Now we define x_i as follows

$x_i = \sup\{y: x_{i-1} < y \leq 1,\ \|f\|^\sigma_{B(x_{i-1},y)} \leq \frac{1}{n}\|f\|_{B(0,1)}\}$.

Suppose $x_m \leq 1$ for some $m \geq 1$. Denote $\Delta_i = (x_{i-1},x_i)$. By (16) we get

$$\sum_{i=1}^m \|f\|^\sigma_{B(\Delta_i)} = \sum_{i=1}^m \int_0^{|\Delta_i|/k} t^{-\alpha\sigma}\|\Delta_t^k f(.)\|^\sigma_{L_\sigma(x_{i-1},x_i-kt)} dt/t$$

$$= \int_0^{1/k} t^{-\alpha\sigma}\Big(\sum_{|\Delta_i|\geq kt} \|\Delta_t^k f(.)\|^\sigma_{L_\sigma(x_{i-1},x_i-kt)}\Big)dt/t$$

$$\leq \int_0^{1/k} t^{-\alpha\sigma}\|\Delta_t^k f(.)\|^\sigma_{L_\sigma(0,1-kt)} dt/t = \|f\|^\sigma_{B(0,1)} ,$$

where the sum is taken over all i such that $|\Delta_i| \geq kt$. Thus we have

(18) $\sum_{i=1}^m \|f\|^\sigma_{B(\Delta_i)} \leq \|f\|^\sigma_{B(0,1)}$.

On the other hand the function $F(v) = \|f\|_{B(u,v)}$, $(u,v)\subset(0,1)$, is obviously continuous and non-decreasing for $v\in[u,1]$. Hence $\|f\|^\sigma_{B(\Delta_i)} = n^{-1}\|f\|^\sigma_{B(0,1)}$ for $i = 1,2,\ldots,m-1$. These equalities and (18) imply that there exists a natural number m ($m \leq n+1$) such that $x_m = 1$.

Also, by the definition of $x_o,x_1,\ldots,x_m$ it follows that

(19) $\|f\|^{\sigma}_{B(\Delta_i)} \leq n^{-1}\|f\|^{\sigma}_{B(0,1)}$ for $i = 1,2,\dots,m$.

Now we apply the estimate (17) to f in each interval Δ_i. According to (19) we get

$$E_{k-1}(f,\Delta_i)_p \leq C\|f\|_{B(\Delta_i)} \leq Cn^{-1/\sigma}\|f\|^{(1)}_{B_{\sigma;k}}$$

for $i = 1,2,\dots,m$ $(m \leq n+1)$. Hence

$$\tilde{S}^k_n(f)_p \leq \Big(\sum_{i=1}^{m} E_{k-1}(f,\Delta_i)^p_p\Big)^{1/p} \leq Cn^{-\alpha}\|f\|^{(1)}_{B^{\alpha}_{\sigma;k}} .$$

In view of (8) the last estimate implies (9). □

<u>Proof of Theorem 2.2.</u> Let $s \in \tilde{S}(k,n)$. Then there exist a partition of $(0,1)$: $0 = x_0<x_1<\dots<x_n = 1$ and polynomials Q_i of degree k-1 such that $s(x) = Q_i(x)$ for $x \in \Delta_i = (x_{i-1},x_i)$, $i = 1,2,\dots,n$.

In order to estimate $\|s\|_{B^{\alpha}_{\sigma;k}}$ we need an estimate of $\omega_k(s,t)_\sigma$ for each $t \in (0,1/k)$. Since $s(x) = Q_i(x)$ for $x \in \Delta_i$ and Q_i is a polynomial of degree k-1, then $\Delta^k_h s(x) = 0$ when $x, x+kh \in \Delta_i$. Hence

(20) $$\omega_k(s,t)^{\sigma}_{\sigma} = \sup_{0\leq h\leq t} \int_0^{1-kh} |\Delta^k_h s(x)|^{\sigma}dx$$

$$\leq C\{\sum_{|\Delta_i|\leq kt} \int_{\Delta_i} |s(x)|^{\sigma}dx + \sum_{|\Delta_i|>kt} \Big(\int_{\Delta'_i}|s(x)^{\sigma}dx + \int_{\Delta''_i} |s(x)|^{\sigma}dx\Big)\},$$

where $\Delta'_i = (x_{i-1},x_{i-1}+kt)$, $\Delta''_i = (x_i-kt,x_i)$, the first sum is taken over all i such that $|\Delta_i| \leq k.t$ and the second sum - over all i such that $|\Delta_i| > k.t$.

Next we shall make use of the following simple inequalities:

(21) $$\|s\|_{L_\sigma(\Delta_i)} \leq |\Delta_i|^{\alpha}\|s\|_{L_p(\Delta_i)} ,$$

(22) $$\|s\|_{L_\sigma(\Delta'_i)} \leq Ct^{1/\sigma}|\Delta_i|^{-1/p}\|s\|_{L_p(\Delta_i)}, \text{ when } |\Delta_i| > kt,$$

(23) $$\|s\|_{L_\sigma(\Delta''_i)} \leq Ct^{1/\sigma}|\Delta_i|^{-1/p}\|s\|_{L_p(\Delta_i)} \text{ when } |\Delta_i| > kt, \text{ where}$$

$C = C(\alpha,p,k)$.

The inequality (21) is an immediate consequence of Hölder's inequality.

It is well known that for each polynomial Q of degree k-1 and for each interval Δ the following inequalities hold

$$|\Delta|^{1/q-1/p}\|Q\|_{L_p(\Delta)} \leq \|Q\|_{L_q(\Delta)} \leq C|\Delta|^{1/q-1/p}\|Q\|_{L_p(\Delta)} ,$$

where $0 < p \leq q \leq \infty$ and $C = C(k,p)$. These inequalities imply (22) and (23). Indeed, we have

$$\|s\|_{L_\sigma(\Delta_i')} \leq |\Delta_i'|^{1/\sigma}\|s\|_{L_\infty(\Delta_i')} \leq (kt)^{1/\sigma}\|s\|_{L_\infty(\Delta_i)}$$
$$\leq Ct^{1/\sigma}|\Delta_i|^{-1/p}\|s\|_{L_p(\Delta_i)} ,$$

whence (22) follows. Inequality (23) can be proved similarly.

Combining (20) with (21)-(23) we get

$$\omega_k(s,t)_\sigma^\sigma \leq C\{\sum_{|\Delta_i|\leq kt}|\Delta_i|^{\alpha\sigma}\|s\|_{L_p(\Delta_i)}^\sigma + \sum_{|\Delta_j|>kt} t|\Delta_i|^{-\sigma/p}\|s\|_{L_p(\Delta_i)}^\sigma\}.$$

Now we are ready to estimate $\|s\|_{B^\alpha_{\sigma;k}}$. Applying above estimate for $\omega_k(s,t)_\sigma$ we get

$$\|s\|^\sigma_{B^\alpha_{\sigma;k}} = \int_0^{1/k}(t^{-\alpha}\omega_k(s,t)_\sigma)^\sigma dt/t \leq C\int_0^{1/k} t^{-\alpha\sigma-1}(\sum_{|\Delta_i|\leq kt}|\Delta_i|^{\alpha\sigma}\|s\|^\sigma_{L_p(\Delta_i)}$$
$$+\sum_{|\Delta_i|>kt} t|\Delta_i|^{-\sigma/p}\|s\|^\sigma_{L_p(\Delta_i)})dt \leq C\sum_{i=1}^n\{|\Delta_i|^{\alpha\sigma}\|s\|^\sigma_{L_p(\Delta_i)}\int_{|\Delta_i|/k}^{\infty} t^{-\alpha\sigma-1}dt$$
$$+|\Delta_i|^{-\sigma/p}\|s\|^\sigma_{L_p(\Delta_i)}\int_0^{|\Delta_i|/k} t^{-\alpha\sigma}dt\} \leq C\sum_{i=1}^n\|s\|^\sigma_{L_p(\Delta_i)}$$
$$\leq C(\sum_{i=1}^n\|s\|^p_{L_p(\Delta_i)})^{\sigma/p}n^{1-\sigma/p} = Cn^{\alpha\sigma}\|s\|^\sigma_{L_p(0,1)},$$

where we have used a discrete variant of Hölder's inequality. The estimate (10) is proved.

It remains to show that the estimate (10) does not hold in the case $p=\infty$. Suppose $k \geq 2$ and $0 < \varepsilon < \frac{1}{4k}$. Consider the spline

$$S_\varepsilon(x) = \begin{cases} 0 & \text{for } x\in[0,1/2] \\ \int_{1/2}^{x} B(t)dt & \text{for } x\in[1/2,1/2+\varepsilon] \\ 1 & \text{for } x\in[1/2+\varepsilon,1], \end{cases}$$

where B is an arbitrary B-spline of degree k-2 with k knots in $(1/2,1/2+\varepsilon)$ such that supp $B\subset(1/2,1/2+\varepsilon)$ and $\int_{1/2}^{1/2+\varepsilon}B(x)dx = 1$. It is readily seen that $s_\varepsilon\in S(k,k+1)$, $\|s_\varepsilon\|_C = 1$ and $\omega_k(s_\varepsilon,\delta)_{1/\alpha} > C(k,\alpha)\delta^\alpha$ for $\varepsilon/2 \leq \delta \leq 1/2k$, $\alpha > 0$, hence $\|s_\varepsilon\|_{B^\alpha_{1/\alpha;k}} > C(k,\alpha)\ln 1/\varepsilon$ and ε may tend to zero. Consequently, the estimate (10) does not hold in the case $p = \infty$. □

The proof of relation (12) of Corollary 2.2 is similar to the proof of Corollary 1.2. The relation (13) follows from (12) since $(L_p, B^\alpha_{\sigma;k})_{\gamma/\alpha,\sigma_1} = B^\gamma_{\sigma_1;k}$, see the article of R.A. DeVore and V.A. Popov in this volume.

Now we consider the spline approximation in uniform metric. The following inequalitites of Jackson and Bernstein type hold:

Theorem 2.6 (Jackson type inequality). If f is absolutely continuous on $[0,1]$ and $f' \in B^{\alpha-1}_{1/\alpha;k-1}$, where $\alpha > 1$ and $k \geq 2$, then

(24) $S^k_n(f)_C \leq Cn^{-\alpha} \| f' \|_{B^{\alpha-1}_{1/\alpha,k-1}}$ for $n = 1,2,\ldots$, where $C = C(\alpha,k)$.

Theorem 2.7 (Bernstein type inequality). If $s \in S(k,n)$, $k \geq 2$, $n \geq 1$ and $\alpha > 1$, then

(25) $\| s' \|_{B^{\alpha-1}_{1/\alpha;k-1}} \leq Cn^\alpha \| s \|_C$,

where $C = C(\alpha,k)$.

Denote by $\overline{B}^\alpha_{\sigma;k}$ the set of all absolutely continuous functions f on $[0,1]$ such that $f' \in B^\alpha_{\sigma;k}$ with quasi-norm $\| f \|_{\overline{B}^\alpha_{\sigma;k}} = \| f' \|_{B^\alpha_{\sigma;k}}$.

According to Proposition 1.1, Theorem 2.6 and 2.7 imply the following direct and converse theorems:

Theorem 2.8 (direct theorem). If $f \in C_{[0,1]}$, $\alpha > 1$ and $k \geq 2$, then we have

$$S^k_n(f)_C \leq C(\alpha)K(f,n^{-\alpha};C,\overline{B}^{\alpha-1}_{1/\alpha;k-1}) \text{ for } n = 1,2,\ldots$$

Theorem 2.9 (converse theorem). If $f \in C_{[0,1]}$, $\alpha > 1$ and $k \geq 2$, then

$$K(f,n^{-\alpha};C,\overline{B}^{\alpha-1}_{1/\alpha;k-1}) \leq C(\alpha,k)n^{-\alpha}\Big(\sum_{\nu=1}^{n} \frac{1}{\nu}(\nu^\alpha S^k_\nu(f)_C)^{1/\alpha}\Big)^\alpha \text{ for } n = 1,2,\ldots$$

Corollaries similar to Corollaries 1.1 and 1.2 follow from Theorems 2.8 and 2.9. We formulate only the following

Corollary 2.3. If $0 < \gamma < \alpha$, $\alpha > 1$, $0 < q \leq \infty$ and $k \geq 2$, then

$$S^\gamma_{q,k}(C) = (C, \overline{B}^{\alpha-1}_{1/\alpha;k-1})_{\gamma/\alpha,q}$$

with equivalent quasi-norms.

In order to prove Theorem 2.6 we shall use the following trivial lemma:

Lemma 2.1. If f is absolutely continuous on $[0,1]$ and $k \geq 2$ then

(26) $\tilde{S}^k_{2n}(f)_C \leq n^{-1}.\tilde{S}^{k-1}_n(f')_1$ for $n = 1,2,\ldots$

Proof. Suppose $s \in \tilde{S}(k-1,n)$ and

$\|f'-s\|_{L(0,1)} = \tilde{S}_n^{k-1}(f')_1$. Clearly, there exist points $0 = u_0 < u_1 < \dots < u_{2n} = 1$ such that: i) s is an algebraic polynomial of degree k-2 in each interval (u_{i-1}, u_i) and ii) $\int_{u_{i-1}}^{u_i} |f'(x)-s(x)|dx \leq n^{-1}.S_n^{k-1}(f')_1$. Put $\varphi(x) = s(u_{i-1}) + \int_{u_{i-1}}^{x} s(t)dt$ for $x \in [u_{i-1}, u_i)$, $i = 1,2,\dots,2n$ and $\varphi(1) = \varphi(1-0)$. Obviously, $\varphi \in \tilde{S}(k,2n)$ and

$\|f-\varphi\|_{C[u_{i-1},u_i)} \leq \int_{u_{i-1}}^{u_i} |f'(t) - S(t)|dt \leq n^{-1}\tilde{S}_n^{k-1}(f')_1$, hence (26) follows. □

Proof of Theorem 2.6. By Theorem 2.1 we have

$$S_n^{k-1}(f')_1 \leq Cn^{-\alpha+1}\|f'\|_{B^{\alpha-1}_{1/\alpha;k-1}} .$$

Combining this estimate with Lemma 2.1 and (8) we establish estimate (24). □

Proof of Theorem 2.7. Let $s \in \tilde{S}(k,n)$, $k \geq 2$, $n \geq 1$. By Theorem 2.2 we have

$$\|s'\|_{B^{\alpha-1}_{1/\alpha;k-1}} \leq C.n^{\alpha-1}\|s'\|_{L_1} .$$

On the other hand obviously

$$\|s'\|_{L_1} \leq Cn\|s\|_C .$$

Combining these two estimates we obtain (25). □

Remarks 2.1. The problem of characterization of the free knots spline approximation has been considered by many authors although in other terms, see [2], [3], [4], [18], [19]. J. Peetre was the first who realized that Besov spaces play an essential role in the theory of spline approximation. At the Conference on approximation theory in Kiev in 1983 Yu. Brudnyi announced a relation similar to (13) in Corollary 2.2 but without any indication of proof, see also [5].

Usually the number k in the definition of Besov space $B^{\alpha}_{p,q;k}$ is being connected to the smooth index α (usually $k = [\alpha] + 1$). In our case there is no need of such restriction because generally $p < 1$ and the space $B^{\alpha}_{p,q;k}$ is non-trivial when $0 < \alpha < k-1 +1/p$. The fact that k is not depending on α allows us to use Theorems 2.3 and 2.4 for arbitrary orders of approximation $O(n^{-\gamma})$, $0 < \gamma < \infty$, instead of $0 < \gamma \leq k$ as usual. Spaces $B^{\alpha}_{p,q;k}$ are considered by P. Oswald [8]. A problem arises to investigate the K-functional and the interpolation spaces generated by L_p and Besov spaces.

The results of this section are announced in [17].

3. Direct and converse theorems for rational approximation in L_p $(1 < p < \infty)$ metric

Now, using the recently proved relations between rational and spline approximations we shall characterize rational approximation in L_p $(1 < p < \infty)$ by K-functional generated by L_p and appropriate Besov space.

Denote by $R_n(f)_p$ the best approximation of f in $L_p(0,1)$ by means of all rational functions of degree at most n. We have proved the following result:

Theorem 3.1 [16]. If $f \in L_p(0,1)$, $1 \leq p < \infty$, $k \geq 1$ and $\alpha > 0$, then

$$R_n(f)_p \leq Cn^{-\alpha} \sum_{\nu=1}^{n} \nu^{\alpha-1}\tilde{S}_\nu^k(f)_p$$

for $n \geq \max\{1,k-1\}$, where $C = C(p,k,\alpha)$.

In the opposite direction A.A. Pekarski [13] obtained:

Theorem 3.2 (A.A. Pekarski [13]). If $f \in L_p(0,1)$, $1 < p \leq \infty$, $k \geq 1$ and $\sigma = (k+1/p)^{-1}$, then

$$(27) \quad \tilde{S}_n^k(f)_p \leq C.n^{-k}\{\sum_{\nu=0}^{n} \frac{1}{\nu+1}((\nu+1)^k R_\nu(f)_p)^\sigma\}^{1/\sigma}$$

for $n = 1,2,\ldots$, where $C = C(k,p)$.

Theorem 3.1 and 3.2 completely characterize rational approximation in L_p $(1 < p < \infty)$ by spline approximation in the same metric. They show that there is no substantial difference between rational functions and splines as tools for approximation in L_p $(1 < p < \infty)$ for orders of approximation not better than $O(n^{-k})$. The following corollary follows immediately by Theorems 3.1 and 3.2:

Corollary 3.1. If $f \in L_p(0,1)$, $1 < p < \infty$, then $R_n(f)_p = O(n^{-\gamma})$ iff $\tilde{S}_n^k(f)_p = O(n^{-\gamma})$ provided $0 < \gamma < k$.

Now, when we consider rational approximation there is no need the number k in the definition of space $B_{\sigma;k}^\alpha$ to be independent. As usual, we shall denote $B_\sigma^\alpha = B_{\sigma;k}^\alpha$ and $\|f\|_{B_\sigma^\alpha} = \|f\|_{B_{\sigma;k}^\alpha}$, where $k = [\alpha] + 1$

Combining Theorems 3.1 and 3.2 together with Theorems 2.3 and 2.4 we obtain:

Theorem 3.3 (direct theorem). If $f \in L_p(0,1)$, $1 \leq p < \infty$, $\alpha > 0$ and $\sigma = (\alpha+1/p)^{-1}$, then

(28) $R_n(f)_p \leq CK(f,n^{-\alpha};L_p,B^\alpha_\sigma)$ for $n \geq \max\{1,[\alpha]\}$, where $C = C(\alpha,p)$.

Theorem 3.4 (converse theorem). If $f \in L_p(0,1)$, $1 < p < \infty$, $\alpha > 0$ and $\sigma = (\alpha+1/p)^{-1}$, then

(29) $$K(f,n^{-\alpha};L_p,B^\alpha_\sigma) \leq Cn^{-\alpha}\{\sum_{\nu=0}^{n} \frac{1}{\nu+1}((\nu+1)^\alpha R_\nu(f)_p)^\lambda\}^{1/\lambda}$$

for $n = 1,2,\ldots$, where $\lambda = \min\{1,\sigma\}$, $C = C(\alpha,p)$.

Consequences similar to Corollaries 1.1 and 1.2 follow from Theorems 3.3 and 3.4. We shall formulate only the analogue of Corollary 1.2.

Denote

$$R^\gamma_q(L_p) = \{f \in L_p(0,1): \|f\|_{R^\gamma_q(L_p)} = \{R_0(f)^q_p + \sum_{\nu=0}^{\infty} (2^{\nu\gamma}R_{2^\nu}(f)_p)^q\}^{1/q} < \infty\},$$
$0 < q < \infty$,

and

$$R^\gamma_\infty(L_p) = \{f \in L_p(0,1): \|f\|_{R^\gamma_\infty(L_p)} = \sup_n n^\gamma R_n(f)_p < \infty\}.$$

Corollary 3.2. If $1 < p < \infty$, $0 < \gamma < \alpha$, $0 < q \leq \infty$ and $\sigma = (\alpha+1/p)^{-1}$, then $R^\gamma_q(L_p) = (L_p,B^\alpha_\sigma)_{\gamma/\alpha,q}$ with equivalent quasi-norms.

In particular, if $1 < p < \infty$, $\gamma > 0$ and $\sigma_1 = (\gamma+1/p)^{-1}$, then

(30) $R^\gamma_{\sigma_1}(L_p) = B^\gamma_{\sigma_1}$ with equivalent quasi-norms.

Proof of Theorems 3.3 and 3.4. In what follows we shall denote $k = [\alpha] + 1$.

By Theorems 3.1 and 2.3 we have

$$R_n(f)_p \leq C.n^{-\beta} \sum_{\nu=1}^{n} \nu^{\beta-1}\tilde{S}^k_\nu(f)_p \quad \text{and}$$

$$\tilde{S}^k_\nu(f)_p \leq CK(f,\nu^{-\alpha};L_p,B^\alpha_\sigma) \leq C(\tfrac{n}{\nu})^\alpha K(f,n^{-\alpha};L_p,B^\alpha_\sigma).$$

Choose $\beta = \alpha+1$. Then combining the last estimates we establish (28).

The estimate (27) in Theorem 3.2 can be written in the form

$$\tilde{S}^k_{2^m}(f)_p \leq C.2^{-mk}\{R_0(f)^\sigma_p + \sum_{\nu=0}^{m} (2^{\nu k}R_{2^\nu}(f)_p)^\sigma\}^{1/\sigma}$$

and the estimate (11) in Theorem 2.4 - in the form

$$K(f,2^{-m\alpha};L_p,B^\alpha_{\sigma;k}) \leq C2^{-m\alpha}\{\sum_{\nu=0}^{m} (2^{\nu\alpha}\tilde{S}^k_{2^\nu}(f)_p)^\lambda\}^{1/\lambda}.$$

Combining those two estimates and applying Hardy's inequality (7) and the fact that $\alpha < k$ we get

$$K(f,2^{-m\alpha};L_p,B_\sigma^\alpha)$$

$$\leqslant C2^{-m\alpha}\{\sum_{\nu=0}^{m}(2^{\nu(\alpha-k)}(R_o(f)_p^\sigma+\sum_{\mu=0}^{\nu}(2^{\mu k}R_{2^\mu}(f)_p)^\sigma)^{\lambda/\sigma}\}^{1/\lambda}$$

$$\leqslant C2^{-m\alpha}\{R_o(f)_p^\lambda+\sum_{\mu=0}^{m}(2^{\mu\alpha}R_{2^\mu}(f)_p)^\lambda\}^{1/\lambda}$$

which implies (29). □

The proof of Corollary 3.2 is similar to that of Corollaries 1.2 and 2.2. We leave the details.

Remarks 3.2. V. Peller [14], [15] was the first who characterized classes of functions which have a certain order of approximation by rational functions in BMO. A.A. Pekarski [11], [12] proved pairs of adjusted inequalities of Jackson and Bernstein type and characterized the rational approximation in H^p and BMO. In the Conference in Kiev in 1983 Yu. Brudnyi announced result similar to relation (30) in Corollary 3.2 but again without proof, see also [5].

References

[1] Bergh, J., J. Löfström, Interpolation spaces, Springer Ver., Grundleren, vol. 223, Berlin, 1976.

[2] Bergh, J., J. Peetre, On the Spaces V_p ($0 < p \leqslant \infty$), Bolletino U.M.I. (4)10 (1974), 632-648.

[3] Brudnyi, Yu., Piecewise polynomial approximation and local approximations, Soviet Math. Dokl. 12 (1971) 1591-1594.

[4] Brudnyi, Yu., Spline approximation and functions of bounded variation, Soviet Math. Dokl. 15 (1974), 518-521.

[5] Brudnyi, Yu., Rational approximation and embedding theorems, Doklady AN USSR, 247 (1979), 681-684 (Russian).

[6] Butzer, P., K. Scherer, Approximationsprozesse und interpolationsmethoden, B.I. Hochschulskripten, Mannheim, 1968.

[7] Oswald, P., Ungleichungen vom Jackson-Typ für die algebraische beste Approximation in L_p, J. of Approx. Theory 23 (1978) 113-136.

[8] Oswald, P., Spline approximation in L_p ($0 < p < 1$) metric, Math. Nachr. Bd. 94 (1980), 69-96 (Russian).

[9] Peetre, J., New Thoughts on Besov Spaces, Duke Univ. Math. Series, Duke Univ., Durham, 1976.

[10] Peetre, J., G. Sparr, Interpolation of normed Abelian groups, Ann. Mat. Pura Appl. 92 (1972), 217-262.

[11] Pekarski, A., Bernstein type inequalities for the derivatives of rational functions and converse theorems for rational approximation, Math. Sbornik 124(166) (1984), 571-588 (Russian).

[12] Pekarski, A., Classes of analytic functions determined by rational approximation in H_p, Math. Sbornik 127(169) (1985), 3-20 (Russian).

[13] Pekarski, A., Estimates for the derivatives of rational functions in $L_p[-1,1]$, Math. Zametki, 39(1986), 388-394 (Russian).

[14] Peller, V., Hankel operators of the calss G_p and their applications (Rational approximation, Gaussian processes, Operator majorant problem), Math. Sbornik 113(155) (1980), 538-582 (Russian).

[15] Peller, V., Description of Hankel operators of the class σ_p for $p < 1$, investigation of the order of rational approximation and other applications, Math. Sbornik 122(164) (183), 481-510 (Russian).

[16] Petrushev, P., Relation between rational and spline approximations in L_p metric, J. of Approx. Theory (to appear).

[17] Petrushev, P., Direct and converse theorems for spline approximation and Besov spaces, C.R. Acad. bulg. Sci. 39 (1986), 25-28.

[18] Popov, V.A., Direct and converse theorems for spline approximation with free knots, C.R. Acad. bulg. Sci. 26 (1973), 1297-1299.

[19] Popov, V.A., Direct and converse theorems for spline approximation with free knots in L_p, Rev. Anal. Numér. Théorie Approximation 5 (1976), 69-78.

Interpolation of Some Analytic Families of Operators

Y. SAGHER
Department of Mathematics, Statistics, and Computer Science
University of Illinois at Chicago

Introduction

The idea of analytic families of operators has had far-reaching consequences in analysis, and the existence of an interpolation theorem for such families is a major advantage of complex interpolation theory. In the context of L_p theory, $1 \leq p$, it is natural to define analyticity of a family of operators in terms of dual spaces, and that is how the theory developed, with an early exception [6].

When one wishes to extend the theory to H_p spaces, $0 < p \leq \infty$, complications occur. There are two interpolation theorems for analytic families. One by A.P. Calderón and A. Torchinsky [1], considers roughly four possibilities: the domain spaces and the range spaces can be either Banach spaces or H_p spaces. The second theorem, by S. Janson and P. Jones [5], is very general but the range spaces have to have rich duals. Moreover the applications of the theorem to concrete cases requires verification of some non-trivial conditions.

In this note we consider an interpolation theorem for some analytic families of operators. It is tailored to handle multiplers of Fourier Series, which is a major application of the theory. A second result in this note (Theorem 1) treats a problem which has been ignored in the application of the theory. When the continuity of the operators is tested in practice, it is done so not for the full intersection of the domain spaces but on a dense subset: trigonometric polynomials, finite linear combinations of atoms or the like. In the context of the general theory one has to justify this restriction. We do this below.

The results of this paper generalize to families of quasi-Banach spaces. We keep the 2-space setting since it avoids some technicalities.

Complex Interpolation

Denote $S = \{z/0 < Rez < 1\}$, $\overline{S} = \{z/0 \leq Rez \leq 1\}$.

$$\varphi(z) = \frac{e^{i\pi(z-1/2)} - 1}{e^{i\pi(z-1/2)} + 1}$$

maps S conformally onto $|w| < 1$. A function $f(z)$ will be said to belong to $N^+(S)$ iff $f \circ \varphi^{-1}$ is in the Nevanlina class N^+ of the unit disc. It can be shown that $f \in N^+(S)$ has non-tangential limits almost everywhere on ∂S and that if we denote the limit function by $f(\varsigma)$ then $log|f(\varsigma)|$ is integrable with respect to $P(z,\varsigma)d\varsigma$ where $P(z,\varsigma)$ is Poisson's kernel for S. Further

$$log|f(z)| \leq \int_{\partial S} log|f(\varsigma)|P(z,\varsigma)d\varsigma.$$

Early works on analytic families of operators [4], [7], imposed conditions of admissible growth:

$$log|f(x+iy)| \leq Ae^{\alpha|y|}, \quad \alpha < \pi.$$

If $f(z)$ has admissible growth however, taking $h(z) = expcos\beta(z-1/2)$, $\alpha < \beta < \pi$, we have $h(z) \in N^+(S)$, $|f(z)| \leq C|h(z)|$ so that $f \in N^+(S)$.

Continue now with the definition of the interpolation norms. Let (A_0, A_1) be an interpolation couple of quasi-Banach spaces, and let $A \subset A_0 \cap A_1$. Denote by $\mathcal{G}(A)$ the space of functions of the form $f(z) = \sum_k \psi_k(z)a_k$, where $a_k \in A$, $\psi_k \in N^+(S)$. We use the convention that sums $\sum_k$ are over a finite number of terms. Define for $0 < s < 1$ and $0 < p \leq \infty$

$$|f|_{\mathcal{G}_s^p} = \left(\int_{-\infty}^{\infty} |f(iy)|_{A_0}^p P_0(s,y)\, dy + \int_{-\infty}^{\infty} |f(1+iy)|_{A_1}^p P_1(s,y)\, dy\right)^{\frac{1}{p}}$$

and

$$|f|_{\mathcal{G}_s^0} = exp\left[\int_{-\infty}^{\infty} log|f(iy)|_{A_0} P_0(s,y)\,dy + \right.$$

$$\left. + \int_{-\infty}^{\infty} log|f(1+iy)|_{A_1} P_1(s,y)\,dy\ \right].$$

Note that $\mathcal{G}^\infty$ does not depend on s. One can show as in the Banach-space case that for $a \in A_0 \cap A_1$, $0 < s < 1$,

$$inf\{|f|_{\mathcal{G}_s^p} /\ f(s) = a, f \in \mathcal{G}(A_0 \cap A_1)\}$$

does not depend on p, $0 \leq p \leq \infty$. Denote this quantity by $|a|_{[A_0,A_1]_s}$.

When the spaces are implied by the context we will write simply $\mathcal{G}$, $|a|_s$ etc. We will also write $|f|_{\mathcal{G}}$ for $|f|_{\mathcal{G}_s^0}$.

We clearly have $|a|_s \leq |a|_{A_0 \cap A_1}$. When A_j are Banach spaces then also $|a|_{A_0+A_1} \leq |a|_s$.

Very frequently in applications, one has the operators defined on a dense subset of $A_0 \cap A_1$. This motivates the following:

THEOREM 1. *If A is dense in $A_0 \cap A_1$ then for $a \in A$ one has*

$$|a|_s = inf\{|f|_{\mathcal{G}} /\ f(s) = a,\ f \in \mathcal{G}(A)\}.$$

PROOF: Let $a \in A$. $|a|_s = inf\{|a - g(z)|_{\mathcal{G}}\ /\ g \in \mathcal{G}(A_0 \cap A_1), g(s) = 0\}$. If $g(z) = \sum_k \psi_k(z) a_k$, $g(s) = 0$ we can assume $\psi_k(s) = 0$ for all k : if not, assume $\psi_1(s) \neq 0$, then $-\psi_1(z) a_1 = \frac{1}{\psi_1(s)} \sum_{k \geq 2} \psi_1(z) \psi_k(s) a_k$ and so $g(z) = \sum_{k\geq 2} \left[\psi_k(z) - \psi_1(z) \frac{\psi_k(s)}{\psi_1(s)}\right] a_k = \sum_k \varphi_k(z) a_k$, and $\varphi_k(s) = 0$, for all k . Assume therefore $\psi_k(s) = 0$ all k . We can also assume $\psi_k \in H^\infty(S)$ (see proposition 2.5 in [2]). We may assume, by passage to equivalent quasi-norms, that there exist $0 < r_j \leq 1$ so that $|\ |_{A_j}^{r_j}$ are subadditive. Let $r = \min\{r_0, r_1\}$; we have $|\sum c_k|_{A_j}^r \leq \sum |c_k|_{A_j}^r$. Choose now $b_k \in A$ so that $\sum |\psi_k(z)|_{H^\infty(S)}^r |a_k - b_k|_{A_0 \cap A_1}^r < \epsilon$. We then have:

$$|a - \sum_k \psi_k(j+iy) b_k|_{A_j}^r = |a - \sum_k \psi_k(j+iy) a_k + \sum \psi_k(j+iy)(b_k - a_k)|_{A_j}^r$$

$$\leq |a - \sum_k \psi_k(j+iy)a_k|_{A_j} + \epsilon.$$

so that $|a - \sum_k \psi_k(z)b_k|^r_{\mathcal{G}\infty} \leq |a - \sum \psi_k(z)a_k|^r_{\mathcal{G}\infty} + \epsilon$. Therefore

$$inf\{|a - \sum_k \psi_k(z)b_k|_{\mathcal{G}} /\ \psi_k(z) \in N^+(S),\ b_k \in A\} =$$

$$= inf\{|a - \sum_k \psi_k(z)b_k|_{\mathcal{G}\infty} /\ \psi_k(z) \in H^\infty(S),\ b_k \in A\} =$$

$$= inf\{|a - \sum_k \psi_k(z)a_k|_{\mathcal{G}\infty} \ /\ \psi_k(z) \in H^\infty(S),\ a_k \in A_0 \cap A_1\} =$$

$$= inf\{|a - \sum_k \psi_k(z)a_k|_{\mathcal{G}} /\ \psi_k(z) \in N^+(S),\ a_k \in A_0 \cap A_1\} =$$

$$= |a|_s.$$

and theorem is proved.

THEOREM 2. *Let* (A_0, A_1) , (B_0, B_1) *be two interpolation couples of quasi-Banach spaces. Let* $A \subset A_0 \cap A_1$ *be dense in* $A_0 \cap A_1$, *and let* $T_z : A \to B_0 + B_1$ *be a family of linear operators parametrized by* $z \in \bar{S}$. *Assume*

(1) *For each* $a \in A$ *the range of* $\{T_z a\}_{z \in \bar{S}}$ *is finite dimensional and there exist.*
 $b_k = b_k(a) \in B_0 + B_1$ *and* $\psi_k(z) = \psi_k(z,a) \in N^+(S)$ *so that* $T_z a = \sum_k \psi_k(z) b_k$.

(2) (B_0, B_1) *is an* (L_{p_0}, L_{p_1}) *precursor, i.e. there exists a measure space* $(X, \sum, \mu)$ *and an operator* $\mathcal{M} : B_0 + B_1 \to M^+(X, \sum, \mu)$ *(non-negative measurable functions on* $(X, \sum, \mu)$*) such that:*

(2a) *(Sublinearity): for any* $b_0, \ldots, b_n \in B_0 + B_1$ *there exists a set* $E = E(b_0, \ldots, b_n)$, $\mu(E) = 0$, *such that for all* $\lambda_k \in \mathbf{C}$, $m \notin E$, $\mathcal{M}(\sum \lambda_k b_k)(m) \leq \sum |\lambda_k| \mathcal{M} b_k(m)$.

(2b) $|\mathcal{M}b|_{L_{p_j}} \leq |b|_{B_j}$.

(3) $|T_{j+iy}a|_{B_j} \leq K^j_y |a|_{A_j} \quad j = 0,1$ *and* $log K^j_y \in L(\partial S, dP)$.
 Then, for all $a \in A$, $0 < s < 1$, *we have:*

$$|\mathcal{M}T_s a|_{L_p(s)} \leq K_s |a|_{[A_0, A_1]_s}$$

where $\frac{1}{p(s)} = \frac{1-s}{p_0} + \frac{s}{p_1}$, $log K_s \leq \int_{\partial S} log K_\varsigma P(s,\varsigma)d\varsigma$.

PROOF.: Let $a \in A$. By lemma 1, $|a|_s = inf\{|f(z)|_{\mathcal{G}} / \ f(s) = a, f \in \mathcal{G}(A)\}$. Fix $f(z) = \sum_k a_k\varphi_k(z)$, $\varphi_k \in N^+(S)$, $a_k \in A$ and $f(s) = a$. $T_z f(z) = \sum_k \varphi_k(z) \sum_l \psi_l(z, a_k)b_l(a_k) = \sum_{k,l} \phi_{k,l}(z)b_l(a_k)$. $\phi_{k,l} \in N^+(S)$. Since $b_l(a_k) \in B_0 + B_1$, $\mathcal{M}b_l(a_k)(m) \in L_{p_0} + L_{p_1}$, and so is finite for a.e. m . By the sublinearity of $\mathcal{M}$ therefore, we can find a set E , $\mu(E) = 0$, so that for $m \notin E$, $b \to \mathcal{M}b(m)$ is a semi-norm on the span of $\{b_l(a_k)\}$. Our next claim is that $(\varsigma, m) \to \mathcal{M}T_\varsigma f(\varsigma)(m)$ is a measurable function on $\partial S \times X$. If $\phi_{k,l}(\varsigma)$ are bounded, this has been shown in [3]. Let therefore $g(\varsigma) = \frac{1}{1+\sum_{k,l}|\phi_{k,l}(\varsigma)|}$. $(\varsigma, m) \to \mathcal{M}g(\varsigma)T_\varsigma f(\varsigma)(m)$ is measurable and so $(\varsigma, m) \to \mathcal{M}T_\varsigma f(\varsigma)(m) = \frac{1}{|g(\varsigma)|}\mathcal{M}g(\varsigma)T_\varsigma f(\varsigma)(m)$ is measurable too. Since $b \to \mathcal{M}b(m)$ is a seminorm on the span of $b_l(a_k)$, we can find a linear functional defined on this span, so that $lf(s) = \mathcal{M}T_s f(s)(m)$, and $|l\ b| \leq \mathcal{M}b(m)$ for all b in this span. $lf(z) \in N^+(S)$ so that

$$log \mathcal{M}T_s f(s)(m) = log\, l f(s) \leq \int_{\partial S} log|lf(\varsigma)|P(s,\varsigma)d\varsigma \leq$$
$$\leq \int_{\partial S} log \mathcal{M}T_\varsigma f(\varsigma)(m) P(s,\varsigma)d\varsigma .$$

Using Hölder's inequality and then the limiting case of Minkowski's inequality, see [6], one then gets:

$$\frac{1}{p_s}log\left[\int_X \mathcal{M}^{p_s}T_s f(s)(m)d\mu(m)\right] \leq$$

$$\leq \int_{-\infty}^{\infty} log|\mathcal{M}T_{iy}f(iy)(.)|_{p_0} P_0(s,y)dy +$$
$$+ \int_{-\infty}^{\infty} log|\mathcal{M}T_{1+iy}f(1+iy)(.)|_{p_1} P_1(s,y)dy \leq$$

$$\leq \int_{-\infty}^{\infty} log|T_{iy}f(iy)|_{B_0} P_0(s,y)dy + \int_{-\infty}^{\infty} log|T_{1+iy}f(1+iy)|_{B_1} P_1(s,y)dy \leq$$

$$\leq log K_s + \int_{-\infty}^{\infty} log|f(iy)|_{A_0} P_0(s,y)dy + \int_{-\infty}^{\infty} log|f(1+iy)|_{A_1} P_1(s,y)dy \leq$$

$$\leq log K_s + log|f|_{\mathcal{G}}$$

and finally, taking infinum over all $f \in \mathcal{G}(A)$,

$$|\mathcal{M}T_s a|_{L_p(s)} \leq K_s |a|_{[A_0,A_1]_s}$$

and the proof is complete.

REFERENCES

(1) *A.P. Calderón and A. Torchinsky*. Parabolic maximal functions associated with a distribution. ADVANCES IN MATHEMATICS **16**(1975) pp. 1-64. II. ADVANCES IN MATHEMATICS **24**(1977) pp. 101-171.

(2) *R.R. Coifman, M.Cwikel, R.Rochberg, Y.Sagher and G.Weiss*. A theory of complex interpolation for families of Banach spaces. ADVANCES IN MATHEMATICS **43**(1982) pp. 203-229.

(3) *M. Cwikel, M.Milman, and Y.Sagher* Complex interpolation of some quasi-Banach spaces. JOURNAL OF FUNCTIONAL ANALYSIS, **65**(1986), pp. 339-347.

(4) *I.I. Hirschman Jr.* A convexity theorem for certain groups of transformations. JOURNAL D'ANALYSE **2**(1952), pp. 209-218.

(5) *S.Janson and P.Jones*. Interpolation between H_p spaces, the complex method. JOURNAL OF FUNCTIONAL ANALYSIS, **48**(1982), pp. 58-80.

(6) *Y.Sagher*. On analytic familes of operators. ISRAEL JOURNAL OF MATHEMATICS **7**(1969) pp. 350-356.

(7) *E.M. Stein*. Interpolation of linear operators. TRANSACTIONS AM. MATH. SOC. **83**(1956), pp. 482-492.

FUNCTION SPACES ON LIE GROUPS AND ON ANALYTIC MANIFOLDS

Hans Triebel
Sektion Mathematik
Universität Jena
DDR-6900 Jena
German Democratic Republic

1. Introduction and outline of methods

This paper is the continuation of /11/ and /9,10/, which, in turn, are based on /8/. In these papers we developed the theory of the spaces F^s_{pq} and B^s_{pq} on Lie groups, complete Riemannian manifolds, and the euclidean n-space R^n from a unified point of view, cf. also the survey /12/. The spaces $F^s_{pq}(R^n)$ and $B^s_{pq}(R^n)$ with $-\infty < s < \infty$, $0 < p \leq \infty$, $0 < q \leq \infty$ ($p < \infty$ in the case of the F-spaces) cover many well-known classical function spaces: Sobolev spaces, Bessel-potential spaces, Hölder-Zygmund spaces, Besov-Lipschitz spaces and (inhomogeneous) Hardy spaces. As an outgrowth of /8/ (but, at least implicitly, contained in /7,2.3.6/) we have a completely local theory of the spaces $F^s_{pq}(R^n)$ which can be described as follows. Let Z^n be the lattice with integer-valued components. Let $\psi(x)$ be a C^∞ function in R^n with a compact support and with $\sum \psi(x-j) = 1$ if $x \in R^n$, where the sum is taken over all $j \in Z^n$. Then we have

$$\| f \mid F^s_{pq}(R^n) \|^p \sim \sum_{j \in Z^n} \| \psi(\cdot - j) f \mid F^s_{pq}(R^n) \|^p \tag{1}$$

(equivalent quasi-norms). This property (together with an obvious counterpart if $p = \infty$) is the basis to extend the theory of the spaces F^s_{pq} from R^n to n-dimensional complete Riemannian manifolds M with a bounded geometry and a positive injectivity radius, cf. /9,10/. Let

G be a n-dimensional connected Lie group (or a Lie group consisting of finitely many connected components) which we furnish with a left-invariant Riemannian metric. In /11/ we reduced on this way the theory of the spaces $F^s_{pq}(G)$ to the theory of the spaces $F^s_{pq}(M)$. Both for the manifold M and the Lie group G the spaces B^s_{pq} can be incorporated afterwards via real interpolation. However the aim of the papers /9, 10,11/ is at least twofold. First, of course, we justified the roughly sketched way to introduce spaces F^s_{pq} and B^s_{pq} on manifolds and Lie groups. But secondly, and maybe more important, we gave intrinsic descriptions of these spaces in natural terms on manifolds and Lie groups and compared these spaces with other possibilities (Sobolev spaces, Hölder spaces, characterizations via derivatives and differences). For the purpose of the present paper and its motivation it is useful to outline some of the key ideas. Let k be a function on the real line such that $k(|y|)$ with $y \in R^n$ is a C^∞ function in R^n with a support contained in the unit ball in R^n. Then we introduce the means

$$k(t,f)(x) = \int_{R^n} k(|y|)\, f(x+ty)\, dy, \qquad x \in R^n, \quad t > 0, \tag{2}$$

which make sense for any tempered distribution $f \in S'(R^n)$ (appropriate interpretation). If k_o and k_N are two appropriate kernels (the conditions will be described later on in detail) then

$$F^s_{pq}(R^n) = \Big\{ f \,\Big|\, f \in S'(R^n),\ \ \| f \mid F^s_{pq}(R^n) \|^{k_o,k_N}_{\varepsilon,r} =$$

$$\| k_o(\varepsilon,f) \mid L_p(R^n) \| + \Big\| \Big(\int_0^r t^{-sq}\, |k_N(t,f)(\cdot)|^q \frac{dt}{t} \Big)^{\frac{1}{q}} \Big| \ L_p(R^n) \Big\| < \infty \Big\} \tag{3}$$

where either $-\infty < s < \infty$, $0 < p < \infty$, $0 < q \leqq \infty$ or $-\infty < s < \infty$, $p = q = \infty$ (modification of (3) if $q = \infty$). Furthermore $\varepsilon > 0$ and $r > 0$. This assertion has to be understood in the sense of equivalent quasi-norms. In any case, (3) shows that the spaces $F^s_{pq}(R^n)$ can be characterized via the means (2) where the behaviour with respect to $x \in R^n$ and $t > 0$ is measured separately. A similar characterization can be given for the spaces $B^s_{pq}(R^n)$. It is one of the main aims of the papers /9,10,11/ to extend these characterizations, or at least these equivalent quasi-norms, from R^n to the above manifold M or the above Lie group G. One has to look for intrinsic counterparts of the means (2). Let M be the above complete Riemannian manifold M (with bounded geometry and positive injectivity radius). Then

$$k(t,f)(P) = \int_{T_P M} k(|X|)\, f(c(P,X,t))\, dX, \quad P \in M,\ 0 < t < r, \tag{4}$$

is the counterpart of (2), where $T_P M$ is the tangent space in the point $P \in M$ and $c(P,X,t)$ stands for the geodesic with $c(P,X,0) = P$ and $\frac{dc}{dt}(P,X,0) = X$ (the parameter t may be interpreted as the arc length). Furthermore, $|X|$ and dX refer to the Riemannian metric in the point $P \in M$. If one inserts the means (4) in (3) (with M instead of R^n), where $\varepsilon > 0$ and $r > 0$ are small, and k_0 and k_N are appropriate functions, then one obtains an equivalent quasi-norm in $F^s_{pq}(M)$. This is one of the main results in /9/. Let G be the above Lie group. Then it is easy to equip G with a left-invariant Riemannian metric. The result is a Riemannian manifold which has all the desired properties in order to apply the theory sketched above. We have the means (4) where $c(P,X,t)$ are the Riemannian geodesics connected with the introduced left-invariant Riemannian metric. However this looks unnatural. One would strongly prefer to replace the Riemannian geodesics $c(P,X,t)$ by the Lie geodesics $x \cdot \exp tX$ where $x \in G$, $X \in \mathfrak{g}$ (the Lie algebra) and $t > 0$. In other words, the prefered means look as follows,

$$k(t,f)(x) = \int_{\mathfrak{g}} k(|X|)\, f(x \cdot \exp tX)\, dX, \quad x \in G,\ 0 < t < r, \tag{5}$$

where the Lie algebra $\mathfrak{g}$ is equipped with an euclidean metric. Then again one has similar equivalent quasi-norms on $F^s_{pq}(G)$ as in (3). A proof is given in /11/. The basic idea is to follow the detailed proof in /9/ for complete Riemannian manifolds and to look for the necessary replacements. There is essentially only one point which caused serious trouble. This is the proof of Theorem 2(i) in /9/ for large values of s, cf. /9, 4.2, pp. 318/319/. We used two facts:

(i) In local charts the Riemannian geodesic c(t) with the components $c^i(t)$ satisfies the differential equations

$$\frac{d^2c^j(t)}{dt^2} + \Gamma^j_{ik}\left(c(t)\right)\frac{dc^i(t)}{dt}\,\frac{dc^k(t)}{dt} = 0\,,\quad j=1,\dots,n, \qquad (6)$$

where Γ^j_{ik} stands for the Christoffel symbols with respect to the Levi-Cività derivation (summation convention).

(ii) If g_{ij} is the Riemannian metric then $g_{ij}\,\frac{dc^i}{dt}\,\frac{dc^j}{dt}$ is constant along a Riemannian geodesic.

In /11/ we described the substitutes of (i) and (ii). The equations (6) can be replaced by the full Campbell-Baker-Hausdorff formula. On the other hand, (ii) was necessary to ensure uniform remainder estimates. This can be done now on the basis of the analyticity of all the involved operations on the Lie group.

The aim of the present paper is to offer a second way to reduce function spaces on Lie groups to those ones on complete Riemannian manifolds. For this purpose we modify the construction in /9,10/ somewhat. We beginn with an analytic complete Riemannian manifold. Beside the Levi-Cività derivation we assume that we have a second derivation. Let (6) be now the corresponding system of differential equations with respect to this new derivation and let c(P,X,t) be the corresponding geodesics. Then is comes out that one obtains equivalent quasi-norms of type (3) based on these newly interpreted means (4). Maybe this result is not so interesting for its own sake. But it allows a quick access to the corresponding assertions for the spaces F^s_{pq} on a Lie group G: The Lie group G is furnished with a left-invariant Riemannian metric and, in addition, with the left-invariant covariant derivation independent of the left-invariant Riemannian metric. The crucial point is that the geodesics with respect to this left-invariant covariant derivation coincide with the Lie geodesics from (5). This completes the reduction.

In the following sections we give precise definitions and results. Furthermore we return to the above vague outline of our method, mostly in order to provide the necessary references. We restrict our attention to the spaces F^s_{pq} because the considerations for the spaces B^s_{pq} remain essentially unchanged and there are no new results and new methods as far as the latter spaces are concerned, compared with /9,10, 11/.

2. Definitions and results

2.1. Spaces on R^n

Let R^n be the real euclidean n-space and let $B = \{y \mid |y| < 1\}$ be the unit ball in R^n. Let k_o and k be two functions defined on the real line such that both $k_o'(y) = k_o(|y|)$ and $k'(y) = k(|y|)$ with $y \in R^n$ are C^∞ functions on R^n with

$$\text{supp } k_o \subset B \quad \text{and} \quad \text{supp } k \subset B.$$

Let

$$\widehat{k'}(0) \neq 0 \quad \text{and} \quad \widehat{k_o'}(y) \neq 0 \quad \text{for all } y \in R^n,$$

where $\widehat{k_o'}$ and $\widehat{k'}$ stand for the Fourier transform of k_o' and k', respectively. If $N = 1,2,\dots$ is a natural number then we put

$$k_N' = \left(\sum_{j=1}^{n} \frac{\partial^2}{\partial x_j^2} \right)^N k' .$$

Of course, $k_N'(y) = k_N(|y|)$ with $y \in R^n$ is also rotation-invariant. If $N = 0,1,2,\dots$ then the means $k_N(t,f)(x)$ have been defined in (2) (with k_N instead of k). If a is a real number then we put $a_+ = \max(a,0)$. Finally

$$\| h \mid L_p(R^n) \| = \left(\int_{R^n} |h(x)|^p \, dx \right)^{\frac{1}{p}}, \qquad 0 < p \leq \infty ,$$

with the usual modification if $p = \infty$.

Definition 1. Let k_N be the above functions. Let $0<\varepsilon<\infty$, $0<r<\infty$ and $-\infty<s<\infty$. Let either $0<p<\infty$, $0<q\leq\infty$ or $p=q=\infty$. Let N be a natural number with $2N>\max(s,n(\frac{1}{p}-1)_+)$. Then $F^s_{pq}(R^n)$ is given by (3) (with the usual modification if $q=\infty$).

Remark 1. The theory of these spaces has been developed in /7/. However the above definition is a consequence of /8/, cf. also the recent survey /12/. All these spaces are quasi-Banach spaces. They are independent of k_0 and k, as well as of the numbers ε, r and N (equivalent quasi-norms). These spaces cover Sobolev spaces, Bessel-potential spaces and (inhomogeneous) Hardy spaces. Similarly one can introduce the spaces $B^s_{pq}(R^n)$: one has to change the L_q and the L_p spaces in (3). However we restrict ourselves to the spaces of type F^s_{pq}.

2.2. Spaces on analytic manifolds: definitions

Let M be a connected n-dimensional analytic manifold, equipped with an analytic Riemannian metric, a real twice covariant analytic tensor field g such that g_P for every $P\in M$ is a positive definite bilinear form, i. e.

$$g_P(X,Y)=g_P(Y,X) \quad \text{and} \quad g_P(X,X)>0 \quad \text{if} \quad 0\neq X\in T_PM$$

and $Y\in T_PM$. Here, as usual, T_PM stands for the tangent space in the point $P\in M$. The exponential map, $\exp_P\colon T_PM\longrightarrow M$, has the usual meaning. It yields the analytic normal geodesic coordinates (where T_PM is identified with R^n). In particular, if $r>0$ is small (in dependence on P) then $\exp_P$ is a diffeomorphic map from

$$B_P(r)=\{X\mid X\in T_PM,\ |X|<r\} \quad \text{onto} \quad \Omega_P(r)=\exp_P B_P(r), \tag{7}$$

where $|X|=\sqrt{g_P(X,X)}$. We assume that M has a positive injectivity radius $r_0>0$, i. e. the just explained mapping property holds independently of $P\in M$ for all r with $0<r<r_0$, and r_0 is the largest number with this property. Finally we assume that M has a bounded geometry:

Let g_{ij} be the components of the metric tensor in normal geodesic coordinates (with respect to a given point $P \in M$). Let $0 < r < r_0$. Then there exist positive numbers c and c_α, which are independent of the chosen point $P \in M$, such that

$$\det g_{ij} \geq c \quad \text{and} \quad |D^\alpha g_{ij}| \leq c_\alpha \tag{8}$$

holds for all multi-indices α and all points in the ball $B_P(r)$ from (7). For manifolds with these properties we have a resolution of unity which is similar to those one used in (1). If $\delta > 0$ is sufficiently small and the natural number L is sufficiently large then there exist a covering $M = \bigcup_{j=1}^{J} \Omega_{P_j}(\delta)$ with the geodesic balls from (7) such that a fixed ball of this covering has a non-empty intersection with at most L of the other balls of this covering. Here J may be a natural number (compact case) or may be ∞ (non-compact case). Furthermore there exists an associated resolution of unity $\{\psi_j\}$ with the following properties:

(i) $\psi_j \in C^\infty(M)$, $0 \leq \psi_j \leq 1$, $\sum_{j=1}^{J} \psi_j = 1$ on M,

(ii) $\operatorname{supp} \psi_j \subset \Omega_{P_j}(\delta)$, $j = 1,2,\dots$,

(iii) for any multi-index α there exists a positive number b_α with

$$|D^\alpha(\psi_j \circ \exp_{P_j})(x)| \leq b_\alpha, \quad x \in B_{P_j}(r), \quad 0 < \delta < r < r_0. \tag{9}$$

<u>Remark 2</u>. More detailed versions of these conditions and notations may be found in /9,10,12/, where we dealt with C^∞ manifolds instead of analytic manifolds. Furthermore we refer to /4,5/ (injectivity radius), /2,6/ (bounded geometry) and /1/ (resolution of unity).

Let $D'(M)$ be the collection of all complex-valued distributions on the above manifold M.

<u>Definition 2</u>. Let M be the above analytic Riemannian manifold (connected, complete, with positive injectivity radius, with bounded geometry). Let $\psi = \{\psi_j\}$ be the above resolution of unity. Let $\delta > 0$ be sufficiently small. Let $-\infty < s < \infty$. Let either $0 < p < \infty$, $0 < q \leq \infty$

or $p = q = \infty$. Then

$$F^s_{pq}(M) = \left\{ f \mid f \in D'(M),\ \|f \mid F^s_{pq}(M)\|^{\psi} = \left(\sum_{j=1}^{J} \|\psi_j f \circ \exp_{P_j} \mid F^s_{pq}(R^n)\|^p \right)^{\frac{1}{p}} < \infty \right\} \qquad (10)$$

(modification if $p = \infty$).

Remark 3. Again $T_{P_j}M$ is identified with R^n. In particular, $(\Omega_{P_j}(r), \exp^{-1}_{P_j})$ with $0 < \delta < r < r_0$ are the corresponding normal geodesic coordinates, cf. (7). Tacitly it is assumed that $\psi_j f \circ \exp_{P_j}$ is extended outside of $B_{P_j}(r)$ by zero. Hence $\psi_j f \circ \exp_{P_j} \in S'(R^n)$, and (10) makes sense.

Remark 4. If $M = R^n$, then the above defined space $F^s_{pq}(M)$ coincides with the previously defined space $F^s_{pq}(R^n)$. This is essentially a consequence of (1).

Remark 5. It comes out that the spaces $F^s_{pq}(M)$ are well defined. They are independent of the local charts $\{(\Omega_{P_j}(r), \exp^{-1}_{P_j})\}$ and the chosen resolution of unity ψ . Furthermore $F^s_{pq}(M)$ is a quasi-Banach space (Banach space if $p \geq 1$ and $q \geq 1$). Cf. /9,10/.

2.3. Spaces on analytic manifolds: equivalent quasi-norms

Let M be again the above analytic manifold. It is our aim to describe equivalent quasi-norms in the sense of (3) but with the help of the means (4). This has been done in /9,10/ where $c(P,X,t)$ stood for the Riemannian geodesics connected with the Levi-Civitá derivation, and hence with the given metric tensor on M. Now we generalize this concept (however we need now the analyticity of the underlying manifold, in contrast to /9,10/ where we dealt with C^{∞} manifolds). Now we assume that M carries a second structure, an analytic covariant derivation, which may be different from the Levi-Civitá derivation. The corresponding geodesics are denoted by $c(P,X,t)$. They obey in local charts the differential equations (6), cf. /5,1.5.1/. Let $(\Omega_P(r), \exp_P^{-1})$ be the above normal geodesic coordinates (with respect to the Levi-Cività derivation!), $0<r<r_0$. Then we assume that the Christoffel symbols Γ^j_{ik} (of this additional covariant derivation) are analytic in $(\Omega_P(r), \exp_P^{-1})$ and that there exist two positive numbers b and c with

$$|D^{\alpha}\Gamma^j_{ik}(P)| \leq c\,\alpha!\, b^{|\alpha|} \quad \text{in} \quad (\Omega_P(r), \exp_P^{-1}) \quad \text{for all } P\in M. \tag{11}$$

This is the counterpart of (8). It ensures a uniform analyticity of these new Christoffel symbols. (For the Christoffel symbols

$$\frac{1}{2} g^{jl}\left[\partial_i\, g_{kl} + \partial_k\, g_{il} - \partial_l\, g_{ik}\right]$$

of the Levi-Cività derivation we have only a somewhat weaker assertion at hand which comes from (8). But this is completely sufficient for this special derivation.). Now the means (4) must be understood with respect to this additional covariant derivation with $|X| = \sqrt{g_P(X,X)}$, where dX stands for the Riemannian volume element induced on T_PM by $g_P(X,Y)$ (for more details we refer to /9,12/). Finally, $L_p(M)$ and $\|\cdot|L_p(M)\|$ with $0<p\leq\infty$ must be understood with respect to the invariant Riemannian volume element on M.

<u>Theorem 1</u>. Let M be the above manifold with its double analytic structure (the metric and the independent covariant derivation). Let $-\infty< s<\infty$. Let either $0<p<\infty$, $0<q\leq\infty$ or $p=q=\infty$. Let $\varepsilon>0$ and $r>0$ be sufficiently small and let the natural number N be sufficiently large (in dependence on M,s,p,q). Let k_N be the functions from 2.1. Let $k_N(t,f)(\cdot)$ be the means in the sense of (4) with respect to the additional covariant derivation. Then

$$\|k_0(\varepsilon,f)\,|\,L_p(M)\| + \left\|\left(\int_0^r t^{-sq}|k_N(t,f)(\cdot)|^q \frac{dt}{t}\right)^{\frac{1}{q}} \,\Big|\, L_p(M)\right\| \tag{12}$$

is an equivalent quasi-norm in $F^s_{pq}(M)$ (modification if $q=\infty$).

<u>Remark 6</u>. This theorem is the counterpart of (3). For the special case of the Levi-Cività derivation we gave a detailed proof in /9,10/,

even under the weaker assumption that M is C^∞. In the introduction we have given a rough outline of the method and the necessary changes. We return to this point in 3.1 below.

2.4. Spaces on Lie groups: definitions

Let G be a n-dimensional connected Lie group. (It is not really necessary to suppose that G is connected. It would be sufficient to assume that G consists of finitely many connected components. The assumption that G is connected is convenient: we have a direct access to what has been said above about analytic manifolds). Let e be the unit element and let $\mathfrak{g} = T_e G$ be the corresponding Lie algebra. Let g be a real positive-definite symmetric bilinear form on $\mathfrak{g}$. Let $L_a: x \longrightarrow ax$ be the left translation on G, where $a \in G$ and $x \in G$. Then $g_a = (L_{a^{-1}})_a^* g$ with $a \in G$ (pull back operation) generates a left-invariant analytic Riemannian metric which satisfies all the conditions of 2.2. In particular we can apply Definition 2, where we now replace M by G. The corresponding spaces are denoted by $F^s_{pq}(G)$: It is easy to see that these spaces are independent of the chosen left-invariant Riemannian metric. On the other hand if one replaces the left-invariant metric by a right-invariant metric then the corresponding spaces of F^s_{pq}-type are different, in general, from the just introduced spaces $F^s_{pq}(G)$. If G is compact (or abelian) then these two types of spaces coincide. We developed the theory of these spaces in /11/, cf. also /12/.

2.5. Spaces on Lie groups: equivalent quasi-norms

Let G again be the above Lie group. We can apply Theorem 1 if we identify the covariant derivation with the Levi-Cività derivation: As we remarked above in this special case (8) is sufficient, instead of the sharper assumption (11), cf. also 3.1 below. In other words, the means in (12) are given by (4), where $c(P,X,t)$ stands for the Riemannian geodesics with respect to the constructed left-invariant Riemannian metric. However it seems to be more natural to ask for equivalent quasi-norms of type (12) where the means are given by (5). As usual $L_p(G)$ and $\|\cdot|L_p(G)\|$ with $0<p\leq\infty$ must be understood with respect to a fixed left-invariant Haar measure on G (which may be identified with the Riemannian volume element of the above left-invariant Riemannian metric).

Theorem 2. Let G be the above Lie group. Let $-\infty<s<\infty$. Let either $0<p<\infty$, $0<q\leq\infty$ or $p=q=\infty$. Let $\varepsilon>0$ and $r>0$ be sufficiently small and let the natural number N be sufficiently large (in dependence on G,s,p,q). Let k_N be the functions from 2.1. Let $k_N(t,f)$ be the means in the sense of (5). Then

$$\|k_o(\varepsilon,f)\mid L_p(G)\| + \left\|\left(\int_0^r t^{-sq}|k_N(t,f)(\cdot)|^q \frac{dt}{t}\right)^{\frac{1}{q}} \mid L_p(G)\right\| \qquad (13)$$

is an equivalent quasi-norm in $F^s_{pq}(G)$ (modification if $q=\infty$).

Remark 7. This theorem is one of the main results of /11/. The proof in /11/ was based on the full Campbell-Baker-Hausdorff formula as a substitute of (6), cf. Sect. 1. The main aim of the present paper is to make clear that Theorem 2 is a special case of Theorem 1.

3. Outline of proofs

3.1. Analytic manifolds

In order to prove Theorem 1 one can follow the detailed proof given in /9/, where the nasty additional assumption (15') in /9,p. 306/ is not necessary. This was shown in /10/ on the basis of /2/. We refer in this connection also to /3/. As we outlined in Section 1 there is essentially only one critical point. In /9, formulas (68),

(69) on p. 319/ we expanded the geodesics (now with respect to the additional covariant derivation)

$$c^i(x,X,t) = x^i + tX^i + \sum_{\ell=2}^{2L-1} \frac{t^\ell}{\ell!} \frac{d^\ell c^i}{dt^\ell}(x,X,0) + \frac{t^{2L}}{(2L)!} \frac{d^{2L} c^i}{dt^{2L}}(x,X,\vartheta^i t)$$

$$= x^i + tX^i + \sum_{2 \le |\alpha| \le 2L-1} t^{|\alpha|} b^i_\alpha(x)\, X^\alpha + t^{2L} R^i_{2L}(x,X,t), \tag{14}$$

where the second line comes from (6) and (11). As in /9/ we have the necessary uniform estimates for the functions $b^i_\alpha(x)$. The uniform estimate of the remainder terms $R^i_{2L}(x,X,t)$ in /9/ was based on the assertion (ii) formulated in Section 1, which is not available now. However (6) and the uniform analyticity of the Christoffel symbols Γ^j_{ik}, cf. (11), yield the necessary estimates for $R^i_{2L}(x,X,t)$. In the same way one can overcome a second minor difficulty on p. 324, formula (93) in /9/. Again one can use the uniform analyticity of the geodesics. All the other arguments can be taken over from /9/.

Remark 8. In rough terms: In /9/ (and /10/) M is a C^∞ manifold with the above properties (complete, connected, with positive injectivity radius and with bounded geometry). If we wish to replace the natural Riemannian geodesics by other line systems, geodesics with respect to an additional covariant derivation, then we are forced to strenghten the assumption: In addition M is analytic and we have the uniform analyticity condition (11).

3.2. Lie groups

Let G be the above Lie group equipped with a (uniform analytic) left-invariant Riemannian metric, cf. 2.4. As an additional structure we introduce the left-invariant covariant derivation in the sense of /5, 1.7.7/, cf. also /4, p. 102/. It comes out that the corresponding geodesics coincide with the left translates of the 1-dimensional subgroups of G, i. e. with the Lie geodesics $x \cdot \exp tX$, cf. /5, 1.7.10

and 1.7.13/ or /4, pp. 102 and 104/. Furthermore, (11) is satisfied. Now Theorem 2 is a special case of Theorem 1.

Remark 9. In general the left-invariant covariant derivation on G is not symmetric. However there exists also a symmetric covariant derivation on G with the desired properties, cf. /5, 1.7.12/.

References

1. T. Aubin. Nonlinear Analysis on Manifolds. Monge-Ampère Equations. Springer, New York, Heidelberg, Berlin, 1982.
2. J. Cheeger, M. Gromov, M. Taylor. Finite propagation speed, kernel estimates for functions of the Laplace operator, and the geometry of complete Riemannian manifolds. J. Differential Geometry 17 (1982), 15 - 53.
3. E. B. Davies. Kernel estimates for functions of second order elliptic operators. Preprint, London, 1986.
4. S. Helgason. Differential Geometry, Lie Groups, and Symmetric Spaces. Academic Press, New York, San Francisco, London, 1978.
5. W. Klingenberg. Riemannian Geometry. W. de Gruyter, Berlin, New York, 1982.
6. D. Michel. Estimées des coefficients du Laplacien d'une variété Riemannienne. Bull. Sci. Math. 102 (1978), 15 - 41.
7. H. Triebel. Theory of Function Spaces. Birkhäuser, Boston, 1983.
8. H. Triebel. Characterizations of Besov-Hardy-Sobolev spaces: a unified approach. J. Approximation Theory
9. H. Triebel. Spaces of Besov-Hardy-Sobolev type on complete Riemannian manifolds. Ark. Mat.
10. H. Triebel. Characterizations of function spaces on a complete Riemannian manifold with bounded geometry. Math. Nachr.
11. H. Triebel. Function spaces on Lie groups, the Riemannian approach. J. London Math. Soc.
12. H. Triebel. Einige neuere Entwicklungen in der Theorie der Funktionenräume. Jber. Deutsch. Math.-Verein.

EQUIVALENT NORMALIZATIONS OF SOBOLEV AND NIKOL'SKIĬ SPACES IN DOMAINS. BOUNDARY VALUES AND EXTENSION

S.K.Vodop'yanov
Novosibirsk State University
Novosibirsk, 630090, Soviet Union

The main result of this work consists of some new equivalent normalizations of Sobolev and Nikol'skiĭ spaces in domains, which include geometrical characteristics of the domain in an explicit way. The geometry of the domain is determined by the modulus of continuity defining the function space. Two applications of this result are given.

The first one deals with traces (boundary values) of a function space to a boundary of the domain of definition. In the case of an arbitrary domain the boundary is realized as the completion of the domain of definition with respect to the corresponding metric. The elements of the function spaces are extended by continuity to the boundary and these traces belong to some function space. It is proved that such a characterization of traces is reversible.

The other application concerns some new necessary and sufficient extendability conditions for differentiable functions across the boundary of the domain of definition. They are formulated in terms of equivalence of the corresponding metrics in the domain and in the surrounding space.

1. Taylor expansion and equivalent normalizations of Sobolev and Nikol'skiĭ spaces

For an arbitrary $\alpha \in (0,1]$ we define an inner α-metric $d_{\alpha,G}(x,y)$ on a domain $G \subset R^n$ as follows

$$d_{\alpha,G}(x,y) = \inf_{\gamma} \sum_{i=1}^{m} |x_i - y_i|^{\alpha} .$$

where the infimum is taken over all polygons γ consisting of segments $[x_{i-1}, x_i] \subset G$, $x_0 = 0$, $x_m = y$. It is evident that the 1-metric coincides with the infimum of the lengths of rectifiable curves connecting points $x, y \in G$, and therefore it is the inner metric of the domain in the commonly accepted sense (see [1]). We denote by G_α the metric space $(G, d_{\alpha,G})$.

The elements of the function space $Lip(l, G_\alpha)$, $l \in R_+ = \{x: x>0\}$, $l \in (k, k+1]$, $k = 0,1,2,\ldots$, are L_∞-functions $f: G \to R$, whose weak partial derivatives, denoted by $D^j f$, also belong to L_∞ for $|j| \leq k$.

The norm in $Lip(l,G_\alpha)$ is defined by

$$||f||_{Lip(l,G_\alpha)} = \sum_{|j|\le k} \{||D^j f||_{L_\infty(G)} + \sup_{x,y\in G} |R_j(x,y)|/d_{\alpha,G}(x,y)^{(l-|j|)/\alpha}\},$$

where $\alpha = l-k$ and

$$D^j f(x) = \sum_{|j+s|\le k} D^{j+s} f(y)(x-y)^s/s! + R_j(x,y), \quad 0 \le |j| \le k.$$

We use the usual multi-index notation, $j = (j_1,j_2,\dots,j_n)$, $s = (s_1,s_2,\dots,s_n)$, $s! = s_1!s_2!\dots s_n!$, $j = j_1 + j_2 +\dots+ j_n$ and $x^s = x_1^{s_1} x_2^{s_2} \dots x_n^{s_n}$.

Two function spaces coincide if the operator of embedding of one of the spaces into the other is a bounded isomorphism.

Theorem 1. Let G be an arbitrary domain in R^n. The following function spaces coincide.

$$Lip(l,G_\alpha) = W^l_\infty(G) \text{ for } l \in N \text{ and } Lip(l,G_\alpha) = H^l_\infty(G) \text{ for } l \notin N.$$

Here the space $W^l_\infty(G)$ $(H^l_\infty(G))$ is a member of the scale of the Sobolev (Nikol'skiĭ) spaces $W^l_p(G)$ [2] ($H^l_p(G)$ [3]). Let us recall that the elements of $W^l_p(G)$, $l \in N$, $p \in [1,\infty]$ ($H^l_p(G)$, $l \in R_+$, $l \in (k,k+1)$, $k = 0,1,2,\dots$, $p \in [1,\infty]$) are L_p-runctions f whose weak partial derivatives, denoted by $D^j f$, also belong to L_p for $|j| \le l$ ($|j|\le k$). The norm in $W^l_p(G)$ $(H^l_p(G))$ is defined by

$$||f||_{W^l_p(G)} = ||f||_{L_p(G)} + ||\nabla_l f||_{L_p(G)}$$

$$(||f||_{H^l_p(G)} = ||f||_{L_p(G)} + \sup_{h\in R^n} ||\Delta_h \nabla_k f||_{L_p(G)}/|h|^\alpha),$$

where $\nabla_l f = \{D^j f\}$, $|j| = l$ ($\alpha = l-k$, $\Delta_h g(x) = g(x+h) - g(x)$ if the segment $[x,x+h] \subset G$, and $\Delta_h g(x) = 0$ if the segement $[x,x+h] \not\subset G$). The Nikol'skiĭ spaces $H^l_p(G)$ again belong to the scale of the Besov spaces $B^l_{p,q}(G)$, $q \in [1,\infty]$: $H^l_p(G) = B^l_{p,\infty}(G)$ [3,4].

Proof of Theorem 1. If $f \in Lip(l,G_\alpha)$, $l \in N$, then any point $x \in G$ is the centre of some ball B in which all the derivatives of function f of the order $l-1$ are bounded and satisfy a Lipschitz condition. This implies that the function f has bounded generalized derivatives of order l. Thus the boundedness of the embedding $Lip(l,G_\alpha) \to W^l_\infty(G)$, $l \in N$, is proved. The continuity of the embedding

$$\mathrm{Lip}(1,G_\alpha) \to H^l_\infty(G), \quad l \notin N,$$

is evident.

Now we shall prove the continuity of the reciprocal embedding. Let us consider any function f of a Sobolev or a Nikol'skii space with norm equal to 1. The estimates $|D^j f(x)| \leq C$, $|j| \leq l$, $x \in G$, are proved in [3]. Therefore it remains to establish the inequalities

$$|R_j(x,y)| \leq C_j d_{\alpha,G}(x,y)^{(l-|j|)/\alpha}, \tag{1.1}$$

where $\alpha = l-k$, $k \in N$, $k < l \leq k+1$, $|j| \leq k$ and the constant C_j equals $n^{(k-|j|)/2}$. We shall prove the estimate (1.1) by induction. If $|j| = k$, then $R_j(x,y) = D^j f(x) - D^j f(y)$. We have the following estimate for the difference of the derivative $D_j f$, $|j| = k$, along a polygon consisting of segments $[x_{i-1}, x_i] \subset G$, $i = 0,1,\dots,m$, $x_0 = 0$, $x_m = y$:

$$|D^j f(x)-D^j f(y)| \leq \sum_{i=1}^{m} |D^j f(x_i)-D^j f(x_{j-1})| \leq \sum_{i=1}^{m} |x_i-x_{i-1}|^{\alpha}, \tag{1.2}$$

where $\alpha = 1$ if $f \in W^l_\infty(G)$, and $\alpha = l-k$ if $f \in H^l_\infty(G)$. Now minimizing the right hand side of (1.2), we obtain the estimate

$$|R_j(x,y)| = |D^j f(x)-D^j f(y)| \leq d_{\alpha,G}(x,y) = d_{\alpha,G}(x,y)^{(l-|j|)/\alpha}, \tag{1.3}$$

where $|j| = k$ and α depends on the choice of function space made in the above mentioned way.

The inequalities (1.3) constitute the basis for the induction. Let us suppose that the estimate (1.1) in which $C_j = n^{(k-|j|)/2}$, is proved for all j such that $C \leq p < |j| \leq k$, $p = 0,1,2,\dots,k-1$. For any multi-index j, $|j| = p$, we shall estimate $R_j(x,y)$, where the points $x,y \in G$ are fixed. Since the value $R_j(y,y)$ in question is equal to 0,

$$|R_j(x,y)| \leq \sup_{\xi\in G,\, d_{\alpha,G}(\xi,y)\leq d_{\alpha,G}(x,y)+\varepsilon} |\nabla R_j(\xi,y)| \sum_i |a_i|, \tag{1.4}$$

where the segments a_i form a polygon with endpoints x and y for which

$$\sum_i |a_i|^\alpha \leq d_{\alpha,G}(x,y) + \varepsilon.$$

Since

$$|\nabla R_j(\xi,y)|^2 = \sum_{|j+k|=p+1} R_{j+k}(\xi,y)|^2$$

and

$$\sum_i |a_i| \leq [\sum_i |a_i|^{\alpha}]^{1/\alpha} \leq [d_{\alpha,G}(x,y) + \varepsilon]^{1/\alpha}$$

by the reciprocal Minkowski inequality (see [5]), we obtain from (1.4)

$$|R_j(x,y)| \leq n^{1/2} n^{[k-(p+1)]/2} d_{\alpha,G}(x,y)^{[1-(p+1)]/\alpha} d_{\alpha,G}(x,y)^{1/\alpha} =$$

$$= n^{(k-p)/2} d_{\alpha,G}(x,y)^{(1-p)/\alpha}.$$

This proves the inequalities (1.1) for $p = |j|$.

2. Traces of functions of Sobolev and Nikol'skiĭ spaces to a boundary of the domain of definition

Let us consider the metric space $G_\alpha = (G, d_{\alpha,G})$, $\alpha \in (0,1]$. Let $\tilde{G}_\alpha$ be the completion of G_α with respect to the metric d_α,G. It is easy to verify that any function of $\mathrm{Lip}(1,G_\alpha)$ extends by the continuity to the completion $\tilde{G}_\alpha$. Indeed, if $|j| = k$, then $|f^{(j)}(x)-f^{(j)}(y)| \leq$ $\leq R_j(x,y) \leq Md_{\alpha,G}(x,y)$, where $f^{(j)}(x) = D^j f(x)$. Hence we have an extension by continuity of the functions $f^{(j)}(x)$, $|j| = k$, to the completion $\tilde{G}_\alpha$. Let us now suppose that the functions $f^{(j)}(x) = D^j f(x)$, $0 < p \leq |j| \leq k$, are already extended to $\tilde{G}_\alpha$. The extension to $\tilde{G}_\alpha$ of the functions $f^{(j)}(x) = D^j f(x)$, $|j| = p-1$, follows from the expansion

$$f^{(j)}(x) - f^{(j)}(y) = \sum_{|j+s|\leq k} f^{(f+s)}(y)(x-y)^s/s! + R_j(x,y)$$

and the inequalities $|R_j(x,y)| \leq Md_{\alpha,G}(x,y)^{[1-(p-1)]/\alpha}$, $|x-y|^\alpha \leq$ $\leq d_{\alpha,G}(x,y)$. Since the functions $\mathrm{Lip}(1,G_\alpha)$ extend uniquely to $\tilde{G}_\alpha$ it is natural to introduce the function class $\mathrm{Lip}(1,\tilde{G}_\alpha)$, consisting of the jets $\{f^{(j)}(x)\}$, $|j| \leq k$, defined on $\tilde{G}_\alpha$, as the extension by continuity of the collections of derivatives of functions $\mathrm{Lip}(1,G_\alpha)$ with respect to the metric $d_{\alpha,G}(x,y)$.

In order to describe the class $\mathrm{Lip}(1,\tilde{G}_\alpha)$ in an invariant way, let us define the mapping $i_\alpha:\tilde{G}_\alpha \to \bar{G}$ as the extension by continuity of the identical mapping $i_\alpha:G_\alpha \to G$, $i_\alpha(x) \Rightarrow x$, $x \in G_\alpha$. This extension exists by the inequality $|i_\alpha(x)-i_\alpha(y)|^\alpha \leq d_{\alpha,G}(x,y)$. The elements of the class $\mathrm{Lip}(1,\tilde{G}_\alpha)$ are the jets $\{f^{(j)}\}$, $0 \leq |j| \leq k$, consisting of the continuous functions defined on $\tilde{G}_\alpha$. The norm in the space $\mathrm{Lip}(1,\tilde{G}_\alpha)$ is

$$||f||_{\mathrm{Lip}(1,\tilde{G}_\alpha)} = \sum_{|j|\leq k} \{\sup |f^j(x)| + \sup |R_j(x,y)|/d_{\alpha,G}(x,y)^{(1-|j|)/\alpha}\}. \tag{2.1}$$

where the supremum is taken over all points $x,y \in G_\alpha$, $\alpha = l-k$,

$$f^{(j)}(x) = \sum_{|j|\le k} f^{(j+s)}(y)(i_\alpha(x)-i_\alpha(y))^s/s! + R_j(x,y).$$

It is easy to verify that at the points $x \in G_\alpha$ we have $f^{(j)}(x) = D^j f(x)$, $0 \le |j| \le k$.

Thus we have proved the following assertion.

Proposition 1. Let G be an arbitrary domain in R^n. There exists a natural isometry of the spaces $Lip(l,G_\alpha)$ and $Lip(l,\tilde{G}_\alpha)$: any element of $Lip(l,G_\alpha)$ extends uniquely to $\tilde{G}_\alpha$ with respect to the inner α-metric $d_{\alpha,G}(x,y)$ and is an element of $Lip(l,\tilde{G}_\alpha)$ on $\tilde{G}_\alpha$.

Further the set $\partial G_\alpha = \tilde{G}_\alpha \setminus G_\alpha$ will be called the α-boundary of the domain G. It is natural to consider the restrictions of the elements of $Lip(l,\tilde{G}_\alpha)$ to ∂G_α "as boundary values or traces of the functions of $Lip(l,G_\alpha)$ to the α-boundary of the domain G". To give a rigorous definition of the concept of the trace we denote by $Lip(l,\partial G_\alpha)$ the function space whose elements are jets $\{f^{(j)}\}$, $0 \le |j| \le k$, consisting of continuous functions defined on $(\partial G_\alpha, d_{\alpha,G})$ with the finite norm (2.1), where the supremum is taken over all points $x,y \in \partial G_\alpha$. The trace operator $tr_l: Lip(l,G_\alpha) \to Lip(l,\partial G_\alpha)$ is defined as the composition of the isometry $i: Lip(l,G_\alpha) \to Lip(l,\tilde{G}_\alpha)$ given in Proposition 1 and the restriction of the elements of $Lip(l,\tilde{G}_\alpha)$ to ∂G_α, which are contained in the class $Lip(l,\partial G_\alpha)$.

Let us formulate the above as the following Theorem.

Theorem 2. Let G be an arbitrary domain in R^n. There exist bounded trace operators definied by the continuity with respect to the metric $d_{\alpha,G}$

$$tr_l: W^l_\infty(G) \to Lip(l,\partial G_1), \quad \alpha = 1, \ l \in N,$$

$$tr_l: H^l_\infty(G) \to Lip(l,\partial G_\alpha), \quad \alpha = l-k, \ l \in (k,k+1), \ k = 0,1,\dots .$$

The characterization of the trace of the functions in the Sobolev or Nikol'skiĭ classes given by Theorem 2, is reversible.

Theorem 3. Let G be an arbitrary domain in R^n. There exist linear bounded extension operators

$$ext_k: Lip(l,\partial G_1) \to W^l_\infty(G), \quad k = l-1, \ l \in N,$$

$$ext_k: Lip(l,\partial G_\alpha) \to H^l_\infty(G), \ l \in (k,k-1), \ k = 0,1,\dots, \ \alpha = l-k,$$

in the sense that $tr_l \circ ext_k$ is the identical mapping.

Proof. The mapping $i_\alpha: (G_\alpha, d_{\alpha,G}) \to (G, |x-y|^\alpha)$ has the following property: if U is a convex set such that $U \subset G$, then for any points

$x,y \in U$ we have $|i_\alpha(x)-i_\alpha(y)|^\alpha = d_{\alpha,G}(x,y)$. Hence it follows that we have the inequality

$$d_{\alpha,G}(x,\partial G_\alpha) = |i_\alpha(x)-\partial G|^\alpha, \quad x \in G,$$

where $|i_\alpha(x)-\partial G|^\alpha = \inf\{|x-y|^\alpha : y \in \partial G\}$, $d_{\alpha,G}(x,\partial G_\alpha) = \inf\{d_{\alpha,G}(x,y): y \in G_\alpha\}$. This relation allows the space $G_\alpha = (G,d_{\alpha,G})$ to be partitioned by "Whitney cubes" in the same way as the domain G is partitioned by Whitney cubes with respect to the Euclidean metric in the Whitney extension theorem (see [6]). We underline that all the properties of the partition needed to realize the sketch of the proof of a Whitney type extension theorem (see [6]) with respect to the inner α-metric $d_{\alpha,G}$ hold.

3. Extension of differentiable function

An inner α-metric $d_{\alpha,G}(x,y)$, $\alpha \in (0,1]$, in a domain $G \subset R^n$ is called locally equivalent to the α-metric $d_\alpha(x,y) = |x-y|^\alpha$ if there exist numbers $r > 0$ and $M > 0$ such that for all points $x,y \in G$ with $|x-y|^\alpha < r$ the inequality

$$d_{\alpha,G}(x-y) \le M|x-y|^\alpha \tag{3.1}$$

is valid. Let us indicate by $d_{\alpha,G}(x,y) \approx_{loc} d_\alpha(x,y)$ that the metrics $d_{\alpha,G}(x,y)$ and $d_\alpha(x,y)$ are locally equivalent in the domain G, and let $M_\alpha(r)$ be the least constant for which inequality (3.1) is valid under the condition $|x-y|^\alpha < r$, $x,y \in G$. Let us also set

$$M_\alpha = \lim_{r\to 0} M_\alpha(r), \tag{3.2}$$

$$r_\alpha = \sup\{r: M_\alpha(r) < \infty\}. \tag{3.3}$$

From the definition it follows at once that the inner α-metric $d_{\alpha,G}(x,y)$ is locally equivalent to the α-metric $d_\alpha(x,y)$ in the domain G if and only of the value M_α is finite.

Let us consider two seminormed spaces $F(G)$ and $N(R^n)$ of functions defined on a domain $G \subset R^n$ respectively in Euclidean space R^n. The mapping ext: $F(G) \to N(R^n)$ is called an extension operator if $(\text{ext } f)|_G = f$ for all the functions $f \in F(G)$ and if

$$||\text{ext}|| = \sup\{||\text{ext } f||_{N(R^n)}/||f||_{F(G)} : f \in F(G)\} < \infty.$$

Theorem 4. If in the domain $G \subset R^n$ the inner α-metric $d_{\alpha,G}(x,y)$ is locally equivalent to the α-metric $d_\alpha(x,y) = |x-y|^\alpha$, $\alpha = 1-k$, $k < l \le k+1$, $l \in R_+$, $k = 0,1,\ldots,$ then there exists a linear bounded extension operator

$$\mathrm{ext}_k\colon W^l_\infty(G) \to W^l_\infty(R^n) \quad \text{for} \quad l \in N \quad \text{or} \quad \mathrm{ext}_k\colon H^l_\infty(G) \to H^l_\infty(R^n) \quad \text{for} \quad l \notin N.$$

For the norm of the extension operator the following upper estimate holds

$$\|\mathrm{ext}_k\| \le \gamma[\max(M_\alpha, 1/r_\alpha)]^{1/\alpha}.$$

where the constant γ does not depend on the domain G, and M_α, r_α are defined in (3.2)-(3.3).

Proof. According to Theorem 1 and Proposition 1 we have the coincidence of spaces

$$W^l_\infty(G) = \mathrm{Lip}(l, \tilde G_\alpha) \quad \text{for} \quad l \in N \quad \text{and} \quad H^l_\infty(G) = \mathrm{Lip}(l, \tilde G_\alpha) \quad \text{for} \quad l \notin N.$$

Therefore we shall prove the existence of a bounded extension operator

$$\mathrm{ext}_k\colon \mathrm{Lip}(l, \tilde G_\alpha) \to \mathrm{Lip}(l, R^n), \quad l \in R_+,\ k < l \le k+1,\ k = 0,1,\dots,$$

in the sense that $\mathrm{ext}_k f|_G = f|_G$ for an arbitrary function $f \in \mathrm{Lip}(l, \tilde G_\alpha)$, assuming that in the domain G we have $d_{\alpha,G}(x,y) \approx_{loc} d_\alpha(x,y)$.

To prove this we have to introduce the space $\mathrm{Lip}(l,F)$ where F is a closed set in R^n, $l \in R_+$, $k < l \le k+1$, $k = 0,1,\dots$ (see [6]). The elements of $\mathrm{Lip}(l,F)$ are jets $\{f^{(j)}\}$, $0 \le |j| \le k$, consisting of bounded functions such that $f^{(0)} = f$ and

$$f^{(j)}(x) = \sum_{|j+s| \le k} f^{(j+s)}(y)(x-y)^s/s! + R_j(x,y), \tag{3.4}$$

where

$$|f^{(j)}(x)| \le M, \quad |R_j(x,y)| \le M|x-y|^{l-|j|}, \quad x,y \in F,\ |j| \le k. \tag{3.5}$$

The norm of an element of $\mathrm{Lip}(l,F)$ is equal to the smallest value M for which the inequalities (3.5) hold. It is easy to verify that $f^{(j)}(x) = D^j f(x)$ in $x \in G$, where $G = \mathrm{Int}\, F$.

According to a Whitney type extension theorem (see [6]) there exists a linear bounded extension operator

$$\widetilde{\mathrm{ext}}_k\colon \mathrm{Lip}(l,F) \to \mathrm{Lip}(l,R^n),$$

whose norm does not depend on the closed set $F \subset R^n$. To prove Theorem 4, it is sufficient to establish an embedding

$$i\colon \mathrm{Lip}(1,\tilde{G}_\alpha) \to \mathrm{Lip}(1,\bar{G}) \tag{3.6}$$

such that for each function $f \in \mathrm{Lip}(1,\tilde{G}_\alpha)$ the relation $(if)(x) = f(x)$, $x \in G$, is true. The desired extension operator will be the coposition of the mappings in the diagram

$$\mathrm{Lip}(1,\tilde{G}_\alpha) \xrightarrow{\;i\;} \mathrm{Lip}(1,\bar{G}) \xrightarrow{\;\widetilde{\mathrm{ext}}_k\;} \mathrm{Lip}(1,R^n).$$

Thus $\mathrm{ext}_k = \widetilde{\mathrm{ext}}_k \circ i\colon \mathrm{Lip}(1,\tilde{G}_\alpha) \to \mathrm{Lip}(1,R^n)$.

To find an upper estimate for the norm of operator ext_k it is sufficient to know the contribution of the geometry of the space G_α to the norm of the operator i, since

$$||\mathrm{ext}_k|| \le ||i||\;||\widetilde{\mathrm{ext}}_k|| \tag{3.7}$$

and the norm of operator $\widetilde{\mathrm{ext}}_k$ does not depend on the domain G.

Now let us prove the boundedness of the embedding (3.6). Let $||f||_{\mathrm{Lip}(1,\tilde{G}_\alpha)}$ be equal to 1. Since $|f^{(j)}(x)| = |D^j f(x)| \le 1$, $x \in G$, $|j| \le k$, it remains to prove the inequalities

$$|R_j(x,y)| \le C|x-y|^{1-|j|}, \quad x,y \in G. \tag{3.8}$$

Indeed it follows from (3.8) that the functions $f^{(j)}$ extend by continuity to G, that for the extended functions the inequalities (3.5) hold at all points $x,y \in G$, and that the embedding (3.6) is bounded.

To prove (3.8) we denote by r_m the largest number such that for points $x,y \in G$, $0 < |x-y|^\alpha < r_m$, we have the inequality

$$d_{\alpha,G}(x,y) \le mM_\alpha |x-y|^\alpha, \quad m \in N, \tag{3.9}$$

Since $|f^{(j)}(x)| \le 1$, the estimate for the remainders $R_j(x,y)$, $0 \le |j| \le k$, in $x,y \in G$, $|x-y|^\alpha \ge r_m$,

$$|R_j(x,y)| \le [C(n,k)/r_m^{(1-|j|)/}]\,|x-y|^{1-|j|} \tag{3.10}$$

follows from the expansion (3.4). If $|x-y|^\alpha < r_m$, $x,y \in G$, then we use (3.9) to estimate the remainders $R_j(x,y)$ of a function $f \in \mathrm{Lip}(1,G)$ in (3.4). We obtain

$$|R_j(x,y)| \le d_{\alpha,G}(x,y)^{(1-|j|)/\alpha} \le (mM_\alpha)^{1/\alpha}|x-y|^{1-|j|}. \tag{3.11}$$

The inequalities (3.10) and (3.11) together constitute the estimate (3.8).

It follows that the embedding (3.6) is continuous. The upper bound of the embedding is obtained from $\widetilde{(3}.10)$ and (3.11) in $|j| = 0$. For the norm of the operator $ext_k = \widetilde{ext}_k \circ i$ we have the following estimate:

$$||ext_k|| \leq \gamma \max(mM_\alpha, 1/r_m)^{1/\alpha}, \tag{3.12}$$

where the constant γ depends only on n and l. To finish the proof of Theorem 4 it remains to minimize the inequality (3.12) over all r_m such that

$$d_{\alpha,G}(x,y) \leq mM_\alpha |x-y|^\alpha, \quad m \in N,$$

in $x,y \in G$, $0 < |x-y|^\alpha < r_m$.

In the case of the Sobolev spaces $(\alpha = 1)$ Theorem 4 was proved by another method in [7].

There exists a relation between the α-metrics resulting from the reciprocal Minkowski inequality (see [5]): for any numbers α and β, $0 < \alpha < \beta < 1$, the inequalities

$$d_{1,G}(x,y) \leq d_{\beta,G}(x,y)^{1/\beta} \leq d_{\alpha,G}(x,y)^{1/\alpha}, \tag{3.13}$$

where x,y are points from a domain G, hold. It follows that the relation $d_{\alpha,G}(x,y) \approx_{loc} d_\alpha(x,y)$ is stronger than the relation $\alpha_{\beta,G}(x,y) \approx_{loc} d_\beta(x,y)$. For given numbers α and β, $0 < \alpha < \beta \leq 1$, we can construct an example of a domain $G \subset R^2$ such that $d_{\beta,G}(x,t) \approx_{loc} d_{\beta,G}(x,y)$, but $d_{\alpha,G}(x,y) \not\approx_{loc} d_\alpha(x,y)$.

4. Necessary extension conditions for differentiable functions

The following result shows that the conditions of Theorem 4 are sharp.

Theorem 5. If extension operators

$$ext_1 : W_p^1(G) \longrightarrow W_p^1(R^n), \quad p > n \quad (\alpha = 1) \quad \text{or}$$

$$ext_1 : H_p^1(G) \longrightarrow H_p^1(R^n), \quad lp > n \quad (\alpha = 1 < 1),$$

exist then in the domain G the inner α-metric $d_{\alpha,G}(x,y)$ is locally equivalent to the α-metric $d_\alpha(x,y) = |x-y|^\alpha$ and

$$||ext_\alpha|| \geq \begin{cases} KM_\alpha^{1-n/\alpha p}, & \alpha p \in (n,\infty), \\ \max(M_\alpha, r_\alpha^{-1} + 1), & p = \infty, \end{cases}$$

where the constant K does not depend on the domain G.

The assertion of Theorem 5 in the case of a bounded domain and $\alpha = 1$, $p = \infty$ is proved in [7]. In [8] Theorem 5 is formulated assuming that $\alpha = 1$, $p \in (2,\infty]$ and that the domain of definition is a simply connected plane domain.

Proof. In the first part of this proof we shall construct the following function.

Lemma 1. For any two points $x,y \in G$, $G \subset R^n$, there exists a function $f \in Lip(1,G_1)$, $1 \in (0,1]$, with the properties:

1) $0 \leq f(u) \leq 1$, $u \in G$,; $f(x) = 1$, $f(y) = 0$;

2) $|f(u)-f(v)| \leq d_{1,G}(u,v)/d_{1,G}(x,y)$, $u,v \in G$;

3) the support of the function f is contained in the Euclidean ball with the radius $R = d_{1,G}(x,y)^{1/1}$ and the centre at the point x;

4) if $1 = 1$, then $|\nabla f| \leq d_{1,G}(x,y)^{-1}$.

Proof. It follows at once that the function defined by the formula $f(u) = d_{1,G}(u,G_x)/d_{1,G}(x,y)$, where $G_x = \{v \in G: d_{1,G}(x,v) \geq d_{1,G}(x,y)\}$ satisfies the conditions 1,2,4 of Lemma 1. If $u \notin B(x,R)$, then $d_{1,G}(x,y) = R^1 < |x-u|^1 \leq d_{1,G}(x,u)$. From here follows property 3.

In the hypothesis of Theorem 5 the extension operator ext_1 is bounded. Therefore for each function $f \in W^1_p(G)$ $(f \in H^1_\rho(G))$ we obtain the inequality

$$||ext_1 f||_{W^1_p(R^n)} \leq ||ext_1||\cdot||f||_{W^1_p(G)} \tag{4.1}$$
$$(||ext_1 f||_{H^1_p(R^n)} \leq ||ext_1||\cdot||f||_{H^1_p(G)}).$$

Let f be the function in Lemma 1 constructed with the points $x,y \in G$. Before substituting it in (4.1), we first make some observations on the case $1 \in (0,1)$.

If $h \in R^n$ is any vector, $h \geq R = d_{1,G}(x,y)^{1/1}$, then it is evident that $|f(u+h)-f(u)|/|h|^1<2/|h|^1 < 2/R^1$ and since $\text{supp } f \subset B(x,R)$ we have $\text{supp } \Delta_h f \subset B(x-h,R) \cup B(x,R)$. Thus in the case in question we have the estimate

$$||\Delta_h f||_{L_p(G)}/|h|^1 \leq 4^{n-p}/R^{1-n/p}.$$

In the case $|h| < R$ by Lemma 1 we have the inequality $|f(u+h)-f(u)|/$ $/|h|^1 < d_{1,G}(u+h,u)/|h|^1 d_{1,G}(x,y) = 1/R^1$ for the segment $[u,u+h] \subset G$, and the inclusion $\text{supp } \Delta_h f \subset B(x,2R)$. Therefore for $|h| < R$ the following inequality holds

$$\|\Delta_h f\|_{L_p(G)}/|h|^l \leq 2^{n/p}R^{n/p}/R^l = 2^{n/p}/R^{l-n/p}.$$

For $l = 1$ we obtain $\|\nabla_h f\|_{L_p(G)} \leq 1/R^{l-n/p}$.

Now substituting the function f in (4.1) we obtain the relation

$$\|\mathrm{ext}_l f\| \leq \|\mathrm{ext}_l\| \begin{cases} [\min(|G|^{1/p}, v_n^{1/p}R^{n/p}) + 4^{n/p}/R^{l-n/p}] & \text{if } p < \infty, \\ [1 + d_{l,G}(x,y)^{-1}] & \text{if } p = \infty. \end{cases} \tag{4.2}$$

where $v_n = |B(0,1)|$ and $|A|$ denotes the Lebesgue measure of a measurable set A. For the left side of (4.2) we have the inequality

$$\gamma/|x-y|^{l-n/p} \leq \begin{cases} \|\mathrm{ext}_l f\|_{W_p^l(R^n)} & \text{for } l = 1, \\ \|\mathrm{ext}_l f\|_{H_p^l(R^n)} & \text{for } l < 1, \end{cases} \tag{4.3}$$

where the constant γ does not depend on the choice of the points x, y. Now combining (4.2) and (4.,3) we obtain the estimate

$$\gamma/|x-y|^{l-n/p} \leq \|\mathrm{ext}_l\| \begin{cases} [\min(|G|^{1/p}, v_n^{1/p}R^{n/p}) + 4^{n/p}/R^{l-n/p}] & \text{if } p < \infty, \\ [1 + d_{l,G}(x,y)^{-1}] & \text{if } p = \infty, \end{cases} \tag{4.4}$$

The assertion of Theorem 5 now follows for $p = \infty$.

In the case of finite p we first prove that

$$\lim_{r \to 0} \sup \{d_{l,G}(x,y) : 0 < |x-y|^l < r,\ x, y \in G\} = 0. \tag{4.5}$$

Supposing the opposite we have two sequences $\{x_k\}$, $\{y_k\}$ of points of the domain G such that $|x_k - y_k| \to 0$ and $\underline{\lim}_{k\to\infty} d_{l,G}(x_k, y_k) > 0$. For every pair of points x_k, y_k, $k \in N$, we take a function satisfying the conditions of Lemma 2, and substitute it in (4.4). It follows that

$$\gamma(|x-y|^{l-n/p} < \|\mathrm{ext}_l\| [\min(|G|^{1/p}, v_n^{1/p}R_k^{n/p}) + 4^{n/P}/R_k^{l-n/p}] \tag{4.6}$$

(here $R_k = d_{l,G}(x_k, y_k)^{1/l}$). It the measure of the domain G is finite then the assertion of Theorem 5 follows from (4.6).

In the case $|G| = \infty$ instead of the sequence of functions f_k it is necessary to consider the sequence $g_k(u) = f_k(u)\varphi_k(u)$, $u \in G$, where $\varphi_k(u)$ is the restriction of the function $\varphi(u - x_k)$ to the domain G (here $\varphi \in C^\infty(R^n)$, $\varphi(u) = 1$ in $u \in B(0,1)$, $\varphi(u) = 0$ in $u \notin B(0,2)$). If $|x_k - y_k| < 2$ then $g_k(x_k) = 1$, $g_k(y_k) = 0$ and besides, since supp $g_k \subset B(x_k,2)$, the norms $\|g_k\|_{W^1_p(G)}$ ($\|g_k\|_{H^\ell_p(G)}$) are bounded by a constant C not depending on k. From here instead of (4.6) we obtain the inequality

$$\gamma/|x_k-y_k|^{1-n/p} < C\,\|\mathrm{ext}_1\|,$$

contradicting the assumption that $\lim_{k\to\infty} d_{1,G}(x_k,y_k) > 0$ for $\lim_{k\to\infty}|x_k-y_k| = 0$. Thus (4.5) is proved.

Now substituting R in (4.4) by $d_{1,G}(x,y)$ we obtain

$$\gamma < \|\mathrm{ext}_1\|\{|x-y|^{1-n/p}\min[|G|^{1/p},v_n^{1/p}d_{1,G}(x,y)^{n/1p}] + + 4^{n/p}/M_1(r)^{(1-n/p)/1}\}.$$

Passing here to the limit for $r \to 0$, we obtain

$$4^{-n/p}\gamma M_1^{(1-n/p)/1} \leq \|\mathrm{ext}_1\|, \tag{4.7}$$

where M_1 is defined in (3.2). The local equivalence of the metrics $d_{1,G}(x,y)$ and $d_1(x,y) = |x-y|^1$ in the domain G follows also from the inequality (4.7). The proof of Theorem 5 is finished.

REFERENCES

1. Aleksandrov, A.D.: Inner geometry of convex surfaces. Moscow: Gostekhizdat 1948. [German translation of the enlarged version: Die innere Geometrie der konvexen Flächen. Berlin: Akademie-Verlag 1955].
2. Sobolev, S.L.: Applications of functional analysis in mathematical physics. Izd. Leningrad. Univ. 1950 [English translation: Providence: Amer. Math. Soc. 1963].
3. Nikol'skiĭ, S.M.: Approximation of functions of several variables and embedding theorems. Moscow: Nauka 1969 [English translation: Berlin: Springer-Verlag 1975].
4. Besov, O.V., Il'in, V.P., Nikol'skiĭ, S.M.: Integral representation of functions and embedding theorems. Moscow: Nauka 1975 [English translation: Vol. 1-2, New York: John Wiley & Sons 1979].
5. Hardy, G.H., Littlewood, J.E., Polya, G.: Inequalities. Cambridge: Cambridge University Press 1934.
6. Stein, E.M.: Singular integrals and differentiability properties of functions. Princeton: Princeton University Press 1970.
7. Konovalov, V.N.: Description of the traces of some classes of functions of several variables. Preprint nr. 84.21. Kiev: Institute of mathematics 1984 (In Russian).

8. Vodop'yanov, S.K.: Geometrical properties of domains and lower bounds for the norm of the extension operator. Vsesojuziǐ seminar molodikh utchenikh po aktual'nim voprosam kompleksnogo analiza. Tashkent, Sept. 1985. Abstracts, pp. 23-24 (In Russian).

Remarks on Interpolation of Subspaces

Robert Wallstén

Matematiska Institutionen
Thunbergsvägen 3
S-752 38 Uppsala, Sweden

1. INTRODUCTION.

Let (A_0, A_1) be a compatible of Banach spaces and $M \subset A_0$ a closed subspace. Then, what is $(M, A_1)_{\theta,q}$? Is the inclusion $(M, A_1)_{\theta,q} \subset (A_0, A_1)_{\theta,q}$ closed, or, equivalently, are the norms on $(M, A_1)_{\theta,q}$ and $(A_0, A_1)_{\theta,q}$ equivalent on $(M, A_1)_{\theta,q}$?

A special case of a result by Grisvard and Seeley (see [4], p.320) says:

$$[L^2(\Omega),\ H_0^0(\Omega))]_\theta = \begin{cases} H^\theta(\Omega) & \theta < \frac{1}{2} \\ H_0^\theta(\Omega) & \theta > \frac{1}{2} \end{cases}$$

if $\Omega \subset \mathbb{R}^n$ is a bounded C^∞-domain and $H_0^\theta(\Omega) = \{u \in H^\theta(\Omega) | u|\partial\Omega = 0\}$, where H^θ means the usual Sobolev space. For $n = 1$ the same result has been obtained, with different methods, by Kellogg [2].

Triebel [3] has shown that for any positive integer n there are separable Hilbert spaces H_0, H_1 and H_2 such that $H_1 \subset H_0$, H_2 is a closed subspace of H_1, $\dim(H_1/H_2) = n$ and the norms of $(H_0, H_1)_{1/2,2}$ and $(H_0, H_2)_{1/2,2}$ are not equivalent to each other on H_2.

The case we consider in the following is when A_0 and A_1 are the L^1- and L^∞-spaces of complex-valued measurable functions on a σ-finite and non-atomic measure space. M is assumed to be of codimension 1.

It turns out there are $0 \le \theta_0 \le \theta_1 \le 1$ such that

$$(M,L^{\infty})_{\theta,q} = \begin{cases} \{f \in L^{\frac{1}{1-\theta}q} \mid \langle f,g\rangle = 0\} & \theta < \theta_0 \\ L^{\frac{1}{1-\theta}q} & \theta > \theta_1 \end{cases}$$

where $g \in L^{\infty}$ is such that $M = \mathrm{Ker}(g)$. We also show, by example, that it may happen that $\theta_0 < \theta_1$, in fact $\theta_0 = 0$ and $\theta_1 = 1$. Then, if $\theta_0 < \theta < \theta_1$, $(M,L^{\infty})_{\theta,q}$ is not a closed subspace of $(L^1,L^{\infty})_{\theta,q}$.

Remark. The notation used in the following coincides with that in Bergh-Löfström [1], where also the relevant facts from interpolation theory can be found.

2. The norm on $(M,A_1)_{\theta,q}$.

LEMMA 1. Let $\bar{A} = (A_0,A_1)$ be a compatible couple of Banach spaces such that $A_0 \cap A_1$ is dense in both A_0 and A_1. Let $M \subset A_0$ be a closed subspace of codimension 1. Then

$$\|a\|_{(M,A_1)_{\theta,q}} \sim \|a\|_{\bar{A}_{\theta,q}} + \Phi_{\theta,q}\left(\frac{|\langle a_0(t),g\rangle|}{K(t^{-1},g\,A_0^*,A_0^*)}\right)$$

where $g \in A_0^*$ and $a_0(t) \in A_0$ are such that $M = \mathrm{Ker}(g)$ and

$$K(t,a,A_0,A_1) \sim \|a_0(t)\|_{A_0} + t\|a_1(t)\|_{A_1}, \qquad t > 0.$$

Proof. We use the triangle inequality and obtain, for every $h \in A_0 \cap A_1$ with $\langle h,g\rangle \neq 0$,

$$K(t,a,M,A_1) \leq \left\|a_0(t) - \frac{\langle a_0(t),g\rangle}{\langle h,g\rangle}h\right\|_{A_0} + t\left\|a_1(t) + \frac{\langle a_0(t),g\rangle}{\langle h,g\rangle}h\right\|_{A_1} \leq$$

$$\leq \|a_0(t)\|_{A_0} + \frac{|\langle a_0(t),g\rangle|}{|\langle h,g\rangle|}\|h\|_{A_0} + t\|a_1(t)\|_{A_1} + t\frac{|\langle a_0(t),g\rangle|}{|\langle h,g\rangle|}\|h\|_{A_1} =$$

$$= \|a_0(t)\|_{A_0} + t\|a_1(t)\|_{A_1} + |\langle a_0(t),g\rangle|\left(\frac{|\langle h,g\rangle|}{\|h\|_{A_0} + t\|h\|_{A_1}}\right)^{-1}$$

$$\sim K(t,a,A_0,A_1) + |\langle a_0(t),g\rangle|\left(\frac{|\langle h,g\rangle|}{J(t,h)}\right)^{-1}.$$

Taking infimum for $h \in A_0 \cap A_1$ yields

$$K(t,a,M,A_1) \leq C\left(K(t,a,A_0,A_1) + \frac{|\langle a_0(t),g\rangle|}{K(t^{-1},g,A_0^*,A_1^*)}\right)$$

and half of the statement is proved by the definition of the norm on $(M,A_1)_{\theta,q}$.

The other half follows if we can prove

$$\frac{|\langle a_0(t),g\rangle|}{K(t^{-1},g,A_0^*,A_1^*)} \le C\cdot K(t,a,M,A_1) ,$$

since $K(t,a,A_0,A_1) \le K(t,a,M,A_1)$.

Fix t. If $\langle a_0(t),g\rangle = 0$ nothing needs to be proved. Assume $\langle a_0(t),g\rangle \ne 0$ and let b_0 and b_1 be such that

$$K(t,a,M,A_1) \sim \|b_0\|_{A_0} + t\|b_1\|_{A_1} .$$

Set $h = a_0(t) - b_0 = b_1 - a_1(t)$. Then $\frac{\langle a_0(t),g\rangle}{\langle h,g\rangle} h \in A_0 \cap A_1$ and

$$K(t,a,M,A_1) \sim \left\|a_0(t) - \frac{\langle a_0(t),g\rangle}{\langle h,g\rangle} h\right\|_{A_0} + t\left\|a_1(t) + \frac{\langle a_0(t),g\rangle}{\langle h,g\rangle} h\right\|_{A_1} \ge$$

$$\ge |\langle a_0(t),g\rangle| \left(\frac{|\langle h,g\rangle|}{\|h\|_{A_0} + t\|h\|_{A_1}}\right)^{-1} - (\|a_0(t)\|_{A_0} + t\|a_1(t)\|_{A_1}) \ge$$

$$\ge |\langle a_0(t),g\rangle| \left(\sup_{h\in A_0\cap A_1} \frac{|\langle h,g\rangle|}{J(t,h)}\right)^{-1} - C\cdot K(t,a,A_0,A_1) =$$

$$= |\langle a_0(t),g\rangle|\, K(t^{-1},g,A_0^*,A_1^*)^{-1} - C\cdot K(t,a,A_0,A_1)$$

and the desired inequality follows. ∎

3. The spaces $(L^1_{\perp g}, L^\infty)_{\theta,q}$.

In the sequel we assume that $g \in L^\infty$, and $M \subset L^{pq}$ will be denoted $L^{pq}_{\perp g}$ if $M = \mathrm{Ker}(g)$ and $g \in (L^{pq})^*$.

LEMMA 2. If the norms on $(L^1_{\perp g},L^\infty)_{\theta,q}$ and $(L^1,L^\infty)_{\theta,q}$ are equivalent to each other and $q < \infty$, then either

(i) $g \in \left(L^{\frac{1}{1-\theta}q}\right)^*$ and $(L^1_{\perp g},L^\infty)_{\theta,q} = L^{\frac{1}{1-\theta}q}_{\perp g}$ or

(ii) $g \notin \left(L^{\frac{1}{1-\theta}q}\right)^*$ and $(L^1_{\perp g}, L^\infty)_{\theta,q} = L^{\frac{1}{1-\theta}q}$.

Proof. Elementary Banach space arguments. ∎

LEMMA 3. $(L^{p_0q_0}_{\perp g}, L^{p_1q_1}_{\perp g})_{\theta,q} = \left[(L^{p_0q_0}, L^{p_1q_1})_{\theta,q}\right]_{\perp g}$.

Proof. This is a corollary of Theorem 1, p. 118 in [4]. ∎

THEOREM. There are $\theta_0, \theta_1 \in [0,1]$ such that

$$(L^1_{\perp g}, L^\infty)_{\theta,q} = \begin{cases} L^{\frac{1}{1-\theta}q}_{\perp g}, & \theta < \theta_0 \\ L^{\frac{1}{1-\theta}q}, & \theta > \theta_1 \end{cases}$$

and $(L^1_{\perp g}, L^\infty)_{\theta,q}$ is not a closed subspace of $L^{\frac{1}{1-\theta}q}$ if $\theta_0 < \theta < \theta_1$. θ_0 and θ_1 are given by

$$\theta_0 = \sup\{\theta \geq 0 \mid g \in L^{\frac{1}{\theta}\infty} \text{ and } \sup_{f \in L^{\frac{1}{1-\theta}}} \int_1^\infty g^*(t) \int_1^t \frac{s^{-\theta}}{K(s,g)} ds\, f^*(t)dt / \|f\|_{L^{\frac{1}{1-\theta}1}} < \infty\}$$

$$\theta_1 = \inf\{\theta \leq 1 \mid g \notin L^{\frac{1}{\theta}\infty} \text{ and } g^*(t) \int_t^\infty \frac{s^{-\theta}}{K(s,g)} ds \in L^{\frac{1}{\theta}\infty}\}.$$

Proof. By Lemma 3 and the reiteration theorem it follows that the set of θ for which $(L^1_{\perp g}, L^\infty)_{\theta,q} = L^{\frac{1}{1-\theta}q}_{\perp g}$ is an interval. Thus, the first statement holds with θ_0 equal to the right endpoint. Similarly, we we find θ_1. That $(L^1_{\perp g}, L^\infty)_{\theta,q} \subset L^{\frac{1}{1-\theta}q}$ is not closed for $\theta_0 < \theta < \theta_1$ is a consequence of Lemma 2. It remains to verify the expressions for θ_0 and θ_1.

To begin with, note that

$$K(t,f,L^1,L^\infty) = \int_0^t f^*(s)ds \sim \|f_0(t)\|_{L^1} + t\|f_1(t)\|_{L^\infty}, \quad \text{where}$$

$$f_0 = f \cdot \chi_{E^f_t}, \quad f_1 = f - f_0, \quad \{x \in X \mid |f(x)| > f^*(t)\} \subset E^f_t \subset \{x \in X \mid |f(x)| \geq f^*(t)\}$$

and $\mu(E^f_t) = t$. The existence of such an E^f_t is implied by the Hausdorff maximality theorem.

By Lemma 1 and the fact that $(L^1, L^\infty)_{\theta,q} = (L^1, \overline{L^1 \cap L^\infty})_{\theta,q}$, where $\overline{L^1 \cap L^\infty}$ means the L^∞-closure, we have to estimate

$$\Phi_{\theta,q}\left(\frac{t}{K(t,g,L^1,L^\infty)} \left| \int_{E_t^f} fg \right|\right)$$

and it is no restriction to set $q = 1$. It will be convenient to deal with the cases $g \in L^1$ and $g \notin \overline{L^1 \cap L^\infty}$ separately.

If $g \in L^1$ and $f \in (L^1_{\perp g}, L^\infty)_{\theta,q}$ then Lemma 2 implies $\langle f,g \rangle = 0$. We have, by Hölder's inequality,

$$\frac{t}{K(t,g)} \left| \int_{E_t^f} fg \right| \leq \frac{t}{C \cdot t} \|g\|_{L^\infty} K(t,f) = C \cdot K(t,f), \qquad t < 1,$$

and

$$\frac{t}{K(t,g)} \left| \int_{E_t^f} fg \right| = \frac{t}{K(t,g)} \left| \int_{(E_t^f)^c} fg \right| \leq \frac{t}{K(1,g)} f^*(t) \|g\|_{L^1} \leq C \cdot K(t,f),$$

$$t \geq 1.$$

If $g \notin \overline{L^1 \cap L^\infty}$, then

$$\frac{t}{K(t,g)} \left| \int_{E_t^f} fg \right| \leq \frac{1}{C \cdot t} \|g\|_{L^\infty} K(t,f) = C \cdot K(t,f), \qquad t > 0.$$

Hence $(L^1_{\perp g}, L^\infty)_{\theta,q} = L^{\frac{1}{1-\theta}q}_{\perp g}$ and $(L^1_{\perp g}, L^\infty)_{\theta,q} = L^{\frac{1}{1-\theta}q}$ hold respectively for $0 < \theta < 1$ and $1 \leq q \leq \infty$, in accord with the theorem. We now turn to the general case.

Assume $g \notin L^{\frac{1}{\theta}\infty}$ and $g^*(s) \int_s^\infty \frac{t^{-\theta}}{K(t,g)} dt \in L^{\frac{1}{\theta}\infty}$. Then, using Fubini's theorem, we have

$$(\Phi_{\theta,1} \frac{t}{K(t,g)} \left| \int_{E_t^f} fg \right|) \leq \int_0^\infty \frac{t^{1-\theta}}{K(t,g)} \int_0^t f^*(s) g^*(s) ds \frac{dt}{t} =$$

$$= \int_0^\infty g^*(s) \int_s^\infty \frac{t^{-\theta}}{K(t,g)} dt\, f^*(s) ds \leq C \|f\|_{L^{\frac{1}{1-\theta}1}}.$$

By Lemma 2,

$$(L^1_{\perp g}, L^\infty)_{\theta,q} = L^{\frac{1}{1-\theta}q}.$$

The condition $g^*(s)\int_s^\infty \frac{t^{-\theta}}{K(t,g)}\,dt \in L^{\frac{1}{\theta}\infty}$ is also necessary. To see this pick $f \in (L^1_{1g}, L^\infty)_{\theta,1}$ and let

$$M_{n,k} = E^g_{\frac{k+1}{2^n}} - E^g_{\frac{k}{2^n}} ,$$

where n and k range over the non-negative integers. Define

$$f_n(x) = \begin{cases} f^*(\frac{k+1}{2^n}) & x \in M_{n,k} \\ 0 & \text{otherwise} \end{cases}$$

and

$$f^{**}(x) = \frac{\overline{g(x)}}{|g(x)|} \lim_{n\to\infty} f_n(x) .$$

Then $(f^{**})^* = f^*$ and

$$\Phi_{\theta,1}\left(\frac{t}{K(t,g)} \Big| \int_{E_t^{f^{**}}} f^{**}g \Big| \right) = \int_0^\infty \frac{t^{1-\theta}}{K(t,g)} \int_0^t f^*(s)g^*(s)ds \frac{dt}{t} =$$

$$= \int_0^\infty g^*(s) \int_s^\infty \frac{t^{-\theta}}{K(t,g)}\,dt\, f^*(s)ds,$$

which is bounded by $C\|f\|_{L^{\frac{1}{1-\theta}}}$ only if the non-increasing function $g^*(s)\int_s^\infty \frac{t^{-\theta}}{K(t,g)}\,dt \in L^{\frac{1}{\theta}\infty}$, because the L^{pq}-spaces are rearrangement invariant. The formula for θ_1 follows.

Assume now $g \in L^{\frac{1}{\theta}\infty}$ and rewrite

$$_{\theta,1}(\varphi(t)) = \int_0^1 t^{-\theta}\varphi(t)\frac{dt}{t} + \int_1^\infty t^{-\theta}\varphi(t)\frac{dt}{t} .$$

The first term can be **estimated** as in the case $g \in L^1$. If $f \in (L^1_g, L^\infty)_{\theta,1}$ then, by Lemma 2, $\langle f,g\rangle = 0$. We have

$$\frac{1}{K(t,g)}\Big|\int_{E_t^f} fg\Big| = \frac{1}{K(t,g)}\Big|\int_{(E_t^f)^c} fg\Big| = \frac{1}{K(t,g)}\Big|\int_{(E_t^f)^c\cap(E_t^g)^c} fg + \int_{(E_t^f)^c\cap E_t^g} fg\Big|$$

$$\leq \frac{1}{K(t,g)}\Big|\int_{(E_t^f)^c\cap(E_t^g)^c} fg\Big| + f^*(t)$$

and

$$\int_1^{\infty} \frac{t^{1-\theta}}{K(t,g)} \left| \int_{(E_t^f)^c \cap (E_t^g)^c} f\, g \right| \frac{dt}{t} \le \int_1^{\infty} \frac{t^{-\theta}}{K(t,g)} \int_t^{\infty} f^*(s) g^*(s)\, ds\, dt =$$

$$= \int_1^{\infty} g^*(s) \int_1^{s} \frac{t^{-\theta}}{K(t,g)}\, dt\, f^*(s)\, ds\ .$$

By Lemma 2 $\quad (L^1_{\perp g}, L^{\infty})_{\theta,q} = L^{\frac{1}{1-\theta}\, q}_{\perp g}$ if this integral is bounded by $C \cdot \|f\|_{L^{\frac{1}{1-\theta}\, 1}}$ or, equivalently,

$$\sup_{f \in L^{\frac{1}{1-\theta}\, 1}} \frac{\int_1^{\infty} g^*(s) \int_1^{s} \frac{t^{-\theta}}{K(t,g)}\, dt\, f^*(s)\, ds}{\|f\|_{L^{\frac{1}{1-\theta}\, 1}}} < \infty\ .$$

This condition is also necessary, if $(L^1_{\perp g}, L^{\infty})_{\theta,1} = L^{\frac{1}{1-\theta}\, 1}_{\perp g}$.

To see this, rearrange f as before and let

$$\alpha = \frac{\left| \int_{(E_1^g)^c} f^{**}\, g \right|}{\int_0^1 g^*(s)\, ds}\ .$$

If $\alpha < f^*(1)$ then there is $t_0 < 1$ such that

$$\int_{E^g_{t_0}} f^{**} g = \int_{(E^g_{t_0})^c} f^{**} g\ .$$

Set

$$I(x) = \begin{cases} -1 & x \in E^g_{t_0} \\ 1 & x \notin E^g_{t_0} \end{cases}$$

Then $\|f^{**} \cdot I\|_{L^{\frac{1}{1-\theta}\, 1}} = \|f\|_{L^{\frac{1}{1-\theta}\, 1}}$, $\langle f^{**} \cdot I, g\rangle = 0$ and

$$\int_1^{\infty} \frac{t^{1-\theta}}{K(t,g)} \left| \int_{(E_t^{f^{**}\cdot I})^c} f^{**} \cdot I \cdot g \right| \frac{dt}{t} = \int_1^{\infty} g^*(s) \int_1^{s} \frac{t^{-\theta}}{K(t,g)}\, dt\, f^*(s)\, ds\ .$$

If $\alpha \ge f^*(1)$, let

$$f'(x) = \begin{cases} -\alpha \cdot \dfrac{\overline{g(x)}}{|g(x)|} & x \in E_1^g \\ f^{**}(x) & x \in E_1^g \end{cases}$$

Then $\|f'\|_{L^{\frac{1}{1-\theta}1}} \leq C \cdot \|f\|_{L^{\frac{1}{1-\theta}1}}$, $\langle f', g\rangle = 0$ and

$$\int_1^{\infty} \frac{t^{-\theta}}{K(t,g)} \left| \int_{(E_t^{f'})^c} f'g \right| \frac{dt}{t} = \int_1^{\infty} g^*(s) \int_1^s \frac{t^{-\theta}}{K(t,g)} dt\, f^*(s) ds .$$

The formula for θ_0 follows. ∎

5. Examples.

I. Let $X = (0,\infty)$ with Lebesgue measure and let f and g be positive and non-increasing. Then

$$\Phi_{\theta,q}\left(\frac{t^{1-\theta}}{K(t,g)} \left| \int_{E_t^f} f g \right|\right) = \left(\int_0^{\infty}\left(\frac{t^{1-\theta}}{K(t,g)} \int_0^t f^*(s) g^*(s) ds\right)^q \frac{dt}{t}\right)^{1/q} \geq$$

$$\geq C\left(\sum_{\nu\in\mathbb{Z}}\left(\frac{2^{\nu(1-\theta)}}{K(2^{\nu},g)} \int_0^{2^{\nu}} f^*(s) g^*(s) ds\right)^q\right)^{1/q} \geq$$

$$\geq C\left(\sum_{\nu\in\mathbb{Z}}\left(\frac{2^{\nu(1-\theta)}}{K(2^{\nu},g)} \sum_{\mu\leq\nu} f^*(2^{\mu}) g^*(2^{\mu}) 2^{\mu}\right)^q\right)^{1/q} \geq$$

$$\geq C\left(\sum_{\nu\in\mathbb{Z}} \frac{2^{\nu q(1-\theta)}}{K(2^{\nu},g)^q} \sum_{\mu\leq\nu} f^*(2^{\mu})^q g^*(2^{\mu})^q 2^{\mu q}\right)^{1/q} =$$

$$= C\left(\sum_{\mu\in\mathbb{Z}} \sum_{\nu\geq\mu} \frac{2^{\nu q(1-\theta)}}{K(2^{\nu},g)^q} f^*(2^{\mu})^q g^*(2^{\mu})^q 2^{\mu q}\right)^{1/q} \geq$$

$$\geq C\left(\sum_{\mu\in\mathbb{Z}} \int_{2^{\mu}}^{\infty} \left(\frac{s^{1-\theta}}{K(s,g)}\right)^q \frac{ds}{s} f^*(2^{\mu})^q g^*(2^{\mu})^q 2^{\mu q}\right)^{1/q} \geq$$

$$\geq C\left(\int_0^{\infty} g^*(t)^q t^{\theta q} \int_t^{\infty} \left(\frac{s^{1-\theta}}{K(s,g)}\right)^q \frac{ds}{s} \left(t^{1-\theta} f^*(t)\right)^q \frac{dt}{t}\right)^{1/q} .$$

Let

$\theta = \frac{1}{2}$, $q = 2$ and $g(t) = (1+t)^{-\frac{1}{2}}$. If the inclusion $(L^1_{\perp g}, L^{\infty})_{\frac{1}{2},2} \subset L^2$

were closed, then, since $g \notin (L^2)^* = L^2$, we would have $(L^1_{\perp g}, L^\infty)_{\frac{1}{2},2} = L^2$ according to Lemma 2. But this is impossible, for the last integral is infinite.

For $\theta \neq \frac{1}{2}$ we have by our theorem $\theta_0 = \theta_1 = \frac{1}{2}$ and

$$(L^1_{\perp g}, L^\infty)_{\theta,q} = \begin{cases} L^{\frac{1}{1-\theta}q}_{\perp g} & \theta < \frac{1}{2} \\ L^{\frac{1}{1-\theta}q} & \theta > \frac{1}{2} \end{cases}$$ ∎

II. Let $X = (0,\infty)$ with Lebesgue measure, take $t_1 > e$, let $t_{n+1} = e^{t_n}$ for $n \geq 1$ and set

$$g(t) = \max\{(\log t_n)^{-1} \mid t_n > t\}.$$

Then $g \notin L^{\frac{1}{\theta}\infty}$ for any $0 < \theta < 1$, hence $\theta_0 = 0$. It remains to check the condition

$$\sup t^\theta \; g^*(t) \int_t^\infty \frac{s^{-\theta}}{K(s,g)} ds < \infty .$$

Simple estimates give $K(t_n, g) \sim t_n/\log t_n$. Let $t = \frac{1}{2} t_n$. Then

$$t^\theta g^*(t) \int_t^\infty \frac{s^{-\theta}}{K(s,g)} ds \geq C\, t_n^\theta \cdot \frac{1}{\log t_n} \int_{t_n}^{\frac{t_n^2}{\log t_n}} \frac{s^{-\theta}}{\frac{t_n}{\log t_n} + \frac{s}{t_n}} ds \geq$$

$$\geq C\, t_n^{\theta-1} \int_{t_n}^{\frac{t_n^2}{\log t_n}} s^{-\theta} ds \geq C\left(\frac{t_n}{\log t_n}\right)^{1-\theta} \to \infty \quad \text{as} \quad n \to \infty .$$

Thus $\theta_1 = 1$ and it follows that for $0 < \theta < 1$ $(L^1_{\perp g}, L^\infty)_{\theta,q}$ is not a closed subspace of $L^{\frac{1}{1-\theta}q}$. ∎

Remark. In the case of higher co-dimension, statements corresponding to Lemmas 1-3 hold. However, explicit calculation for arbitrary annihilators are out of question. On the other hand, by considering annihilators with disjoint supports, examples analogous to I, but with several jumps and infinite co-dimension at the jump points, can be found.

References

1. J. Bergh and J. Löfström, Interpolation Spaces. Springer-Verlag, Berlin 1976.

2. R.B. Kellogg, Interpolation between subspaces of a Hilbert space. Technical Report, Univ. of Maryland, 1971.

3. H. Triebel, Allgemeine Legendresche Differentialoperatoren. Ann. Scuola Norm. Pisa 24(1970).

4. H. Triebel, Interpolation Theory, Function Spaces, Differential Operators. VEB Deutscher Verlag der Wissenschaften, Berlin 1978.

SPECTRAL ANALYSIS IN SPACES OF CONTINUOUS FUNCTIONS

Yitzhak Weit

Department of Mathematics
University of Haifa

The problem of spectral analysis in spaces of continuous functions is intimately connected with problems in complex variable function theory.

The "two-circles" theorem of Zalcman [12] and similar results concerning the characterization of analytic and harmonic functions, Pompeiu's problem [2,5,12,13,14] are based on Schwartz's theorem on mean periodic functions of one variable [8].

Recently, C.A. Berenstein and R. Gay obtained a local version of the "two-circles" theorem [3]. Using, basically, their methods we may introduce the following localization of the results in [11]. Here, $C(D)$ denotes the space of continuous functions in the open unit disc and $\omega_n = \exp[2\pi i/(n+1)$.

Theorem: *The function* $f \in C(D)$ *satisfies*

$$\sum_{k=0}^{n} \omega_{n+1}^{k} f(z + \omega_{n+1}^{k} \xi) = 0$$

for $|\xi| = r$ *for some fixed* $r < \frac{1}{2}$ *and for all* z *with* $|z| < 1-r$, *if and only if,* f *is a polynomial (in* z*) of degree not exceeding* n.

For harmonic polynomials we have

Theorem: *The function* $f \in C(D)$ *satisfies*

$$\frac{1}{n+1} \sum_{k=0}^{n} f(z + \omega_n^k \xi) = f(z)$$

for $|\xi| = r$ *for some* $r < \frac{1}{2}$ *and for all* z *with* $|z| < 1-r$, *if and only if,* f *is a harmonic polynomial of degree not exceeding* n.

The problem of spectral analysis was investigated on symmetric spaces [1,4,9] and on some locally compact groups [6,7,10].

By Gurevich [7] spectral analysis fails to hold for $\mathbb{R}^N$, $N \geq 2$. However, the following is an analogue of Schwartz's theorem for the motion group $M(3) = SO(3) \rtimes \mathbb{R}^3$. Let Γ denote the set of all matricial coefficients of the irreducible unitary spherical representations of $SO(3)$.

Theorem: *Every two-sided translation-invariant closed subspace of* $C(M(3))$ *contains a function of the form* $\varphi(\sigma,x) = (\Sigma m_{i,j}) \otimes e_\lambda(\sigma,x)$ *where* $m_{i,j} \in \Gamma$, $\lambda \in \mathbb{C}^3$ *and* e_λ *is the exponential* $e_\lambda(x) = e^{\lambda \cdot x}$. *(Here* $\lambda \cdot x = \Sigma_{i=1}^{3} \lambda_i x_i$ *where* $\lambda = (\lambda_i) \in \mathbb{C}^3$, $x = (x_i) \in \mathbb{R}^3$).

Proof: Let V denote the closed subspace generated by the two-sided translates of a function $f \in C(M(3))$, $f \neq 0$. V contains all functions h such that

$$h(\sigma,x) = f(\sigma'\sigma\sigma'', \sigma'x + \sigma w + v)$$

where $\sigma',\sigma'' \in SO(3)$ and $w,v \in \mathbb{R}^N$. By the Peter-Weyl theorem, and by the completeness of Γ there exists a matrix coefficient $m_{i,j} \in \Gamma$ of an irreducible unitary spherical representation π of $SO(3)$, such that the function g defined as

$$g(\sigma,x) = \int_{SO(3)} f(\sigma\sigma',x)\overline{m_{i,j}}(\sigma')d\sigma' = \sum_{i=1}^{r} m_{i,j} \otimes g_i(\sigma,x)$$

is non-zero and belongs to V, where

$$g_i(x) = \int_{SO(3)} f(\sigma',x)\overline{m_{i,j}}(\sigma')d\sigma' \quad , \quad i = 1,2,\ldots,r$$

and r is the rank of π.

V contains with g its left translates of the form

$$g(\sigma'\sigma,x) = \sum_{i=1}^{r} m_{i,j} \otimes g_i(\sigma'\sigma,x)$$

$$= \sum_{i=1}^{r} m_{i,j} \otimes \left(\sum_{s=1}^{r} m_{s,i}(\sigma')g_s \right)(\sigma,\sigma'x) \ .$$

Let H_i, $i = 1,2,\ldots,r$ denote the solid spherical harmonics related to the spherical representation π.

It follows that $\tilde{g} \in V$ where

$$\tilde{g}(\sigma,x) = \int_{SO(3)} g(\sigma'\sigma,x)d\sigma'$$

$$= \sum_{i=1}^{r} m_{i,j} \otimes P_i(\sigma,x)$$

and P_i are functions of the form

$$P_i(x) = \sum_{k=1}^{r} f_k(|x|)\overline{H}_k\left(\frac{x}{|x|}\right) .$$

To ensure that $\tilde{g} \neq 0$ one notices that provided $g_1 \neq C$ there exists $r > 0$ such that $g_1 * (H_j \mu_r)$ is not identically zero. (Here, μ_r denotes the normalized Lebesgue measure of $\{x \in \mathbb{R}^N : \|x\| = r\}$. Thus we have to replace g by a suitable left translate $g_\tau(\sigma,x) = g(\sigma,x+\tau)$. To complete the proof we need the following result which is proved for $N = 2$ in [10]:

Lemma: *Let* $f \in C(\mathbb{R}^N)$, $f \neq 0$ *where*

$$f(x) = \sum_{k=1}^{r} f_k(|x|)\overline{H}_k\left(\frac{x}{|x|}\right) .$$

Then the translation-invariant closed subspace generated by f contains an exponential function.

To this end, by using a vector-valued spectral analysis result for $P = (P_1,P_2,\ldots,P_r)$ [10], we complete the proof of the theorem.

Remark: The problem of spectral analysis in $C(\mathbb{R}^N)$ under the action of the Poincaré group is open. As pointed out by J. Peetre this problem is connected to the question of irreducibility of the representation of the group on eigen-spaces of the d'Alembertian $\Box = \partial_1^2 - \partial_2^2 - \partial_3^2 - \partial_4^2$.

References

1. C.A. Berenstein, Spectral synthesis on symmetric spaces, to appear in Contemporary Mathematics.
2. C.A. Berenstein, An inverse spectral theorem and its relation to the Pompeiu problem, J. Analyse Math. 37 (1980), 128-144.

3. C.A. Berenstein and R. Gay, A local version of the two-circles theorem, to appear.

4. C.A. Berenstein and L. Zalcman, Pompeiu's problem on symmetric spaces, Comment. Math. Helvetici 55 (1980), 593-621.

5. L. Brown, B.M. Schreiber and B.A. Taylor, Spectral synthesis and the Pompeiu problem, Ann. Inst. Fourier 23 (1973), 125-154.

6. L. Ehrenpreis and F.I. Mautner, Some properties of the Fourier transform on semisimple Lie groups II, Trans. Amer. Math. Soc. 84 (1957), 1-55.

7. D.I. Gurevich, Counterexamples to a problem of L. Schwartz, Funct. Anal. Appl. 197 (1975), 116-120.

8. L. Schwartz, Théorie générale des fonctions moyenne-périodiques, Ann. of Math. 48 (1947), 857-928.

9. A. Wawrzrynczyk, Spectral analysis and synthesis on symmetric spaces, preprint.

10. Y. Weit, On Schwartz's theorem for the motion group, Ann. Inst. Fourier 30 (1980), 91-107.

11. Y. Weit, A characterization of polynomials by convolution equations, J. London Math. Soc. (2) 23 (1981), 455-459.

12. L. Zalcman, Analyticity and the Pompeiu problem, Arch. Rational Mech. Anal. 47 (1972), 237-254.

13. L. Zalcman, Mean values and differential equations, Israel J. Math. 14 (1973), 339-352.

14. L. Zalcman, Offbeat integral geometry, Amer. Monthly 87 (1980), 161-175.

LIPSCHITZ SPACES AND INTERPOLATING POLYNOMIALS ON SUBSETS OF EUCLIDEAN SPACE

Peter Wingren
Department of Mathematics
University of Umeå
S-901 87 Umeå, Sweden

0. Introduction

This paper deals with definitions of Lipschitz spaces by means of interpolating polynomials.

Let k be a non-negative integer. Let $\Lambda_\alpha(\mathbb{R}^n)$, $k<\alpha\leq k+1$, be the Lipschitz space which is the subspace of $C^k(\mathbb{R}^n)$ consisting of bounded functions with bounded partial derivatives and where the kth order derivatives have second differences which fulfil the Zygmund-condition $|\Delta_h^2| \leq M\cdot|h|^{\alpha-k}$, where M is a constant and $h \in \mathbb{R}^n$. This function space may be defined for subsets of $\mathbb{R}^n$ (see [10, p.51]); if F is any closed subset of $\mathbb{R}^n$, the space $\Lambda_\alpha(F)$ consists of families of functions where the first function in the family, together with the others, corresponds to a function and its partial derivatives up to order k respectively. A family belongs to $\Lambda_\alpha(F)$ if it can be approximated locally, in a special way, by polynomials of degree at most the integerpart of α.

If we, in the definition of $\Lambda_\alpha(\mathbb{R}^n)$, substitute the second difference Δ_h^2 by the first difference Δ_h then we get the Lipschitz space $\mathrm{Lip}(\alpha,\mathbb{R}^n)$. Correspondingly, if we, in the generalized definition of $\Lambda_\alpha(F)$, change the maximal degree of the polynomials from the integerpart of α to the greatest degree less than α then we get the space $\mathrm{Lip}(\alpha,F)$.

The closed set F is not arbitrary in this work, but belongs to a class of sets, which is of a very general type in relation to the considered problems. We present, for sets from this class, an alternative definition of $\Lambda_{k+1}(F)$ by means of interpolating polynomials; see the Theorem in Section 1. The advantage of the alternative definition, given in the Theorem, is that instead of a fundamentally complicated and sometimes unwieldy characterization we obtain an uncomplicated and easily handled one.

The starting-point for our investigation is a result in [7] obtained by A. Jonsson and H. Wallin. In [7, p.179] $\Lambda_\alpha(F)$, $\alpha=1$, is defined by interpolating polynomials in the case where $F \subset \mathbb{R}^2$ has the following properties:
There are positive constants c_1, c_2 and v, $0<v<\pi/3$, so that, for every $x_0 \in F$ and every $\nu=1,2,\ldots$, there are points $x_1, x_2 \in F$ such that

$$c_1 2^{-\nu} \leq |x_i - x_0| \leq c_2 \cdot 2^{-\nu},\ i=1,2$$

and the angles in the triangle, determined by x_0, x_1 and x_2, are all greater than v.

We generalize the above class of sets to any dimension and the Definition in [7] to any integer α and any dimension. Since the result in [7] and our generalization leads to the solution of an extension problem, we end the Introduction with a few comments concerning this problem.

By forming equivalence classes of families in $\mathrm{Lip}(\alpha,F)$ and $\Lambda_\alpha(F)$, so that a class consists of families which have the same first functions one gets a factor space; see [10, p.69]. When analysing the problem of the existence of a continuous and linear extension operator the above formation is natural. Corresponding to the two formations, i.e. the Lipschitz space and the generated factor space, are two extension problems, one for spaces of families and one for spaces of equivalence classes of families, or expressed in another way: One can look for an extension operator which is linear in families, or one which is linear in the first function of the family. Observe that these two problems, "the family problem" and "the function problem" coincide trivially when the family, by definition, consists of one function, i.e. when $\alpha \leq 1$, but also non-trivially when the first function in the family uniquely determines the others. In all other cases the problems are different.

In case of $\mathrm{Lip}(\alpha,F)$, every α, the family-problem is solved. The solution is in principle the classical Whitney's extension theorem; see [13, pp.63-89]. In the extension operator Whitney uses polynomials which are determined by interpolation in one point. The corresponding function-problem in the general case (F and α arbitrary) is only solved for dimension one. This solution is also a Whitney solution; see [14].

The space $\Lambda_\alpha(F)$ is identical with $\mathrm{Lip}(\alpha,F)$ if α is non-integer and consequently we only comment on the integer case. If α is an integer then the general case (arbitrary F and α) is not even solved for families and it is likely that considerable problems are involved in the construction. An easily observable difficulty, compared with the non-integer case, is that the number of functions in the function families is less than the number needed to determine a one-point-interpolating polynomial of degree $k+1$, (as Whitney does in the non-integer case). However the problem is solved by A. Jonsson (see [4]) for $F \subset \mathbb{R}^1$ and α any non-negative integer; see also [2] and [11] for similar results. Recently Brudnyi and Shvartsman announced (see [1]) that they had succeeded in the case $\alpha=1$ and n (the dimension) arbitrary, i.e. in the case when the family-problem and the function-problem trivially coincide. Lastly, as we hinted at above, A. Jonsson and H. Wallin gave a relatively simple solution for the case $\alpha=1$, $n=2$ and F from the class of sets previously described. Sets from this class give the nontrivial case when the family-problem and function-problem coincide. Our findings also generalize the result in [7], in the case of linear extension operators, to any dimension and any integer index α.

Acknowledgement: I am grateful to Professor Hans Wallin for suggesting the problem and providing valuable criticism.

1. Notation

The usual multiindex notation is used. c is a positive constant whose value may differ each time it appears. $B(x_0,r) = \{x \in \mathbb{R}^n: d(x_0,x) \leq r\}$. $\mathrm{aff}\, A$ is the affine hull of A. The convex hull of A is denoted $\mathrm{conv}\, A$. k is always a non-negative integer. F is always a closed set. The notations, which are an integral part of the method of proving the Theorem, are introduced later; see Proofs.

2. Definitions and results

In this report we generalize the result in [7, p.179], which were reproduced in the Introduction. In [7] the result is proved for $\alpha=1$ and $n=2$, and we generalize it in the sense that we obtain the same result for any non-negative integer α and any n. The generalization is contained in the Theorem below, which is the main outcome in our investigation. The Corollary following the Theorem is an important application. The proofs can be found in the next section. Before we formulate our Theorem, a definition is presented.

DEFINITION 1. A closed subset F of $\mathbb{R}^n$ preserves Markov's inequality if the following condition holds for all positive integers k: For all polynomials P of degree at most k and all balls $B(x_0,r)$ with $x_0 \in F$ and $0<r\leq 1$, we have

$$\max_{F\cap B}|\text{grad } P| \leq c\cdot r^{-1}\max_{F\cap B}|P| \tag{1}$$

with a constant c depending only on F, n and k.

The notation "F preserves Markov's inequality" was first introduced in [8] and [9]. We are now ready to formulate the Theorem.

THEOREM. Let $F \subset \mathbb{R}^n$ preserve Markov's inequality. Then $f \in \Lambda_{k+1}(F)$ if, and only if, for some positive constant M

1) $|f(x)| \leq M, \quad x \in F$ (2)

and

2) if $x_0 \in F$ and r is a real number such that $0<r\leq 1$, then there is a unique polynomial, P_I, which is determined by interpolation and which is of degree at most $k+1$ such that

$$|f(x)-P_I(x)| \leq M\cdot r^{k+1}, \quad x \in B(x_0,r)\cap F. \tag{3}$$

The points at which interpolation takes place are independent of f.

COROLLARY. Let $F \subset \mathbb{R}^n$ preserve Markov's inequality. There then exists a continuous and linear extension operator from $\Lambda_{k+1}(F)$ into $\Lambda_{k+1}(\mathbb{R}^n)$.

At the end of the proof of the Theorem in Section 3 we estimate the norm of the extension operator. The norm depends on $(c^*)^{-(k+1)}$, where c^* is a constant which has to do with the geometric properties of F and can be defined in the following way. If $x_0 \in F$, $0<r\leq 1$, and a_i, $i=0,1,\dots,n$ are $n+1$ affinely independent points in $B(x_0,r)\cap F$ and the inscribed ball in $A = \mathrm{conv}\{a_i\colon i=0,1,\dots,n\}$ has radius $r'(a_0,\dots,a_n)$ then

$$c^* = \inf_{x_0\in F,0\leq r<1} \quad \sup_{\{a_0,\dots,a_n\}\subset F\cap B(x_0,r)} \frac{r'(a_0,\dots,a_n)}{r} . \qquad (4)$$

From Proposition 2 (see Section 3) it follows that F preserves Markov's inequality if, and only if, $c^*>0$.

Finally we also indicate the limits of the method of proof by an example in Remark 3 in Section 3.

It is now necessary to make some comments about the definitions and their consequences and on the way in which the Theorem is proved.

From the fact that F preserves Markov's inequality it follows (see [10, p.40 and p.72] and for the essential fundamentals [3]) that the members of $\Lambda_{k+1}(F)$ can be identified with functions and the following easily handled definition holds.

DEFINITION 2. Let $F \subset \mathbb{R}^n$ be a set preserving Markov's inequality. $f \in \Lambda_\alpha(F)$ if

1) $|f(x)| \leq M,\quad x \in F$ (5)

and

2) for every ball B with centre in F and radius $r\leq 1$ there exists a polynomial P_B of degree at most $[\alpha]$ such that

$$|f(x)-P_B(x)| \leq M\cdot r^\alpha, \quad \text{for } x \in B\cap F. \qquad (6)$$

The norm of f in $\Lambda_\alpha(F)$ is equal to the infimum of all possible constants M.

This is the definition we refer to when we prove the Theorem. The proof is partly based on a geometric characterization of the sets which preserve Markov's inequality, Proposition 2, and partly on a construction of interpolating polynomials. The geometric characterization is made in terms of affinely independent point sets, which each contain $n+1$ points. Starting with such a point set the interpolating polynomial is constructed. The polynomial interpolates the values which correspond to each mixed directed derivative of order up to k in the $n+1$ points in the following sense. We start with n linearly independent directions in point number 0 and then decrease the number of directions with one for every subsequent point. This means that the number of values in point i are $\binom{n-i+k}{k}$. Hence the number of interpolating values altogether are $\sum_{i=0}^{n}\binom{n-i+k}{k} = \binom{n+k+1}{k+1}$, which equals the maximal number of terms in a polynomial of degree $k+1$ in n variables.

Finally we make two remarks.

Remark 1. It is not difficult to see through Proposition 2, that the Theorem is a generalization of the result in [7].

Remark 2. A. Jonsson has proved the Corollary by other methods (see [5]).

3. Proofs

Before proving the Theorem, we make the following preparations:

1. We characterize the sets, which preserve Markov's inequality, in a way which is suitable for the proof of the Theorem.
2. We introduce and use some special notations and we formulate and prove Lemma 1 in terms of these notations.
3. Finally, we define, also in terms of the special notations, the interpolating polynomial which appears in the Theorem.

2.1. Markov's inequality, defined in Section 2, is the basis of the characterization of an important class of subsets of $\mathbb{R}^n$; see [12] and [10]. It was shown in [6, p.7] that the sets which preserve Markov's inequality can be characterized geometrically (see also [10, p.38] for this geometric characterization). We give this characterization in a reformulated version.

PROPOSITION 1. F preserves Markov's inequality iff there exists a constant $c_*>0$ such that

$$(B(x_0,r) \setminus \{x \in \mathbb{R}^n: |b\cdot(x-x_0)| < c_* r\}) \cap F \neq \emptyset \tag{7}$$

for every x_0, r and b such that $x_0 \in F$, $0<r\leq 1$ and $b \in \mathbb{R}^n$, $|b|=1$.

The characterization in Proposition 1 says, roughly speaking, that F preserves Markov's inequality if it is not locally contained in a band which is too narrow.

In our case a slightly different characterization is required.

PROPOSITION 2. F preserves Markov's inequality iff there is a constant $c^*>0$ such that for every $x_0 \in F$ and for every r, $0<r\leq 1$, there are $a_i \in F\cap B(x_0,r)$, $i=0,\ldots,n$ such that the ball inscribed in $\mathrm{conv}\{a_0,\ldots,a_n\}$ has a radius not less than c^*r.

<u>Proof of Proposition 1 $\Leftrightarrow$ Proposition 2.</u>
Proposition 2 $\Rightarrow$ Proposition 1 is trivial.
Proposition 1 implies Proposition 2 is easy to prove. Let $a_0 = x_0$. From Proposition 1 follows that there exists $a_1 \in F\cap B(x_0,r)$ such that $d(a_0,a_1) \geq c_*r$. It is obvious that Proposition 1 gives us the possibility to continue to choose $a_i \in F\cap B(x_0,r)$, $i=2,\ldots,n$, so that $d(a_i,\mathrm{aff}\{a_0,\ldots,a_{i-1}\}) \geq c_*r$. The volume of $\mathrm{conv}\{a_0,\ldots, a_n)$ is greater than $c_*^n\cdot r^n/n!$ and since all a_i are points in $B(x_0,r)$ we conclude that there exists an inscribed ball with radius c^*r for some $c^*=c^*(n,F)$.

2.2. Let δ be the projection which maps $(x_1,x_2,\ldots,x_i)$ on $(x_1,x_2,\ldots,x_{i-1})$ i.e. δ deletes the last coordinate. Π_{n-m} denotes the hyperplane in $\mathbb{R}^{n-m}$, which is determined by the points $\delta^m a_{m+1},\ldots,\delta^m a_n$, where $a_0,a_1,\ldots,a_n$ are the $n+1$ affinely independent points in Proposition 2; observe that it is always possible, though sometimes necessary to first change the names of the coordinates, to write the equation of this hyperplane as

$$x_{n-m} = c_0^{(n-m)}+c_1^{(n-m)}x_1+\ldots+c_{n-m-1}^{(n-m)}x_{n-m-1}, \tag{8}$$

where

$$|c_i^{(n-m)}| \leq 1,\ i=1,2,\ldots,\ n-m-1. \tag{9}$$

If f is a function on $F \subset \mathbb{R}^n$ then we let

$$f_n(x_1,\ldots,x_n) = f(x_1,\ldots,x_n),\ (x_1,\ldots,x_n) \in F\cap B(x_0,r), \tag{10}$$

$$f_{n-m-1}(x_1,\ldots,x_{n-m-1}) = f_{n-m}(x_1,\ldots,x_{n-m-1},c_0^{(n-m)} + c_1^{(n-m)}x_1+\ldots+c_{n-m-1}^{(n-m)}x_{n-m-1}) \tag{11}$$

$m=0,1,\dots,n-2$, if the right side is defined.

S_{n-m}, which will appear in Lemma 1, and P_{n-m}, which is presented later, are defined in a corresponding way.

We would like to point out that, since there is a 1-1 correspondence between points in $\mathbb{R}^{n-m-1}$ and points in $\mathrm{aff}\{a_{m+1},\dots,a_n\}$ via the hyperplanes Π_{n-m} and the function δ, it is equivalent to write either about the value of the function f_{n-m-1} at $(x_1,\dots,x_{n-m-1})$ or about the value of f at the corresponding point in $\mathrm{aff}\{a_{m+1},\dots,a_n\}$. There is also a correspondence between the boundedness of the partial derivatives of $(f_{n-m}-S_{n-m})$, see the lemma below, and the boundedness of the partial derivatives of $(f-S)$,

$$|D^j(f-S)(x)| \le c\cdot M\cdot r^{k+1-|j|}, \quad |j|\le k, \quad x \in F\cap B(x_0,r) \qquad (12)$$

(12) follows from Definition 2 by Taylorexpansion and Markov's inequality. We state the correspondence, to the extent required for our purposes, in the following lemma.

LEMMA 1. If $f \in \Lambda_{k+1}(F)$ and S is a polynomial which approximates f according to Definition 2 and $a_0,\dots,a_n$ are the points in $B(x_0,r)\cap F$ appearing in the characterization in Proposition 2 then

$$|D^j(f_{n-m}-S_{n-m})(\delta^m a_{m-1})| \le c\cdot M\cdot r^{k+1-|j|}, \quad |j|\le k$$

$$\text{and} \quad c = c(k,n,F) \qquad (13)$$

Observe that since F preserves Markov's inequality, the partial derivatives exist and are unique; see Section 2.

<u>Proof of Lemma 1.</u> The formulas (10) and (11) give us the possibility to successively increase the subscript in $f_{n-m}-S_{n-m}$. Thus if we repeatedly apply (10) and (11) and the chain rule, then formula (12) yields (13) and finally (9) yields $c=c(F,n,k)$.

2.3. We now define the interpolating polynomial $P_I(x)$, which appears in the Theorem. The definition of $P_I(x)$ is in n steps and is formulated so that the proof of the Theorem will be easy to follow. In the description below $P_I(x) = P_n(x)$, and we do not always set out the dependent variable when the dependence is obvious.

Step m, $m=0,1,\dots,n-2$. Let Q_{n-m} be the polynomial in $n-m$ variables of degree at most k interpolating f_{n-m} and its partial derivatives of order at most k in the point $\delta^m a_m$. H_{n-m} is a homogenous polynomial in $n-m$ variables of degree $k+1$ and

$$P_{n-m} = Q_{n-m} + H_{n-m} \tag{14}$$

where H_{n-m} is uniquely determined by

$$Q_{n-m} + H_{n-m} = P_{n-m-1} \tag{15}$$

where the equality holds for all $(x_1,\dots,x_{n-m})$ which satisfy (8).

Step $n-1$. P_1 is the polynomial in one variable such that

$$P_1(\delta^{n-1}x) = \sum_{j=0}^{k} \frac{D^j f_1(\delta^{n-1}a_{n-1})}{j!}(\delta^{n-1}(x-a_{n-1}))^j+\beta(\delta^{n-1}(x-a_{n-1}))^{k+1}/(k+1)! \tag{16}$$

where β is a unique real number determined by

$$P_1(\delta^{n-1}(a_n)) = f(a_n) \tag{17}$$

We now state the existence and the uniqueness of P_I in a lemma.

LEMMA 2. There exists one and only one polynomial P_I of degree at most $k+1$ which fulfils the n-steps definition above.

Proof. The construction of P_I in n steps is valid as a proof if we add the following argument. The polynomial H_{n-m} is a homogenous polynomial in $x_{m+1},\dots,x_n$, it is zero at $\delta^m(a_m)$ and the values of H_{n-m} are known on a hyperplane which does not contain $\delta^m(a_m)$. Thus all its values are known and the proof is complete.

The preparatory statements and proofs are now completed and provide us with a simple proof of our main result, the Theorem in Section 2.

Proof of the Theorem. The "if"-part is trivial since it follows from the conditions 1) and 2) in the Theorem that $f \in \Lambda_{k+1}(F)$ by Definition 2.

The "only if"-part of the Theorem is proved by showing that the polynomial P_I, which through Lemma 2, exists and is unique, really approximates f according to condition "2)" in the Theorem.

Since all estimates are made for points in $B(x_0,r)\cap F$ (when f is involved) or on $B(x_0,r)$ when only polynomials are involved, we do not set out this in every one of the formulas which follow.

Suppose that $f \in \Lambda_{k+1}(F)$. Then there exists, by Definition 2, a polynomial S of degree at most $k+1$ such that

$$|f-S| \leq M\cdot r^{k+1}. \tag{18}$$

Since $S=S_n$ and $P_I=P_n$ (see above) we obtain

$$|f-P_I| \leq |f-S| + |S_n-P_n| \leq M\cdot r^{k+1} + |S_n-P_n|. \tag{19}$$

The proof is complete if we show that the last term in (19) is less than $c\cdot M\cdot r^{k+1}$. We estimate this term in n steps.

We expand $S_{n-m}-P_{n-m}$, $m=0,1,\dots,n-1$, in Taylor series at $\delta^m a_m$ and we use the notation $H_{S,n-m}$ for terms of degree $k+1$ which belong to the expansion of S_{n-m},

$$S_{n-m}-P_{n-m} = \sum_{|j|\leq k} \frac{D^j(S_{n-m}-P_{n-m})(\delta^m a_m)}{j!}(\delta^m(x-a_m))^j+H_{S,n-m}-H_{n-m} \tag{20}$$

Before continuing, we would like to make two observations concerning (20). Firstly, we observe that, since the interpolating polynomial is determined so that

$$(P_{n-m}-f_{n-m})(\delta^m a_m) = 0, \quad |j|\leq k \tag{21}$$

then Lemma 1 yields

$$\left|\sum_{|j|\leq k} \frac{D^j(S_{n-m}-P_{n-m})(\delta^m a_m)}{j!}(\delta^m(x-a_m))^j\right| \leq c\cdot M\cdot r^{k+1}, \tag{22}$$

$$x \in B(x_0,r) \cap \operatorname{aff}\{a_{m+1},\dots,a_n\}.$$

Secondly we observe that, from (20), (21) and from the homogeneity of the polynomial $H_{S,n-m}-H_{n-m}$, it follows that if

$$|S_{n-m}-P_{n-m}| \leq c\cdot M\cdot r^{k+1} \quad \text{on} \quad \mathrm{conv}\{a_{m+1},\dots,a_n\} \tag{23}$$

for some $m=m'$

then

the same estimate is true on $\mathrm{conv}\{a_m,\dots,a_n\}$. (24)

A combination of the second observation and formula (11) yields: If (23) holds for some $m=m'>0$, then it holds for $m=m'-1$, and hence by induction, for $m=0$. It is easy to find such an m', if $m=n-1$, then the polynomials in (23) are polynomials in one variable and thus it follows from the construction of P_I and the equations (20) and (22) that (23) is true for $m=n-1$.

Finally, if we expand $S-P_I$ in Taylor series at the centre of the ball inscribed in $\mathrm{conv}\{a_0,\dots,a_n\}$ and use Markov's inequality we can conclude that

$$|S(x)-P_I(x)| \leq c\cdot M\cdot r^{k+1}, \quad x \in B(x_0,r) \tag{25}$$

where $c=c(n,c^*) = c(n)\cdot(c^*)^{-(k+1)}$.

<u>Proof of the Corollary</u>. In the extension operator, given in [16] or [10, p.58], we substitute the polynomials by the interpolating polynomials which exist through the Theorem. Thus the operator becomes linear.

<u>Remark 3.</u> We illustrate the limits of the above method, in the search for a linear and bounded extension operator, by the following example.

Let $F = \{(x_1,x_2)\colon x_1 \geq 0,\ x_2 \geq 0,\ x_2 \leq \frac{1}{x_1}\}$

and

$$f(x_1,x_2) = \begin{cases} x_2\cdot|\log|x_2|| & \text{if} \quad 0<|x_2|\leq 1 \\ 0 & \text{otherwise} \end{cases}$$

$f \in \Lambda_1(F)$ (see e.g. [15, p.14]), but if we in the extension operator in [10] choose interpolating polynomials for the Whitney cubes for large values of x_1 then the extended function does not belong to $\Lambda_1(\mathbb{R}^2)$ since it is unbounded.

REFERENCES

[1] Yu. A. Brudnyi and P.A. Shvartsman, A linear extension operator for a space of smooth functions defined on a closed subset of $\mathbb{R}^n$, Soviet Math. Dokl. 31 (1985), 48-51.

[2] V.K. Dzyadyk and I.A. Shevchuk, Math. USSR Izv. 22 (1983).

[3] G. Glaeser, Étude de quelques algèbres Tayloriennes, J. d'Analyse Math. 6 (1958), 1-125.

[4] A. Jonsson, The trace of the Zygmund class $\Lambda_k(\mathbb{R})$ to closed sets and interpolating polynomials, J. Approx. Theory 44 (1985), 1-13.

[5] A. Jonsson, Markov's Inequality and Local Polynomial Approximation, No. 3, Department of Math., Univ. of Umeå, 1986.

[6] A. Jonsson, P. Sjögren, and H. Wallin, Hardy and Lipschitz Spaces on Subsets of $\mathbb{R}^n$, No. 8, Department of Math., Univ. of Umeå, 1980.

[7] A. Jonsson and H. Wallin, The trace to closed sets of functions in $\mathbb{R}^n$ with second difference of order $O(h)$, J. Approx. Theory 26 (1979), 159-184.

[8] A. Jonsson and H. Wallin, Local Polynomial Approximation and Lipschitz Type Conditions on General Closed Sets, No. 1, Department of Math., Univ. of Umeå, 1980.

[9] A. Jonsson and H. Wallin, Local polynomial approximation and Lipschitz functions on closed sets, Proceedings of the International Conference on Constructive Function Theory, Varna, Bulgaria, 1981.

[10] A. Jonsson and H. Wallin, Function Spaces on Subsets of $\mathbb{R}^n$, Mathematical Reports 2, Part 1, Harwood Academic Publ. (1984).

[11] P.A. Shvartsman, Studies in the theory of functions of several real variables (Yu.A. Brudnyi, editor), Yaroslav. Gos. Univ., Yaroslavl', (1982), 145-168 (Russian).

[12] H. Wallin, Markov's inequality on subsets of $\mathbb{R}^n$, Canadian Mathematical Society, Conference Proceedings volume 3 (1983), 377-388.

[13] H. Whitney, Analytic extensions of differentiable functions defined in closed sets, Trans. Amer. Math. Soc. 36 (1934), 63-89.

[14] H. Whitney, Differentiable functions defined in closed sets, Trans. Amer. Math. Soc. 36 (1934), 369-387.

[15] P. Wingren, Imbeddings and Characterizations of Lipschitz Spaces on Closed Sets, No. 5, Department of Math., Univ. of Umeå, 1985.

[16] L. Ödlund, On the Extension of $g \in \Lambda_\alpha(F)$ to $Eg \in \Lambda_\alpha(\mathbb{R}^n)$, No. 3, Department of Math., Univ. of Umeå, 1982.

PROBLEM SECTION

M. CWIKEL (Haifa)

1. Searching for a variant of Wiener measure which can be applied to interpolation theory. There is a fairly natural way (see e.g. [2]) to generate complex interpolation spaces with respect to an n-tuple or even an infinite family of compatible Banach spaces. The real method was extended long ago to n-tuples by several authors but a way of extending it to infinite families was only described rather recently in [3]. It uses Sparr's spaces $(A_1,A_2,\dots,A_n)_{\theta_1,\theta_2,\dots,\theta_n,p}$ defined for an n-tuple $\bar{A} = (A_1,A_2,\dots,A_n)$ via the K-functional

$$K(t_1,t_2,\dots t_n,a;\bar{A}) = \inf_{a = a_1+a_2+\dots+a_n} t_1\|a_1\|_{A_1}+t_2\|a_2\|_{A_2} +\dots+ t_n\|a_n\|_{A_n}$$

and the idea is to make the transition from spaces defined by n-tuples to spaces defined by an infinite family $\{A_\gamma | \gamma \in \Gamma\}$ by a sort of "integration" theory, much as one can make the transition from integration of simple functions to measurable functions. An important role is played here by consistency relations such as

$$(A_1,A_2,\dots,A_n)_{\theta_1,\theta_2,\dots,\theta_n,p} = (A_1,A_3,\dots,A_n)_{\theta_1+\theta_2,\theta_3,\dots,\theta_n,p}$$

which holds if $A_1 = A_2$. In the passage from n-tuples to infinite families the role of the multi-index $(\theta_1,\theta_2,\dots,\theta_n)$ (whose components must add up to one) is assumed by a probability measure μ on Γ and the consistency relations guarantee that the new interpolation space which we can denote by $(A(\gamma)|\gamma \in \Gamma)_{\mu,p}$ coincides with $(A_1,A_2,\dots,A_n)_{\theta_1,\theta_2,\dots,\theta_n,p}$ if Γ is the disjoint union of sets $\Gamma_1,\Gamma_2,\dots,\Gamma_n$, with $\mu(\Gamma_j) = \theta_j$ and $A(\gamma) = A_j$ for all $\gamma \in \Gamma_j$, $j=1,2,\dots,n$. Since these sorts of consistency conditions are somehow reminiscent of the definition of Wiener measure via cylinder sets this suggests that there ought to be some way of defining a measure ν on a measure space of non negative functions $t = t(\gamma)$, which permits an alternative construction for the spaces obtained in [3], i.e.

$$\|a\|_{(A(\gamma)\,|\,\gamma\in\Gamma)_{\mu,p}} = \left(\int K(t,a;A(\cdot))^p d\nu(t)\right)^{1/p}$$

where $K(t,a;A(\cdot))$ denotes an appropriate infinite dimensional version of the Sparr K-functional; perhaps it could be

$$K(t,a;A(\cdot)) = \inf_{a=\Sigma_{\gamma\in\Gamma}a(\gamma)} \Sigma_{\gamma\in\Gamma} t(\gamma)\|a(\gamma)\|_{A(\gamma)}.$$

Problem: Construct the measure ν . (N.B. It will probably depend on p .)

2. Video games and complex interpolation. It would surely be fun, and should also give considerable insight into properties of the complex interpolation method if we could draw profiles of the unit balls of the two dimensional spaces A_0 and A_1 on a computer screen, punch in a desired value of θ and then very soon afterwards obtain a plot or picture of the unit ball of $[A_0,A_1]_\theta$, and indeed this is fairly easy to do if A_0 and A_1 are lattices since then $[A_0,A_1]_\theta$ is the complexification of $A_0^{1-\theta}A_1^{\theta}$.

Problem: Write a program which plots unit balls for general two dimensional A_0 and A_1 . You will first have to decide what is the most appropriate way to complexify the spaces over the reals whose unit balls you can hope to draw. Another more difficult preliminary step might be to make a careful analysis of conditions governing the rate of convergence of sequences in the class $G(A_0,A_1)$ to elements in $J(A_0,A_1)$ (as defined in [1]).

Jaak Peetre has suggested also trying a similar project for the more "canonical" of the real interpolation spaces, namely $(A_0,A_1)_{\theta,1;J}$ or $(A_0,A_1)_{\theta,\infty;K}$.

When you also have a version for three dimensional spaces I confess to being very curious to see the result of interpolating an icosahedron with a dodecahedron.

References

[1] Calderón, A.P.: Intermediate spaces and interpolation, the complex method, Studia Math. 24, 113-190 (1964).

[2] Coifman, R.R., Cwikel, M., Rochberg, R., Sagher, Y., Weiss, G.: A theory of complex interpolation for families of Banach spaces, Adv. in Math. 43, 203-229 (1982).

[3] Cwikel, M., Janson, S.: Real and complex interpolation methods for finite and infinite families of Banach spaces; Adv. in Math. (to appear).

J. PEETRE (Lund)

1. Generalization of Guillemin's formula. Let μ be any radial measure on the unit disk $D \subset \underline{C}$ and let P^{μ} be the corresponding "Bergman projection". Set $T_b^{\mu} = P^{\mu}(b\cdot)$ (Toeplitz-Bergman operator with symbol b ; we write S_b for the ordinary Toeplitz(-Szegö) operator, corresponding to the special case when μ is a delta function). Consider also $F(a) = \int_0^1 r^{2a} d\mu(r)$. Then formally

$$T_b^{\mu} = \sum_{k=0}^{\infty} (G_k)\left(\frac{1}{i} \cdot \frac{d}{d\Theta}\right) \circ S_{b_k}$$

where $b_k = (\bar{z}\partial/\partial\bar{z})^k b$ and $G_k = F^{(k)}/F$. It is easy to extend this formula to the case of the unit ball ("Rudin ball") in $\underline{C}^n$. The special case μ = volume (uniform) measure was considered by Guillemin [3]. To which other domains in $\underline{C}^n$ does it extend? I believe that at least general Reinhardt domains will do. Next, to estimate the remainder (in suitable norms, e.g. in appropriate Schatten classes). This is analogous to the problem of estimating the remainder in the asymptotic series ("Leibniz's formula") for the product A o B of two Ψ.D.O. A and B (see any book on Ψ.D.O., e.g. [5]), a special case of which at least was mentioned in [7].

2. G-spaces. The one parameter scale of "diagonal" Besov spaces $B_\sigma^\alpha = B_\sigma^{\alpha\sigma}$, where $1/\sigma - \alpha = \text{const.} = 1/p$, has proved to be of vital interest in several recent investigations: a) Hankel operators - $p = \infty$ (see e.g. [12]), b) rational approximation - p general (see [10], [11]). But what one really should look at in such contexts are the spaces obtained by real interpolation from these spaces; they are then not any longer Besov spaces. More precisely, set

$$G_{\sigma q}^\alpha = (B_{\sigma_0}^{\alpha_0}, B_{\sigma_1}^{\alpha_1})_{\theta q},$$

in the natural relations between the parameters. (By the reiteration theorem the result of the interpolation is independent of the endpoints.) A characterization of the G-spaces is given in [13] and another one is at least outlined in [7]. But, in view of their importance, one should really try to do much more with them. For instance (a question for D. Adams), what about their potential theory (capacities and all that)?

3. By Pekarskiĭ's theorem (cf. problem 2; see also my and Johan Karlsson's report to the Edmonton conference [9]), the scale of Besov spaces B_σ^α, $1/\sigma = \alpha + 1/p$, arises in rational approximation. In an $(\alpha, 1/\sigma)$-diagram they correspond to a straight line forming an angle 45° with the coordinate axes. By contrast, the spaces encountered more classically in, say, polynomial approximation (Jackson, Bernstein) give a horisontal line. Is there an "approximation" description also for the scale of spaces corresponding to a line of arbitrary inclination (see figure)?

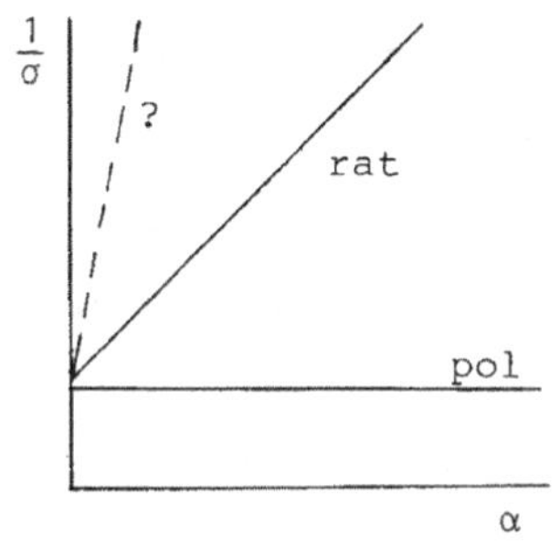

4. <u>Schatten classes of multilinear forms.</u> If one formulates Hankel theory not for operators, but in terms of bilinear forms (cf. e.g. [6]), it is natural to try to incorporate also the case of the corresponding multilinear (Hankel) forms (see [6], Sec. 5). But what is S_p in the multilinear case? The definition of S_p, $0 < p \leq 1$, as p-nuclear forms in the sense of Grothendieck is clear, S_2 consists of the Hilbert-Schmidt forms and S_∞ of either the bounded or the compact forms. This suggests (cf. [8], Lect. V, [6], Sec. 5) to define S_p, $1 < p < \infty$, by either real or complex interpolation between S_1 and S_∞. If $\Theta = 1/2$ one gets back the Hilbert-Schmidt forms. But it is not clear that the real and the complex method will give the same result in general. And what about other characterizations? For instance, by approximation (using "s-numbers")?

5. <u>What does the Schottky double mean for interpolation.</u> If Ω is any finitely connected plane domain bounded by smooth arcs then its Schottky double is a (compact) Riemann surface R obtained by pasting together two copies of Ω. (Example. If $\Omega = D$ then R = Riemann sphere; if Ω = an annulus then R = a torus.) But R carries also a natural projective structure. Cwikel and Fisher [1] have used the uniformization theorem in the context of interpolation (to prove that the passage to multiply connected domains gives the same spaces as the classical Calderón construction with a simply connected domain, usually the disk D or else a strip), which is connected with a different, but in my eyes less natural (in the present context) projective structure. Quite generally, I am asking for a more conscious inclusion of the point of view of algebraic geometry in interpolation. Another instance, where a construct from algebraic geometry intervenes at least in a hidden way, is the paper [2] by Cwikel and Janson, where something appears which may be identified as the Abel map (mapping a curve into its Jacobian variety). Otherwise, the Schottky double is considered e.g. in [4].

References

[1] Cwikel, M., Fisher, S.D.: Complex interpolation spaces on multiply connected domains. Adv. in Math. 48, 286-294 (1983).

[2] Cwikel, M., Janson, S.: Real and complex interpolation methods for finite and infinite families of Banach spaces. Adv. in Math. (to appear).

[3] Guillemin, V.: Toeplitz operators in n-dimensions. Integral Equations Operator Theory 7, 145-205 (1984).

[4] Hawley, N., Schiffer, M.: Half-order differentials on Riemann surfaces. Acta Math. 115, 199-236 (1966).

[5] Hörmander, L.: The analysis of linear partial differential operators III. Grundlehren 274. Berlin etc.: Springer 1985.

[6] Janson, S., Peetre, J., Rochberg, R.: Hankel forms and the Fock space. Technical report. Uppsala 1986. (Submitted to Ann. Math.)

[7] Peetre, J.: Hankel operators, rational approximation and allied questions in analysis. In: Second Edmonton Conference on Approximation Theory. C.M.S. Conference Proceedings 3, pp. 287-332. A.M.S.: Providence 1983.

[8] Peetre, J.: Paracommutators and minimal spaces. In: S.C. Power (ed.), Operators and function theory, pp. 163-224. Dordrecht: Reidel 1985.

[9] Peetre, J., Karlsson, J.; Rational approximation - analysis of the work of Pekarskiǐ. (Report prepared for the Edmonton conference, July 1986.)

[10] Pekarskiǐ, A.A.: Inequalities of Bernstein type for derivatives of rational functions and converse theorems for rational approximation. Mat. Sb. 124, 571-588 (1984) [Russian].

[11] Pekarskiǐ, A.A.: Classes of analytic functions defined by best rational approximations. Mat. Sb. 127, 3-19 (1985) [Russian].

[12] Peller, V.V.: Hankel operators of class S_p and their applications (rational approximation, Gaussian processes, the majorant problem for operators). Mat. Sb. 113, 538-581 (1980) [Russian].

[13] Peller, V.V.: Continuity properties for the averaging projection onto the set of Hankel matrices. Dokl. Akad. Nauk SSSR 278, 275-281 (1984) [Russian].

V.V. PELLER (Leningrad)

1. Let X be a space of functions on the unit circle. The problem is for which X the following property holds:

For any unimodular function u (i.e. $|u| = 1$ a.e.) in X such that $\mathbb{P}_- u \in X$ ($\mathbb{P}_-$ is the orthogonal projection from L^2 onto $H^2_- \stackrel{\text{def}}{=} L^2 \ominus H^2$) we have $u = v\Theta$ for some inner function Θ and $v \in X$.

This question is very important in the problems of spectral characterizations of stationary Gaussian processes with continuous time. The only result familiar to me is T. Wolff's theorem (Duke Math. J., 49 (1982), 321-328) which proves the above property for $X = VMO$.

The above problem for $X = B^{1/p}_{p,p}$ (Besov spaces) seems especially important in connection with stationary processes.

Note that for various spaces X under the additional assumption that the Toeplitz operator T_u on H^2 has dense range, a stronger result is proved in V.V. Peller, S.V. Khrushchev, Uspekhi Matem. Nauk, 37 (1982), 53-124 (=Russian Math. Surveys, 37 (1982), 61-144). Namely, in this case we can conclude that $u \in X$.

2. Given a function φ on the unit circle, consider the function $\check{\varphi}$ of two variables defined by

$$\check{\varphi}(z_1, z_2) = \frac{\varphi(z_1) - \varphi(z_2)}{z_1 - z_2} .$$

This function generates a Schur multiplier $M_{\check{\varphi}}$ on the Schatten - von Neumann class S_p, $0 < p < 2$:

$$M_{\check{\varphi}} R_k = R_{\check{\varphi}k},$$

where

$$(R_k f)(z) = \int k(z,\zeta)f(\zeta)dm(\zeta)$$

(each operator in S_p can be realized as an integral operator, and $M_{\check{\varphi}}$ is the multiplication of the kernel of an operator by $\check{\varphi}$).

Suppose that $M_{\check{\varphi}}$ is bounded on S_p, $1 \le p \le 2$.

Question. Is it true that the lacunary Fourier coefficients of the derivative φ' belong to ℓ^p:

$$\{\widehat{\varphi'}(2^n)\}_{n\geq 0} \in \ell^p \;?$$

For $p = 2$ this is obviously true because in this case $M_{\check{\varphi}}$ is bounded if and only if $\varphi' \in L^\infty$.

For $p = 1$ this is also true because in this case $\varphi \in B^1_{1,1}$ (see V.V. Peller, Funkts. Anal. i ego Pril., 19, N 2 (1985), 37-51) and this implies the result for $p = 1$. For all other p the answer is unknown. I conjecture that it should be positive.

Note that the problem of boundedness of $M_{\check{\varphi}}$ on S_p is important in perturbation theory. It is very closely related to the problem of when $\varphi(U) - \varphi(V) \in S_p$ for any unitary operators U, V satisfying the condition $U - V \in S_p$.

3. This problem is closely connected with the previous one.

Problem. Describe the real interpolation spaces $(B^0_{\infty 1}, L^\infty)_{\theta,q}$ and the complex interpolation spaces $(B^0_{\infty 1}, L^\infty)_{[\theta]}$.

This problem is also important in connection with the above problem in perturbation theory. Indeed, if $\varphi' \in B^0_{\infty 1}$ then $M_{\check{\varphi}}$ is bounded on S_1 (see V.V. Peller, Funkts. Anal. i ego Pril., 19, N 2

(1985), 37-51) and as mentioned in the previous problem $M_{\check{\varphi}}$ is bounded on S_2 for $\varphi' \in L^\infty$. So, if we knew interpolation spaces between $B^0_{\infty 1}$ and L^∞, we could obtain sharp sufficient conditions on φ in order that $M_{\check{\varphi}}$ be bounded on S_p for $1 < p < 2$. (Note that $M_{\check{\varphi}}$ is bounded on S_p if and only if it is bounded on $S_{p'}$, where $1/p + 1/p' = 1$).

4. Let φ be a function analytic in the unit disc. Consider the Hankel operator Γ_φ acting from the Hardy class H^q to the Hardy class H^r which is defined by

$$\Gamma_\varphi z^n = \sum_{k \geq 0} \hat{\varphi}(n+k) z^k.$$

Suppose that $1 < r < \infty$ and $q > r$. Then it is well known (and easily verifiable) that Γ_φ is bounded from H^q to H^r if and only if $\varphi \in H^p$ where $1/p = 1/r - 1/q$.

The problem is whether there is an analogue of the theorem of Adamyan, Arov, and Krein (Matem. Sbornik, 86 (1971), 34-75 = Mat. USSR Sbornik, 15 (1971), 31-73). Namely, is it true that

$$\operatorname{dist}(\Gamma_\varphi, \{T : \operatorname{rank} T \leq n\})$$

is equivalent to

$$\operatorname{dist}(\Gamma_\varphi, \{\Gamma_\psi : \operatorname{rank} \Gamma_\psi \leq n\})?$$

Note that for $r = s$ the norm of Γ_φ is equivalent to the norm of φ in BMO and so the above question has an affirmative answer.

If the answer is positive for $q > r$, it would be possible to relate the question of description of approximative numbers of Hankel operators to the problem of rational approximation in the norm of H^p.

R. ROCHBERG (St. Louis)

The complex interpolation spaces $[A_0,A_1]_\Theta = A_\Theta$ make sense even if A_1 is a Banach space and A_0 is only a quasi-Banach space. Under what conditions will (some of) the intermediate spaces be Banach spaces? The example of Lebesgue spaces, $A_0=L^{1/2}$, $A_1=L^1$ and $A_\Theta=L^p$, $1/2 < p < 1$, shows that some conditions are necessary to get positive conclusions. Possibly these conditions involve the degree of convexity of A_1. Again, the example of Lebesgue spaces, this time with $A_0=L^p$ for small p and $A_1=L^2$ shows that some control of the lack of convexity of A_0 is needed for there to be any hope of a uniform result. Here is a very specific version of this general question. Suppose that the vector spaces A_0 and A_1 are $\mathbb{C}^2$. Suppose that A_0 has the $\ell^{1/2}$ norm and that A_1 has a norm given by an inner product (but not necessarily the standard one); that is $\|v\|_{A_1}^2 = (\Omega v,v)$ for some positive definite 2 by 2 matrix Ω. Is there a Θ_0, <u>independent of Ω</u> so that A_Θ is a Banach space for $\Theta_0 \leq \Theta < 1$?